PROJECT MANAGEMENT

PROJECT MANAGEMENT
Engineering, Technology, and Implementation

AVRAHAM SHTUB
Department of Industrial Engineering
Tel Aviv University

JONATHAN F. BARD
Department of Mechanical Engineering
Graduate Program in Operations Research & Industrial Engineering
University of Texas at Austin

SHLOMO GLOBERSON
School of Business Administration
Tel Aviv University

Prentice Hall International Series
in Industrial and Systems Engineering
W. J. Fabrycky and J. H. Mize, Series Editors

PRENTICE HALL, Englewood Cliffs, NJ 07632

Library of Congress Cataloging-in-Publication Data

Shtub, Avraham.
    Project management : engineering, technology, and implementation /
  Avraham Shtub, Jonathan F. Bard, Shlomo Globerson.
        p.   cm.
    Includes bibliographical references and index.
    ISBN 0-13-556458-1
    1. Engineering--Management. 2. Industrial project management.
  I. Bard, Jonathan F. II. Globerson, Shlomo,   . III. Title.
  TA190.S58 1994
  658.4'04--dc20   93-32078
                                                             CIP

Acquisitions editor: MARCIA HORTON
Project manager: JENNIFER WENZEL
Cover designer: DELUCA DESIGN
Buyer: DAVE DICKEY
Copy editor: ZEIDERS AND ASSOCIATES
Editorial assistant: DOLORES MARS

 © 1994 by Prentice-Hall, Inc.
A Paramount Communications Company
Englewood Cliffs, New Jersey 07632

Printed in the United States of America

10 9 8 7 6 5

ISBN   0-13-556458-1

Prentice-Hall International (UK) Limited, *London*
Prentice-Hall of Australia Pty. Limited, *Sydney*
Prentice-Hall Canada Inc., *Toronto*
Prentice-Hall Hispanoamericana, S.A., *Mexico*
Prentice-Hall of India Private Limited, *New Delhi*
Prentice-Hall of Japan, Inc., *Tokyo*
Simon & Schuster Asia Pte. Ltd., *Singapore*
Editora Prentice-Hall do Brasil, Ltda., *Rio de Janeiro*

Contents

13. COMPUTER SUPPORT FOR PROJECT MANAGEMENT   537

Preface

We all deal with projects in our daily lives. In most cases, organization and management simply amount to constructing of a list of tasks and executing them in sequence. But when the information available is limited or imprecise, and when cause-and-effect relationships are uncertain, a more considered approach is called for. This is especially true when the stakes are high and time is pressing. Getting the job done right the first time is essential. This means doing the upfront work thoroughly, even at the cost of lengthening the initial phases of the project. Shaving expenses at the outset with the intent of leaving time and money for revisions later might seem like a good idea but could have consequences of painful proportions. Seasoned managers will tell you that it is more cost-effective in the long run to add five extra engineers at the beginning of a project than to have to add 50 toward the end.

The quality revolution in manufacturing has brought this point home. Companies in all areas of technology have come to learn that quality cannot be inspected into a product; it must be built in. Recalling the 1980s, the global competitive battles of that time were won by companies that could achieve cost and quality advantages in existing, well-defined markets. In the 1990s, these battles are being won by companies that can build and dominate new markets. Planning is a critical part of this process and is the foundation of project management.

Projects may involve dozens of firms and hundreds of people who need to be managed and coordinated. They need to know what has to be done, who is to do it, when it should be done, how it will be done, and what resources will be used. Proper planning is the first step in communicating these intentions. The problem is made difficult by what can be characterized as an atmosphere of uncertainty, chaos, and conflicting goals. To ensure teamwork, all major participants and stakeholders should be involved at each stage of the process.

But how to achieve this efficiently, within budget, and on schedule? Our primary objective in writing this book has been to answer this question from the perspective of the project manager. We do this by identifying the components of modern project management and showing how they relate to the basic phases of a project, starting with conceptual design and advanced development, and continuing through detailed design, production, and termination. We take a practical approach, drawing on our collective experience in the electronics, information services, and aerospace industries.

Over the years, there have been numerous books written with a similar objective in mind. We acknowledge their contribution and have endeavored to build on their strengths. In so doing, we have placed the stress on integrative concepts rather than isolated methodologies. We have relied on simple models to convey ideas and have intentionally avoided detailed mathematical formulations and solution algorithms—aspects of the field better left to other parts of the curriculum. Nevertheless, we do present some of the more important mathematical programming models arising in project management, and provide references for those readers who have a need or interest in further study. The availability of powerful, commercial codes now bring the solution of these models within reach of the project team.

To ensure that project participants work toward the same end and hold the same expectations, short- and long-term goals must be identified and communicated on a continuing bases. The project plan is the vehicle by which this is accomplished, and once approved, becomes the basis for monitoring, controlling, and evaluating progress at each phase of the project's life cycle. To help the project manager in this effort, various software packages have been developed; the most common run interactively on microcomputers and have full functional and report-generating capabilities. A unique feature of this book is that it comes with a disk containing SuperProject Expert, one of the most sophisticated project management software packages on the market. The "educational" version included here has all the functionality of the original version, with the exception that projects are limited to 50 tasks. In our experience, even the most timid users are able to take advantage of SuperProject Expert's main features after only a few hours of hands-on instruction.

A second objective in writing this book has been to fill a void between texts aimed at low- to midlevel managers and those aimed at technical personnel with strong analytic skills but little training in, or exposure to, organizational issues. Those teaching engineering or business students at both the late undergraduate and early graduate levels should find it suitable. In addition, the book is intended to serve as a reference for the practitioner who is new to the field or who would like to gain a surer footing in project management concepts and techniques.

The core material, including most of the underlying theory, can be covered in a one-semester course. At the end of Chapter 1, we outline the book's contents. Chapter 2 deals with economic issues, such as cash flow, time value of money, and depreciation, as they relate to projects. With this material and some supplementary notes, coupled with the evaluation methods and multiple-criteria decision-making techniques discussed in Chapters 3 and 4, respectively, it should be possible to teach a combined course in project management and engineering economy. This is the direction in which many undergraduate engineering programs are now headed after many years of industry prodding. Young engineers are often thrust into leadership roles without adequate preparation or training in project management skills.

Writing a textbook is a collaborative effort involving many persons whose names do not always appear on the cover. In particular, we would like to thank Professor R. Balachandra of Northeastern University, Professor Ted Klastorin of the University of Washington, and Professor Gavriel Salvendy of Purdue University for their careful reading of the first draft and their constructive and informative comments. Much appreciation is also due Professor Robert Parsons of Northeastern University for his many contributions in the early stages of development.

With regard to production, we are indebted to the Owen Graduate School of Management at Vanderbilt University for providing support for the first author during his stay. Most of the original manuscript and later corrections were typed by Sandra Hughes of

Vanderbilt, whose patience and dedication proved invaluable in meeting deadlines. Finally, special thanks go to Dr. Lisa Judge, Jewel Bard, and Kinneret Dekel for editing and proofreading the text and to Warren Sharp, Michael Kowal, Siwate Rojanasoothon, and Kishore Sarathy for checking exercise solutions.

Avraham Shtub
Jonathan F. Bard
Shlomo Globerson

List of Abbreviations

ACWP	actual cost of work performed		**EV**	earned value
AHP	analytic hierarchy process		**FMS**	flexible manufacturing system
AOA	activity-on-arrow		**GDSS**	group decision support system
AON	activity-on-node		**GERT**	graphical evaluation and review technique
B/C	benefit/cost		**IRR**	internal rate of return
BCWP	budgeted cost of work performed		**LCC**	life-cycle cost
BCWS	budgeted cost of work scheduled		**LOB**	line of balance
CBS	cost breakdown structure		**LRC**	linear responsibility chart
CCB	change control board		**MARR**	minimum attractive (acceptable) rate of return
CE	certainty equivalent; concurrent engineering		**MAUT**	multiattribute utility theory
C-E	cost-effectiveness		**MBO**	management by objectives
CER	cost estimating relationship		**MIS**	management information system
CDR	critical design review		**MTBF**	mean time between failures
CI	cost index; consistency index; criticality index		**MTTR**	mean time to repair
CPM	critical path method		**NASA**	National Aeronautics and Space Administration
CV	cost variance		**NPV**	net present value
C/SCSC	cost/schedule control systems criteria		**OBS**	organizational breakdown structure
DSS	decision support system		**PDR**	preliminary design review
ECR	engineering change request			
EMV	expected monetary value			

PERT	program evaluation and review technique	**R&D**	research and development
		RFP	request for proposal
PW	present worth	**SI**	schedule index
QA	quality assurance	**SOW**	statement of work
QFD	quality function deployment	**TQM**	total quality management
		WBS	work breakdown structure

1

Introduction

1.1 NATURE OF PROJECT MANAGEMENT

Many of the most difficult engineering challenges of recent decades have been to design, develop, and implement new systems of a type and complexity never before attempted. Examples include the construction of vast petroleum production facilities in the North Sea off the coast of Great Britain, the development of the manned space program in both the United States and the former Soviet Union, and the worldwide installation of fiber optic lines for broadband telecommunications. The creation of these systems with performance capabilities not previously available, and within acceptable schedules and budgets, has required the development of new methods of planning, organizing, and controlling events. This is the essence of project management.

Succinctly, a project is an organized endeavor aimed at accomplishing a specific nonroutine or low-volume task. Although projects are not repetitive, they may take significant amounts of time and, for our purposes, are sufficiently large or complex to be recognized and managed as separate undertakings. Consequently, teams have emerged as the

1

way of supplying the needed talents. But the use of teams complicates the flow of information and places additional burdens on management to communicate with, and coordinate the activities of, the participants.

In general, the amount of time an individual or an organizational unit is involved in a project may vary considerably. Someone in operations may work only with other operations personnel on a project, or may work with a team comprised of specialists from various functional areas to study and solve a specific problem, or to perform a secondary task.

Management of a project differs in several ways from management of a typical enterprise. The objective of a project team is to accomplish its prescribed mission and disband. Few firms are in business to perform just one job and then disappear. Since a project is intended to have a finite life, employees are seldom hired with the intent of building a career with the project. Instead, a team is pulled together on an ad hoc basis from among people who normally have assignments in other parts of the organization. They may be asked to work full time on the project until its completion; or they may be asked to work only part time, such as two days a week, on the project and spend the rest of the time at their usual assignments. A project may involve a short-term task that lasts only a matter of days, or it may run for years. After completion, the team normally disperses and its members return to their original jobs.

1.2 RELATIONSHIP BETWEEN PROJECTS AND OTHER PRODUCTION SYSTEMS

Operations and production management contains three major classes of systems: (1) those designed for mass production, (2) those designed for batch (or lot) production, and (3) those designed for undertaking nonrepetitive projects common to construction and new product development. Each of these classes may be found in both the manufacturing and service sectors.

Mass production systems are typically designed around the specific processes used to assemble a product or perform a service. Their orientation is fixed and their applications are limited. Resources and facilities are composed of special-purpose equipment designed to perform the operations required by the product or the service in an efficient way. By laying out the equipment to parallel the natural routings, material handling and information processing are greatly simplified. Quite frequently, material handling is automated and the use of conveyors and monorails is extensive. The resulting system is capital intensive, very efficient in the processing of large quantities of specific products or services for which relatively little management and control are necessary. However, these systems are very difficult to alter should a need arise to produce new or modified products, or to provide new services. As a result, they are most appropriate for operations that experience a high rate of demand (e.g., several hundred thousand units annually) as well as high aggregate demand (e.g., several million units throughout the life cycle of the system).

Batch-oriented systems are used when several products or services are processed in the same facility. When the demand rate is not high enough or when long-run expectations do not justify the investment in special-purpose equipment, an effort is made to design a more flexible system on which a variety of products or services can be processed. Because the resources used in such systems have to be adjusted (set up) when production switches from one product to another, jobs are typically scheduled in batches to save setup time. Flexibility is achieved by using general-purpose resources that can be adjusted to handle different processes. The complexity of operations planning, scheduling, and control is greater than in mass production systems, as each product has its own routing (sequence of operations). To simplify planning, resources are frequently grouped together based on the type of processes that they perform. Thus batch-oriented systems contain organizational units that specialize in a function or a process, as opposed to product lines that are found in mass production systems. Departments such as metal cutting, painting, testing, and packaging/shipping are typical examples from the batch-oriented manufacturing sector, while word processing centers and diagnostic laboratories are examples from the service sector.

In the batch-oriented system, it is particularly important to pay attention to material handling needs, as each product has its specific set of operations and routings. Material handling equipment such as forklifts are used to move in-process inventory between departments and work centers. The flexibility of batch-oriented systems makes them attractive for many organizations.

In recent years, flexible manufacturing systems have been quick to gain acceptance in some industrial settings. With the help of microelectronics and computer technology, these systems are designed to achieve mass production efficiencies in low-demand environments. They work by reducing setup times and automating material handling operations but are extremely capital intensive. Hence they cannot always be justified when product demand is low or when labor costs are minimal. Another approach is to take advantage of local economies of scale. Group technology cells, which are based on clustering similar products or components into families processed by dedicated resources of the facility, are one way to implement this approach. Higher utilization rates and greater throughput can be achieved by processing similar components on dedicated machines.

By way of contrast, systems that are subject to very low demand (no more than a few units) are substantially different from the first two mentioned. Because of the nonrepetitive nature of these systems, past experience may be of limited value, so little learning takes place. In this environment, extensive management effort is required to plan, monitor, and control the activities of the organization. Project management is a direct outgrowth of these efforts.

It is possible to classify organizations based on their production orientation as a function of volume and batch size. This is illustrated in Fig. 1-1. The borderlines between mass production, batch-oriented, and project-oriented systems are hard to define. In some organizations where the project approach has been adopted, several units of the same product (a batch) are produced, while other organizations use a batch-oriented system that produces small lots (the just-in-time approach) of very large volumes of products. To better understand the transition between the three types of systems, consider an electronics firm that assembles printed circuit boards in small batches in a job shop. As demand for the boards picks up, a decision is made to develop a flow line for assembly. The design and implementation of this new line is a project.

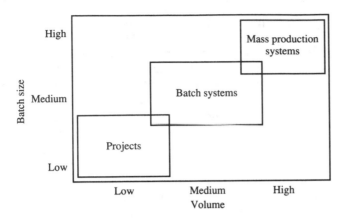

Figure 1-1   Classification of production systems.

1.3 CHARACTERISTICS OF PROJECTS

Although the Manhattan project—the development of the first atomic bomb—is considered by many to be the first instance where modern project management techniques were used, ancient history is replete with examples. Some of the better known ones include the construction of the Egyptian pyramids, the conquest of the Persian Empire by Alexander the Great, and the building of the Temple in Jerusalem. In the 1960s formal project management methods received their greatest impetus with the Apollo program and an array of large, dynamic construction projects.

Today, activities such as the transportation of American forces in Operation Desert Shield, the pursuit of new treatments for AIDS, and the development of the U.S. Space Station are examples of three projects with which most of us are familiar. Additional examples of a more routine nature include:

- Selecting a software package
- Developing a new office plan or layout
- Implementing a new decision support system
- Introducing a new product to the market
- Designing an airplane, supercomputer, or work center
- Opening a new store
- Constructing a bridge, dam, highway, or building
- Relocating an office or a factory
- Performing major maintenance or repair
- Starting up a new manufacturing or service facility
- Producing and directing a movie

1.3.1 Definitions and Issues

As the list above suggests, a project may be viewed or defined in several different ways: for example, as "the entire process required to produce a new product, new plant, new system, or other specified results" (Archibald 1976) or as "a narrowly defined activity which is planned for a finite duration with a specific goal to be achieved" (General Electric 1977). Further, Obrien (1974) states that project management occurs when management gives emphasis and special attention to the conduct of nonrepetitive activities for the purpose of meeting a single set of goals.

By implication, project management deals with a one-time effort to achieve a focused objective. How progress and outcomes are measured, though, depends on a number of critical factors. Typical among these are technology (specifications, performance, quality), time (due dates, milestones), and cost (total investment, required cash flow), as well as profits, resource utilization, and market acceptance.

These factors and their relative importance are major issues in project management. Based on a well-defined set of goals, it is possible to develop appropriate performance measures and to select the organizational structure, required resources, and people that will team up to achieve these goals. Figure 1-2 summarizes the underlying processes. As illustrated, most projects are initiated by a need. A new need may be identified by a customer, the marketing department, or any member of the organization. When management is convinced that the need is genuine, goals may be defined and the first steps taken toward putting together a project team. Most projects have several goals covering such aspects as technical and operational requirements, delivery dates, and cost, and should be ranked according to their relative importance.

Based on these rankings and a derived set of performance measures for each goal, the technological concept (or initial design) is developed along with a schedule and a budget for the project. The next step is to integrate the design, the schedule, and the budget into a project plan specifying what should be done, by whom, at what cost, and when. As the plan is implemented the actual accomplishments are monitored and recorded. Adjustments, aimed at keeping the project on track, are made when deviations or overruns appear. When the project terminates its success is evaluated based on the predetermined goals and performance measures. Figure 1-3 compares two projects with these points in mind. In project 1, a "design to cost" approach is taken. Here the budget is fixed and the technological goals are clearly specified. Cost, performance, and schedule are all given equal weight. In project 2, the technological goals are paramount and must be achieved, even if it means compromising the schedule and the budget in the process.

The first situation is typical of standard construction and manufacturing projects where a contractor agrees to supply a system or a product in accordance with a given schedule and budget. The second situation is typical of "cost plus fixed fee" projects where the technological uncertainties argue against a contractor committing to a fixed cost and schedule. This situation is most common in a research and development environment.

A well-designed organizational structure is required to handle projects due to their uniqueness, variety, and limited life span. In addition, special skills are required to manage them successfully. Taken together, these skills and organizational structures have been the

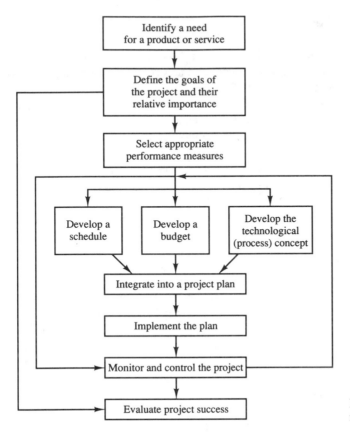

Figure 1-2 Major processes in project management.

catalyst for the development of the project management discipline. Many of the accompanying tools and techniques, though, are equally applicable in the manufacturing and service sectors.

Because projects are characterized by a "one time only" effort, learning is limited and most operations never become routine. This results in a need for extensive management involvement throughout the life cycle of the project. In addition, the lack of continuity leads to a high degree of uncertainty.

1.3.2 Risk and Uncertainty

In project management it is common to refer to very high levels of uncertainty as sources of risk. Risk is present in most projects, especially in the research and development (R&D) environment. Without trying to sound too pessimistic, it is prudent to assume that what can go wrong will go wrong. Principal sources of uncertainty include random variations in component and subsystem performance, inaccurate or inadequate data, and the inability to forecast satisfactorily due to lack of prior experience. Specifically, there may be:

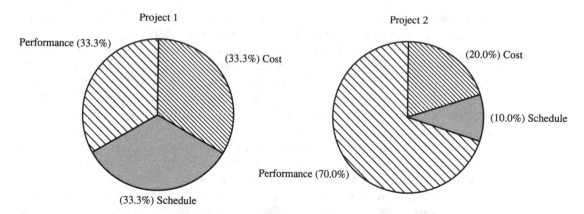

Figure 1-3   Relative importance of goals.

1. *Uncertainty in scheduling.* Changes in the environment that are impossible to forecast accurately at the outset of a project are likely to have a critical impact on the length of certain activities. For example, subcontractor performance or the time it takes to obtain a long-term loan are bound to influence the length of various subtasks. Methods are needed to deal with problematic or unstable time estimates. Probability theory and simulation have both been used successfully for this purpose, as discussed in Chapter 7.

2. *Uncertainty in cost.* Limited information on the duration of activities makes it difficult to predict the amount of resources needed to complete them on schedule. This translates directly into an uncertainty in cost. In addition, the expected hourly rate of resources and the cost of materials used to carry out project tasks may also possess a high degree of variability.

3. *Technological uncertainty.* This form of uncertainty is typically present in R&D projects where new (not well tested and approved) technologies, methods, equipment, and systems are developed or employed. Technological uncertainty may affect the schedule, the cost, and the ultimate success of the project. The integration of familiar technologies into one system or product may cause technological uncertainty as well. The same applies to the development of software and its integration with hardware.

There are other sources of uncertainty, including those of an organizational and political nature. New regulations might affect the market for a project, while the turnover of personnel and changes in the policies of one or more of the participating organizations may disrupt the flow of work.

To gain a better understanding of the effects of uncertainty, consider the three projects mentioned earlier. The transportation of American Armed Forces in Operation Desert Shield faced extreme political and logistical uncertainties. In the initial stages, none of the planners had a clear idea of how many troops would be needed or how much time was available to put the troops in place. Also, it was unknown whether permission would be granted to use NATO air bases or even to fly over European and Middle Eastern countries, or how much tactical support would be forthcoming from U.S. allies.

The development of a treatment for AIDS is an ongoing project fraught with technological uncertainty. Hundreds of millions of dollars have already been spent with little progress toward a cure. As expected, researchers have taken many false steps, and many promising paths have turned out to be dead ends. Lengthy trial procedures and duplicative efforts have produced additional frustration. If success finally comes, it is unlikely that the original plans or schemes will have predicted its form.

The design of the U.S. Space Station is an example where virtually every form of uncertainty is present. Politicians continue to play havoc with the budget, while special interest groups (both friendly and hostile) push their individual agendas; schedules get altered and rearranged; software fails to perform correctly; and the needed resources never seem to be available in adequate supply. Inflation, high turnover rates, and scaled-down expectations take their toll on the internal workforce, as well as the legion of subcontractors.

Taking its cue from Parkinson, the *American Production and Inventory Control Society* has fashioned the following laws in an attempt to explain the consequences of uncertainty on project management.

LAWS OF PROJECT MANAGEMENT

1. No major project is ever installed on time, within budget, or with the same staff that started it. Yours will not be the first.
2. Projects progress quickly until they become 90% complete, then they remain at 90% complete forever.
3. One advantage of fuzzy project objectives is that they let you avoid the embarrassment of estimating the corresponding costs.
4. When things are going well, something will go wrong.
   - When things just cannot get any worse, they will.
   - When things appear to be going better, you have overlooked something.
5. If project content is allowed to change freely, the rate of change will exceed the rate of progress.
6. No system is ever completely debugged. Attempts to debug a system inevitably introduce new bugs that are even harder to find.
7. A carelessly planned project will take three times longer to complete than expected; a carefully planned project will take only twice as long.
8. Project teams detest progress reporting because it vividly manifests their lack of progress.

1.3.3 Phases of a Project

A project passes through a life cycle that may vary with size and complexity, and the style established by the organization. The names of the various phases may differ, but they typically include those shown in Fig. 1-4. To begin, there is a *conceptual design* phase during which

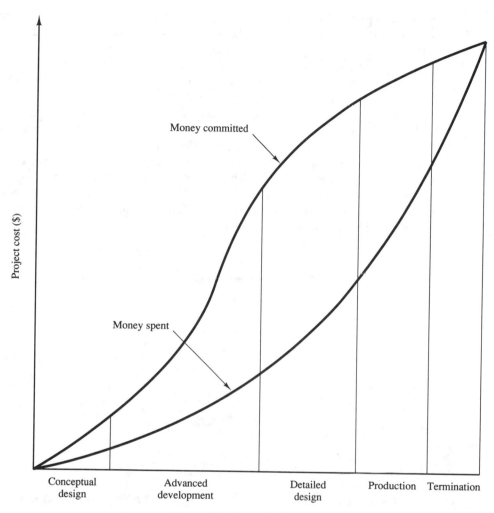

Figure 1-4   Project life cycle.

the organization realizes that a project may be needed, or receives a request from a customer to propose a plan to perform a project. Next there is an *advanced development or preliminary system design* phase where the project manager (and perhaps a staff if the project is complex) plans the project to a level of detail sufficient for initial scheduling and budgeting. If the project is approved, it will enter a more *detailed design* phase, a *production* phase, and a *termination* phase.

In Fig. 1-4 the five phases in the life cycle of a project are presented as a function of time. The cost during each phase depends on the specifics, but usually the majority of the budget is spent during the production phase. However, most of this budget is committed during the advanced development phase and the detailed design phase before the actual work takes place. Management plays a vital role during the conceptual design phase, the advanced development phase, and the detailed design phase. The importance of this involvement in defining goals, selecting performance measures, and planning the project cannot be overemphasized. Pressures to start the "real work" on the project, that is, to begin the production phase as early as possible, may lead to high cost and schedule risks due to the commitment of resources without adequate planning.

In most cases, a work breakdown structure (WBS) is developed during the conceptual design phase. The WBS is a document that divides the project into major hardware, software, data, and service elements. These elements are further divided and a list is produced identifying all tasks that must be accomplished to complete the project. The WBS helps define the work to be performed and provides a framework for planning, budgeting, monitoring, and control. Therefore, as the project advances, schedule and cost performance can be compared to plans and budgets. Table 1-1 shows an abbreviated WBS for an orbital space laboratory vehicle.

TABLE 1-1   PARTIAL WORK
BREAKDOWN STRUCTURE
FOR SPACE LABORATORY

Index	Work element
1.0	Command module
2.0	Laboratory module
3.0	Main propulsion system
3.1	Fuel supply system
3.1.1	Fuel tank assembly
3.1.1.1	Fuel tank casing
3.1.1.2	Fuel tank insulation
4.0	Guidance system
5.0	Habitat module
6.0	Training system
7.0	Logistic support system

The detailed project definition, as reflected in the WBS, is examined during the advanced development phase to determine the skills necessary to achieve the project's goals. Depending on the planning horizon, personnel from other parts of the organization may be used temporarily to accomplish the project. However, previous commitments may limit the availability of these resources. Other strategies might include hiring new personnel or subcontracting various work elements, as well as leasing equipment and facilities.

1.3.4 Organizing for a Project

A variety of structures are used by organizations to perform project work. The actual arrangement may depend on the proportion of the company's business that is project oriented, the scope and duration of the underlying tasks, the capabilities of the available personnel, preferences of the decision makers, and so on. The following five possibilities range from no special structure to a totally separate project organization.

1. *Functional organization.* Many companies are organized as a hierarchy with functional departments that specialize in a particular type of work, such as engineering and sales (see Fig. 1-5). These departments are often broken into smaller units that focus on special areas within the function. Upper management may divide a project into work tasks and assign them to the appropriate functional units. The project is then budgeted and managed through the normal management hierarchy.

2. *Project coordinator.* A project may be handled through the organization as described above, but with a special appointee to coordinate it. The project is still funded through the normal channels and the functional managers retain responsibility and authority for their portion of the work. The coordinator meets with the functional managers and provides direction and impetus for the project and may report its status to higher management.

3. *Matrix organization.* In a matrix organization a project manager is responsible for completion of the project and is often assigned a budget. The project manager essentially contracts with the functional managers for completion of specific tasks and coordinates project efforts across the functional units. The functional managers assign work to employees and coordinate work within their areas. These arrangements are depicted schematically in Fig. 1-6.

4. *Project team.* A particularly significant project (development of a new product or business venture) that will have a long duration and require the full-time efforts of a group

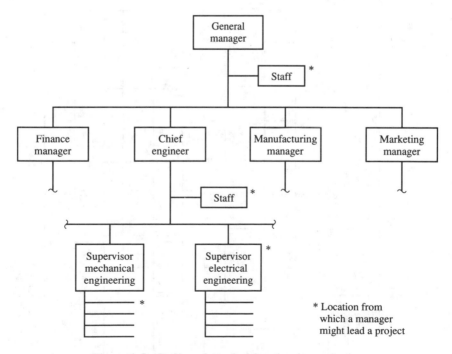

Figure 1-5   Portion of a typical functional organization.

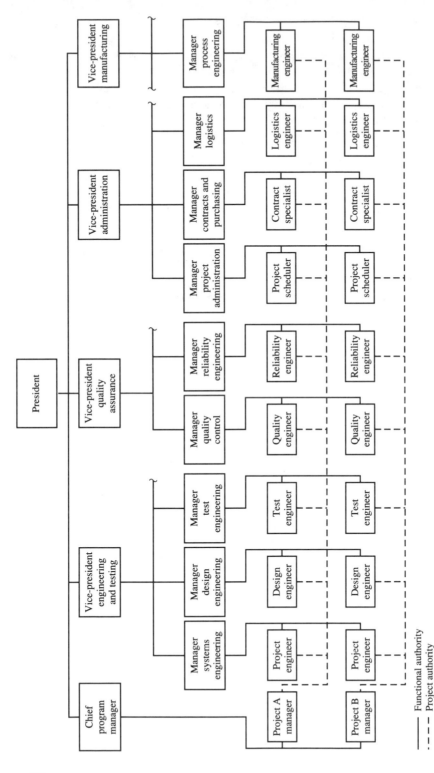

Figure 1-6 Typical matrix organization.

Bard et al. 1.6

——— Functional authority
- - - Project authority

may be supervised by a project team. Full-time personnel are assigned to the project and are physically located with other team members. The project has its own management structure and budget as though it were a separate division of the company.

5. *Projectized organization.* When the project is extremely complex, of long duration, and involves a number of disparate organizations, it is advisable to give one person complete control of all the elements necessary to accomplish the stated goals. For example, when Rockwell International was awarded two multimillion-dollar contracts (the Apollo command and service modules, and the second stage of the Saturn launch vehicle) by NASA, two separate programs were set up in different locations of the organization. Each program was under a division vice-president and had its own manufacturing plant and staff of specialists. Such an arrangement takes the idea of a self-sufficient project team to an extreme and is known as a *projectized* organization.

Table 1-2 enumerates some advantages and disadvantages of the functional and projectized organizations. Companies that are frequently involved in a series of projects and occasionally shift around personnel often elect to use a matrix organization. This type of organization provides the flexibility to assign employees to one or more projects. In this arrangement, project personnel maintain a permanent reporting relationship that connects vertically to the supervisor, who directs the scope of their work. At the same time, each person assigned to a project has a horizontal reporting relationship to the manager of a particular project, who coordinates his or her participation in that project. Pay and career advancement are developed within a particular discipline even though a person may be assigned to different projects. At times this dual reporting relationship can give rise to a host of personnel problems.

Regardless of the structure, a project will usually draw on many of the organization's administrative support groups. It is unnecessary to duplicate existing services in procurement, legal, human resources, logistics, or most other support areas. In addition,

TABLE 1-2 ADVANTAGES AND DISADVANTAGES OF TWO ORGANIZATIONAL STRUCTURES

Functional organization	Projectized organization
Advantages	
Efficient use of technical personnel	Good project schedule and cost control
Career continuity and growth for technical personnel	Single point for customer contact
Good technology transfer between projects	Rapid reaction time possible
Good stability, security, and morale	Simpler project communication
	Training ground for general management
Disadvantages	
Weak customer interface	Uncertain technical direction
Weak project authority	Inefficient use of specialists
Poor horizontal communications	Insecurity regarding future job assignments
Discipline (technology) rather than program oriented	Poor crossfeed of technical information between
Slower work flow	projects

the availability of microcomputers and project management software makes it possible to tie all of these functions together.

1.4 PROJECT MANAGER

The presence of uncertainty coupled with limited experience and hard-to-find data makes project management a combination of art, science, and most of all, logical thinking. A good project manager must be familiar with a large number of disciplines and techniques. Breadth of knowledge is particularly important because most projects have technical, financial, marketing, and organizational aspects that inevitably conspire to derail the best of plans.

The role of the project manager may start at different points in the life cycle of a project. Some managers are involved from the beginning, helping to select the project, form the team, and negotiate the contracts. Others may begin at a later stage and be asked to execute plans that they did not have a hand in developing. At some point, though, most project managers deal with the basic issues: scheduling, budgeting, resource allocation, resource management, human relations, and negotiations.

It is essential and perhaps the most difficult part of the project manager's job to pay close attention to the entire picture without losing sight of the critical details, no matter how slight. The project manager has to trade off different aspects of the project each time a decision is called for. Questions like: "How important is the budget relative to the schedule?" and "Should more resources be acquired to avoid delays at the expense of a budget overrun, or should a slight deviation in performances be tolerated as long as the project is kept on schedule and on budget?" are common.

Some skills can be taught, whereas others come only with time and experience. We will not dwell on these but simply point them out as we attempt to define fundamental principles and procedures. Nevertheless, one of our basic aims is to highlight the practical aspects of project management and to show how modern organizations can function more effectively by adopting them. In so doing, we hope to provide all members of the project team with a comprehensive view of the field.

1.4.1 Basic Functions

The Project Management Institute identifies six basic functions that the discipline must address.

1. Manage the project's scope by defining the goals and work to be done in sufficient detail to facilitate understanding and corrective action, should the need arise.
2. Manage the human resources involved in the project.
3. Manage communications to see that the appropriate parties are informed and have sufficient information to keep the project on track.
4. Manage time by planning and meeting a schedule.

5. Manage quality so that the project's results are satisfactory.
6. Manage costs so that the project is performed at the minimum practical cost and within budget, if possible.

Managing a project is a complex and challenging assignment. Since projects are one-of-a-kind endeavors, there is little in the way of experience, normal working relationships, or established procedures to guide the participants. A project manager may have to coordinate many diverse efforts and activities to achieve the project goals. Persons from various disciplines and from various parts of the organization who have never worked together may be assigned to the project for different spans of time. Subcontractors who are unfamiliar with the organization may be brought in to carry out major tasks. The project may involve thousands of interrelated activities performed by persons employed by any one of several different subcontractors or by the sponsoring organization.

For these and other reasons, it is important that the project leaders have an effective means of identifying and communicating the planned activities and their interrelationships. A computer-based scheduling and monitoring system is usually essential. Network techniques such as CPM (*critical path method*) and PERT (*program evaluation and review technique*) are likely to figure prominently in such systems. CPM was developed in 1957 by J. E. Kelly of Remington-Rand and M. R. Walker of Dupont to aid in scheduling maintenance shutdowns of chemical plants. PERT was developed in 1958 under the sponsorship of the U.S. Navy Special Projects Office, as a management tool for scheduling and controlling the Polaris missile program. Collectively, their value has been demonstrated time and again during both the planning and the execution phases of projects.

1.4.2 Characteristics of Effective Project Managers

The project manager is responsible for assuring that tasks are completed on time and within budget, but often has no formal authority over those actually performing the work. He or she must therefore have a firm understanding of the overall job, and rely on negotiation and persuasion skills to influence the array of contractors, functionaries, and specialists assigned to the project. The skills that a typical project manager needs are summarized in Fig. 1-7; the complexity of the situation is depicted in Fig. 1-8, which shows client, subcontractor, and top management interactions.

The project manager is a lightning rod, frequently under a storm of pressure and stress. He must deal effectively with the changing priorities of his client, the anxieties of his own management ever fearful of cost and schedule overruns, and the divided loyalties of the personnel assigned to his team. The ability to trade off conflicting goals and to find the optimal balance between conflicting pressures is probably the most important skill of the job.

In general, project managers need enthusiasm, stamina, and an appetite for hard work to withstand the onslaught of technical and political concerns. Where possible, they should have seniority and position in the organization commensurate with that of the functional managers with whom they must deal. But whether they are coordinators within a functional structure, or managers in a matrix structure, they will frequently find their

Figure 1-7   Important skills for the project manager.

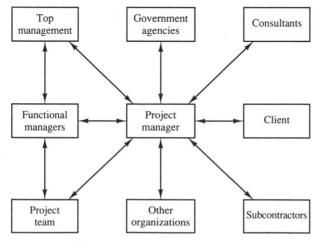

Figure 1-8   Major interactions of project participants.

formal authority incomplete. Therefore, they must have the blend of technical, administrative, and interpersonal skills as illustrated in Fig. 1-7 to furnish effective leadership.

1.5 COMPONENTS, CONCEPTS, AND TERMINOLOGY

The definition of a project fits a large variety of individual and organizational endeavors. Remodeling a house, writing a paper, building a new facility, introducing a new computer, and developing a training program are just a few examples. Although each project has a unique set of goals, there is enough commonality at a generic level to permit the development of a unified framework for planning and control. Project management techniques are designed to handle the common processes and problems that arise during a project's life cycle. This does not mean, however, that one versed in such techniques will be a successful manager. Experts are needed to collect and interpret data, negotiate contracts, arrange for resources, and deal with the welter of technical and organizational issues that impinge on both the cost and schedule.

The following list contains the major components of a "typical" project.

- *Project initiation, selection, and definition*
  Identification of needs [e.g., by *quality function deployment* (QFD)]
  Development of (technological) alternatives
  Evaluation of alternatives
  Selection of the "most promising" alternatives
  Estimation of the *life-cycle cost* of the promising alternatives
  Assessment of risk
  Development of a *configuration baseline*
  "Selling" the configuration and getting approval

- *Project organization*
  Selection of participating organizations
  Structuring the work content of the project into a *work breakdown structure* (WBS)
  Development of the project organizational structure and associated communication and reporting facilities
  Allocation of WBS elements to participating organizations

- *Analysis of activities*
  Definition of the project's major tasks
  Development of a list of activities required to complete the project's tasks
  Development of precedence relations among activities
  Development of a network model
  Development of higher-level network elements (hammock activities, subnetworks)
  Development of milestones
  Updating of the network and its elements

- *Project scheduling*
  Development of a calendar
  Estimation of activity durations
  Estimation of activity performance dates
  Monitoring actual progress and milestones
  Updating the schedule

- *Resource management*
  Definition of resource requirements
  Acquisition of resources
  Allocation of resources among projects/activities
  Monitoring of actual resource use and cost

- *Technological management*
  Development of a configuration management plan
  Identification of technological risks
  Configuration control
  Risk management and control
  Total quality management (TQM)

- *Project budgeting*
  Estimation of direct and indirect costs
  Development of a cash flow forecast
  Development of a budget
  Monitoring actual cost

- *Project execution and control*
  Development of data collection systems
  Development of data analysis systems
  Execution of activities
  Data collection and analysis
  Detection of deviations in cost, configuration, schedule, and quality
  Development of corrective plans
  Implementation of corrective plans
  Forecasting of project cost at completion

- *Project termination*
  Evaluation of project success
  Recommendation for improvements in project management practices
  Analysis and storage of information on actual cost, actual duration, actual performances, and configuration

Each of these activities is discussed in detail in subsequent chapters. Presently, we give an overview with the intention of introducing important concepts and the relationships among them. We also mention some of the tools developed to support the management of each activity.

1. *Project initiation, selection, and definition.* This process starts with identifying a need for a new service, product, or system. The trigger can come from any number of sources, including a current client, line personnel, or a proposed request from an outside organization. If the need is considered important and feasible solutions exist, the need is translated into technical specifications through such techniques as QFD (quality function deployment). Next, a study of an alternative approach may be initiated. Each alternative is evaluated based on a predetermined set of performance measures, and the most promising alternatives are put on a candidate list. Following this, an effort is made to estimate the costs and returns associated with the most suitable candidates. Cost estimates for development, production (or purchasing), maintenance, and operations form the basis of a "life-cycle cost" model that provides a framework for selecting the "optimal" alternative.

Due to uncertainty, most of the estimates are likely to be problematic. A risk assessment may be required if high levels of uncertainty are present. The risk associated with an unfavorable outcome is defined as the probability of that outcome multiplied by the cost associated with it. Major risk drivers should be identified early in the process, and contingency plans should be prepared to handle unfavorable events if and when they occur.

Once an alternative is chosen, design details are fleshed out during the concept formulation and definition phase of the project. Preliminary design efforts end with a configuration baseline. This configuration (the principal alternative) has to satisfy the "customer's" needs and be accepted and approved by management. A well-structured selection and evaluation process in which all relevant parties are involved increases the probability of management approval. A generic flow diagram for the processes of project initiation selection and definition is presented in Fig. 1-9.

2. *Project organization.* Many entities, ranging from private firms and research laboratories to public utilities and government agencies, may participate in a particular project. In the advanced development phase, it is common to define the work content [statement of work (SOW)] as a set of tasks, and to array them hierarchically in a treelike form known as the *work breakdown structure* (WBS). The relationship between participating organizations known as the *organizational breakdown structure* (OBS) is similarly represented.

In the OBS, the lines of communication between and within organizations are defined, and procedures for work authorization and report preparation and distribution are established. Finally, lower-level WBS elements are assigned to lower-level OBS elements to form work packages and a responsibility matrix is constructed, indicating which organizational unit is responsible for which WBS element.

At the end of the advanced development phase, a more detailed cost estimate and a long-range budget proposal are prepared and submitted for management approval. A positive response signals the go-ahead for detailed planning and organizational design. This includes the next five functions.

3. *Analysis of activities.* To assess the need for resources and to prepare a detailed schedule, it is necessary to develop a detailed list of activities that are to be performed. These activities should be aimed at accomplishing the WBS tasks in a logical, economically sound, and technically feasible manner. Each task defined in the initial planning phase may consist of one or more activities. Feasibility is assured by introducing precedence relations among activities. These relations can be represented graphically in the form of a network model.

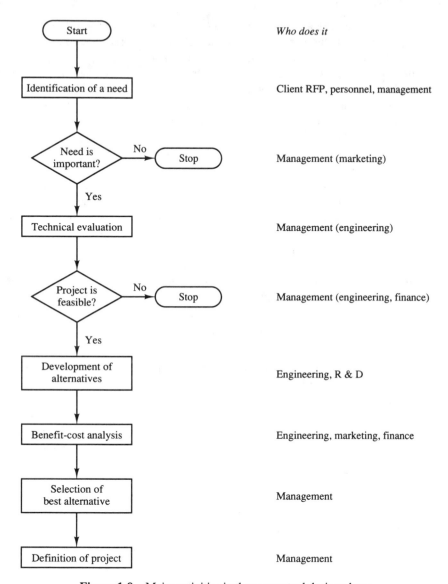

Figure 1-9   Major activities in the conceptual design phase.

The termination of important activities is defined as milestones and is represented in the network model as nodes. Milestones provide feedback in support of project control and form the basis for budgeting, scheduling, and resources management. As progress is made, the model has to be updated to account for the inclusion of new activities in the WBS, the successful completion of tasks, and any changes in design, organization and schedule due to uncertainty, new needs, or new technological and political developments.

4. *Project scheduling.* The expected execution dates of activities are important from both a financial (acquisition of the required funds) and operational (acquisition of the required resources) point of view. Scheduling of project activities starts with the definition of a calendar specifying the working hours per day, working days per week, holidays, and so on. The expected duration of each activity is estimated and a project schedule is developed based on the calendar, precedence relations among activities, and the expected duration of each activity. The schedule specifies the start and ending dates of each activity and the accompanying slack or leeway. This information is used in budgeting and resource management. The schedule is used as a basis for work authorization and as a baseline against which actual progress is measured. It is updated throughout the life cycle of the project to reflect actual progress.

5. *Resource management.* Activities are performed by resources so that before any concrete steps can be taken, requirements have to be identified. This means defining one or more alternatives for meeting the estimated needs of each activity (the duration of an activity may be a function of the resources assigned to perform it). Based on the results, and in light of the project schedule, total resource requirements are estimated. These requirements are the basis of resource management and resource acquisition plans.

In the case where resource requirements exceed expected availability, schedule delays may occur unless the deficit is made up through additional acquisitions or by subcontracting. Alternatively, it may be possible to reschedule activities (especially those with slack) so as not to exceed expected resource availability. Other considerations, such as minimizing fluctuations in resource usage and maximizing resource utilization, may be applicable as well.

During the execution phase, resources are allocated periodically to projects and activities in accordance with a predetermined timetable. However, because actual and planned use may differ, it is important to monitor and compare progress to plans. Low utilization as well as higher than planned costs or consumption rates indicate problems and should be brought to the immediate attention of management. Large discrepancies may call for significant alterations in the schedule.

6. *Technological management.* Once the primary candidates are evaluated and a consensus forms, the approved configuration is adopted as a baseline. From the baseline, plans for project execution are developed, tests to validate operational and technical requirements are designed, and contingency plans for risky areas are formulated. Changes in needs or in the environment may trigger modifications to the configuration. Technological management deals with execution of the project to achieve the approved baseline. Principal functions include the evaluation of proposed changes, the introduction of approved changes into the configuration baseline, and the development of a total quality management program. The latter involves the continuous effort to prevent defects, to improve the process, and to guarantee a final result that fits the specifications of the project and the expectations of the client.

7. *Project budgeting.* Money is the most common resource used in a project. Resources have to be acquired and suppliers have to be paid. Overhead costs have to be assigned and subcontractors have to be put on the payroll. The preparation of a budget is an important management activity that results in a time-phased plan summarizing expected expenditures, income, and milestones.

The budget is derived by estimating the cost of activities and resources. Since the schedule of the project relates activities and resource use to the calendar, the budget is also related to the same calendar. With this information, a cash flow analysis can be performed and the feasibility of the predicted outlays can be tested. If the resulting cash flow or the resulting budget are not acceptable, the schedule should be modified. This is frequently done by delaying activities that have slack.

Once an acceptable budget is developed, it serves as the basic financial tool for the project. Credit lines and loans can be arranged and the cost of financing the project can be assessed. As work progresses, information on actual cost is accumulated and compared to the budget. This comparison forms the basis for controlling costs.

The sequence of activities performed during the detailed design phase is summarized in Fig. 1-10.

8. *Project execution and control.* The activities described so far comprise the first steps in initializing and preparing a project for execution. A feasible schedule that inte-

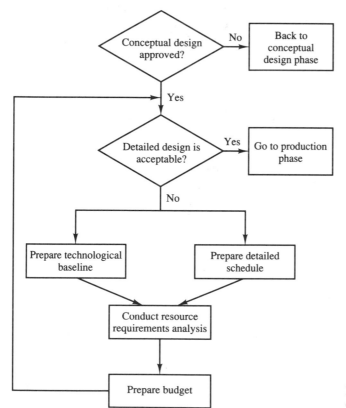

Figure 1-10   Major activities in the detailed design phase.

grates task deadlines, budget considerations, resource availability, and technological requirements, while satisfying the precedence relations among activities, provides a good starting point for a project.

It is important, however, to remember that successful implementation of the initial schedule is subject to unexpected or random effects that are difficult (or impossible) to predict. In situations where all resources are under the direct control of management, and activated according to plan, unexpected circumstances or events may sharply divert progress from the original plan. For those resources not under complete management control, much higher levels of uncertainty may exist. Sources of uncertainty, such as a downturn in the economy, labor unrest, technology breakthroughs or failures, and new environmental regulations, may always be a cause for alarm. These uncertainties should be monitored by keeping check on the effect they have on the project's progress.

Project control systems are designed with three purposes in mind: (1) to detect current deviations and to forecast future deviations between actual progress and the project plans, (2) to trace the source of these deviations, and (3) to support management decisions aimed at putting the project back on the desired course.

Project control is based on the collection and analysis of the most recent performance data. Thus actual progress, cost, resource use, and technological achievements should be monitored continually. The information gleaned from this process is compared to the updated plans across all aspects of the project. Because deviations in one area (e.g., schedule overrun) may affect the performance and deviations in other areas (e.g., cost overrun), it is important to look at things from a systems point of view.

In general, all operational data collected by the control system are analyzed, and if deviations are detected, a scheme is devised to put the project back on course. The existing plan is modified accordingly, and steps are taken to monitor its implementation.

During the life cycle of the project, a continuous effort is made to update original estimates of completion dates and costs. These updates are used by management to evaluate the success of the project and the efficiency of the participating organizations. These evaluations form the basis of management forecasts regarding the expected success of the project at each stage of its life cycle.

Special control systems are necessary when several projects are being conducted simultaneously. If the deviations in cost are positive for some projects and negative or zero for others, the total budget might not be affected due to the former canceling the latter. Nevertheless, if the deviations are all negative, and actual cost tends to be higher than the expected cost, the entire organization might find itself in a dilemma. These risks constitute the need for a control system so that the decision makers can be forewarned of any problems that could delay the project.

Schedule deviations might have a similar effect since payments are usually based on actual progress. If a schedule overrun occurs and payments are delayed, cash flow difficulties might result. Schedule overruns might also cause excess load on resources due to the accumulation of work content. A well-designed control system in the hands of a well-trained project manager is the best way of counteracting the negative effects of uncertainty that we can recommend.

9. *Project termination.* A project does not necessarily terminate as soon as its technical objectives are met. Management should strive to learn from past experience to improve the handling of future projects. A detailed analysis of the original plan, the modifications made over time, the actual progress, and the relative success of the project should be conducted. The underlying goal is to identify procedures and techniques that were not effective and to recommend ways of improving operations. An effort aimed at identifying missing or redundant managerial tools should also be initiated; new techniques for project management should be adopted where necessary, and obsolete procedures and tools discarded.

Information on the actual cost and duration of activities, and the cost and utilization of resources, should be stored in well-organized databases to support the planning effort in future projects. Only by striving for continuous improvement through programs based on past experience is competitiveness likely to persist in an organization. Policies, procedures, and tools must be updated on a regular basis.

1.6 LIFE CYCLE OF A PROJECT: STRATEGIC AND TACTICAL ISSUES

Because of the degree to which projects differ in their principal attributes, such as length, cost, type of technology used, and sources of uncertainty, it is difficult to generalize the operational and technical issues they each face. It is possible, however, to discuss some strategic and tactical issues that are relevant to many types of projects. The framework for the discussion is the project life cycle or the major phases through which a "typical" project progresses. An outline of these phases is depicted in Fig. 1-11 and elaborated on by Cleland and King (1983), who identify the long-range (strategic) and medium-range (tactical) issues that management must consider. A synopsis follows.

1. *Conceptual design phase.* In this phase, an organization (client, contractor, subcontractor) initiates the project and evaluates potential alternatives. A client organization may start by identifying a need or a deficiency in existing operations and issuing a request for proposal (RFP). If the soliciting organization is a government agency, the RFP will be published in the *Commerce Business Daily*. Potential contractors can study the RFPs and develop a strategy for responding.

The selection of projects at the conceptual design phase is a strategic decision based on the established goals of the organization, needs, ongoing projects, and long-term commitments and objectives. In this phase, expected benefits from alternative projects, assessment of cost and risks, and estimates of required resources are some of the factors weighed. Important action items include the initial "make or buy" decisions for components and equipment, the development of contingency plans for high-risk areas, and the preliminary selection of subcontractors and other team members that will participate in the project.

In addition, upper management must consider the technological aspects, such as availability and maturity of the required technology, its performance, and expected usage

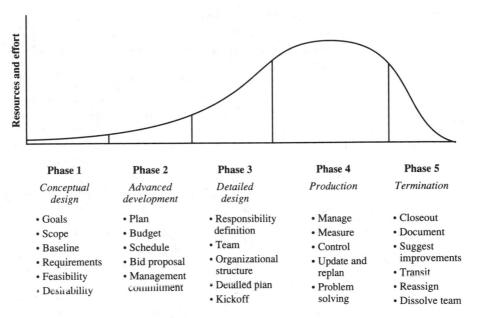

Phase 1	Phase 2	Phase 3	Phase 4	Phase 5
Conceptual design	*Advanced development*	*Detailed design*	*Production*	*Termination*
• Goals	• Plan	• Responsibility definition	• Manage	• Closeout
• Scope	• Budget	• Team	• Measure	• Document
• Baseline	• Schedule	• Organizational structure	• Control	• Suggest improvements
• Requirements	• Bid proposal	• Detailed plan	• Update and replan	• Transit
• Feasibility	• Management commitment	• Kickoff	• Problem solving	• Reassign
• Desirability				• Dissolve team

Figure 1-11   Project life cycle.

in subsequent projects. Environmental factors related to government regulations, potential markets, and competition must also be analyzed.

The selection of projects is based on a variety of goals and performance measures, including expected cost, profitability, risk, and potential for follow-on assignments. Once a project is selected and its conceptual design is approved, work begins on the second phase, where many of the details are ironed out.

2. *Advanced development phase.* In this phase the organizational structure of the project is formed by weighing the tactical advantages and disadvantages of each possible arrangement mentioned in Section 1.3.4. Once a decision is made, lines of communication and procedures for work authorization and performance reporting are established. This leads to the framework in which the project is executed.

3. *Detailed design phase.* This is the phase in a project's life cycle where comprehensive plans are prepared. These plans consist of:

- Product and process design
- Final performance requirements
- Detailed breakdown of the work structure
- Scheduling information
- Blueprints for cost and resource management
- Detailed contingency plans for high-risk activities
- Budgets
- Expected cash flows

In addition, and most important, procedures and tools for executing, controlling, and correcting the project are developed. When this phase is completed, implementation can begin since the various plans should cover all aspects of the project in sufficient detail to support work authorization and execution.

The success of a project is highly correlated with the quality and depth of the plans prepared during this phase. A detailed design review of each plan and each aspect of the project is therefore conducted prior to approval. A sensitivity analysis of environmental factors that contribute to uncertainty may also be needed. This analysis is typically performed as part of "what-if" studies using expert opinions and simulation as supporting mechanisms.

In most situations, the resources committed to the project are defined during the initial phases of its life cycle. Although these resources are used later, the strategic issues of how much to spend and at what rate are addressed here.

4. *Production phase.* The fourth life-cycle phase involves the execution of plans and in most projects dominates the others in effort and duration. The critical strategic issue here relates to maintaining top management support, while the critical tactical issues center on the flow of communications within and among the participating organizations. At this level the focus is on actual performance and changes in the original plans. Modifications can take different forms—in the extreme case, a project may be canceled. More likely, though, the scope of work, schedule, and budget will be adjusted as the situation dictates. Throughout this phase, management's task is to assign work to the participating parties, to monitor actual progress, and to compare it to the baseline plans. The establishment and operation of a well-designed communications and control system are therefore necessary.

Support of the product or system throughout its entire life (logistic support) requires management attention in most engineering projects where an operational phase is scheduled to follow implementation. The preparation for logistic support includes documentation, personnel training, maintenance, and initial acquisition of spare parts. Neglecting this activity or giving it only cursory attention can doom an otherwise successful venture.

5. *Termination phase.* In this phase, management's goal is to consolidate what it has learned and translate this knowledge into ongoing improvements in the process. Current lessons and experience serve as the basis for improved practice. But while successful projects can provide valuable insights, failures can teach us even more. Unless we learn from our mistakes we are bound to repeat them, making the task of continuous improvement little more than an empty exercise. Databases that store and support the retrieval of information on cost, schedules, resource utilization, and so on, are assets of an organization. Readily available, accurate information is a key factor in the success of future projects.

6. *Operational phase.* The operational phase may or may not be included in the project life cycle. In many instances, termination takes place before the end of the operational phase, so the project is considered finished even when production and deployment continue to occur. Once the production phase is completed and the final results are accepted (normally, it is the responsibility of the quality control group to evaluate the system by performing acceptance tests based on the original specifications and other technological information), the product enters its operational phase. The operational phase is frequently carried out by organizations other than those involved in the earlier life-cycle stages and may be outside the scope of the project. From the project manager's point of view, this phase is the most crucial because it is here that a judgment is made as to whether or not the project has achieved its operational goals.

Strategic issues such as long-term relationships with customers, as well as customer service and satisfaction, have a strong influence on upper management's attitudes and decisions. Therefore, the project manager should be particularly aware of the need to open and maintain lines of communication between all parties, especially during this phase.

In some projects the transition from the production phase to the operational phase is gradual. This is the case when a project involves the production of a series of similar units. The first few units might be placed in operation while the remainder are still in various stages of completion. Special communications and control techniques (e.g., line of balance) have been developed for this type of situation to help the project manager carry out his or her duties throughout the life cycle of the project.

1.7 FACTORS AFFECTING THE SUCCESS OF A PROJECT

The project manager must handle a wide variety of tasks simultaneously. Decisions have to be made continuously at all levels regarding resource use, schedule adjustments, budgeting, organizational communications, technical problems, and human relations. There is a need to identify all the basic issues, whether they be strategic, tactical, or operational, and to establish a priority scheme for helping the project manager focus on those that are critical. Of course, defining what is critical will vary from one phase of the project to another.

This issue was addressed in a study by Pinto and Slevin (1987) aimed at finding those factors that contribute most to a project's success, and at measuring their significance over the life cycle. They found the following 10 factors to be of primary importance. Additional insights are provided by Balachandra (1984) regarding new product development.

1. *Project mission and goals.* Well-defined and intelligible understanding of the project goals are the basis of planning and executing the project. Understanding the goals and performance measures used in the evaluation is important for good coordination of efforts and building organizational support. Therefore, starting at the conceptual design phase of the project life cycle, the overall mission should be defined and explained to team members, contractors, and other participants.

2. *Top management support.* The competition for resources, coupled with the high levels of uncertainty typically found in the project environment, often lead to conflict and crisis. The continuous involvement of top management throughout the life cycle of the project increases their understanding of its mission and importance. This awareness, if translated into support, may prove invaluable in resolving problems when crises and conflicts arise or when uncertainty strikes. Therefore, continued, solid communication between the project manager and top management is a catalyst for the project to be a success.

3. *Project planning.* The translation of the project mission, goals, and performance measures into a workable (feasible) plan is the link between the conceptual design phase and the production phase. A detailed plan that covers all aspects of the project—technical, financial, organizational, scheduling, communication, and control—is the basis of implementation. Planning does not end when execution starts because deviations from the original plans during implementation may call for replanning and updating from one period to the next. Thus planning is a dynamic and continuous process that links changing goals and performance to the final results.

4. *Client consultation.* The ultimate user of the project is the final judge of its success. A project that was completed on time according to the technical specifications and within budget but was never (or rarely) used can certainly be classified as a failure. In the conceptual design phase of the project life cycle, client input is the basis for setting the mission and establishing goals. In subsequent phases continuous consultation with the client can help in correcting errors previously made in translating goals into performance measures. However, due to changing needs and conditions, a mission statement that represented the client's needs in the conceptual design phase accurately may no longer be valid in the planning or implementation phases. As discussed in Chapter 6, the configuration management system provides the link between existing plans and change requests issued by the client, as well as the project team.

5. *Personnel issues.* Satisfactory achievement of technical goals without violating schedule and budgetary constraints does not necessarily constitute a complete success, even if the customer is satisfied. If relations among team members, between team members and the client, or between team members and other personnel in the organization are poor and morale problems are frequent, project success is doubtful. Well-motivated teams with a sufficient level of commitment to the project and a good relationship with the client are the key determinants of project success.

6. *Technical issues.* Understanding the technical aspects of the project and ensuring that members of the project team possess the necessary skills are important responsibilities of the project manager. Inappropriate technologies or technical incompatibility may affect all aspects of the project, including cost, schedule, actual performances, and morale.

7. *Client acceptance.* Ongoing client consultation during the project life cycle increases the probability of success regarding user acceptance. In the final stages of implementation, the client has to judge the resulting project and decide whether or not it is acceptable. A project that is rejected at this point must be viewed as a failure.

8. *Project control.* The continuous flow of information regarding actual progress is a feedback mechanism that allows the project manager to cope with uncertainty. By comparing actual progress to current plans, the project manager can identify deviations, anticipate problems, and initiate corrective actions. Lower-than-planned achievements in technical areas as well as schedule and cost deviations detected early in the life cycle can help the project manager focus on the important issues. Plans can be updated or partially adjusted to keep the project on schedule, on budget, and on target with respect to its mission.

9. *Communication.* The successful transition between the phases of a project's life cycle and good coordination between participants during each phase requires a continuous exchange of information. In general, communication in the project team, with other parts of the organization, and between the project managers and the client is made easier if lines of authority are well defined. The organizational structure of the project should specify the communication channels and the information that should flow through each one. In addition, it should specify the frequency at which this information should be generated and transmitted. The formal communication lines, in addition to a positive working environment that enhances informal communication within the project team, contribute to the success of a project.

10. *Troubleshooting.* The control system is designed to identify problem areas, and if possible, to trace their source through the organization. Since uncertainty is always a likely culprit, the development of contingency plans is a valuable preventive step. The availability of prepared plans and procedures for handling problems can reduce the effort required for dealing with them should they actually occur.

The foregoing list of factors affecting the success of a project is broad enough to provide a starting point for discussing the issues surrounding project management. Of course, the individual nature of projects gives rise to a unique set of factors that go beyond any attempts to generalize. Specific examples, usually in the form of case studies drawn from our collective experience, are given throughout the book to provide additional insights.

1.8 ABOUT THE BOOK: PURPOSE AND STRUCTURE

The book is designed to bridge the gap between theory and practice by presenting the tools and techniques most suited for modern project management. A principal goal is to give managers, engineers, and technology experts a larger appreciation of their roles by defining a common terminology and by explaining the interfaces between the underlying disciplines.

Theoretical aspects are covered at a level appropriate for a senior undergraduate course or a first-year graduate course in either an engineering or an MBA program. Special attention is paid to the use and evaluation of specific tools with respect to their real-world applicability. Whether the book is adopted for a course, or is read by practitioners who want to learn the "tools of the trade," we have tried to present the subject matter in a concise and fully integrated manner.

The book is structured along functional lines and offers an in-depth treatment of the economic aspects of project selection and evaluation, the technological aspects of configuration management, and the various issues related to budgeting, scheduling, and control. By examining these functions and their organizational links, a comprehensive picture emerges of the relationship that exists between project planning and implementation.

The end of each chapter contains a series of discussion questions and exercises designed to stimulate thought and to test the readers' grasp of the material. In some cases the intent is to explore supplementary issues in a more open-ended manner. Also included at the end of each chapter is a team project centering on the design and construction of a solid waste disposal facility known as a *thermal transfer plant*. As readers go from one chapter to the next, they are asked to address a particular aspect of project management as it relates to the planning of this facility. Each of the remaining 13 chapters deals with a specific component of project management or a specific phase in the project life cycle. A short description of Chapters 2 through 14 follows.

In *Chapter 2* we address the economic aspects of projects and the quantitative techniques developed for analyzing a specific alternative. The long-term perspective is first presented via the concept of time value of money. Investment evaluation criteria based on the payback period, net present value, and internal rate of return are discussed. Next, the short-term perspective is considered via cash flow analysis and its effect on a single project or a portfolio of projects. The notion and impacts of risk are introduced, followed by some concepts common to decision making, such as expected monetary value, utility theory, break-even analysis, and diminishing returns. Specific decisions such as buy, make, rent, or lease are also elaborated.

The selection of a project from a list of available candidates and the selection of a particular configuration for a specific project are two key management decisions. The purpose of *Chapter 3* is to present the basic techniques that can be used to aid in this process. Checklists and scoring models are the simplest and first to be introduced. This is followed by a presentation of the formal aspects of benefit-cost and cost-effectiveness analysis. Issues related to risk, and how to deal with them, tie all the material together. The chapter closes with a comprehensive treatment of decision trees. The strengths and weakness of each methodology are highlighted and examples are given to demonstrate the computations.

It is rare that any decision is made on the basis of one criterion alone. To deal more rigorously with situations where many objectives, often incommensurate, must be juggled simultaneously, a value model is needed that goes beyond simple checklists. In *Chapter 4* we introduce two of the most popular such models for combining incommensurate objectives into a single measure of performance. Multiattribute utility theory (MAUT) is the first presented. Basic theory is discussed along with the guiding axioms. Next, the concepts and assumptions behind the analytical hierarchy process (AHP) are detailed. A case study contained in the appendix documents the results of a project aimed at comparing the two approaches and points out the relative advantages of each.

The integrative notions of the organizational breakdown structure and the work breakdown structure are introduced in *Chapter 5*. The former combines several organizational units residing in one or more organizations by defining communication channels for work authorization, performance reports, and assigning general responsibility for tasks. Questions related to the selection of the most appropriate organizational structure are addressed, and the advantages and disadvantages of each are presented.

Next, the work breakdown structure of projects is discussed. This structure combines hardware, software, and services performed in a project into a hierarchical framework. It further facilitates identification of the critical relationships that exist among various project components.

Subsequently, the combined OBS–WBS matrix is introduced, where each element in the lowest WBS level is assigned to an organizational unit at the lowest level of the OBS. This type of integration is the basis for detailed planning and control, as explained in subsequent chapters. We close with a discussion of human resources and the special needs they engender for the project manager.

In *Chapter 6* the process by which the technological configuration of projects is developed and maintained is discussed. We show how tools like benefit-cost analysis and MAUT can be used to select the best technological alternative from a set of potential candidates. Procedures used to handle engineering change requests via configuration management and configuration control are presented. Finally, the integration of total quality management into the project, and its relationship to configuration test and audit, are highlighted.

Network analysis has played an important role in project scheduling over the last 30 years. In *Chapter 7* we introduce the notions of activities, precedence relations, and task times, and show how they can be combined in an analytic framework to provide a mechanism for planning and control. The idea of a calendar and the relationship between activities and time are presented, first by Gantt charts, and then by network models of the AOA/AON (activity-on-arrow/activity-on-node) type. This is followed by a discussion of precedence relations, feasibility issues, and the concepts of milestones, hammock activities, and subnetworks. Finally, uncertainty is introduced along with the PERT approach to estimating the critical path and the use of Monte Carlo simulation to gain a deeper understanding of a project's dynamics.

In *Chapter 8* we deal with the budget as a tool by which organizational goals, policies, and constraints are transformed into an executable plan that relates task completions and capital expenditures to time. Techniques commonly used for budget development, presentation, and execution are discussed. Issues also examined are the relationship between the duration and timing of activities and the budget of a project, cash flow constraints and liabilities, and the interrelationship among several projects performed by a single organizational unit.

Chapter 9 opens with a discussion on the type of resources used in projects. A classification scheme is developed according to resource availability, and performance measures are suggested for assessing efficiency and effectiveness. Some general guidelines are presented as to how resources should be used to achieve better performance levels. The relationship between resources, their cost, and the project schedule is analyzed, and mathematical models for resources allocation and leveling are described.

The integration of life-cycle cost (LCC) analysis into the project management system is covered in *Chapter 10*. LCC concepts and the treatment of uncertainty in the analysis are discussed, as well as classification schemes for cost components. The steps required in building LCC models are outlined and explained to facilitate their implementation.

The idea that the cost of new product development is only a fraction of the total cost of ownership is a central theme of the chapter. The total life-cycle cost is determined largely in the early phases of a project, when decisions regarding product design and process selection are being made. Some of the issues discussed in this context include cost estimation and risk evaluation. The concept of the cost breakdown structure and how it is used in planning is also presented.

The execution of a project is frequently subject to unforeseen difficulties that force deviation from the original plans. The focus of *Chapter 11* is on project control—a function that depends heavily on the early detection of such deviations. The integration of OBS and

WBS elements serves as a basis for the control system. Complementary components include a mechanism for tracing the source of each deviation, and a forecasting procedure for assessing their implications if no corrective action is taken. Cost and schedule control techniques such as the earned value approach are presented, and the concept of cost/schedule control systems criteria (C/SCSC) is made clear.

Engineering projects where new technologies are developed and implemented are subject to high levels of uncertainty. In *Chapter 12* we define R&D projects and highlight their unique characteristics. The typical goals of such projects are discussed and measures of success are suggested. Techniques for handling risk, including the idea of parallel funding, are presented. The need for rework or repetition of some activities is discussed, and techniques based on the *Q*-GERT approach to scheduling R&D projects are outlined. The idea of a portfolio is introduced and tools used for portfolio management are discussed. A case study that involves screening criteria, project selection and termination criteria, and the allocation of limited resources is contained in the appendix.

A wide variety of software has been developed to assist the project manager in performing his or her job. In *Chapter 13* we discuss the basic functions and range of capabilities associated with these packages. A classification system is devised and a process by which the most appropriate package can be selected for a project or an organization is outlined. Finally, a list of selected packages is presented together with a summary of the advantages and disadvantages of each.

Most people find it difficult to give up a habit or to abandon an arrangement in which they feel comfortable or secure. Nevertheless, breaks must be made. Long-run goals should not be compromised for the sake of preserving the familiarity of the status quo. In *Chapter 14* the need to terminate a project in a planned, orderly manner is discussed. The process by which information gathered in past projects can be stored, retrieved, and analyzed is presented. Post-mortem analysis is suggested as a vehicle by which continuous improvement can be achieved in an organization. The goal is to show how projects can be terminated so that the collective experience and knowledge can be transferred to future endeavors.

It goes without saying that the huge body of knowledge in the area of project management cannot be condensed into a single book. Over the last 30 years much has been written on the subject in technical journals, textbooks, company reports, and trade magazines. In an effort to cover some of this material, a bibliography of important works is provided at the end of each chapter. The interested reader can further his or her understanding of a particular topic by consulting these references.

The final and unique component of the book is the inclusion of a DOS disk containing an educational version of SuperProject Expert. SuperProject was developed by Computer Associates, Inc. in the mid-1980s, and remains one of the most powerful and widely used project management software products on the market. The educational version has all the functionality of the original commercial version with the exception of a 50-task limit on projects. Recently, Computer Associates has significantly enhanced the basic program to include greater graphics capabilities and a set of features enabling the user to manage several projects at once.

TEAM PROJECT*

Thermal Transfer Plant

Introduction

To exercise the techniques used for project planning and control, and to implement these techniques using the software package provided with this book, the reader is encouraged to work out each aspect of the *thermal transfer plant* case study. At the end of each chapter a short description of the relevant components of the thermal transfer plant is provided along with an assignment. If possible, the assignment should be done in groups of three or four to develop the interpersonal and organizational skills necessary for teamwork.

Not all the information required for each assignment is given. Before proceeding, it may be necessary for the group to research a particular topic and to make some logical assumptions. Accordingly, there is no "correct solution" to compare recommendations and conclusions. Each assignment should be judged with respect to the availability of information and the force of the underlying assumptions.

Total Manufacturing Solutions, Inc.

Total Manufacturing Solutions, Inc. (TMS) designs and integrates manufacturing and assembly plants. Their line of products and services includes the selection of manufacturing and assembly processes for new or existing products, the design and selection of manufacturing equipment, facilities design and layout, the integration of manufacturing and assembly systems, and the training of personnel and startup management teams. The broad range of services that TMS provides to its customers makes it a rather unique and successful organization. Its headquarters are in Nashville, Tennessee, with branches in New York and Los Angeles.

TMS began operations in 1965 as a consulting firm in the areas of industrial engineering and operations management. In the late 1970s, the company started its design and integration business. Recently, it began promoting just-in-time systems and group technology-based manufacturing facilities. The organization structure of TMS is depicted in Fig. 1-12; financial data are presented in Tables 1-3 and 1-4.

TMS employs some 500 people, 300 of whom are in the Nashville area, 100 in New York, and 100 in Los Angeles. About 50% of these are industrial, mechanical, and electrical engineers, and about 10% also have M.B.A. degrees, mostly with operations management concentrations. The other employees are technicians, support personnel, and managers. Some information on labor costs follows.

*The authors wish to thank Warren Sharp and Ian St. Maurice for their help in writing this case study.

Engineers $50,000/year
Technicians $25/hour
Administrators $35,000/year
Other $10/hour

These rates do not include fringe benefits or overhead. Moreover, bear in mind that individual salaries are a function of experience, position, and seniority within the company.

In the last 10 years, TMS averaged 20 major projects annually. Each project consisted of the design of a new manufacturing facility, the selection, installation, and integration of equipment, and the supervision of startup activities. In addition, TMS experts are consultants to more than 100 clients, many of whom own TMS-designed facilities.

The broad technical basis of TMS in the areas of mechanical, electrical, and industrial engineering, coupled with its wide-ranging experience, are its most important assets. Management feels that the company is an industry leader in automatic assembly, material

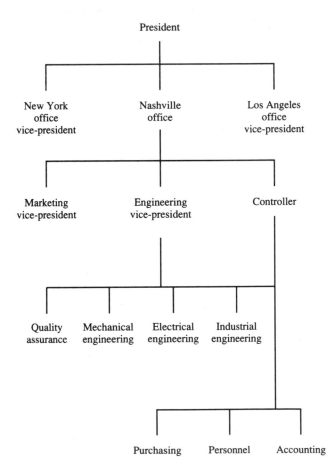

Figure 1-12 Simplified organization chart.

TABLE 1-3 TMS FINANCIAL
DATA: INCOME STATEMENT

Income statement ($1,000) Year ending December 31, 1993	
Net sales	47,350
Cost of goods sold	
Direct labor	26,600
Overhead	6,000
	32,600
Gross Profit	14,750
General and administrative	5,350
Marketing	4,900
	10,250
Profit before taxes	4,500
Income tax (32%)	1,440
Net profit	3,060

TABLE 1-4 TMS FINANCIAL
DATA: BALANCE SHEET

Balance sheet ($1,000) Year ending December 31, 1993	
Assets	
Current assets	
Cash	1,100
Accounts receivables	1,500
Inventory	12
Other	3
Total current assets	2,615
Net fixed assets	325
Total assets	2,940
Liabilities	
Current liabilities	
Notes payable	35
Accounts payable	137
Accruals	90
Total current liabilities	262
Long-term debt	50
Capital stock and surplus	1,300
Earned surplus	1,328
Net worth	2,628
Total liabilities	2,940

handling, industrial robots, command and control, and computer-integrated manufacturing. TMS is using subcontractors mainly in software development, and when necessary, for fabrication, since it does not have any shops or manufacturing facilities.

Recently, management has decided to expand its line of operations and services into the area of recycling and waste management. New regulations in many states are forcing the designers of manufacturing plants to analyze and solve problems related to waste generation and disposal.

Your team has been selected by TMS–Nashville to work on this new line of business. Your first assignment is to analyze the needs and opportunities in your geographical area. Based on a literature search and conversations with local manufacturers, environmentalists, and politicians, making whatever assumptions you feel are necessary, write a report and prepare a presentation that answers the following questions:

1. How well does this new line of business fit into TMS operations? What are the existing or potential opportunities?

2. How should a waste management project be integrated into TMS's current organizational structure?

3. What are the problems that TMS might encounter should it embark on this project? How might these problems affect the project? How might they affect TMS's other business activities?

4. If a project is approved in waste management, what would its major life-cycle phases be?

Any assumptions regarding TMS's financial position and borrowing power, personnel, previous experience, and technological capabilities relating, for example, to computer-aided design should be stated explicitly.

DISCUSSION QUESTIONS

1. Explain the difference between a project and a batch-oriented production system.

2. Describe three projects, one whose emphasis is on technology, one with emphasis on cost, and one with emphasis on scheduling.

3. Identify a project that is "risk free." Explain why this project is not subject to risk (low probability of undesired results, low cost of undesired results, or both).

4. In the text it is stated that a project manager needs a blend of technical, administrative, and interpersonal skills. What attributes do you believe are desirable in an engineering specialist working on a project in a matrix organization?

5. Write a job description for a project manager.

6. Identify a project with which you are familiar, and describe its life-cycle phases and between 5 and 10 of the most important activities in each phase of its life cycle.

7. Find a recent news article on an ongoing project, evaluate the management's performance, and explain how the project could be better organized and managed.

8. Analyze the factors affecting the success of projects as a function of the project's life cycle. Explain in which phase of the life cycle each factor is most important, and why.

9. In a matrix management structure, the person responsible for a specific activity on a specific project has two bosses. What considerations in a well-run matrix organization reduce the resulting potential for conflict?

10. Outline a strategy for effective communication between project personnel and the customer (client).

11. Select a project and discuss what you think are the interfaces between the engineers and managers assigned to the project.

12. The project plan is the basis for monitoring, controlling, and evaluating the project's success once it has started. List the principal components or contents of a project plan.

EXERCISES

1-1. What type of production system would be associated with the following processes?
   (a) A production line for window assemblies
   (h) A special order of 150 window assemblies
   (c) Supplying 1,000 window assemblies per month throughout the year

1-2. You decided to start a self-service restaurant. Identify the stages of this project and the type of production system involved in each stage, from startup until the restaurant is running well enough to sell.

1-3. Select two products and two services and describe the needs that generated them. Give examples of other products and services that could satisfy those needs equally well.

1-4. You have placed an emergency order for materials from a company that is located 2,000 miles away. You were told that it will be shipped by truck and will arrive within 48 hours. Discuss the issues surrounding the probability that the materials will reach you on time. How would things change if shipment were by rail?

1-5. Your plumber recommends that you replace your cast iron pipes with copper pipes. He claims that although the price for the job is $7,000, he has to add $2,000 for unforeseen expenses. Discuss his proposal.

1-6. In statistical analysis, the coefficient of variation is considered to be a measure of uncertainty. It is defined as the ratio of the standard deviation to the mean. Select an activity, say driving from your home to school, generate a frequency distribution for that activity and calculate its mean and the standard deviation. Analyze the uncertainty.

1-7. Specify the type of uncertainties involved in completing each of the following activities successfully.
   (a) Writing a term paper on a subject that does not fall within your field of study
   (b) Undertaking an anthropological expedition in an unknown area
   (c) Driving to the airport to pick up a friend
   (d) Buying an item at an auction

1-8. Your professor told you that the different departments in the school of business administration are organized in a matrix structure. Functional areas include organizational behavior, mathematics (operations research and statistics), and computer science. Develop an organization chart that depicts these functions along with the management, marketing, accounting, and finance departments. What is the product of a business school? Who is the customer?

1-9. Provide an organizational structure for a school of business administration that reflects either a functional orientation or a product orientation.

1-10. Assume that a recreational park is to be built in your community and that the city council has given you the responsibility of selecting a project manager to lead the effort. Write a job description for the position. Generate a list of relevant criteria that can be used in the selection process, and evaluate three fictitious candidates (think about three of your friends).

1-11. Write an RFP soliciting proposals for preparing your thesis. The RFP should take into account the need for tables, figures, and multiple revisions. Make sure that it adequately describes the nature of the work and what you expect so there will be no surprises once a contract is signed.

1-12. Explain how you would select the best proposal submitted in Exercise 1-11. That is, what measures would you use, and how would you evaluate and aggregate them with respect to each proposal?

1-13. The following list of activities is relevant to almost any project. Identify the phase in which each is typically performed, and order them in the correct sequence.
- Developing the network
- Selecting participating organizations
- Developing a calendar
- Developing corrective plans
- Executing activities
- Developing a budget
- Designing a project
- Recommending improvement steps
- Monitoring actual performance
- Managing the configuration
- Allocating resources to activities
- Developing the WBS
- Estimating the life-cycle cost
- Getting the customer's approval to the design
- Establishing milestones
- Estimating the activity duration

1-14. Drawing from your personal experience, give two examples for each of the following situations.
   (a) The original idea was attractive but not sufficiently important to invest in.
   (b) The idea was compelling but was not technically feasible.
   (c) The idea got past the selection process but was too expensive to implement.
   (d) The idea was successfully transformed into a completed project.

1-15. List two projects that either you or your organization are involved with that are currently in each of the various life-cycle phases.

1-16. Select five national projects that were completed successfully and identify the factors affecting their success. Discuss the attending risks, uncertainty, schedule, cost, technology, and resources usage.

1-17. Identify five projects that have failed and discuss the reasons for their failure.

REFERENCES

Elements of Project Management

BALACHANDRA, R., "Critical Signals from Making the Go/No Go Decisions in New Product Development," *Journal of Product Innovation Management*, Vol. 2 (1984), pp. 92–100.

GENERAL ELECTRIC CORPORATION, "Guidelines for Use of Program/Project Management in Major Appliance Business Group" (March 1977), in D. J. Cleland and W. R. King (editors), *System Analysis and Project Management*, McGraw-Hill, New York, 1983.

O'BRIEN, J. B. "The Project Manager: Not Just a Firefighter," *S.A.M. Advanced Management Journal*, Jan. 1974.

PINTO, J. K., and D. P. SLEVIN, "Critical Factors in Successful Project Implementation," *IEEE Transactions on Engineering Management*, Vol. EM-34, No. 1 (1987), pp. 22–27.

SCHMITT, T., T. D. KLASTORIN, and A. SHTUB, "Production Classification System: Concepts, Models and Strategies," *International Journal of Production Research*, Vol. 23, No. 3 (1985), pp. 563–578.

Books on Project Management

ARCHIBALD, R. D., *Managing High-Technology Programs and Projects*, Wiley, New York, 1976.

BADIRU, A. B., *Project Management Tools for Engineering and Management Professionals*, IIE Press, Norcross, GA, 1991.

BURTON, V. D. (editor), *Project Management: Methods and Studies*, North-Holland, Amsterdam, 1985.

CLELAND, D. I., *Project Management, Strategic Design and Implementation*, TAB Books, Blue Ridge Summit, PA, 1990.

CLELAND, D. I., and W. R. KING (editors), *Systems Analysis and Project Management*, Third Edition, McGraw-Hill, New York, 1983.

CLELAND, D. I., and W. R. KING (editors), *Project Management Handbook*, Second Edition, Van Nostrand Reinhold, New York, 1988.

GILBREATH, R. D., *Winning at Project Management: What Works, What Fails, and Why*, Wiley, New York, 1986.

GROOL, M. C., J. VISSER, W. J. VRIETHOFF, and G. WIJNEN (editors), *Project Management in Progress: Tools and Strategies for the 90s*, North-Holland, Amsterdam, 1986.

KERZNER, H., *Project Management: A Systems Approach to Planning, Scheduling and Control*, Van Nostrand Reinhold, New York, 1989.

LOCK, D., *Project Management*, Gower Publishing Company, Aldershot, Hants, England, 1990.

MEREDITH, J. R., and S. J. MANTEL, JR., *Project Management: A Managerial Approach*, Second Edition, Wiley, New York, 1989.

MODER, J. J., C. R. PHILLIPS and E. W. DAVIS, *Project Management with CPM, PERT and Precedence Diagramming*, Third Edition, Van Nostrand Reinhold, New York, 1983.

RANDOLPH, W. A., and Z. B. POSNER, *Effective Project Planning and Management: Getting the Job Done*, Prentice Hall, Englewood Cliffs, NJ, 1988.

ROMAN, D. D., *Managing Projects: A Systems Approach*, Elsevier, New York, 1986.

SOUDER, W. E., *Project Selection and Economic Appraisal,* Van Nostrand Reinhold, New York, 1984.

SPINNER, M., *Elements of Project Management: Plan, Schedule, and Control*, Prentice Hall, Englewood Cliffs, NJ, 1981.

SPINNER, M., *Improving Project Management Skills and Techniques*, Prentice Hall, Englewood Cliffs, NJ, 1989.

STOWINSKY, R., and J. WEGLARZ (editors), *Advances in Project Scheduling*, Elsevier, New York, 1989.

WIEST, D. L., and F. K. LEVY, *A Management Guide to PERT/CPM* (2nd edition), Prentice Hall, Englewood Cliffs, NJ, 1977.

APPENDIX 1A

Engineering versus Management

1A.1 NATURE OF MANAGEMENT

Practically everyone has some conception of the meaning of the word *management* and to some extent understands that it requires talents that are distinct from those needed to perform the work being managed. Thus a person may be a first-class engineer but unable to manage a high-tech company successfully. Similarly, a superior journeyman may make an inferior foreman. We have all read about cases in which an enterprise failed, not because the owner did not know the field, but because he was a poor manager. To cite just one example, Thomas Edison was perhaps the foremost inventor of the last century, but he lost control of the many businesses that grew from his inventions because of his inability to plan, and to direct and supervise others.

So what exactly is management, and what does a good manager have to know? Although there is no simple answer to this question, there is general agreement that to a large extent, management is an art grounded in application, judgment, and common sense. More precisely, it is the art of getting things done through other people (Dale 1965, Souder 1980). To work effectively through others, a manager must be able to perform competently the seven functions listed in Table 1A-1. Of those, planning, organizing, staffing, directing, and controlling are fundamental. If any of these five functions is lacking, the management process will not be effective. Note that these are necessary but not sufficient functions for success. Getting things done through people requires the manager also to be effective at motivating and leading others.

The relative importance of the seven functions listed in Table 1A-1 may vary with the level of management. Top management success requires an emphasis on planning, organizing, and controlling. Middle-level management activities are more often concerned with staffing, directing, and leading. Lower-level managers must excel at motivating and leading others.

1A.2 DIFFERENCES BETWEEN ENGINEERING AND MANAGEMENT

Many people start out as engineers and over time work their way up the management ladder. As Table 1A-2 shows, the skills required by a manager are very different from those normally associated with engineering (Badawy 1982, Souder 1984). Engineering involves

TABLE 1A-1 FUNCTIONS OF MANAGEMENT

Function	Description
Planning	The manager must first decide what must be done. This means setting short- and long-term goals for the organization and determining how they will be met. Planning is a process of anticipating problems, analyzing them, estimating their likely impacts, and determining actions that will lead to the desired outcomes, objectives, or goals.
Organizing	Establishing interrelationships between people and things in such a way that human and material resources are effectively focused toward achieving the goals of the enterprise. Organizing involves grouping activities and people, defining jobs, delegating the appropriate authority to each job, specifying the reporting structure and interrelationships between these jobs, and providing the policies or other means for coordinating these jobs with each other.
Staffing	In organizing, the manager establishes positions and decides which duties and responsibilities properly belong to each. Staffing involves appraising and selecting candidates, setting the compensation and reward structure for each job, training personnel, conducting performance appraisals, and performing salary administration. Turnover in the workforce and changes in the organization make it an ongoing function.
Directing	Because no one can predict with certainty the problems or opportunities that will arise, duties must naturally be expressed in general terms. Managers must guide and direct subordinates and resources toward the goals of the enterprise. This involves explaining, providing instructions, pointing out proper directions for the future, clarifying assignments, orienting personnel in the most effective directions, and channeling resources.
Motivating	A principal function of lower management is to instill in the workforce a commitment and enthusiasm for pursuing the goals of the organization. Motivating refers to the interpersonal skills to encourage outstanding human performance in others and to instill in them an inner drive and a zeal to pursue the goals and objectives of the various tasks that may be assigned to them.
Leading	This means encouraging others to follow the example set for them, with great commitment and conviction. Leading involves setting examples for others, establishing a sense of group pride and spirit, and instilling allegiance.
Controlling	Actual performance will normally differ from the original plans, so checking for deviations and taking corrective actions is a continuing responsibility of management. Controlling involves monitoring achievements and progress against the plans, measuring the degree of compliance with the plans, deciding when a deviation is significant, and taking actions to realign operations with the plans.

hands-on contact with the work. Managers are always one or more steps removed from the shop floor and can only influence output and performance through others. An engineer can derive personal satisfaction and gratification in his or her own physical creations, and from the work itself. Managers must learn to be fulfilled through the achievements of those they supervise. Engineering is a science. It is characterized by precision, reproducibility, proven theories, and experimentally verifiable results. Management is an art. It is characterized by intuition, studied judgments, unique events, and one-time occurrences. Engineering is a world of things; management is a world of people. People have feelings, sentiments, and motives that may cause them to behave in unpredictable or unanticipated ways. Engineering is based on physical laws, so that most events occur in an orderly, predictable fashion (Souder 1980).

TABLE 1A-2   ENGINEERING VERSUS MANAGEMENT

What engineers do	What managers do
Minimize risks, emphasize accuracy and mathematical precision	Take calculated risks, rely heavily on intuition, take educated guesses, and try to be "about right"
Exercise care in applying sound scientific methods, on the basis of reproducible data	Exercise leadership in making decisions under widely varying conditions, based on sketchy information
Solve technical problems based on their own individual skills	Solve techno-people problems based on skills in integrating the talents and behaviors of others
Work largely through their own abilities to get things done	Work through others to get things done

1A.3 TRANSITION FROM ENGINEER TO MANAGER

Engineers are often propelled into management out of economic considerations or a desire to take on more responsibility. Some organizations have a dual career ladder that permits good technical people to remain in the laboratory and receive the same financial rewards that attend supervisory promotions. This type of program has been most successful in research intensive environments such as those found at the IBM Research Center in Yorktown Heights and AT&T Bell Laboratories.

Nevertheless, when an engineer enters management, new perspectives must be acquired and new motivations must be found. He or she must learn to enjoy leadership challenges, detailed planning, helping others, taking risks, making decisions, working through others, and using the organization. In contrast to the engineer, the manager achieves satisfaction from directing the work of others (not things), exercising authority (not technical knowledge), and conceptualizing new ways to do things (not doing them). Nevertheless, experience indicates that the following three critical skills are the ones that engineers find most troublesome to acquire: (1) learning to trust others, (2) learning how to work through others, and (3) learning how to take satisfaction in the work of others.

The step from engineering to management is a big one. To become successful managers, engineers usually must develop new talents, acquire new values, and broaden their point of view. This takes time, on-the-job and off-the-job training, and careful planning. In short, engineers can become good managers only through effective career planning.

ADDITIONAL REFERENCES

BADAWY, M. K., *Developing Managerial Skills in Engineers and Scientists*, Van Nostrand Reinhold, New York, 1982.

DALE, E., *Management: Theory and Practice*, McGraw-Hill, New York, 1965.

MOORE, D. C., and D. S. DAVIES, "The Dual Ladder: Establishing and Operating It," *Research Management*, Vol. 20, No. 4 (1977), pp. 21–27.

SOUDER, W. E., *Management Decision Methods for Managers of Engineering and Research*, Van Nostrand Reinhold, New York, 1980.

2

Engineering Economic
Analysis

2.1 PROBLEM OF PROJECT EVALUATION

The design of a system represents a decision about how resources will be transformed to achieve a given set of objectives. The final design is a choice of a particular combination of resources and a blueprint for using them; it is selected from among other combinations that would accomplish the same objectives, but perhaps with different cost and performance consequences. For example, the design of a commercial aircraft represents a choice of structural materials, size and location of engines, spacing of seats, and so on; the same result could be achieved in any number of ways.

A design must satisfy a host of technical considerations and constraints since only some things are possible. In general, it must conform to the laws of natural science. To continue with the aircraft example, there are limits to the strength of metal alloys or composites, and to the thrust attainable from jet engines. The creation of a good design for a system requires solid technical knowledge and competence. Engineers may take this fact to be self-evident, but it often needs to be stressed to upper management and political leaders, who may be motivated by what a proposed system might accomplish rather than by costs and the limitations of technology.

Economics and value must also be taken into account in the choice of design; the best configuration cannot be determined from technical qualities alone. Moreover, value per dollar spent tends to dominate the final choice of a system. As a general rule, the engineer must pick from among many possible configurations, each of which may appear equally effective from a technical point of view. The selection of the best configuration is determined by comparing the costs and relative values associated with each. The choice between constructing an aircraft of aluminum or titanium is generally a question of cost, as both can meet the required standards. For more complex systems, political or other values may be more important than costs. In planning an airport for a city, for instance, it is usually the case that several sites will be judged suitable. The final choice hinges on societal decisions regarding the relative importance of accessibility, congestion, and other environmental and political impacts, in addition to cost.

The centrality of economic considerations in the design of engineering systems needs to be stressed. The recognition that economic theory is essential to engineering practice is relatively new and thus relatively limited. This new relevance results from the evolution in engineering from the design of components and mechanisms to the design and analysis of systems.

As engineers have become increasingly involved with interoperability and integration of systems, they must deal with new issues and incorporate new methods into their analyses. Traditionally, engineering education and practice have been concerned with detailed design. At that level technical problems dominate, with economics taking a back seat. In designing an engine for example, the immediate task—and the trademark of the engineer—is to make the device work properly. At the systems level, however, economic considerations are likely to be critical. Thus the design of a transportation system generally assumes that engines to power vehicles will be available and focuses attention on such issues as whether service can be provided at a price low enough to generate sufficient traffic to make the enterprise worthwhile.

Systems optimization, an integral component of the field known as "operations research," is the prime example of how economic theory has been incorporated into engineering systems analysis. Many key concepts have been adapted with little modification. The production function, the basic model of the way resources are transformed to achieve objectives, comes directly from economics (Sage 1983). So does marginal analysis, which is the basic method for solving the resource allocation, or design, problem.

2.1.1 Need for Economic Analysis

The purpose of an economic evaluation is to determine whether any project or investment is financially desirable. Specifically, an evaluation addresses two sorts of questions:

- Is any individual project worthwhile? That is, does it meet our minimum standards?
- Given a list of projects, which is the best? How does each project rank or compare to the others on the list?

This chapter shows how both these questions should be answered when dealing strictly with cash flows. Chapters 3 and 4 add qualitative considerations to the discussion.

In practice, economic evaluations are difficult to perform correctly. This is in great part due to the fact that those responsible for carrying out the analyses—middle-level managers or staff—necessarily have a limited view of their organization's activities and cannot realistically take all the appropriate factors into account. The result is that most evaluations are done on the basis of assumptions often out of date or otherwise inaccurate for the situation at hand.

Conceptual difficulties are another source of error and confusion. A number of the standard criteria for evaluation contain biases that make them inappropriate and even quite wrong for particular kinds of situations. Our intention is to focus on these issues so that the practitioner will be able—to the extent possible within the constraints imposed by the employer or client—to select the most suitable criteria.

The heart of the presentation centers on the various possible criteria that may be used for evaluation. These include net present value, rate of return, and payback period. Each method is discussed in detail and then compared to the others. The role that taxes and depreciation play in several common business situations, such as make versus buy, or rent versus own, are also examined. The chapter concludes with a discussion of utility theory that can be used to help decision makers deal with uncertain outcomes.

2.1.2 Time Value of Money

Many projects, particularly large systems, evolve over long periods. Costs incurred in one period may generate benefits for many years to come. The evaluation of whether these projects are worthwhile must therefore compare benefits and costs that occur at quite different times.

The essential problem in evaluating projects over time comes from the fact that money has a time value. A dollar now is not the same dollar a year from now. The money represents the same nominal quantity, to be sure, but a dollar later does not have the same usefulness or buying power that it has today. The problem is one of compatibility. Because of this value differential, we cannot estimate total benefits (or costs) simply by adding dollar amounts that are realized in different periods. To make a valid comparison, we need to translate all cash flows into comparable quantities. For purposes of definition, two amounts of money or a series of monies at different points in time are said to be equivalent if they are equal to each other at a given point in time for a given interest rate.

From a mathematical point of view the solution to the evaluation problem is simple. It consists of employing a handful of formulas that depend on only two parameters: the duration or "life" of the project, n, and the discount rate, i. These formulas are built into many pocket calculators, and are routinely embedded in spreadsheet programs available on personal computers. Their results have traditionally been arrayed in extensive tables, which are standard features in engineering handbooks (see, e.g., Newnan 1991, White et al. 1989). In the next three sections, we present these essential formulas and examine their use.

From a practical point of view, the analytic solutions are quite delicate and must be interpreted with care. The values generated by the formulas are sensitive to their two parameters, which are rarely known with certainty. The results are therefore somewhat arbitrary, implying that the problem of evaluating projects over time is a mixture of art and science.

2.1.3 Discount Rate, Interest Rate, and Minimum Attractive Rate of Return

A dollar now is worth more than a dollar in the future because it can be used productively between now and then. At a personal level, for example, you can place money in a savings account and get a greater amount back after some period. In the economy at large, businesses and governments can use money to build plants, manufacture products, grow food, educate people, and undertake other worthwhile activities.

Moreover, any given amount of money now is typically worth more than the same amount in the future because of inflation. As prices go up due to inflation, the current buying power of the dollar erodes. The discount rate represents the way money now is worth more than money later. It determines by how much any future amount is discounted, that is, reduced to make it correspond to an equivalent amount today. The discount rate is thus the key factor in the evaluation of projects over time. It is the parameter that permits us to compare costs and benefits incurred at different instances in time.

The discount rate is generally specified as a rate, given as some percentage per year. Normally, this rate is assumed to be constant for any particular evaluation. Because we usually have no reason to believe that it would change in any known way, we take it to be constant over time when looking at any project. It may, however, be quite different for various individuals, companies, or government, and may also vary among persons or groups as circumstances change. Baumol (1968) discusses the effect of the discount rate on social choice, and De Neufville (1990) indicates how to select an appropriate value for both public- and private-sector investments.

The discount rate is similar to what we think of as the prevailing interest rate but is actually quite a different concept. It is similar in that both can be stated as a percentage per period, and both can indicate a connection between money now and money later. The difference is that the discount rate represents real change in value to a person or group, as determined by their possibilities for productive use of the money and the effects of inflation. By contrast, the interest rate narrowly defines a contractual arrangement between a borrower and a lender. This distinction implies a general rule: *discount rate* > *interest rate*. Indeed, if people were not getting more value from the money they borrow than the interest they pay for it, they would be silly to go to the trouble of incurring the debt.

Having said all this, it is common in the engineering economic literature to use the terms *discount rate* and *interest rate* interchangeably. A third term, *minimum attractive rate of return*, has the same meaning. In the remainder of the book, we follow convention and take all three terms to be synonymous unless otherwise indicated.

2.2 COMPOUND INTEREST FORMULAS

Whenever the interest charge for any period is based on the remaining principal to be repaid plus any accumulated interest charges up to the beginning of that period, the interest is said to be compound. Basic compound interest formulas and factors assuming discrete (lump-sum) payments and discrete interest periods are discussed in this section. The notation used to present the concepts is as follows:

i = effective interest rate per interest period, sometimes referred to as the discount rate or minimum attractive rate of return (MARR); given as a decimal number in the formulas below (i.e., 12% is equivalent to 0.12)

n = number of compounding periods

P = present sum of money (equivalent worth of one or more cash flows at a point in time called the present)

F = future sum of money (equivalent worth of one or more cash flows at a point in time called the future)

A = end-of-period cash flow (or equivalent end-of-period value) in a uniform series continuing for n periods (sometimes called "annuity")

G = uniform increase or decrease in end-of-period cash flows or amounts (the arithmetic gradient)

The compound interest formulas follow.

Single-payment compound amount factor

$$(F/P,\ i\%,\ n) = (1 + i)^n$$

Single-payment present worth factor

$$(P/F,\ i\%,\ n) = \frac{1}{(1 + i)^n} = \frac{1}{(F/P,\ i\%,\ n)}$$

Uniform series compound amount factor

$$(F/A,\ i\%,\ n) = \frac{(1 + i)^n - 1}{i}$$

Uniform series sinking fund factor

$$(A/F,\ i\%,\ n) = \frac{i}{(1 + i)^n - 1} = \frac{1}{(F/A,\ i\%,\ n)}$$

Uniform series present worth factor

$$(P/A,\ i\%,\ n) = \frac{(1 + i)^n - 1}{i(1 + i)^n}$$

Uniform series capital recovery factor

$$(A/P,\ i\%,\ n) = \frac{i(1 + i)^n}{(1 + i)^n - 1} = \frac{1}{(P/A,\ i\%,\ n)}$$

Arithmetic gradient present worth factor

$$\star \ (P/G, \ i\%, \ n) = \frac{(1 + i)^{n} - in - 1}{i^{2}(1 + i)^{n}}$$

Arithmetic gradient uniform series factor

$$\star \ (A/G, \ i\%, \ n) = \frac{(1 + i)^{n} - in - 1}{i(1 + i)^{n} - i}$$

2.2.1 Present Worth, Future Worth, Uniform Series, and Gradient Series

Figure 2-1 is a diagram that shows typical placements of P, F, A, and G over time for n periods with interest at $i\%$ per period. Upward-pointing arrows usually indicate payments or disbursements, and downward-pointing arrows receipts or savings. As depicted in the figure, the following conventions apply in using the discrete compound interest formulas and corresponding tables:

1. A occurs at the end of the interest period.
2. P occurs one interest period before the first A.
3. F occurs at the same point in time as the last A, and n periods after P.
4. There is no G cash flow at the end of period 1; hence the total gradient cash flow at the end of period n is $(n - 1)G$.

Most economic analyses involve conversion of estimated or given cash flows to some point or points in time, such as the present, per annum, or the future. The specific calculations are best illustrated with the help of examples.

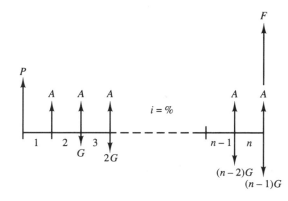

Figure 2-1 Standard cash flow diagram indicating points in time for P, F, A, and G.

Example 2-1

Suppose that a $20,000 piece of equipment is expected to last 5 years and then result in a $4,000 salvage value. If the minimum attractive rate of return (interest rate) is 15%, what are the following values?

(a) Annual equivalent (cost)

(b) Present equivalent (cost)

Solution   Figure 2-2 shows all the cash flows.

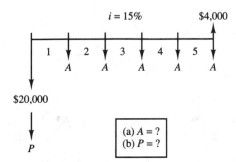

$20,000

(a) $A = ?$
(b) $P = ?$

P

Figure 2-2   Cash flow diagram for
Example 2-1.

(a) A = $-\$20,000(A/P, 15\%, 5) + \$4,000(A/F, 15\%, 5)$
      $= -20,000(0.2983) + 4,000(0.1483) = -\$5,373$

[*Note:* $5,373 is sometimes called the *annual cost* (AC) or *equivalent uniform annual cost*
(EUAC).]

(b) P = $-\$20,000 + \$4,000(P/F, 15\%, 5)$
      $= -20,000 + 4,000(0.4972) = -\$18,011$

Alternatively, it is possible to solve part (b) by exploiting the results obtained from part (a) as
follows:

$$P = A(P/A, 15\%, 5)$$
$$= -\$5,373(3.3522) = -\$18,011$$

■

Example 2-2 (Deferred Uniform Series and Gradient Series)

Suppose that a certain savings is expected to be $10M at the end of year 3 and to increase $1M
each year until the end of year 7. If the MARR is 20%, what are the following values?

(a) Present equivalent (at beginning of year 1)

(b) Future equivalent (at end of year 7)

Solution   Once again, the first step is to draw the cash flow diagram. Figure 2-3 shows the
gradient beginning at the end of year 3 and the unknowns to be calculated (dashed arrows). In
the solution, subscripts are used to indicate a point or points in time.

(a) $A_{3\text{-}7}$ = $\$10M + \$1M(A/G, 20\%, 5)$
        $= \$10M + \$1M(1.6405) = \$11.64M$

      P_2 = $A_{3-7}(P/A, 20\%, 5)$
        $= 11.64M(2.9906) = \$34.81M$

      P_0 = $F_2(P/F, 20\%, 2)$
        $= \$34.81(0.6944) = \$24.17M$

Notice that in the last calculation, the value of P_2 is substituted for F_2.

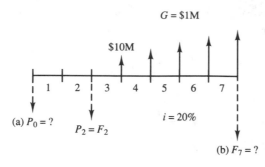

(a) $P_0 = ?$

(b) $F_7 = ?$

Figure 2-3 Cash flow diagram for Example 2-2 showing deferred uniform and gradient series.

(b) (Skipping intermediate calculations):

$$F_7 = [\$10M + \$1M(A/G, 20\%, 5)](F/A, 20\%, 5)$$
$$= [10M + 1M(1.6405)](7.4416) = \$86.62M$$

Alternatively, one can utilize part (a) results to obtain F_7 as follows:

$$F_7 = P_0(F/P, 20\%, 7)$$
$$= \$24.17M(3.5832) = \$86.62M$$

■

Example 2-3 (Repeating Cycle of Payments)

Suppose that the equipment in Example 2-1 is expected to be replaced three times with identical equipment, making four life cycles of 5 years each. To compare this investment correctly with another alternative that can serve 20 years, what are the following values when MARR = 15%?

(a) Annual equivalent (cost)

(b) Present equivalent (cost)

Solution Figure 2-4 shows the costs involved. The key to this type of problem is to recognize that if the cash flows repeat each cycle, the annual equivalent for one cycle will be the same for all other cycles.

(a) We will demonstrate a slightly different way to get the same answer as in Example 2-1.

$$A = [-\$20M + \$4M(P/F, 15\%, 5)](A/P, 15\%, 5)$$
$$= [-20M + 4M(0.4972)](0.2983) = -\$5,373$$

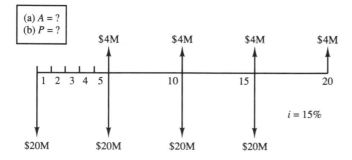

Figure 2-4 Cash flow diagram for Example 2-3.

(b) $P = -\$5,373(P/A, 15\%, 20)$
$$= -5,373(6.2593) = -\$33,629$$

■

2.2.2 Chain Rule and Inflation

Two final points should be made about the use of the compound interest formulas. The first is that the chain rule applies. For example, if you want to find P given F, instead of solving $P = F(P/F, i\%, n)$, you can solve $P = F(A/F, i\%, n)(P/A, i\%, n)$ should it be more convenient for some reason.

The second point concerns inflation, a condition in the economy characterized by rising prices for goods and services. An inflationary trend makes future dollars have less purchasing power than present dollars. This helps long-term borrowers at the expense of lenders, since a loan of present dollars may be repaid in the future with dollars of lesser value.

In the analysis, one approach used to compensate for inflation is to first convert all cash flows into year-0 dollars. If the inflation rate is, say f, this can be done by discounting future dollars to the present, as follows:

$$\text{year-0 dollars} = [(1 + f)^{-n}][\text{year-}n \text{ dollars}]$$

We would now proceed as before with the analysis. Alternatively, one may compute an equivalent interest rate,

$$i_e = i + f + i \times f$$

and use it in conjunction with the present worth factors to compute the present worth of future cash flows in year-0 dollars.

2.2.3 Treatment of Risk

Risk comes in many forms. If a new product is being developed, the probability of commercial success is a major consideration. If a new technology is being pursued, we must constantly reevaluate the probability of technical success and the availability of critical personnel and resources. Once a product is ready for the market, such factors as financing, contractual obligations, reliability of suppliers, and strength of competition must be brought into the equation.

In the private sector, projects that are riskier than others are forced to pay higher interest rates to attract capital. A speculative new company will have to pay the banks several percentage points more for its borrowing than will established, prime customers. Private companies, which always run the risk of bankruptcy, have to pay more than the government. This extra amount of interest is known as the *risk premium*, and as a practical matter, is already included in the discount rate.

When a particular project faces uncommon technical or commercial risks, the evaluation process should address them directly. Decision analysis (Chapter 3), coupled with the use of multiple-criteria methodologies (Chapter 4), is the preferred way to appraise projects with a high component of risk.

2.3 COMPARISON OF ALTERNATIVES

The essence of all economic evaluation is a discounted cash flow analysis. The first step in every situation is to lay out the estimated cash flows, the sequence of benefits (returns) and costs (payments) over time. These are then discounted back to the present, using the methods shown in Section 2.2, either directly or indirectly in the case of the rate-of-return and payback-period methods.

The relative merits of the available alternatives are determined by comparing the discounted cash flows of benefits and costs. In general, a project is considered to be worthwhile if its benefits exceed its costs. The relative ranking of the projects is then determined by one of several evaluation criteria. The methods of evaluation differ from each other principally in the way they handle the results of the discounted cash flow analysis. The present value method focuses on the difference between the discounted benefits and costs, the ratio methods involve various comparisons of these qualities, and the internal rate-of-return method tries to equalize them. The question of what one does with the results of the discounted cash flows is the central problem of economic evaluation.

Most methods presume that the discount rate to be used in the cash flow analysis is known. This is often a reasonable assumption, since many companies or agencies require that a specific rate be used for all their economic evaluations. In many instances, however, the discount rate must be determined.

In carrying out an evaluation, it will be essential to remember that correct selection of the discount rate may be crucial. Its choice can easily change the ranking of projects, making one or another appear best depending on the rate used. This is because lower rates make long-term projects, with benefits in the distant future, appear much more attractive relative to short-term projects with immediate benefits than they would be if a higher rate were used.

To see this, suppose that your organization has the choice of two storage and retrieval systems, one that requires a human operator and one that is fully automated. Both will last for 10 years. The human-assisted system costs $10,000 and requires $4,200 per year of labor. The automated system has an initial cost of $18,000 and consumes an additional $3,000 per year in power. The decision is a question of whether the benefits of the annual savings ($4,200 − $3,000 = $1,200 a year) justify the additional initial cost of $8,000. Is the net present value (NPV) of the upgrade to the more expensive alternative positive?

If the discount rate were zero, implying that future benefits are not discounted, the upgrade is clearly worthwhile.

$$\text{NPV}(i = 0\%) = (\$1,200/\text{yr})(10 \text{ yrs}) - \$8,000 = \$4,000$$

Conversely, if the discount rate were large, future benefits would be heavily discounted. For infinite i,

$$\text{NPV}(i = \infty) = \$1,200(0) - \$8,000 = -\$8,000$$

so the project is not worthwhile.

The variation of the net present value with the discount rate is summarized as follows:

i (%)	0	5	10	15	∞
NPV(i%)	4,000	1,264	−632	−1,976	−8,000

The critical value of i, below which the more expensive system is preferred, is around $8\frac{1}{2}$%, as determined by interpolation.

As this example shows, the choice of the discount rate can steer an analysis in one direction or another. Powerful economic and political forces allied with a particular technology may encourage this. When the U.S. Federal Highway Administration promulgated a regulation in the early 1970s that the discount rate for all federally funded highways would be zero, this was widely interpreted as a victory for the cement industry over asphalt interests. Roads made of concrete cost significantly more than those made of asphalt, but require less maintenance and less frequent replacement.

2.3.1 Defining Investment Alternatives

Every evaluation deals with two distinct sets of projects or alternatives: the explicit and the implicit. The explicit set consists of the opportunities that are to be considered in detail; they are the focus of the analysis. The implicit set, which can only be defined imprecisely, is important because it provides the frame of reference for the evaluation and defines the minimum standards.

Explicit set of alternatives.   This is a limited list of the potential projects that could actually be chosen. The list is usually defined by a manager concerned with a particular issue; for example:

- An official of the department of highways, responsible for maintenance and construction of roads
- A manager of a computer center, proposing to acquire new equipment
- An investment officer for a bank, presenting a menu of opportunities for construction loans

The lists of projects suggested by each of the preceding situations illustrates two characteristics typical of the explicit choices considered by an evaluation. The explicit set is:

1. Limited in scope, in that it includes only a portion of the projects that might be in front of the organization as a whole. Thus the manager of the computer center is only competent in, and only considers various ways to improve the information systems; whether money should be spent on developing a new product or replacing the central heating is literally not his or her department.

2. Limited in number, being only a fraction of all the projects that could be defined over the next several years. Usually, the explicit list deals only with the immediate choices, not the ones that could occur during the next budget or decision period.

The fact that the sets of projects we consider explicitly are limited poses a basic problem for any organization. The difficulty is that any procedure that analyzes separate sets of projects independently can quite easily lead to a list of recommended choices that are not the best ones for the organization as a whole. For example, consider a company with a computer department, a research laboratory, and a manufacturing plant: if we evaluate the projects proposed by each group, we can determine the best computer, the best instrument, and the best machine tool to buy. But this plan may not be in the best interests of the company. It is quite possible that the second-best machine tool is a better investment than the best instrument, or that none of the computers is worthwhile financially.

The issue is: How does an organization ensure that the projects selected by its components are best for the organization as a whole? In addressing this question we must recognize that the obvious answer, of considering all possible projects simultaneously, is neither practical nor even possible. A large number of analyses could be done, but the level of computation is not the real obstacle.

An analysis of all alternatives at once is not practical because it would be extremely difficult for any group in an organization to be sufficiently knowledgeable both to generate the possible projects for all departments and to estimate their benefits and costs. They simply would not have sufficient knowledge of the topic, region, or clients. Further, the analysis of all alternatives at once is not even conceptually feasible because we are unable to predict what options will be available in the future. We can therefore never be sure that the projects we select from a current list, however comprehensive it may be, will include all the opportunities that will occur over the life of the projects and that might otherwise be selected. Some degree of suboptimization is unavoidable.

To reduce the possibility for suboptimization it is necessary to create some means to make the evaluation of any set of explicit alternatives less dependent on other evaluations. This can be done by creating a substitute for the entire list of innumerable possible alternatives. This is the role of the implicit alternatives.

Implicit set of alternatives.   This set generally represents all the projects that have been available previously and that might be available in the reasonable future. Because it refers in part to unknown prospects, it can never be described in detail. It thus indicates inexactly what could be done instead of the immediate, explicit alternatives.

The implicit set of alternatives is of interest because it establishes the minimum standards for deciding if any explicit project is worthwhile. To illustrate, consider the situation in which a person has been making investments and has consistently been able to choose possibilities that provide yearly profits of 12% or more, and has rejected all others with lesser returns. Faced now with the problem of evaluating an explicit set of specific proposals, this person will naturally turn to past experience for guidance. If the investment possibilities have not changed fundamentally, the person may assume that there are continued possibilities—the implicit set of alternatives—for earning 12% or more as before and should correctly conclude that any explicit choice can be worthwhile only if its profitability equals or exceeds the 12% implicitly available elsewhere.

The minimum standards suggested by the implicit alternatives can be stated in several ways. An obvious and common way is to stipulate a minimum attractive rate of return. *MARR*
Minimum standards of profitability can also be expressed quite differently, however. In business, they are typically stated in terms of the highest number of periods that will be required for the benefits to equal the initial investment (the maximum payback period; see *PAYBACK PERIOD* Section 2.4.6). Minimum standards can also be defined in terms of minimum ratios of benefits to costs (Section 3.4).

Organizations use minimum standards for the economic acceptability of projects as the way to reduce the possibility of suboptimization. These standards force each department or group to take into account the global picture. They cannot, for example, choose projects unless they are at least as good as others available elsewhere in the organization.

2.3.2 Steps in the Analysis

A systematic procedure for comparing investment alternatives can be outlined as follows:

1. Define the alternatives.
2. Determine the study period.
3. Provide estimates of the cash flows for each alternative.
4. Specify the interest rate (MARR).
5. Select the measure(s) of effectiveness (i.e., the criteria for judging success).
6. Compare the alternatives.
7. Perform sensitivity analyses.
8. Select the preferred alternative(s).

The study period defines the period of time over which the analysis is to be performed. It may or may not be the same as the useful lives of the equipment involved. In general, if the study period is less than the useful life of an asset, an estimate of its salvage value should be provided in the final period; if the study period is longer than the useful life, estimates of cash flows are needed for subsequent replacements of the asset.

Whenever alternatives having different lives are to be compared, the study period is usually one of the following:

1. The organization's "planning horizon"
2. The life of the shortest-lived alternative
3. The life of the longest-lived alternative
4. The lowest common multiple of the lives of the alternatives

If the study period for both alternatives is forced to be the same by using item 1, 2, or 3 above or for any other reason, the so-called *coterminated assumption* is said to be used, and whatever cash flows are thought appropriate are considered within that study period. When the study period is chosen to be item 4 above, alternatives normally are assumed to satisfy the following so-called *repeatability assumptions*:

1. The period of needed service is either indefinitely long or a common multiple of the lives.
2. What is estimated to happen in the first life cycle will happen in all succeeding life cycles, if any, for each alternative.

In the next three subsections illustrating economic analysis methods, when alternatives have different lives and nothing is indicated to the contrary, the repeatability assumptions are used. These assumptions are commonly adopted for computational convenience or by default.

2.4 EQUIVALENT WORTH METHODS

For purposes of analysis, *equivalent worth methods* convert all relevant cash flows into equivalent (present, annual, or future) amounts at the MARR. If a single project is under consideration, it is acceptable (earns at least the MARR) if its equivalent worth is greater than or equal to zero; otherwise, it is not acceptable. These methods all assume that recovered funds (net cash inflows) can be reinvested at the MARR.

If two or more mutually exclusive alternatives are being compared and receipts or savings (cash inflows) as well as costs (cash outflows) are known, that project should be chosen which has the highest net equivalent worth, as long as that equivalent worth is greater than or equal to zero. If only costs are known or considered (assuming that all alternatives have the same benefits), that project which has the lowest total equivalent of those costs should be chosen. Because all three equivalent worth methods give completely consistent results, the choice of which to use is a matter of computational convenience and preference for the form in which the results are expressed.

2.4.1 Present Worth Method

The term *present worth* (PW) denotes a lump-sum amount at some early point in time (often the present) that is equivalent to a particular schedule of receipts and/or disbursements under consideration. If receipts and disbursements are considered, the term can best be expressed as present worth—cost, otherwise known as net present value.

Example 2-4

Consider the following two mutually exclusive alternatives and recommend which one (if either) should be implemented.

	Machine	
	A	B
Initial cost	$20,000	$30,000
Life	5 years	10 years
Salvage value	$4,000	0
Annual receipts	$10,000	$14,000
Annual disbursements	$4,400	$8,600
Minimum attractive rate of return = 15%		
Assume 10-year study period and repeatability		

Solution (using PW method)

	Machine	
	A	B
Annual receipts		
$10,000(P/A, 15%, 10)	$50,188	
$14,000(P/A, 15%, 10)		$70,263
Salvage value at end of		
year 10 = $4,000(P/F, 15%, 10)	989	0
Total PW of cash inflow	51,177	70,263
Annual disbursements		
$4,400(P/A, 15%, 10)	−22,083	
$8,600(P/A, 15%, 10)		−43,162
Initial cost	−20,000	−30,000
Replacement:		
($20,000 − $4,000)(P/F, 15%, 5)	−7,955	0
Total PW of cash outflow	−50,038	−73,162
Net PW (NPV)	$1,139	−$2,899

Thus project A has the higher net present value and represents the better economic choice. The fact that the net present value of project B is negative makes it unacceptable in any case.

■

2.4.2 Annual Worth Method

Annual worth (AW) is merely an annualized measure for assessing the financial desirability of a proposed undertaking. It is a uniform annual series of money over a certain period of time that is equivalent in amount to a particular schedule of receipts and/or disbursements under consideration. If disbursements only are considered, the term is usually expressed as *annual cost* (AC) or *equivalent uniform annual cost* (EUAC). The examples in this section include both cash inflows and outflows.

Calculation of capital recovery cost.   The *capital recovery* (CR) *cost* for a project is the equivalent uniform annual cost of the capital that is invested. It is an annual amount that covers the following two items:

1. Depreciation (loss in value of the asset)
2. Interest (minimum attractive rate of return) on invested capital

Consider an alternative requiring a lump-sum investment P and a salvage value S at the end of n years. At interest $i\%$ per year, the annual equivalent cost can be calculated as

$$CR = P(A/P, i\%, n) - S(A/F, i\%, n)$$

There are several other formulas for calculating the CR cost. Probably the most common is

$$CR = (P - S)(A/P, i\%, n) + S(i\%)$$

One might want to reverse signs so that a cost is negative, as is done in the following example, which includes CR costs in an AW comparison.

Example 2-5

Given the same machines A and B as used to demonstrate the net PW method in Example 2-4, we now compare them by the net AW method.

	Machine	
	A	B
Initial cost	$20,000	$30,000
Life	5 years	10 years
Salvage value	$4,000	0
Annual receipts	$10,000	$14,000
Annual disbursements	$4,400	$8,600
Minimum attractive rate of return = 15%		
Assume repeatability		

Solution (using AW method)

	Machine	
	A	B
Annual receipts	$10,000	$14,000
Annual disbursements	−4,400	−8,600
CR amount = −$20,000(A/P, 15%, 5)	−5,966	
+$4,000(A/F, 15%, 5)	+593	
−$30,000(A/P, 15%, 10)		−5,978
Net AW	$227	−$578

Thus project A, having the higher net annual worth which also is greater than or equal to $0, is the better economic choice. A shortcut for calculating the net AWs given the net PWs calculated in the preceding section is

$$AW(A) = \$1,139(A/P, 15\%, 10) = \$227$$
$$AW(B) = -\$2,889(A/P, 15\%, 10) = -\$578$$

One significant computational shortcut when comparing alternatives with different lives by the present worth method and assuming repeatability is first to calculate AWs as above and then calculate the PWs for the lowest common multiple-of-lives study period. Thus

$$PW(A) = \$227(P/A, 15\%, 10) = \$1,139$$
$$PW(B) = -\$578(P/A, 15\%, 10) = -\$2,899$$

■

2.4.3 Future Worth Method

The *future worth* (FW) measure of merit is a lump-sum amount at the end of the study period which is equivalent to the cash flows under consideration.

Example 2-6

Given the same machines A and B (Examples 2-4 and 2-5), determine which is better based on FWs at the end of the 10-year study period.

Solution (using FW method)   Rather than calculating FWs of all the types of cash flows involved (as was done for the PW solution above), shown below are shortcut solutions based on (a) PWs and (b) AWs calculated previously:

(a) $FW(A) = \$1,139(F/P, 15\%, 10) = \$4,608$
   $FW(B) = -\$2,899(F/P, 15\%, 10) = -\$11,728$
(b) $FW(A) = \$227(F/A, 15\%, 10) = \$4,608$
   $FW(B) = -\$578(F/A, 15\%, 10) = -\$11,735$

Not surprisingly, we have once again found that alternative A is preferred. The ratios of the numbers produced by each of the equivalent worth methods will always be the same. For the machines (above), $FW(A)/FW(B) = PW(A)/PW(B) = AW(A)/AW(B) = -0.393$.

■

Example 2-7 (Different Useful Lives; Fixed-Length Study Period)

Suppose that two measurement instruments are being considered for a certain industrial laboratory. Following are the principal cost data for one life cycle of each alternative:

	Instrument	
	M1	M2
Investment	$15,000	$25,000
Life	3 years	5 years
Salvage value	0	0
Annual disbursements	$4,400	$8,600
Minimum attractive rate of return = 20%		
Assume no repeatability		

Solution   The calculations will be done using the PW method and MARR = 20% for the following two cases:

1. If the study period is taken to be 3 years, we need a salvage value for alternative M2 at the end of the third year. Assuming it to be, say, $6,000, the following results are obtained:

	Instrument	
	M1	M2
Investment	$15,000	$25,000
Annual disbursements		
$8,000(A/P, 20%, 3)	16,852	
$5,000(A/P, 20%, 3)		10,533
Salvage: −$6,000(A/F, 20%, 3)	———	−3,472
Net PW (NPV)	$31,852	$32,061

Thus the first alternative is slightly better. Note that "+" is used for costs.

2. If the study period is taken to be 5 years, we need estimates of what will happen after the first life cycle of alternative M1. Let us assume that it can be replaced at the beginning of the fourth year for $18,000 and that the annual disbursements will be $9,000 for years 4 and 5. Furthermore, it will have a $7,000 salvage value at the end of year 5. In this case, we obtain:

	Instrument	
	M1	M2
Investment	$15,000	$25,000
Annual disbursements		
$8,000(A/P, 20%, 3)	16,852	
$9,000(P/A, 20%, 2)(P/F, 20%, 3)	7,975	
$5,000(A/P, 20%, 5)		14,953
Additional investment: $18,000(P/F, 20%, 3)	10,147	
Salvage: −$7,000(P/F, 20%, 5)	−2,813	———
Net PW (NPV)	$47,413	$39,953

Thus alternative M2 has a slightly lower net PW and hence is better with the new assumption.

■

2.4.4 Discussion of PW, AW, and FW Methods

Some academics assert that the net present worth methods—and in particular, the net present value criterion—should be used in all economic analyses. This prescription should be resisted. NPV (and its equivalents) only provides a good comparison between projects when they are strictly comparable in terms of level of investment or total budget. This condition is rarely met in the real world. The practical consequence is that net present values are used primarily for the analysis of investments, particularly of specific sums of money, rather than for the evaluation of projects, which come in many different sizes.

The advantage of the net present worth criteria is that they focus attention on quantity of money, which is what the evaluation is ultimately concerned with. Net PW, AW, and FW differ in this respect from the other criteria of evaluation, which rank projects by ratios and hence do not directly address the bottom-line question of maximizing profit.

One disadvantage of NPV is that its precise meaning is difficult to explain; practical-minded people such as businesspersons thus tend to avoid the concept. NPV suggests profit, but in fact is not profit in any usual sense of the term. In ordinary language, profit is the difference between what we receive and what we pay out. As an example, consider an investment now for a lump sum of revenue later. This is, crudely,

$$\text{profit} = \text{money received} - \text{money invested}$$

More precisely, if we had to borrow money to make the original investment, the profit would be net of interest paid for n periods:

$$\text{profit} = \text{money received} - (\text{money invested})(F/P, i\%, n)$$

where i is the interest rate. This profit can also be placed in present value terms using the appropriate discount rate, r, for the organization concerned. Note that it is now important to make the distinction between the discount rate and the interest rate:

$$\text{present value of profit} = (\text{money received})(P/F, r\%, n)$$
$$- (\text{money invested})(F/P, i\%, n)(P/F, r\%, n)$$

In this last calculation, it turns out that because the discount rate is not, in general, equal to the interest rate, NPV $\neq$ present value of profit. Thus even when the net present value equals zero, the project may be profitable, as understood in common language. A project with NPV $= 0$ is simply not advantageous compared to other alternatives available to the organization. Net present value thus indicates "extra profitability" beyond the minimum. As a reader, you probably found it difficult to grasp this subtlety; this demonstrates the point that NPV may be a questionable measure to use.

Another difficulty with the net present worth criteria is that they give no indication of the scale of effort required to achieve the result. To see this, consider the problem of evaluating projects P1 and P2 below.

Project	Benefit	Cost	NPV	NPV as percent of cost
P1	$2,002,000	$2,000,000	$2,000	0.1
P2	2,000	1,000	1,000	100

If one considers only NPV, project P1 appears best. Most investors would consider that an absurd choice, however, because of the difference in scale between the projects. Taking scale into account, P2 presumably gives a much better return than P1: the money saved by investing in the former rather than the latter can be invested elsewhere for a return greater than that offered by P1. In any case, net present value by itself is not a good criterion for ranking projects.

Formally, the essential conditions for net worth to be an appropriate criterion for the evaluation and ranking of projects are that

- we have a fixed budget to invest.
- projects require the same investment.

These conditions do not hold with any regularity. On the contrary, it is most often the case that the list of projects consists of a variety of possibilities with varying costs. A central problem in the evaluation and choice of systems is to delimit their size and budget. Analysis of net worth is not particularly helpful in those contexts.

2.4.5 Internal Rate-of-Return Method

The internal rate-of-return (IRR) method involves the calculation of an interest rate that is compared against a minimum standard of desirability (i.e., the MARR). As we will see, it is the interest rate for which the net present value of a project is zero. The concept is that the internal rate of return expresses the real return on any investment (i.e., "return on investment"). For evaluation, the idea is that projects should be ranked from the highest IRR down.

The internal rate of return is used increasingly by sophisticated business analysts. The advantage of this criterion is that it overcomes two difficulties inherent in the calculation of both net present value and benefit-cost ratios. That is:

1. It eliminates the need to determine, indeed to argue about, the appropriate MARR.
2. Its rankings cannot be manipulated by the choice of a MARR.

It also focuses attention directly on the rate of return of each project, an attribute that cannot be understood from either the net present value or the benefit-cost ratio.

The IRR is known by other names, such as *profitability index, investor's method return, discounted cash flow return*, and so on. We will demonstrate its use for a single project and then for the comparison of mutually exclusive projects.

IRR method for single project.   The most common method of calculation of the IRR for a single project involves finding the interest rate, $i\%$, at which the present worth of the cash inflow (receipts or cash savings) equals the present worth of the cash outflow (disbursements or cash savings forgone). That is, one finds the interest rate at which PW of cash inflow equals PW of cash outflow; or at which PW of cash inflow minus PW of cash outflow equals 0; or at which PW of net cash flow equals 0. The IRR could also be calculated by using the same procedures applied to either AW or FW.

The calculations normally involve trial and error until the correct interest rate is found or can be interpolated. Closed-form solutions are not available because the equivalent worth factors are a nonlinear function of the interest rate. The procedure is described below for several situations. (When both cash inflows and outflows are involved, the convention of using a "+" sign for inflows and a "−" sign for outflows will be followed.)

Example 2-8

Given the same machine A as in Section 2.4.1, find the IRR and compare it to a MARR of 15%.

	Machine A
Initial cost	$20,000
Life	5 years
Salvage value	$4,000
Annual receipts	$10,000
Annual disbursements	$4,400

[Handwritten note: NPV = 0 MEANS CASH INFLOW = CASH OUTFLOW, ∴ ANY % RATE > IRR WILL YIELD PROFIT (OR MORE PROFIT DEPENDING ON HOW INFLOW IS COUNTED)]

Solution   Expressing the NPV of cash flow and setting it equal to zero results in the following:

$$\overset{P}{} \qquad \overset{A}{} \qquad \overset{S}{}$$

$$\text{NPV}(i\%) = -\$20{,}000 + (\$10{,}000 - \$4{,}400)(P/A, i\%, 5) + \$4{,}000(P/F, i\%, 5) = 0$$

Try $i = 10\%$:

$$\text{NPV}(10\%) = -\$20{,}000 + \$5{,}600(P/A, 10\%, 5) + \$4{,}000(P/F, 10\%, 5) = \$3{,}713 > 0$$

Try $i = 15\%$:

$$\text{NPV}(15\%) = -\$20{,}000 + \$5{,}600(P/A, 15\%, 5) + \$4{,}000(P/F, 15\%, 5) = \$730 > 0$$

Try $i = 20\%$:

$$\text{NPV}(20\%) = -\$20{,}000 + \$5{,}600(P/A, 20\%, 5) + \$4{,}000(P/F, 20\%, 5) = -\$1{,}196 < 0$$

Since we have both a positive and a negative NPV, the desired answer is bracketed. Linear interpolation can be used to approximate the unknown interest rate, i, as follows:

$$\frac{i\% - 15\%}{20\% - 15\%} = \frac{\$730 - 0}{\$730 - (-\$1{,}196)}$$

so

$$i\% = 15\% + \frac{\$730}{\$730 + \$1{,}196}(20\% - 15\%)$$

Solving gives $i\% = 16.9\%$.* Now, because 16.9% is greater than the MARR of 15%, the project is justified. A plot of NPV versus interest rate is given in Fig. 2-5.

---

*A more exact calculation gives $i\% = 16.47\%$, but we will use 16.9% for the remainder of the chapter.

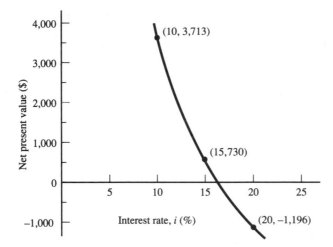

Figure 2-5   Relationship between NPV and IRR for Example 2-8.

■

Because the *P/A* and *P/F* factors are nonlinear functions of the interest rate, the linear interpolation (above) causes an error, but the error is usually inconsequential in economic analyses. The narrower the range of rates over which the interpolation is done, the more accurate are the results. Finally, note that as the trial interest rate is increased, the corresponding NPV decreases.

IRR method for comparing mutually exclusive alternatives.   When comparing alternatives by any rate-of-return (ROR) method when at most one alternative will be chosen, there are three main principles to keep in mind:

1. Any alternative whose IRR is less than the MARR can be discarded immediately.
2. Each increment of investment capital must justify itself (by sufficient ROR on that increment).
3. Compare a higher investment alternative against a lower investment alternative only if that lower investment alternative is justified.

The usual approach when using a ROR method is: choose the alternative requiring the highest investment for which each increment of investment capital is justified. This choice assumes that the organization wants to invest any capital needed as long as the capital is justified by earning a sufficient ROR on each increment of capital. In general, a sufficient ROR is any value greater than or equal to the MARR. The internal rate of return on the incremental investment for any two alternatives can be found by:

1. Finding the rate at which the PW (or AW or FW) of the net cash flow for the difference between the two alternatives is equal to zero; or

2. Finding the rate at which the PWs (or AWs or FWs) of the two alternatives are equal

Example 2-9

Suppose that we have the same machines, A and B, as considered in Section 2.4.1. In addition, machines C and D are mutually exclusive alternatives also to be included in the comparison by the IRR method. Relevant data and the solution are presented below. Repeatability of the alternatives is assumed.

| | | Machine | | |
	A	B	C	D
Initial cost	$20,000	$30,000	$35,000	$43,000
Life	5 years	10 years	5 years	5 years
Salvage value	$4,000	0	$4,000	$5,000
Annual receipts	$10,000	$14,000	$20,000	$18,000
Annual disbursements	$4,400	$8,600	$9,390	$5,250
Net annual receipts − disbursements	$5,600	$5,400	$10,610	$12,750
IRR	16.9%	12.4%	17.9%	17.1%

Solution   As a first step, the alternatives are arranged in order of increasing initial investment because the calculations regarding an increment must be completed before one knows which increment to consider next. The symbol Δ means "increment," and A → B means "the increment in going from alternative A to alternative B." Recall that an increment of investment is justified if the IRR on that increment (i.e., ΔIRR) is ≥15%. The least expensive alternative is always compared to the "do nothing" option.

	A	A → B*	A → C	A → D
Δinvestment	$20,000	$10,000	$15,000	$8,000
Δsalvage	$4,000	−$4,000	$0	$1,000
Δ(annual receipts − disbursements)	$5,600	−$200	$5,010	$2,140
ΔIRR	16.9%	0%	20%	13.3%
Is Δinvestment justified?	Yes	No	Yes	No

*Analysis must include $16,000 replacement lost for alternative A at end of year 5.

The analysis above indicates that alternative C would be chosen because it is associated with the largest investment for which each increment of investment capital is justified. The analysis was performed without considering the IRR on the total investment for each alternative. However, if we look at the individual IRRs, we see that IRR(B) = 12.4% < 15% = MARR, so alternative B could have been discarded.

■

In choosing alternative C, one increment of investment was justified as follows:

Increment	Incremental investment	Internal rate of return on increment, ΔIRR (%)
A	$20,000	16.9
A → C	15,000	20.0
Total investment	$35,000	

Coincidentally, alternative C had the largest IRR, which seems intuitive but is not always the case. If the MARR were, say, 12%, alternative D would have been selected. As a general rule, though, if the most expensive alternative has the highest IRR, it will always turn out to be preferred.

In Example 2-9, because the useful lives of A and B are different and repeatability is assumed, one should closely examine the cash flows for A → B (B minus A) for the lowest common multiple of lives. For the 10-year period, Σ (positive cash flows) = $16,000 (replacement cost) = Σ (negative cash flows) = $10,000 + $4,000 + 10($200). Thus ΔIRR = 0%; any $i\%$ > 0 would produce a negative NPV.

Occasionally, situations arise where a single positive interest rate cannot be determined from the cash flow; that is, solving for NPV = 0 yields more than one solution. Descartes' rule of signs indicates that multiple solutions can occur whenever the cash flow series reverses sign (from net outflow to net inflow, or vice versa) more than once over the study period. This is demonstrated in the following example.

Example 2-10 (No Single IRR Solution)

The Converse Aircraft Company has an opportunity to supply a wide-body airplane to Banzai Airlines. Banzai will pay $19 million when the contract is signed and $10 million one year later. Converse estimates its second- and third-year net cash flows at $50 million each during production. Banzai will take delivery of the plane during year 4, and agrees to pay $20 million at the end of that year and the $60 million balance at the end of year 5. Compute the rate of return on this project.

Solution   Computation of NPV at various interest rates, using single-payment present worth factors (e.g., for year 2 and i = 10%, PW = $-50(P/F, 10\%, 2)$ = $-50(0.826)$ = -41.3) is presented:

Year	Cash flow	0%	10%	20%	40%	50%
0	+19	+19	+19	+19	+19	+19
1	+10	+10	+9.1	+8.3	+7.1	+6.7
2	−50	−50	−41.3	−34.7	−25.5	−22.2
3	−50	−50	−37.6	−28.9	−18.2	−14.8
4	+20	+20	+13.7	+9.6	+5.2	+4.0
5	+60	+60	+37.3	+24.1	+11.2	+7.9
		NPV = +9	+0.2	−2.6	−1.2	+0.6

CONSTR. [2, 3 bracket annotation beside rows 2 and 3]

The NPV plot for these data is depicted in Fig. 2-6. We see that the cash flow produces two points at which NPV = 0; one at about 10.1% and the other at about 47%. Whenever multiple answers such as these exist, it is likely that neither is correct.

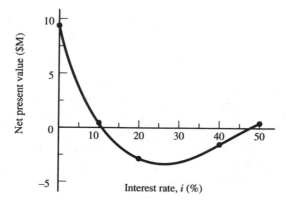

Figure 2-6 NPV plot for more than one change in sign.

An effective way to overcome this difficulty and obtain a "correct" answer is to manipulate cash flows as little as necessary so that there is only one sign reversal in the net cash flow stream. This can be done by using an appropriate interest rate to move lump sums either forward or backward and then solve in the usual manner. To demonstrate, let us assume that all money held outside the project earns 6%. (This value could be considered the external interest rate that Converse faces. If it had to borrow money, the interest rate might be different.) At both year 0 and year 1, there is an inflow of cash resulting from the advance payments by Banzai. The money will be needed later to help pay the production costs. Given an external interest rate of 6%, the $19 million will be invested for 2 years and the $10 million for 1 year. Their compounded amount at the end of year 2 will be

$$\text{future worth at end of year 2} = 19(F/P, 6\%, 2) + 10(F/P, 6\%, 1)$$
$$= 19(1.124) + 10(1.06)$$
$$= 32$$

When this amount is returned to the project, the net cash flow for year 2 becomes $-50 + 32 = -18$. The resulting cash flow for the 5 years is:

Year	Cash flow	0%	8%	10%
0	0	0	0	0
1	0	0	0	0
2	−18	−18	−15.4	−14.9
3	−50	−50	−39.7	−37.6
4	+20	+20	+14.7	+13.7
5	+60	+60	+40.8	+37.3
		NPV = +12	+0.4	−1.5

This cash flow stream has one sign change, indicating that there is either zero or one positive interest rate. By interpolation, we can find the point where NPV = 0:

$$i\% = 8\% + 2\%\frac{0.4}{1.5 + 0.4} = 8\% + 2\%(0.21) = 8.42\%$$

Thus, assuming an external interest rate of 6%, the internal rate of return for the Banzai plane contract is 8.42%.

                                                                            ■

    In many situations we are asked to compare and rank independent investment opportunities rather than a set of mutually exclusive alternatives designed to meet the same need. Portfolio analysis is such an example where the firm is considering a number of different R&D projects and must evaluate the costs and benefits of each. Here the IRR method will always give results that are consistent (regarding project acceptance or rejection) with those obtained from the PW, AW, or FW method. However, the IRR method may give a different ranking regarding the order of desirability than any of the latter when comparing independent investment opportunities.

    As an example, consider Fig. 2-7, depicting the relation of IRR to NPV for two projects, X and Y. The IRR for each project is the interest rate at which the net present value for that project is zero. This is shown for a nominal MARR. For the hypothetical but quite feasible relationship shown in Fig. 2-7, project Y has the higher IRR, while project X has the higher NPV of all IRRs except for the rate at which the net present values are equal. This illustrates the case in which the IRR method does result in a different ranking of alternatives compared to the PW (AW or FW) method. However, because both projects have a NPV greater than zero, the IRR for either is greater than the MARR. The determination of acceptance of both projects is shown consistently by either method. It should be noted

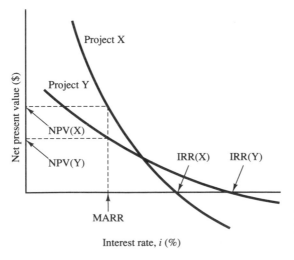

Figure 2-7 Relationship between NPV and IRR for independent investments.

that if X and Y had been mutually exclusive *alternatives* there would have been no inconsistency regarding which to choose provided an incremental IRR analysis was performed.

2.4.6 Payback-Period Method

In its simplest form, the payback period is the number of periods, usually measured in years, it takes for the accruing net *undiscounted* benefits from the investment to equal the cost of the investment. If we assume that the benefits are equal in each future year and that depreciation and income taxes are not included into the calculations, the formula is

$$\text{payback period} = \frac{\text{initial investment}}{\text{annual net undiscounted benefits}}$$

When the benefits differ from year to year, it is necessary to find the smallest value of n such that

$$\sum_{i=1}^{n} B_i \geq P$$

where P is the initial investment and B_i is the annual net benefit in year i.

Example 2-11

The cash flows for two alternatives are as follows:

Alternative	Year					
	0	1	2	3	4	5
A	$-\$2,700$	$+1,200$	$+1,200$	$+1,200$	$+1,200$	$+1,200$
B	$-\$1,000$	$+200$	$+200$	$+1,200$	$+1,200$	$+1,200$

Based on the payback period, which alternative is best?

Solution   *Alternative A:* Because the annual benefits are uniform, the payback period can be computed from the first formula in this section; that is,

$$\frac{\$2,700}{\$1,200/\text{yr}} = 2.25 \text{ years}$$

Alternative B: The payback period is the length of time required for profits or other benefits of an investment to equal the cost of the investment. In the first 2 years, only $400 of the $1,000 cost is recovered. The remaining $600 is recovered in the first half of the third year. Thus the answer is 2.5 years.

Therefore, to minimize the payback period, choose alternative A.   ∎

The great advantage of the payback period is that it is simple and can be applied by anyone. It is thus an excellent mechanism for allowing middle managers and technical staff to choose among proposals without going through a detailed analysis, or to sort through many possibilities before resorting to a more sophisticated approach.

Situations suitable for the use of the payback period are often found in industry. These are projects in which a constant benefit is expected to accrue for an extended period as a result of a particular investment. A typical case would be the purchase of a new robot that would reduce operating expenses each year by a fixed amount, or some insulation or control that would regularly save on energy bills.

The disadvantage of this criterion is that it is crude; it does not clearly distinguish between projects with different useful lives. For any projects with identical useful lives, for which the capital recovery factor will be identical, the payback period gives as good a measure of economic desirability as the net present value or internal rate of return. When the useful lives of projects are different, the capital recovery factors are not the same and the results can be highly misleading, as the following analysis shows.

	Project	
	P1	P2
Investment	$2,000	$2,000
Useful life	3 years	6 years
Annual receipts	1,000	1,000
Payback period, years	2	2.5
NPV at 10%	$487	$1,484
IRR	23.4%	32.7%

In this example, project P1 has a shorter payback period than the alternative P2 and would appear better by this criterion. Yet project P2 is, in fact, more economically desirable for a wide range of discount rates. This is because P2 provides substantial benefits over a much longer period. Thus over a 6-year cycle, P1 would have to be repeated twice for a total cost of $4,000 and benefits of $6,000, whereas P2 would only cost $2,000 and yield returns of $4,800—greater net benefits and a higher net present value for any number of discount factors.

2.5 SENSITIVITY AND BREAK-EVEN ANALYSIS

Many of the data collected in solving a problem represent projections of future consequences and hence may possess a high degree of uncertainty. As the desired result of the analysis is decision making, an appropriate question is: "To what extent do the variations in the data affect the decision?" When small variations in a particular estimate would change the alternative selected, the decision is said to be *sensitive* to the estimate. To better evaluate the impact of any parameter, one should determine the amount of variation necessary in it to effect a change in outcome. This is called *sensitivity analysis*.

This type of analysis highlights the important and significant aspects of a problem. For example, one might be concerned that the estimates for annual maintenance and future salvage value in a facility modernization project vary substantially, depending on the assumptions used. Sensitivity analysis might indicate, however, that the decision is insensitive to the salvage value estimates over the full range of possibilities. At the same time it might show that small changes in annual maintenance expenditures strongly influence the choice of equipment. Under these circumstances, one should place greater emphasis on pinning down the true maintenance costs than on worrying about salvage value estimates.

Succinctly, sensitivity analysis describes the relative magnitude of a particular variation in one or more elements of a problem that is sufficient to alter a particular decision. Closely related is *break-even analysis*, which determines the conditions where two alternatives are equivalent. These two evaluation techniques frequently are useful in engineering problems called *stage construction*. That is, should a facility be constructed now to meet its future full-scale requirements, or should it be constructed in stages as the need for the increased capacity arises? Three examples of this situation are:

- Should we install a cable with 400 circuits now, or a 200-circuit cable now and another 200-circuit cable later?
- A 10-cm water main is needed to serve a new area of homes. Should the 10-cm main be installed now, or should a 15-cm main be installed to provide an adequate water supply later for adjoining areas when other homes are built?
- An industrial firm currently needs a 10,000-m^2 warehouse and estimates that it will need an additional 10,000 m^2 in 4 years. The firm could have a warehouse built now and later enlarged, or have a 20,000-m^2 warehouse built today.

Examples 2-12 and 2-13, adapted from Newnan (1991), illustrate the principles and calculations behind sensitivity and break-even analysis.

Example 2-12

Consider the following situation, where a project may be constructed to full capacity now or may be undertaken in two stages.

Construction costs

Two stage construction	
Construct first stage now	$100,000
Construct second stage n years from now	$120,000
Full-capacity construction	$140,000

Other factors

1. All facilities will last until 40 years from now regardless of when they are installed; at that time they will have zero salvage value.
2. The annual cost of operation and maintenance is the same for both alternatives.
3. Assume that the MARR is 8%.

Plot a graph showing "age when second stage is constructed" versus "costs for both alternatives." Mark the break-even point. What is the sensitivity of the decision to second-stage construction 16 or more years in the future?

Solution   Because we are dealing with a common analysis period, the calculations may be either annual cost or present worth. Present worth calculations appear simpler and are used here:

Construct full capacity now

$$\text{PW of cost} = \$140,000$$

Two-stage construction.   In this alternative, the first stage is constructed now with the second stage to be constructed n years hence. To begin, compute the PW of cost for several values of n (years).

$$\text{PW of cost} = \$100,000 + \$120,000(P/F, 8\%, n)$$

$$n = 5: \quad \text{PW} = 100,000 + 120,000(0.6806) = \$181,700$$
$$n = 10: \quad \text{PW} = 100,000 + 120,000(0.4632) = \$155,600$$
$$n = 20: \quad \text{PW} = 100,000 + 120,000(0.2145) = \$125,700$$
$$n = 30: \quad \text{PW} = 100,000 + 120,000(0.0994) = \$111,900$$

These data are plotted in Fig. 2-8 in the form of a break-even chart. The horizontal axis is the time when the second stage is constructed; the vertical axis represents PW. We see that the PW of cost for two-stage construction naturally decreases as the time for the second stage is deferred. The one-stage construction (full capacity now) option is unaffected by the time variable and hence is a horizontal line on the graph.

The break-even point on the graph is the point at which both alternatives have equivalent costs. We see that if in two-stage construction, the second stage is deferred for 15 years, the PW of that alternative is equal to the PW of the first, which is approximately $137,800.

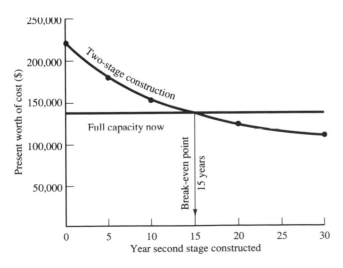

Figure 2-8   Break-even chart for Example 2-12. (From D. G. Newnan, *Engineering Economic Analysis*, copyright © 1991 by Engineering Press, Inc.; reprinted by permission of the publisher.)

Thus year 15 is the break-even point. The plot also shows that if the second stage were needed prior to year 15, one-stage construction, with its smaller PW of cost, would be preferred. If the second stage were not needed until after year 15, the opposite is true.

The decision as to how to construct a project is sensitive to the age at which the second stage is needed only if the range of estimates includes 15 years. For example, if one estimated that the second-stage capacity would be needed sometime over the next 5 to 10 years, the decision is insensitive to that estimate. The more economical thing to do is to build the full capacity now. But if demand for the second-stage capacity were between, say, years 12 and 18, the decision would depend on the estimate of when full capacity would actually be needed.

■

One question posed by Example 2-12 is how sensitive the decision is to the need for the second stage at or beyond 16 years. The graph shows that the decision is insensitive. In all cases for construction on or after 16 years, two-stage construction has a lower PW of cost.

Example 2-13

In this example, we have three mutually exclusive alternatives, each with a 20-year life and no salvage value. Assume that the MARR is 6% and

	A	B	C
Initial cost	$2,000	$4,000	$5,000
Uniform annual benefit	$410	$639	$700

Calculating the NPV of each alternative gives

$$NPV = A(P/A, 6\%, 20)$$
$$NPV(A) = 410(11.470) - 2,000 = \$2,703$$
$$NPV(B) = 639(11.470) - 4,000 = \$3,329$$
$$NPV(C) = 700(11.470) - 5,000 = \$3,029$$

so alternative B is preferred. Now we would like to know how sensitive the decision is to the estimate of the initial cost of B. If B is preferred at an initial cost of $4,000, it will continue to be preferred for any smaller values. But how much higher than $4,000 can the initial cost go up and still have B as the preferred alternative?

Solution   The computations may be performed in several different ways. The first thing to note is that for the three alternatives, B will maximize NPV only as long as its NPV is greater than $3,029. Let x = initial cost of B. Thus we have

$$NPV(B) = 639(11.470) - x > 3,029$$

or

$$x < 7,329 - 3,029 = 4,300$$

implying that B is the best alternative if its initial cost does not exceed $4,300. The break-even chart for the problem is displayed in Fig. 2-9. Because we are maximizing NPV, we see that B

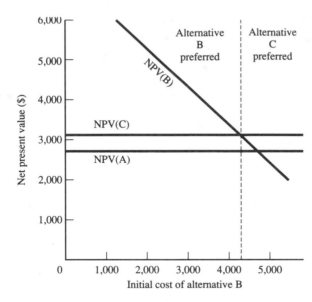

Figure 2-9 Break-even chart for Example 2-13. (From D. G. Newnan, *Engineering Economic Analysis*, copyright © 1991 by Engineering Press, Inc.; reprinted by permission of the publisher.)

is preferred if its initial cost is less than $4,300. At an initial cost above this value, C is preferred. At the break-even point, B and C are equally desirable. For the data given, alternative A is always inferior to alternative C.

■

Sensitivity analysis and break-even point calculations can be very useful in identifying how different estimates affect the decision. It must be recognized, however, that these calculations assume that all parameters except one are held constant, and that the sensitivity of the decision to that parameter is what is being evaluated.

2.6 EFFECT OF TAX AND DEPRECIATION ON INVESTMENT DECISIONS

The discussion thus far referred to investment earnings as cash flows implicitly net of tax consequences. The reason for this is that only the actual cash flow produced by an investment is relevant to the decision process. Earnings before depreciation and taxes do not represent the actual benefits realized by a firm. Consequently, the expected income from an investment must be adjusted to represent the true cash inflow before ranking can take place. Note that depreciation can be viewed as an expense and thus reduces gross income for tax purposes. The procedures and schedules used to compute depreciation in any year are promulgated by the Internal Revenue Service.

Assume that a machine costing $10,000 has a useful life of 5 years and is expected to produce gross earnings of $4,000 each year. With straight-line depreciation (amount per year = (initial cost − salvage value)/(useful life)), no salvage value, and a 40% tax rate, the annual cash flow in each of the 5 years will be:

A.	Gross earnings	$4,000
B.	Depreciation expense	$2,000
C.	Taxable income (A − B)	$2,000
D.	Taxes (40% of C)	−800
E.	Cash flow (A − D)	$3,200

Now, if the MARR for the firm is 10%, the net present value of the investment is

$$\$3,200(P/A, 10\%, 5) - \$10,000 = 3,200(3791) - 10,000 = \$2,131$$

which makes it worthwhile.

2.6.1 Capital Expansion Decision

Problem.   The Leeds Corporation leases plant facilities in which expendable thermocouples are manufactured. Because of rising demand, Leeds could increase sales by investing in new equipment to expand output. The selling price of $10 per thermocouple will remain unchanged if output and sales increase. Based on engineering and cost estimates, the accounting department provides management with the following cost estimates based on an annual increased output of 100,000 units.

Cost of new equipment having an expected life of 5 years	$500,000
Equipment installation cost	$20,000
Expected salvage value	0
New operation's share of annual lease expense	$10,000
Annual increase in utility expenses	$40,000
Annual increase in labor costs	$160,000
Annual additional cost for raw materials	$40,000

The sum-of-the-years' digits (SOYD) method of depreciation will be used and taxes are paid at a rate of 40%. Mr. Leeds' policy is not to invest capital in projects earning less than a 20% rate of return. Should the proposed expansion be undertaken?

Solution.   Compute cost of investment:

Acquisition cost of equipment	$500,000
Equipment installation costs	20,000
Total cost of investment	$520,000

Determine yearly cash flows throughout the life of the investment. The lease expense is a sunk cost. It will be incurred whether or not the investment is made and is therefore irrelevant to the decision and should be disregarded. Annual production expenses to be considered are utility, labor, and raw materials. These total $600,000 per year. Annual sales revenue is $10 × 10,000 units of output, or $1,000,000. Yearly income before depreciation and taxes is thus $1,000,000 gross revenue less $600,000 expenses, or $400,000.

Determine the depreciation charges to be deducted from the $400,000 income each year using the SOYD method (sum-of-the-years' digits $= \Sigma = 1 + 2 + 3 + 4 + 5 = 15$). With SOYD, the depreciation in year j is (initial cost $-$ salvage value) $\times$ $(n - j + 1)/\Sigma, j = 1,...,n$:

Year	Proportion of $500,000 to be depreciated			Depreciation Charge
1	5/15	$\times$	$500,000 =	$166,667
2	4/15	$\times$	500,000 =	133,333
3	3/15	$\times$	500,000 =	100,000
4	2/15	$\times$	500,000 =	66,667
5	1/15	$\times$	500,000 =	33,333
	Accumulated depreciation		=	$500,000

Find each year's cash flow when taxes are 40%. Cash flow is illustrated only for the first year:

Earnings before depreciation and taxes		$400,000
Depreciation expense	$166,667	
Taxable income	$233,333	
Taxes (0.4 $\times$ 233,333)		$-93,332
Cash flow (first year)		$306,668

Determine the present value of the cash flows. Since Leeds demands at least a 20% rate of return on investments, multiply the cash flows by the 20% present value factor $(P/F, 20\%, j)$ for each year j.

Year	Present-value factor		Cash flow		Present value
1	0.833	$\times$	$306,667	=	$255,454
2	0.694	$\times$	293,333	=	203,573
3	0.579	$\times$	280,000	=	162,120
4	0.482	$\times$	266,667	=	128,533
5	0.402	$\times$	253,334	=	101,840
	Total present value of cash flows (discounted at 20%)			=	$851,520

Find whether net present value is positive or negative:

Total present value of cash flows	$851,520
Total cost of investment	520,000
Net present value	$331,520

Decision.   Net present value is positive when returns are discounted at 20%. Therefore, the proposed expansion should be undertaken.

2.6.2 Replacement Decision

Problem.   For 5 years Emetic Pharmaceuticals has been using a machine that attaches labels to bottles. The machine was purchased for $4,000 and is being depreciated over 10 years to a zero salvage value using the straight-line method. The machine can be sold now for $2,000. Emetic can buy a new labeling machine for $6,000 that will have a useful life of 5 years and cut labor costs by $1,200 annually. The old machine will require a major overhaul in the next few months. The cost of the overhaul is expected to be $300. If purchased, the new machine will be depreciated over 5 years to a $500 salvage value using the straight-line method. The company will invest in any project earning more than the 12% cost of capital. The tax rate is 40%. Should Emetic invest in the new machine?

Solution.   Determine the cost of investment:

Price of the new machine		$6,000
Less: Sale of old machine	$2,000	
Avoidable overhaul costs	$300	
Total deductions		−2,300
Effective cost of investment		$3,700

Determine the increase in cash flow resulting from investment in the new machine:

Yearly cost savings = $1,200
Differential depreciation:
   Annual depreciation on old machine:

$$\frac{\text{cost } - \text{ salvage}}{\text{useful life}} = \frac{\$4,000 - \$0}{10} = \$400$$

   Annual depreciation on new machine:

$$\frac{\text{cost } - \text{ salvage}}{\text{useful life}} = \frac{\$6,000 - \$500}{5} = \$1,100$$

   Differential depreciation = $1,100 − $400 = $700

Yearly net increase in cash flow into the firm:

Cost savings		$1,200
Deduct: Taxes at 40%	$480	
Add: Advantage of increase in depreciation		
(0.4 × $700)	$280	
Net deductions		−200
Yearly increase in cash flow		$1,000

Determine the total present value of the investment:

The 5-year cash flow of $1,000 per year is an annuity.
Discounted at 12%, the cost of capital, the present value is:
$1,000 × 3.605 = $3,605.

The present value of the new machine, if sold at its salvage value of $500 at the end of the fifth year, is $500 × 0.567 = $284.

Total present value of the expected cash flows: $3,605 + $284 = $3,889

Determine whether net present value is positive:

$$
\begin{array}{ll}
\text{Total present value} & \$3,889 \\
\text{Cost of investment} & \underline{\hphantom{0}3,700} \\
\text{Net present value} & \$\hphantom{0}189 \\
\end{array}
$$

Decision.   Emetic Pharmaceuticals should make the purchase because the investment will return slightly more than the cost of capital.

Note.   The importance of depreciation has been shown in this example. The present value of the yearly cash flow resulting from operations is *only*

$$
\begin{array}{ccc}
\text{(Cost savings} - \text{taxes)} & \text{(Present value factor)} & \\
(\$1,200 - \$480) & \times \quad 3.605 & = \$2,596 \\
\end{array}
$$

This figure is $1,104 less than the $3,700 cost of the investment. Only a very large depreciation advantage makes this investment worthwhile. The total present value of the advantage is $1,009; that is,

$$
\begin{array}{ccc}
\text{(Tax rate} \times \text{differential depreciation)} & \text{(PV factor)} & \\
(0.4 \times \$700) & \times \quad 3.605 & = \$1,009 \\
\end{array}
$$

2.6.3 Make-or-Buy Decision

Problem.   The GIGO Corporation manufactures and sells computers. It makes some of the parts and purchases others. The engineering department believes that it might be possible to cut costs by manufacturing one of the parts currently being purchased for $8.25 each. The firm uses 100,000 of these parts each year, and the accounting department compiles the following list of annual costs based on engineering estimates.

Fixed costs will increase by $50,000.

Labor costs will increase by $125,000.

Factory overhead, currently running $500,000 per year, is expected to increase 12%.

Raw materials used to make the part will cost $600,000.

Given the estimates above, should GIGO make the part or continue to buy it?

Solution.   Find the total cost per year incurred if the part were manufactured:

Additional fixed costs	$ 50,000
Additional labor costs	125,000
Raw materials cost	600,000
Additional overhead costs = 0.12 × $500,000	60,000
Total cost to manufacturer	$835,000

Find cost per unit to manufacture:

$$\frac{\$835,000}{100,000} = \$8.35 \text{ per unit}$$

Decision.   GIGO should continue to buy the part. Manufacturing costs exceed the present cost to purchase by $0.10 per unit.

Perspective.   The decision to make or buy is arguably the most fundamental component of manufacturing strategy. Should a firm be highly integrated, such as Henry Ford's River Rouge plant, with raw iron ore and coal flowing in one end and a finished Model A rolling out the other? Or should they simply purchase components from capable suppliers and then perform an assembly role much like today's PC manufacturers, such as Compaq and CompuAdd?

Henry Ford's model of vertical integration slipped from favor in the early 1960s when outsourcing became increasingly attractive. Businesses found that outsourcing had certain advantages, potentially allowing them to:

- Convert fixed costs to variable costs, thereby providing flexibility in an economic downturn
- Balance work force requirements
- Reduce capital investment requirements
- Reduce costs via suppliers' economies of scale and lower wage structures
- Accelerate new product development
- Gain access to invention and innovation from suppliers
- Focus resources on high-value-added activities

Nevertheless, recent studies have shown that many make-or-buy decisions have historically been taken with a disproportionate weight placed on unit cost and an insufficient regard for strategic or technical issues (e.g., see Dertouzos et al. 1989). This cost-focused approach has led to competitive disaster for many firms, indeed, entire industries in the United States. The list of those affected by this phenomenon is well known. Some of the most notable include consumer electronics, machine tools, semiconductors, and office equipment. As recently as 1992, General Motors reported more than 8,000 suppliers for direct material alone.

2.6.4 Lease-or-Buy Decision

Problem.   Jeremy Sitzer is a small businessman who has need for a pickup truck in his everyday work. He is considering buying a used truck for $3,000. If he goes ahead, he believes that he will be able to sell it for $1,000 at the end of 4 years, so he will depreciate $2,000 of the truck's value on a straight-line basis. Sitzer can borrow $3,000 from the bank and repay it in four equal annual installments at 6% interest. However, a friend advises him that he may be better off to lease a truck if he can get the same terms from the leasing company that he receives at the bank. Assuming that this is so, should Sitzer buy or lease the truck? Taxes are 40%.

Solution.   Find the cost to buy. The bank loan is an installment loan at 6% interest, so the payments constitute a 4-year annuity. Divide the amount of the loan by the present value factor for a 4-year annuity at 6% [(P/A, 6%, 4) = 3.465] to find the annual payment. Multiply the annual payments by 4 to find the total payment.

$$\frac{\$3,000}{3.465} = \$866 \text{ annual payment}$$

$$4 \times \$866 = \$3,464 \text{ total payment}$$

Next, find the present value of the cost of the loans:

(1) Year	(2) Yearly payment	(3) Interest at 6%	(4) Payment on principal	(5) Remaining balance	(6) Depreciation
1	$866	$180	$686	$2,314	$500
2	866	139	727	1,587	500
3	866	95	771	816	500
4	866	50	816	—	500

(7) Tax-deductible expense (3) + (6)	(8) Tax saving 0.4 × (7)	(9) Cost of owning (2) − (8)	(10) Present value factor	(11) Present value (9) × (10)
$680	$272	$594	0.943	$ 560
639	256	610	0.890	543
595	238	628	0.840	527
550	220	646	0.792	497
		Total present value of payments		$2,127

$$\text{Present value of salvage} = \$1,000 \times 0.792 = \$792$$

$$\text{Present value of cost of loan} = \$2,127 - \$792 = \$1,335$$

Find the cost to lease:

(1)	(2)	(3)	(4)	(5)	(6)
		Tax	Lease cost	Present	Present
	Lease	savings	after taxes	value	value
Year	payment	0.4×866	$(2) - (3)$	factor at 6%	$(4) \times (5)$
1	$866	$346	$520	0.943	$ 490
2	866	346	520	0.890	463
3	866	346	520	0.840	437
4	866	346	520	0.792	411
			Total present value of lease payments		$1,801

Compare present values of cost to buy and cost to lease:

Present value of cost to lease $1,801
Present value of cost to buy   1,335
Advantage of buying $ 466

Decision.   Mr. Sitzer should buy the truck.

Note.   Again, the importance of depreciation should be mentioned. When Sitzer purchases the truck, he gains the accompanying tax advantages of ownership. If the truck were leased, the lessor would depreciate the truck and thereby gain advantage. Sitzer was also aided by being able to reduce the cost of buying by the present value of the salvage (or disposal) value of the truck. In general, depreciation and salvage value reduce the cost of buying. Nevertheless, if an asset is subject to rapid obsolescence, it may be less expensive to lease.

2.7 UTILITY THEORY

Decision theory is concerned with giving structure and rationale to the various conditions under which decisions are made. In general, one must choose between a host of alternatives. These are referred to as actions (or strategies), and each results in a payoff or outcome. If decision makers knew the payoff associated with each action, they would be able to choose the action with the largest payoff. Most situations, however, are characterized by incomplete information, so for a given action, it is necessary to enumerate all probable outcomes together with their consequences and probabilities. The degree of information and understanding that the decision maker has about a particular situation determines how the underlying problem can be approached.

Two persons, faced with the same set of alternatives and conditions, are likely to arrive at very different decisions regarding the most appropriate course of action for them. What is optimal for the first person may not even be an attractive alternative for the second. Judgment, risk, and experience work together to influence attitudes and choices.

Implicit in any decision-making process is the need to construct, either formally or informally, a preference order so that alternatives can be ranked and the final choice made.

For some problems this may be easy to accomplish, as we saw in the preceding sections, where the decision was based on a profit-maximization rule. There, the preference order is adequately represented by the natural order of real numbers ($\leq$ or $\geq$). In more complex situations, where factors other than profit maximization or cost minimization apply, it may be desirable to explore the decision maker's preference structure in an explicit fashion and to attempt to construct a preference ordering directly. An important class of techniques that works by eliciting preference information from the decision maker is predicated on what is known as *utility theory*. This, in turn, is based on the premise that the preference structure can be represented by a real-valued function called a *utility function*.* Once such a function is constructed, selection of the final alternative should be relatively simple. In the absence of uncertainty, an alternative with the highest utility would represent the preferred solution. For the case where outcomes are subject to uncertainty, the appropriate choice would correspond to that which attains the highest expected utility. Thus the decision maker is faced with two basic problems involving judgment:

1. How to quantify (or measure) utility for various payoffs
2. How to quantify judgments concerning the probability of the occurrence of each possible outcome or event

In this section we focus on the first question—of quantifying and exploiting personal preference; the second, subjective probability estimation, falls more appropriately in the realm of elementary statistics and so is not treated here.

2.7.1 Expected Utility Maximization

Assuming the presence of uncertainty, when a decision maker is repeatedly faced with the same problem, experience often leads to a strategy that provides, on average, the best results over the long run. In technical terms, such a strategy is one that maximizes *expected monetary value* (EMV). Notationally, let A be a particular action with possible outcomes $j = 1,...,n$. Also, let p_j be the probability of realizing outcome j with corresponding payoff or return x_j. The expected monetary value of A is calculated as follows:

$$\text{EMV}(A) = \sum_{j=1}^{n} p_j x_j \tag{2-1}$$

For the case where the decision maker is faced with a unique problem, using the EMV criterion might not be such a good idea. In fact, a large body of empirical evidence suggests that it is rarely the criterion selected. To see this, assume that you must select one of the two alternatives in each of the following five situations:

---

*Technically speaking, the term *utility function* is reserved for the case where uncertainty is present. When each alternative has only one possible outcome, the term *value function* is used. In either case, the construction procedure is the same.

Situation 1

a_1: The certainty of receiving $1;

or

a_2: on the flip of a fair coin, $10 if it comes up heads, or $-$1 if it comes up tails.

Situation 2

b_1: The certainty of receiving $100;

or

b_2: on the flip of a fair coin, $1,000 if it comes up heads, or $-$100 if it comes up tails.

Situation 3

c_1: The certainty of receiving $1,000;

or

c_2: on the flip of a fair coin, $10,000 if it comes up heads, or $-$1,000 if it comes up tails.

Situation 4

d_1: The certainty of receiving $10,000;

or

d_2: on the flip of a fair coin, $100,000 if it comes up heads, or $-$10,000 if it comes up tails.

Situation 5

e_1: The certainty of receiving $10,000;

or

e_2: a payment of 2^n, where n is the number of times a fair coin is flipped until heads comes up. If heads appears on the first toss, you receive $2; if the coin shows tails on the first toss and heads on the second, you receive $4; and so on. However, you are given only one chance; the game stops with the first showing of heads.

Most people would probably choose a_2, b_2, c_1, d_1 and e_1. The choices a_2 and b_2 would be those derived from an EMV maximization criterion since $\text{EMV}(a_2) = \frac{1}{2}$10 + \frac{1}{2}$(-$1)$ = $4.5 is greater than the return from the certain choice $a_1 = $1, and $\text{EMV}(b_2) = 450 is greater than $100. Nevertheless, in situations 3 and 4, c_1 would probably be preferred to c_2, even though $\text{EMV}(c_2) = $4,500$ is greater than $1,000, and d_1 would be preferred to d_2 even though $\text{EMV}(d_2) = $45,000$ is greater than $10,000. In situation 5, the EMV of e_2 is infinite; that is,

$$\text{EMV}(e_2) = \tfrac{1}{2}($2) + \tfrac{1}{4}($4) + \tfrac{1}{8}($8) + \cdots$$
$$= $1 + $1 + $1 + \cdots$$
$$= \infty$$

yet e_1 would be preferred to e_2 by practically everyone.

In the first four situations, most persons would tend to change their decision criterion away from maximizing EMV as soon as the thought of losing a large sum of money (say, $1,000) became too painful despite the pleasure to be gained from possibly obtaining a large sum (say, $10,000). At this point, the person faced with such a choice would not be considering EMV but would, instead, be thinking solely of utility. In this sense, *utility* refers to the pleasure (utility) or displeasure (disutility) one would derive from certain outcomes. In essence, we are saying that the person's displeasure from losing $1,000 is greater that the pleasure of winning many times that amount. In situation 5, no prudent person would choose the gamble e_2 over the certainty of a relatively modest amount obtained from e_1. This problem, known as the St. Petersburg paradox, led Daniel Bernoulli to the first investigations of utility rather than EMV as the basis of decision making.

2.7.2 Bernoulli's Principle

Logic, observed behavior, and introspection all indicate that any adequate procedure for handling choice under uncertainty must involve two components: personal valuation of consequences and personal strengths of belief about the occurrence of uncertain events. Bernoulli's principle, as refined by von Neumann and Morgenstern (1947), has the normative justification of being a logical deduction from a small number of axioms that most people find reasonable. The relevant axioms differ slightly depending on whether the decision maker (a) has a single goal, (b) has multiple goals between which he or she can establish acceptable trade-off relations, or (c) has multiple goals that are not substitutable. The first two cases lead to a one-dimensional utility measure (i.e., real number) for each alternative action; the last to a lexicographically ordered utility vector.* We consider only the single-goal case here; multiple goals are taken up in Chapter 4.

Axioms

1. *Ordering*. For the two alternatives A_1 and A_2, one of the following must be true: the person either prefers A_1 to A_2, A_2 to A_1, or is indifferent between them.
2. *Transitivity*. The person's evaluation of alternatives is transitive: if he prefers A_1 to A_2, and A_2 to A_3, then he prefers A_1 to A_3.
3. *Continuity*. If A_1 is preferred to A_2, and A_2 to A_3, there exists a unique probability p, $0 < p < 1$, such that the person is indifferent between outcome A_2 with certainty, or receiving A_1 with probability p and A_3 with probability $(1 - p)$. In other words, there exists a certainty equivalent to any gamble.
4. *Independence*. If A_1 is preferred to A_2, and A_3 is some other prospect, a gamble with A_1 and A_3 as outcomes will be preferred to a gamble with A_2 and A_3 as outcomes if the probability of A_1 and A_2 occurring is the same in both cases.

These axioms relate to choices among both certain and uncertain outcomes. That is, if a person conforms to the four axioms, a utility function can be derived that expresses his

*Given two n-dimensional vectors x and y, if $x_i = y_i$, for $i = 1,...,r-1$, and $x_r > y_r$, the x is said to be lexicographically greater than y.

or her preferences for both certain outcomes (more precisely, we should say "value" function in this case) and the choices in a risky situation. In essence, they are equivalent to assuming that the decision maker is rational and consistent in his or her preferences, and imply Bernoulli's principle, or as it is also known, the *expected utility theorem*.

Expected utility theorem.   Given a decision maker whose preferences satisfy the four axioms, there exists a function U, called a utility function, that associates a single real number or utility index with all risky prospects faced by the decision maker. This function has the following properties:

1. If the risky prospect A_1 is preferred to A_2 (written $A_1 > A_2$), the utility index of A_1 will be greater than that of A_2 [i.e., $U(A_1) > U(A_2)$]. Conversely, $U(A_1) > U(A_2)$ implies that A_1 is preferred to A_2.

2. If A is the risky prospect with a set of outcomes $\{\theta\}$ distributed according to the probability density function $p(\theta)$, the utility of A is equal to the statistically expected utility of A; that is,

$$U(A) = EU(A) \qquad\qquad (2\text{-}2)$$

If $p(\theta)$ is discrete,

$$EU(A) = \sum_\theta U(\theta)p(\theta) \qquad\qquad (2\text{-}3a)$$

and if $p(\theta)$ is continuous,

$$EU(A) = \int_{-\infty}^{\infty} U(\theta)p(\theta)d(\theta) \qquad\qquad (2\text{-}3b)$$

As these equations indicate, only the first moment (i.e., the mean or expected value) of utility is relevant to choice. For a person who accepts the axioms underlying Bernoulli's principle, the variance or other higher moments of utility are irrelevant; the expected value takes full account of all the moments (mean, variance, skewness, etc.) of the probability distribution $p(\theta)$ of outcomes.

3. Uniqueness of the function is defined only up to a positive linear transformation. Given a utility function U, any other function U^* such that

$$U^* = aU + b, \qquad a > 0 \qquad\qquad (2\text{-}4)$$

for scalars a and b, will serve as well as the original function. Thus utility is measured on an arbitrary scale and is a relative measure analogous, for example, to the various scales used for measuring temperature. Concomitantly, because there is no absolute scale for utility and because a person's utility function reflects his or her own personal valuations, it is not possible to compare one person's utility indices with another's.

Bernoulli's principle thus provides a mechanism for ranking risky prospects in order of preference, the most preferred prospect being the one with the highest utility. Hence

Bernoullian or statistical decision theory implies the maximization of utility which by the expected utility theorem, is equivalent to maximization of expected utility. Equations (2-3a) and (2-3b) provide the empirical basis of application of the theory. Two concepts are involved: degree of preference (or utility) and degree of belief (or probability).

2.7.3 Constructing the Utility Function

Utility functions must be assessed separately for each decision maker. To be of use, utility values (i.e., subjective preferences) must be assigned to all possible outcomes for the problem at hand. Usually, we define a frame of reference whose lower and upper bounds represent the worst and best possible outcomes, respectively. In many circumstances, outcomes are nonmonetary in nature. For example, in selecting a portable computer, one weighs such factors as speed, memory, display quality, and weight. It is possible to assign utility values to these outcomes; however, in most business-related problems, a monetary consequence is of major importance. Hence we shall illustrate how to evaluate one's utility function for money, although the same procedure applies to nonmonetary outcomes.

The assessment of a person's utility function involves pinning down, in quantitative terms, subjective feelings that may not have been thought of before in such a precise way. At least four approaches for doing this have been distinguished (Keeney and Raiffa 1976): (1) direct measurement, (2) the von Neumann–Morgenstern (NM) or standard reference contract, (3) the modified NM method, and (4) the Ramsey method.

The first approach involves asking a series of questions of the type: "Suppose that I were to give you an outright gift of $100. How much money would you need to make you twice as happy as the $100 would make you feel?" The answers to a sequence of such questions enable the plotting of a utility curve against whatever arbitrarily chosen utility (value) scale is desired. The drawbacks of this approach are that it is not concerned with uncertainty, and for many people, it cannot be expected to be as precise as the other methods.

The other three approaches deal with the question of risk attitude directly and ask the decision maker to compare certain gambles to sure sums of money, or gambles to gambles. For example, in a new product development problem, a question might be to have the project manager choose between receiving $200,000 for certain versus a gamble (lottery), with equal chances of winning $1,000,000 and losing $500,000. Such a situation might arise if the project manager were faced with selecting one of two technologies: the first being a sure thing, the second being much more risky. Through this type of questioning, one can find some riskless value that would make the project manager indifferent (*Axiom 3*). This value is called the *certainty equivalent* (CE) of the gamble. When the certainty equivalent is less than the expected monetary value (CE < EMV), we say that the decision maker is risk averse. The measurement procedure is continued with different gambles until enough data points are available to plot the utility curve.

In this subsection we discuss the modified NM method, which in our experience, is the most easily understood. The first step in deriving the utility function is to designate two monetary outcomes as reference points. For convenience, we look at the most favorable and least favorable outcomes and then select two values greater than or equal to and less than or equal to these outcomes. The utilities of these extreme points may be selected arbitrarily; however, convention usually assigns them values of 1 and 0, respectively. For example, in the new product development problem given below, the monetary outcomes range from $-\$267,000$ to $\$750,000$. For expediency, we thus might choose extreme values of $-\$500,000$ and $\$1,000,000$, assigning a zero utility to the first and a utility of 1 to the second. That is,

$$U(-\$0.5M) = 0 \quad \text{and} \quad U(\$1M) = 1 \tag{2-5}$$

Once again, the choice of the scale 0 to 1 is arbitrary and just as well could have been -100 to 100, or 34 to 78.

The standard reference contract or NM method is based on the concept of certainty equivalence. If outcome x_1 is preferred to x_2, and x_2 is preferred to x_3, then by continuity there exists a probability p such that

$$pU(x_1) + (1 - p)U(x_3) = U(x_2) \tag{2-6}$$

For specified values of x_1, x_2, and x_3, the utility of x_2 can be determined by questioning to find the value of p at which x_2 is the *certainty equivalent* of the gamble involving x_1 and x_3 (i.e., what value of p will make you indifferent to the gamble of receiving x_2 for certain?), $U(x_1)$ and $U(x_3)$ being given values on an arbitrary scale. For example, if $U(x_1)$ is set at 1 and $U(x_3)$ at 0, then $U(x_2) = p$ [i.e., $p(1) + (1 - p)(0) = U(x_2)$]. By defining the values of p corresponding to an array of values of x_2 between x_1 and x_3, the utility curve may be plotted for values of x in this range.

The difficulty that arises in applying eq. (2-6) is that most people have no experience in specifying probabilities, and consequently become extremely frustrated with the questioning, especially when the appropriate value of p is small, say less than 0.1. To overcome the biases that result, the modified NM method uses neutral probabilities of $p = 0.5 = 1 - p$. Questions are posed to determine the certainty equivalent x_2 for a 50–50 lottery of x_1 and x_3. Thus we have

$$0.5U(x_1) + 0.5U(x_3) = U(x_2) \tag{2-7}$$

If $U(x_1)$ is set at 1 and $U(x_3)$ at 0, then $U(x_2) = 0.5$. In a similar fashion, the CE may be established for the 50–50 lottery of x_1 and x_2, say x_4, which will have a utility of

$$U(x_4) = 0.5U(x_1) + 0.5U(x_2) = 0.75$$

and for the 50–50 lottery of x_2 and x_3, say x_5, which will have a utility of

$$U(x_5) = 0.5U(x_2) + 0.5U(x_3) = 0.25.$$

By such further linked questions, additional points on the utility curve may be established. Now, using eq. (2-5), let's see how we can find the project manager's utility function. To do this, we formulate the following two alternatives: (1) a gamble that offers a 50–50 chance of winning $\$1,000,000$ and losing $\$500,000$, and (2) one that offers a sure amount of money.

Suppose that you have the choice of the gamble (call this scenario B) versus the sure thing (call this A). How much money would the sure thing have to be such that you were indifferent between A and B (i.e., the two alternatives were equally attractive)? Suppose that you said $-\$250,000$. Since you are indifferent to these two options, they must have the same utility, or more properly, the same expected utility. Recall that the expected utility of any set of mutually exclusive outcomes resulting from a decision is the sum of the products of the utility of each outcome and its probability of occurrence. The expected utility of the gamble B is

$$U(\text{B}) = 0.5U(\$1,000,000) + 0.5U(-\$500,000)$$
$$= 0.5(1) + 0.5(0) = 0.5$$

implying that $U(\text{B}) = U(\text{A}) = U(-\$250,000) = 0.5$. The basic concept is depicted in Fig. 2-10.

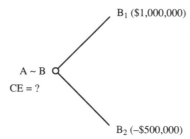

Figure 2-10   Diagram for utility assessment.

We now have three points on the project manager's utility curve. Additional evaluations may be made in a similar manner to obtain a more precise picture. For example, pose an alternative that offers a 50–50 chance of gaining $\$1,000,000$ and losing $\$250,000$. Find the amount that must be offered with certainty to make him indifferent to the gamble. Suppose that he says $\$75,000$. Then

$$U(\$75,000) = 0.5U(\$1,000,000) + 0.5U(-\$250,000)$$
$$= 0.5(1) + 0.5(0.5) = 0.75$$

Next pose the alternative involving a 50–50 chance of losing $\$250,000$ or $\$500,000$. The project manager would clearly consider this gamble unfavorable and would surely be willing to pay some amount to be relieved of the choice (in the same way that one buys insurance to be relieved of risk). Suppose that he was indifferent between the gamble and paying a fixed amount of $\$420,000$. Then

$$U(-\$420,000) = 0.5U(-\$250,000) + 0.5U(-\$500,000)$$
$$= 0.5(0.5) + 0.5(0) = 0.25$$

We now have five points on his utility function, as given in Table 2-1. These can be connected by a smooth curve to approximate the "true" utility function over the entire range from $-\$500,000$ to $\$1,000,000$ (see Fig. 2-11).

TABLE 2-1   ASSESSED UTILITIES
FOR PROJECT MANAGER

Monetary outcome, x	Utility, $U(x)$
$-\$500,000$	0.00
$-\$420,000$	0.25
$-\$250,000$	0.50
$\$75,000$	0.75
$\$1,000,000$	1.00

Note that to be consistent, the project manager should, for example, be indifferent between a gamble C, which offered an equal chance of winning $1,000,000 or losing $500,000, and a second gamble D, which offered an equal chance of winning $75,000 or losing $420,000. That is,

$$U(C) = 0.5U(\$1,000,000) + 0.5U(-\$500,000) = 0.5(1) + 0.5(0) = 0.5$$
$$U(D) = 0.5U(\$75,000) + 0.5U(-\$420,000) = 0.5(0.75) + 0.5(0.25) = 0.5$$

If this is not true, the manager's assessments are inconsistent and should be revised. Similar checks should be performed to gain confidence in the decision maker's responses.

To facilitate the analysis, a number of commercial products such as SmartEdge (Smith et al. 1987), are available. These can be used to guide in the construction of the utility function, assess subjective probabilities, check for inconsistencies in judgment, and rank the alternatives.

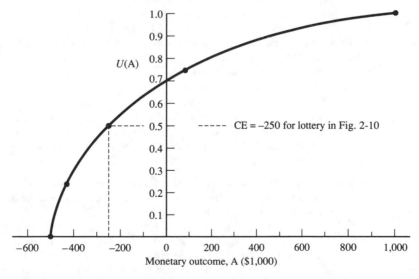

Figure 2-11   Utility function obtained from data in Table 2-1.

2.7.4 Evaluating Alternatives

In the general case, we are given a set of m alternatives $A = \{A_1,...,A_m\}$, where each alternative may result in one of n outcomes or "states of nature." Call these $\theta_1,...,\theta_n$, and denote x_{ij} as the consequence realized if θ_j results when alternative i is selected. Also, let $p_j(\theta_j)$ be the probability that the state of nature θ_j occurs. Then, from eq. (2-3a) we can compute the expected utility of alternative A_i as follows:

$$U(A_i) = \sum_{j=1}^{n} p_j(\theta_j)U(x_{ij}), \qquad i = 1,...,m \tag{2-8}$$

where $x_{ij} \equiv x_{ij}(\theta_j)$ is an implicit function of θ_j. For the deterministic case where $n = 1$, implying that only one outcome is possible, eq. (2-8) reduces to $U(A_i) = U(x_i)$.

Example 2-14 (Selection of New Product Development Strategy)

As project manager of a research and development group you have been assigned the responsibility for coming up with a new switching circuit as a modular component for a laser device. You are given a budget of $300,000 and 3 months to complete the project. Two technical approaches have been identified, one using a circuit incorporating conventional transistors, and another designed around a single integrated chip.

You estimate that a successful conventional circuit design would be worth $478,300 to the company. In contrast, use of a single integrated chip would offer a simpler, more reliable circuit and one that was sufficiently easier to manufacture. Moreover, it would yield an additional cost savings of $150,000 and would be worth an additional $121,700 to the firm over and above any cost savings, for the quantity expected.

You are sure that either of the two approaches could be developed to satisfy the project's specifications given enough time and money. However, within the allotted time and budget, you estimate that there is a 30% chance that the conventional circuit would not meet specifications and a 50% chance that the integrated chip would also fail.

The end result of the project is to be a prototype built in the manufacturing shop from the drawings furnished by you. To work out the design details of the circuit and to identify and resolve unanticipated problems, you plan to design and build a breadboard model. This would take 3 months and cost (in labor, materials, and equipment) $60,000 for the conventional design and $100,000 for the integrated chip. The critical decision you are confronted with is the choice of which design to pursue in construction of a breadboard.

Because you would be within budget, you have the additional option of pursuing the two technical approaches simultaneously. Nevertheless, if you undertake both in parallel, you will incur an additional $107,000 in expenses. What is the best course of action for conducting the development project?

Solution   Let A_1 be the alternative associated with the conventional design, A_2 the alternative associated with the integrated chip, and A_3 the parallel strategy. Note that if the latter is pursued and both breadboards are built, the cost will be $267,000.

The data for this problem are displayed in Table 2-2 in the form of a payoff matrix. For each alternative there are four possible states of nature ($n = 4$), depending on whether the respective breadboard is a success (S) or failure (F). These outcomes, θ_j ($j = 1,...,4$), are indicated in the first row of the table. The probabilities $p_j(\theta_j)$ are computed by taking the product of the two possible outcomes, S or F. For example, $p_1(\theta_1) = \text{Prob}(A_1 \text{ is a success}) \times$

Prob(A_2 is a success) = 0.7×0.5 = 0.35. The monetary consequences of each action for each state of nature are determined by subtracting the costs from the returns. For example, x_{33} represents the payoff where both designs are pursued but only the second succeeds. The cost would be $60,000 for the conventional option + $100,000 for the integrated chip + $107,000 for the duplication of effort = $267,000. The returns to the firm are $478,300 + $121,700 + $150,000 for ease of manufacturability = $750,000. Thus x_{33} = $750,000 − $267,000 = $483,000.

TABLE 2-2 PAYOFF MATRIX FOR NEW PRODUCT
DEVELOPMENT EXAMPLE

θ_j	A_1 : S A_2 : S	S F	F S	F F	
					EMV
$A_i \setminus p_j(\theta_j)$	0.35	0.35	0.15	0.15	($1,000)
A_1	418.3	418.3	−60.0	−60.0	275
A_2	650.0	−100.0	650.0	−100.0	275
A_3	483.0	211.3	483.0	−267.0	275

The last column of Table 2-2 lists the EMV of each alternative. These values were obtained by repeated application of eq. (2-1), and are all equal to $275,000. This suggests that one should be indifferent to all three alternatives. But can this really be the case? You, as the decision maker, might not be willing to tolerate the prospect of losing $100,000 or more (e.g., such a loss might cost you your job or might put the company into a difficult financial position), but you might be willing and able to bear the strain of a $60,000 loss. Hence you would choose A_1 over the other options in that no more than $60,000 could be lost with A_1, whereas $100,000 and $267,000 could be lost with A_2 and A_3, respectively.

If we now approach this problem from a utility theory point of view, where our attitude toward risk is implicitly taken into account in the construction of the utility function, the analysis is more informative. To proceed, the first step is to convert monetary outcomes to "utiles" by using the curve in Fig. 2-11. The results are displayed in Table 2-3, where now we see that A_1 is preferred to A_3, which is preferred to A_2, although only slightly. Evidently, the increased prospect for success with alternative 3 is not sufficiently high to balance the risk of losing $267,000 should both projects fail. Similarly, the $650,000 payoff associated with A_2 is not large enough for this risk-averse decision maker to compensate for the 50% chance of losing $100,000. Nevertheless, because the expected utilities for the three alternatives are so close, additional effort should go into refining the probability, cost, and return estimates.

TABLE 2-3 UTILITY MATRIX FOR NEW PRODUCT
DEVELOPMENT EXAMPLE

θ_j	A_1 : S A_2 : S	S F	F S	F F	
					Expected
$A_i \setminus p_j(\theta_j)$	0.35	0.35	0.15	0.15	utility
A_1	0.90	0.90	0.70	0.70	0.84
A_2	0.95	0.67	0.95	0.67	0.81
A_3	0.92	0.83	0.92	0.49	0.82

■

2.7.5 Characteristics of the Utility Function

The curve derived in Fig. 2-11 increases monotonically from the lower left to the upper right. In other words, it has a positive slope throughout. This is generally the characteristic of utility functions. It simply implies that people ordinarily attach greater utility to larger amounts of money than to smaller amounts (i.e., more is preferred to less). Economists refer to such a psychological trait as a *positive marginal utility for money*.

Three general types of utility functions are depicted in Fig. 2-12. Of course, actual shapes may vary, and the particular application will determine the scale on the horizontal axis. Any number of combinations of the three are possible. The concave-downward shape is illustrative of a person who has a diminishing marginal utility for money, although the marginal utility is always positive (the slope is positive but decreasing as the dollar amount increases—the rate of change of the slope is negative). This type of utility function is indicative of a *risk avoider*, someone who is risk averse. The decreasing slope implies that the utility of a given amount of gain is less than the disutility of an equal amount of loss; also, as the dollar gain increases, it becomes less valuable. This is in part the rationale for a progressive income tax system. A person characterized by such a utility function would prefer a small but certain monetary gain to a gamble whose EMV is greater but may involve a larger but unlikely gain, or a large and not unlikely loss.

The linear function in Fig. 2-12 depicts the behavior of a person who is *neutral* or indifferent to risk. For such a person every increment of, say, $1,000 has an associated constant increment of utility (the slope of the utility curve is positive and constant). That is, she values an additional dollar of income just as highly regardless of whether it is the first dollar or the 100,000th dollar gained. This type of person would use the EMV criterion

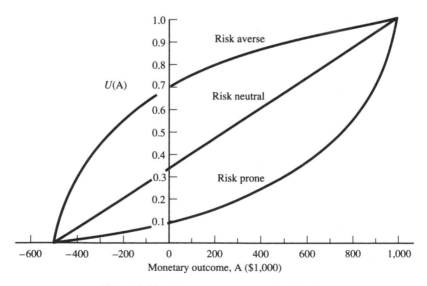

Figure 2-12   Three general types of utility functions.

in making decisions because by so doing she would also maximize expected utility. Government decision making usually proceeds from a risk neutral viewpoint. Referring to Fig. 2-12, the expected utility of each alternative in Example 2-14 is 0.51.

The third curve in Fig. 2-12, which is convex in shape, is that of a *risk seeker*, someone who is risk prone. Note that the slope of the utility function increases as the dollar amount increases. This implies that the utility of a given gain is greater than the disutility of an equivalent loss. A risk-seeking person subjectively values each dollar of gain more highly. This type of person willingly accepts gambles that have a smaller EMV than an alternative payoff received with certainty. He will also take an "unfair" bet in the sense that he will choose an action whose EMV is negative. In the case of such a person, the attractiveness of a possibly large payoff in the gamble tends to outweigh the fact that the probability of such a payoff may indeed be exceedingly small. Persons who persistently buy lottery tickets fall into this category. If the risk-prone curve in Fig. 2-12 is used in Example 2-14, the expected utilities for the three alternatives are 0.155, 0.195, and 0.157, thus reversing the order of preference. Now $A_2 > A_3 > A_1$.

Most people have utility functions whose slopes do not change very much for small changes in money, suggesting risk-neutral attitudes. In considering courses of action, however, in which one of the consequences is very adverse or in which one of the payoffs is very favorable, people can be expected to depart from the maximization of EMV criterion. In fact, most persons are risk seekers for small gains and losses, and risk avoiders when the stakes are high, in either direction. This explains why most of us buy insurance, and stay with secure but often unexciting jobs rather than seeking out risky opportunities that have the probability of making us wildly rich.

For many business decisions, where the monetary consequences may represent only a small fraction of the total assets of the organization, maximization of EMV constitutes a reasonable approximation to the decision-making criterion of maximization of expected utility. In such cases the utility function may be considered linear over the range of possible monetary outcomes.

TEAM PROJECT

Thermal Transfer Plant

Based on your excellent report and presentation, TMS has decided to approve a prototype project in the area of waste management and recycling. Since there is a need for a rotary combustor in one of the company's new plant designs, a decision was made to select this project as a prototype.

Rotary combustors are designed to burn a variety of solid combustible wastes, including municipal, commercial, industrial, and agricultural wastes. The basic component of the combustor is the rotating barrel, made out of alternating carbon steel water tubes and perforated steel bars (Fig. 2-13). The barrel assembly is set at a slope of $-6°$ and is rotated slowly [approximately 10 revolutions per hour (rph)]. Solid waste is charged from the higher end of the barrel and the combustion air comes into the barrel through the perforated holes. As the material burns, it tumbles through the barrel and eventually comes out of the lower end as residue (Fig. 2-14). In the process, heated forced air promotes drying and burning.

Hot gases created inside the barrel convert the boiler water into steam, which is used in the generation of electricity. High thermal efficiency of up to 80% provides maximum energy recovery through heat transfer from all hot surfaces of the combustor/boiler. Simplified moving parts assure ease of operation, maintenance, and servicing, as well as minimal repair costs.

The combustor capacity is targeted for 14 tons/day. Estimated costs are as follows:

Cost of material

Combustor barrel	$10,000
Tires and trunnions	50,000
Chutes	10,000
Drive gears and chain	20,000
Pushers (2)	2,000
Enclosure and insulation	20,000
Rotary water joint	5,000
Hydraulic drive system (includes power unit, cylinders, and combustor drive components)	90,000
Welding materials	30,000

Cost of labor

Combustor barrel fabrication 10 workers, 8 weeks at $50/hr	$160,000
Tire and trunnion installation 5 workers, 1 week	10,000
Chute fabrication 4 workers, 2 weeks	16,000
Gear installation 2 workers, 2 days	1,600
Pusher fabrication 2 workers, 2 weeks	8,000
Enclosure fabrication 10 workers, 4 weeks	80,000
Water joint installation 2 workers, 1 day	800

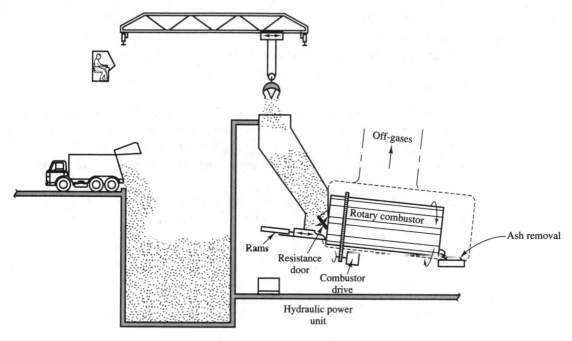

Figure 2-13   Arrangement of rotary combustor.

Other

Design	$15,000
Instrumentation	13,000
Pressure testing	2,500
Preassembly	7,000
Break down and loading for shipment	3,000
Overhead	25%

The following factors contribute to the risk of the project:

Schedule risks

- *NIMBY (not in my back yard).* The construction of thermal transfer plants may prove to be a long, drawn-out affair. In addition to EPA requirements, local opposition must be considered.
- One option for the combustor drive is to use a single large hydraulic motor. There are two manufacturers of this type of motor, one in Sweden and one in Germany.

Cost risks

- Costs due to delays (see first item above).
- Price increases—an entire plant is being built—estimated duration is 2 years.

- Design time—difficult to estimate.
- Fabrication time.
- Overhead is very difficult to estimate and control.

Technological risks

- Hazardous location—a rotary combustor is a furnace and because of fire hazards, mineral-based hydraulic fluid cannot be used. Fire-resistant fluids are an alternative but require that certain hydraulic components be downrated.
- Speed control—the accuracy and degree of variability of speed control for the rams is yet to be determined.
- No satisfactory design exists yet for a rotary water joint.
- Satisfactory seals around the combustor are yet to be developed.

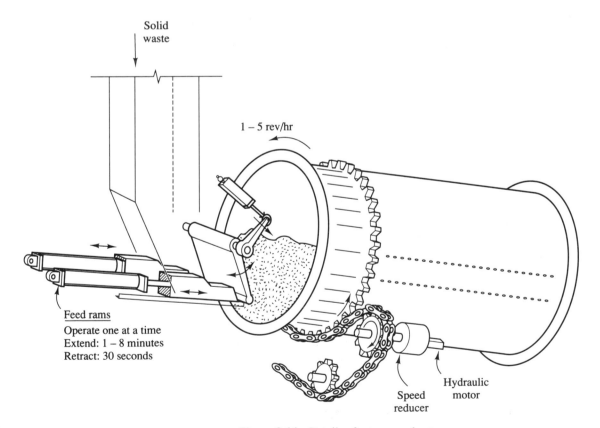

Figure 2-14   Details of rotary combustor.

- Hydraulic leaks at other installations have been a problem, particularly at the rams and at the power unit.
- Instrumentation—there is disagreement as to the sophistication of instrumentation required, particularly on the rams.

TMS's main business is design and consulting; however, it is felt that this new area of operation may present an opportunity for the company to develop manufacturing capabilities. Management has three alternatives under consideration:

(a) Design the rotary combustor at TMS based on customer needs, but subcontract all manufacturing and assembly.
(b) Design the rotary combustor at TMS, subcontracting all manufacturing of parts but assembling the system at TMS facilities.
(c) Design, manufacture, and assemble the combustor at TMS.

Your assignment is to compare the economic aspects, including risks.

1. For each alternative, list the risks involved and their associated costs.
2. Analyze TMS's overall financial position under each alternative.
3. Include projected differences in total expenditures and investments.

In evaluating these alternatives you can make any assumptions necessary. Each should be stated explicitly.

DISCUSSION QUESTIONS

1. What are the shortcomings of engineering economic analysis? What difficulties and uncertainties might one face when performing such an analysis?
2. American businesses have often been criticized for short-term thinking that places too much emphasis on payback period and rate of return. When Honda started making cars in the early 1970s, for example, the chief executive officer stated that the firm would be "willing to accept a rate of return no greater than 2 or 3% for as long as it took to be recognized as the best car maker in the world." In light of the success of many Japanese firms, is the criticism of American business justified?
3. If a firm is short of capital, what action might it take to conserve the capital it has and to obtain more?
4. Explain why the marginal cost for borrowing money increases. Why might the cost also be high for borrowing small amounts?
5. Are there any reasons for using present value analysis rather than future value analysis?
6. Why might a decision maker like to see the payback analysis as well as the rate of return and the net present value?

7. In the 1960s and 1970s the top marginal tax rate for individuals in the United States was 70%; that is, for each dollar a person earned above roughly $100,000, he or she had to pay 70 cents in taxes. It was argued by many economists at the time that this rate was much too high. What do you think are the negative economic and social consequences of such "confiscatory" tax rates?

8. Discuss why the comparison of alternative investment decisions is especially difficult when the investment choices have different useful lives.

9. Break-even analysis is typically simplified by using constant-unit variable costs and revenues. What would you expect realistic costs and revenues to be, and what would a corresponding break-even chart look like?

10. Identify a situation and set of alternatives whose outcomes are not measured on a monetary scale. Assess your utility function for this situation.

11. Give some examples where the axioms underlying Bernoulli's principle are violated.

12. Most countries have a progressive income tax system where each dollar earned in incrementally higher tax brackets is taxed at an increasingly higher rate. Can you explain the rationale for this in terms of utility theory? Do you think that a proportional tax system would be fairer?

13. If you just assessed a corporate executive's utility function for a problem concerning the purchase of a supercomputer, could you use the same utility function for a problem of buying an automobile? a personal computer? Explain.

14. It has been argued that comparable interpersonal utility scales may be established on the basis of equating people's best conceivable situations at the top end and their worst conceivable situations at the bottom end. What's wrong, if anything, with this approach?

15. In situations where wealthy employers bargain over wages and benefits with needy employees on an individual basis the employer usually gives away much less than he actually might have been pressured into, or could have afforded. Can you explain this consequence in terms of utility theory?

EXERCISES

2-1. Construct a diagram illustrating the cash flows involved in the following transactions, from the borrower's viewpoint. The amount borrowed is $2,000 at 10% for 5 years.
   (a) Year-end payment of interest only; repayment of principal at the end of the 5 years
   (b) Year-end repayment of one-fifth of the principal ($400) plus interest on the unpaid balance
   (c) Lump-sum repayment at end of year 5 of principal plus accrued interest compounded annually
   (d) Year-end payments of equal-sized installments, as in a standard installment loan contract

2-2. A firm wants to lease some land from you for 20 years and build a warehouse on it. As your payment for the lease, you will own the warehouse at the end of the 20 years, estimated to be worth $20,000 at that time.
   (a) If $i = 8\%$, what is the present worth of the deal to you?
   (b) If $i = 2\%$ per quarter, what is the present worth of the deal to you?

2-3. In payment for engineering services a client offers you a choice between (1) $10,000 now and (2) a share in the project which you are fairly certain you can cash in for $15,000 five years from now. With $i = 10\%$, which is the most profitable choice?

2-4. Assume that a medium-sized town now has a peak electrical demand of 105 megawatts, increasing at an annually compounded rate of 15%. The current generating capacity is 240 megawatts.

(a) How soon will additional generating capacity be needed on-line?

(b) If the new generator is designed to take care of needs 5 years past the on-line date, what size should it be? Assume that the present generators continue in service.

2-5. Assume that a local government agency asks you to consult regarding acquisition of land for recreation needs for the urban area. The following data are provided:

Urban population 10 years ago	49,050
Urban area population now	89,920
Desired ratio of recreation land in acres per 1,000 population	10 acres/1,000
Actual acres of land now held by local government for recreational purposes	803 acres

(a) Find the annual growth rate in the urban area population by assuming that growth has compounded discretely at year's end over the past 10 years.

(b) How many years from now will it be before the desired ratio of recreation land per 1,000 population is exceeded if no more land is acquired and the population continues to grow at its present rate?

(c) The local government is planning to purchase more land to supply the recreational needs for 10 years past the point in time found in part (b). How many acres of land should they purchase to maintain the desired ratio, assuming that the population growth continues at the same rate?

2-6 A young engineer decides to save $240 per year toward retirement in 40 years.

(a) If he invests this sum at the end of every year at 9%, how much will be accumulated by retirement time?

(b) If by astute investing the interest rate could be raised to 12%, what sum could be saved?

(c) If he deposits one-fourth of this annual amount each quarter ($60 per quarter) in an interest-bearing account earning a nominal annual interest rate of 12%, compounded quarterly, how much could be saved by retirement time?

(d) In part (c), what annual effective interest rate is being earned?

2-7. A lump sum of $100,000 is borrowed now, to be repaid in one lump sum at end of month (EOM) 120, and bearing interest at 1% compounded monthly. No partial repayments will be accepted on this loan. To accumulate the repayment lump sum due, monthly deposits are made into an interest-bearing account. This repayment deposit account bears interest at 0.75% per month from EOM 1 until EOM 48. From EOM 48 until EOM 120 the interest rate changes to 0.5%. Monthly deposits of amount A begin with the first deposit at EOM 1 and continue until EOM 48. Beginning with EOM 49, the deposits are doubled at amount 2A and continued at this level until the final deposit at EOM 120. Draw the cash flow diagram; find the initial monthly deposit amount A.

2-8. A backhoe is purchased for $20,000. The terms are 10% down and 2% per month on the unpaid balance for 60 months.

(a) How much are the monthly payments?

(b) What annual effective interest rate is being charged?

2-9. Your firm owns a large earth-moving machine and has contracts to move earth for $1 per cubic yard. For $100,000 this machine may be modified to increase its production output by an extra 10 yd^3 per hour, with no increase in operating costs. The earth-moving machine is expected to last another 8 years, with zero salvage value at the end of that time. Determine whether or not this investment meets the company objective of earning at least 15% return. Assume that the equipment works 2,000 hours per year.

2-10. Your firm wants to purchase a $50,000 computer, no money down. The $50,000 will be paid off in 10 equal end-of-year payments with interest at 8% on the unpaid balance.

(a) What are the annual end-of-year payments?

(b) What hourly charge should be included to pay off the computer, assuming 2,000 hours of work per year, credited at the end of the year?

(c) Assume that 5 years from now you would like to trade in the computer and purchase a new one. You expect a 5% increase in price each year. What would the new computer cost at the end of year 5?

(d) What is the unpaid balance on the current computer after 5 years?

2-11. A transportation authority asks you to check on the feasibility of financing for a toll bridge. The bridge will cost $2,000,000. The authority can borrow this amount now and repay it from tolls. It will take 2 years to construct and be open for traffic at end of year (EOY) 2. Tolls will be accumulated throughout the third year and will be available for the initial annual repayment at EOY 3. In subsequent years the tolls are deposited at the end of the year. Draw the cash flow diagram. How much must be charged to each car to repay the borrowed funds in 20 equal annual installments (first installment due at EOY 3), with 8% compound interest on the unpaid balance?

Assume 10,000 vehicles per day

2-12. A firm invested $15,000 in a project that appeared to have excellent potential. Unfortunately, a lengthy labor dispute in year 3 resulted in costs that exceeded benefits by $8000. The cash flow for the project is as follows:

Year	0	1	2	3	4	5	6
Cash flow	−15,000	+10,000	+6,000	−8,000	+4,000	+4,000	+4,000

Compute the rate of return for the project, assuming a 12% interest rate on external investments.

2-13. An oil company plans to purchase for $70,000 a piece of vacant land on the corner of two busy streets. The company has four different types of businesses that it installs on properties of this type.

Plan	Cost of improvements[a]	Description
A	$75,000	Conventional gas station with service facilities for lubrication, oil changes, etc.
B	230,000	Automatic car wash facility with gasoline pump island in front
C	30,000	Discount gas station (no service bays)
D	130,000	Gas station with low-cost, quick-car-wash facility

[a]Cost of improvements does not include the $70,000 cost of land.

In each case, the estimated useful life of the improvements is 15 years. The salvage value for each is estimated to be the $70,000 cost of the land. The net annual income, after paying all operating expenses, is projected as follows:

Plan	Net annual income
A	$23,300
B	44,300
C	10,000
D	27,500

If the oil company expects a 10% rate of return on its investments, which plan (if any) should be selected?

2-14. A firm is considering three mutually exclusive alternatives as part of a production improvement program. The relevant data are:

	A	B	C
Installation cost	$10,000	$15,000	$20,000
Uniform annual benefit	$1,625	$1,625	$1,890
Useful life (years)	10	20	20

For each alternative, the salvage value at the end of useful life is zero. At the end of 10 years, alternative A could be replaced by another A with identical cost and benefits. The MARR is 6%. If the analysis period is 20 years, which alternative should be selected?

2-15. Consider four mutually exclusive alternatives that each have an 8-year useful life:

	A	B	C	D
Initial cost	$1,000	$800	$600	$500
Uniform annual benefit	122	120	97	122
Salvage value	750	500	500	0

If the minimum attractive rate of return is 8%, which alternative should be selected?

2-16. A project has the following costs and benefits. What is the payback period?

Year	Costs	Benefits
0	$1,400	
1	500	
2	300	$400
3–10		$300 per year

2-17. A motor with a 200-horsepower output is needed for intermittent use in a factory. A Teledyne motor costs $7,000 and has an electrical efficiency of 89%. An Allison motor costs $6,000 and has an 85% efficiency. Neither motor would have any salvage value, as the cost to remove it would equal its scrap value. The maintenance cost for either motor is estimated at $300 per year. Electric power costs $0.072/kilowatthour (1 hp = 0.746 kW). If a 10%

annual interest rate is used in the calculations, what is the minimum number of hours that the higher-initial-cost Teledyne motor must be used each year to justify its purchase?

SOYD ⟹ **2-18.** Lu Hodler planned to buy a rental property as an investment. After looking for several months, she found an attractive duplex that could be purchased for $93,000 cash. The total expected income from renting both sides of the duplex would be $800 per month. The total annual expenses for property taxes, repairs, gardening, and so on, are estimated at $600 per year. For tax purposes, Lu plans to depreciate the building by the sum-of-the-years'-digits method, assuming that the building has a 20-year remaining life and no salvage value. Of the total $93,000 cost of the property, $84,000 represents the value of the building and $9,000 is the value of the lot (only the former can be depreciated). Assume that Lu is in the 38% incremental income tax bracket (combined state and federal taxes) throughout the 20 years. In this analysis Lu estimates that the income and expenses will remain constant at their present levels. If she buys and holds the property for 20 years, what after-tax rate of return can she expect to receive on her investment, using the assumptions noted below?

(a) The building and lot can be sold at the end of 20 years for the $9,000 estimated value of the lot.

(b) A more optimistic estimate of the future value of the property is that it can be sold for $100,000 at the end of the 20 years.

SL DC ⟹ **2-19.** The effective combined tax rate in an owner-managed corporation is 40%. An outlay of $20,000 for certain new assets is under consideration. It is estimated that for the next 8 years, these assets will be responsible for annual receipts of $9,000 and annual disbursements (other than for income taxes) of $4,000. After this time they will be used only for standby purposes, and no future excess of receipts over disbursements is expected.

(a) What is the prospective rate of return before income taxes?

(b) What is the prospective rate of return after taxes if these assets can be written off for tax purposes in 8 years using straight line depreciation?

(c) What is the prospective rate of return after taxes if it is assumed that these assets must be written off over the next 20 years using straight-line depreciation?

ECONOMIC ⟹ **2-20.** The Coma Chemical Company needs a large insulated stainless steel tank for the expansion
LIFE   of its plant. Coma has located one at a brewery that has just been closed. The brewery offers to sell the tank for $15,000 including delivery. The price is so low that Coma believes that it can sell the tank at any future time and recover its $15,000 investment. The outside of the tank is lined with heavy insulation that requires considerable maintenance. Estimated costs are:

Year	0	1	2	3	4	5
Insulation maintenance cost	$2,000	500	1,000	1,500	2,000	2,500

(a) Based on a 15% before-tax MARR, what is the economic life of the insulated tank; that is, how long should it be kept?

(b) Is it likely that the insulated tank will be replaced by another tank at the end of its computed economic life? Explain.

2-21. The Gonzo Manufacturing Co. is considering the replacement of one of its machine fixtures with a more flexible variety. The new fixture would cost $3,700, have a 4-year useful and depreciable life, and no salvage value. For tax purposes, sum-of-the-years'-digits depreciation would be used. The existing fixture was purchased 4 years ago at a cost of $4,000 and has been depreciated by straight-line depreciation assuming an 8-year life and no salvage value. It could be sold now to a used equipment dealer for $1,000 or be kept in service for another 4 years. It would then have no salvage value. The new fixture would save about $900 per year in operating costs compared to the existing one. Assume a 40% combined state and federal tax rate.

(a) Compute the before-tax rate of return on the replacement proposal of installing the new fixture rather than keeping the old one.

(b) Compute the after-tax rate of return on the proposal.

2-22. The following estimates have been made for two mutually exclusive alternatives; one must be chosen. The before-tax rate of return required is 20%.

	A	B
Installed cost	$120,000	$150,000
Estimated useful life	10 years	10 years
Salvage at retirement	$20,000	$30,000
Annual operating costs	$20,000	$15,000

Try to minimize your computations as you determine which course of action to recommend.

2-23. The following cost estimates apply to independent equipment alternatives A and B. The before-tax rate of return required is 20%.

	A	B
Installed cost	$100,000	$40,000
Operating costs	$5,000 at the end of year 1 and increasing by $1,000 per year for 20 years	$10,000 at the end of year 1 and increasing by $2,000 per year for 10 years
Overhaul costs every 5 years	$10,000	None required
Economic life	20 years	10 years
Salvage value at end of life (just overhauled)	$20,000	$10,000

(a) Compare the net present value of each using a study period of 20 years.

(b) Compare the annual equivalent costs.

2-24. What is the argument for using assessment procedures based on 50–50 gambles as opposed to assessment procedures based on using reference gambles?

2-25. Explain why identification of special attitudes toward risk can simplify the utility assessment process.

2-26. Given the following information, plot four points on the person's preference curve. The maximum payoff is $1,000. The minimum payoff is $0. The certainty equivalent for a 50–50 gamble between $1,000 and $0 is $400. The certainty equivalent for a 50–50 gamble between $400 and $0 is $100.

2-27. As part of a decision analysis, Archie Leach provided the following information.

  ■ He was indifferent between a 50–50 chance at +$10 million and −$10 million, and −$5 million for certain.

  ■ His certainty equivalent for a lottery offering a 0.5 chance at −$5 million and a 0.5 chance at +$10 million was $0.

  ■ He was indifferent between a lottery with a 0.7 chance at +$10 million and a 0.3 chance at $0, and +$5 million for certain.

  Sketch a preference curve for Leach based on this information.

2-28. Refer to Fig. 2-15.

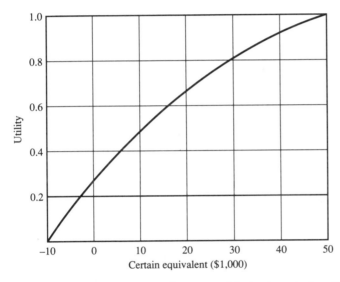

Figure 2-15 Preference curve for risk-averse decision maker.

  (a) Specify a reference gamble that is equivalent (based on this curve) to the certain amount $30,000.

  (b) Specify a 50–50 gamble that is equivalent (based on this curve) to the certain amount $30,000.

2-29. Beverly Silverman had long been promised a graduation present of $10,000 by her father, to be received on graduation day, 3 months hence. Her father had recently offered an alternative gift, and Silverman was trying to decide between the two gifts, since her father had asked for a decision by the following day. The alternative gift would be 1,000 shares of stock in Opera Systems, Inc., a consulting firm with which Silverman was slightly acquainted. On the day she was trying to decide, the stock was selling for $12 per share. Thus it looked to Silverman as if she would be wise to take the stock, since its present value was $12,000. She would not receive the stock until graduation day, however, and she recognized that the stock price 3 months in the future was uncertain. She also recognized that her utility for money was not linear, and that her risk aversion would therefore influence her decision. With these facts in mind, Silverman reached the following conclusions:

  1. She felt that the stock price was more likely to rise than fall in the intervening 3 months, and that it was as likely to be above $14 per share as below that figure when she would receive the stock.

2. She felt that there was only 1 chance in 100 that the stock price would drop to less than $6 per share, and an equal chance that the price would be more than twice its current price on graduation day.

3. She also thought that there was only 1 chance in 5 that the price would be below $10, and that there was 1 chance in 4 that it would be above $16 when she received it.

In considering her preferences, Silverman reached the following conclusions:

1. That her certainty equivalent for a lottery offering a 50–50 chance at zero and $25,000 was $9,000.

2. That her certainty equivalent for a lottery offering a 0.2 chance at $25,000 and a 0.8 chance at zero was $3,000.

3. That his certainty equivalent for a lottery offering a 50–50 chance at $3,000 or $25,000 was $12,000.

4. That her certainty equivalent for a lottery offering a 50–50 chance at $12,000 or $25,000 was $17,000.

Determine the cumulative probability distribution that Silverman has assigned to the stock price. Calculate Silverman's certainty equivalent for the gift of the stock.

2-30. A manager expresses indifference between a certain profit of $5,000 and a venture with a 70% chance of making $10,000 and a 30% chance of making nothing. If the manager's utility scale was set at 1 utile for $0, and 100 for $10,000, what is the utility index for $5,000?

2-31. The manager in Exercise 2-30 is indifferent between a venture that has a 60% chance of making $10,000 and a 40% chance of making $1,000, and a sure investment that yields $5,000. Find the value of $1,000 in utiles for this manager.

2-32. Below are the results of a preference test given to an executive:

1. The executive is indifferent between an investment that will yield a certain $10,000 and a risky venture with a 50% chance of $30,000 profit and a 50% chance of a loss of $1,000.

2. The executive's utility function for money has the following shape:

Money ($)	−1,000	0	5,000	20,000	30,000
Utility	−2	0	10	20	30

A new risky venture is proposed. The possible payoffs are *either* $0 or $20,000. The probabilities of the gain cannot be determined. Find the probability combination of $0 and $20,000 that would make the executive indifferent to the certain $10,000.

2-33. Frances Gumm has an opportunity of investing $3,000 in a venture that has a 0.2 chance of making nothing, a 0.3 chance of making $2,000, a 0.2 chance of making $4,000, and a 0.3 chance of making $6,000. Her utilities for each of the outcomes are 0 for $0, 25 for $2,000, 35 for $4,000, and 40 for making $6,000. Draw Frances's utility curve and advise her on making the investment.

2-34. A plant manager has a utility of 10 for $20,000, 6 for $11,000, 0 for $0, and −10 for a loss of $5,000.

(a) The plant manager is indifferent between receiving $11,000 for certain and a lottery with a 0.6 chance of winning $5,000 and a 0.4 chance of winning $20,000. What is the utility of $5,000 for the manager? Construct the manager's utility curve.

(b) Using this curve, find the "certainty equivalent" for the following gamble (i.e., the amount of cash that will make the manager indifferent to the gamble):

[handwritten left margin: DECISION TREE + UTILITIES]

Payoff	Probability
−$2,000	0.2
0	0.3
3,000	0.4
10,000	0.1

(c) What probability combination of $0 and $20,000 would make the manager indifferent to the certain $11,000?

(d) The manager is facing a decision about buying a new production machine that can bring a net profit of $15,000 (80% chance) or a loss of $1,000 (20% chance); alternatively, the manager can use the old machine and make a $10,000 profit. Use the utility curve to find which alternative the manager should select. Specify all necessary assumptions.

REFERENCES

AU, T., and P. AU, *Engineering Economics for Capital Investment*, Allyn and Bacon, Boston, 1983.

BAUMOL, W. J., "On the Social Rate of Discount," *American Economic Review*, Vol. 57, No. 4 (1968), pp. 778–802.

BUSSEY, L. E., and T. G. ESCENBACK, *Economic Analysis of Industrial Projects*, Second Edition, Prentice Hall, Englewood Cliffs, NJ, 1992.

CANADA, J. R., and W. G. SULLIVAN, *Economic and Multiattribute Evaluation of Advanced Manufacturing Systems*, Prentice Hall, Englewood Cliffs, NJ, 1989.

DeGARMO, E. P., W. G. SULLIVAN, and J. A. BONTADELLI, *Engineering Economy*, Eighth Edition, Macmillan, New York, 1988.

DE NEUFVILLE, R., *Applied Systems Analysis: Engineering Planning and Technology Management*, McGraw-Hill, New York, 1990.

DERTOUZOS, M., R. LESTER and R. SOLOW (editors), *Made in America: Regaining the Productive Edge*, MIT Press, Cambridge, MA, 1989.

ENGLISH, J. M., *Project Evaluation: A Unified Approach for the Analysis of Capital Investments*, Macmillan, New York, 1984.

HUMPHREYS, K. K., *Jelen's Cost and Optimization Engineering*, Third Edition, McGraw-Hill, New York, 1991.

KEENEY, R. L., and H. RAIFFA, *Decisions with Multiple Objectives: Preferences and Value Tradeoffs*, Wiley, New York, 1976.

MILLER, C., and A. P. SAGE, "A Methodology for the Evaluation of Research and Development of Projects and Associated Resource Allocation," *Computers & Electrical Engineering*, Vol. 8, No. 2 (1981), pp. 123–152.

NEWNAN, D.G., *Engineering Economic Analysis*, Fourth Edition, Engineering Press, San Jose, CA, 1991.

SAGE, A. P., *Economic Systems Analysis: Microeconomics for Systems Engineering, Engineering Management, and Project Selection*, North-Holland, Amsterdam, 1983.

SMITH, J. H., A. FEINBERG, and T. LEE, *SmartEdge: A Decision Support System for Expert Knowledge Acquisition and Evaluation*, Haviland-Lee, Northridge, CA, 1987.

VON NEUMANN, J., and O. MORGENSTERN, *Theory of Games and Economic Behavior*, Second Edition, Princeton University Press, Princeton, NJ, 1947.

WHITE, J. A., M. H. AGEE, and K. E. CASE, *Principles of Engineering Economic Analysis*, Third Edition, Wiley, New York, 1989.

3

Project Screening
and Selection

3.1 COMPONENTS OF THE EVALUATION PROCESS

Every new project starts with an idea. Typically, new ideas arrive continuously from a variety of sources, such as customers, suppliers, upper management, and shop floor personnel. For organizations working in the private sector, the details of the steps involved in processing these ideas and the related analyses are highlighted in Fig. 3-1. A similar diagram can be drawn for other classes of organizations.

Depending on the scope and estimated costs, management may simply be interested in determining the merit of the idea, or it may want to determine how best to allocate a budget among a portfolio of projects. If the organization is a consulting firm or an outside contractor, it may want to decide on the most advantageous strategy for responding to requests for proposals (RFPs).

Of course, there are many different types of projects, so the evaluation criteria and accompanying methodology should reflect the peculiar characteristics of the sponsoring or respondent organization. The usual divisions are public sector versus private sector, R&D versus operations, and internal customer versus external customer. Project size, expected duration, underlying risks, and required resources are some of the factors that must weigh on the decision.

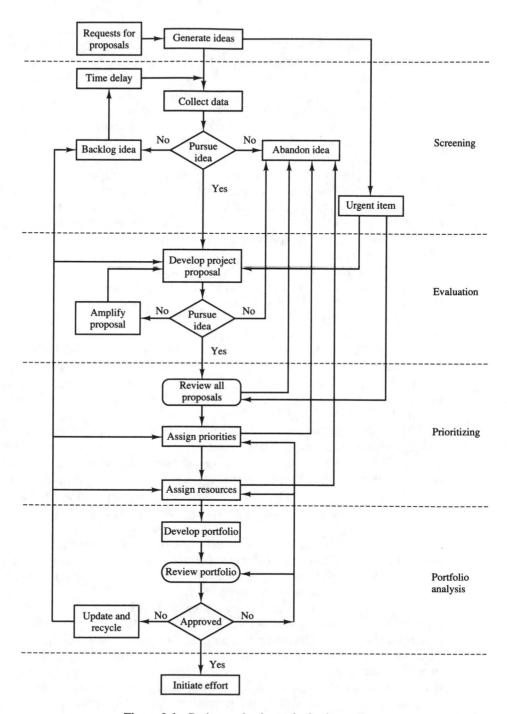

Figure 3-1   Project evaluation and selection process.

Regardless of the source or nature of the customer, screening is usually the first step. Here the proposed project is analyzed in a preliminary fashion in light of the most prominent criteria or prevailing conditions. This should be a quick and inexpensive exercise. The results may suggest, for example, that no further effort is warranted, due to uncertainty in the technology or the lack of the well-defined market. If some promise exists, the project may be temporarily backlogged in deference to more attractive contenders. At some time in the future when conditions are more favorable, it may be desirable to resurrect it. Or the project may be deemed so urgent or beneficial to the organization that it is placed at the top of the priority list. Alternatively, the results of the screening may indicate that the proposed project possesses some merit and deserves further investigation.

If this is the case, a more in-depth analysis should be performed, with the goal of narrowing the uncertainties associated with the project's costs, benefits, and risks. This evaluation process usually involves extensive data collection and reduction, the solicitation of expert opinion, sample computations, and perhaps technological forecasting. As with the screening process, several courses of action might be suggested. The proposal may be rejected or abandoned for lack of merit; it may be backlogged for later retrieval and analysis; or it may be found to be acceptable and placed on a candidate list for a comparative analysis. In some cases it may be initiated immediately.

When the results of the evaluation process indicate that a proposal passes an acceptance threshold but that it is not clearly superior to the other candidates, each should be assessed and ranked competitively. In this prioritizing process, the relative strengths and weaknesses of each candidate are examined carefully and a weighted ranking is obtained. Ideally, the ranking would indicate not only the most preferred project, but the degree to which it is preferred over the other contenders. A number of assessment methodologies are presented in Sections 3.3 to 3.7 and Chapter 4.

If the ranking of a particular proposal is high enough, resources may tentatively be assigned. However, the decision to fund and initiate work on a proposal involves full consideration of the available human and financial resources within the organization. The level of available funds and personnel skill types, and the commitments to the current portfolio of activities, must all be factored into the decision. It may be that the new idea is so meritorious that it should replace one or more of the ongoing projects. If this is the case, some ongoing projects will be terminated or halted temporarily so that resources can be freed up for the new project. Portfolio models have been developed to aid in making these decisions. A portfolio model determines the best way to allocate available resources among the competing alternatives, including the new candidates and the ongoing projects. An example of such a model is presented in Chapter 12.

Obviously, it is justified to use portfolio models only in cases where projects compete for the same resources. However, when a contractor with excess capacity is approached by a potential customer with a project, the selection process is simple; just sign the contract as long as there is a profit in it above a predetermined rate of return. This is a common situation in times of recession. At the other extreme, a contractor may find himself overburdened with projects should he win more than the expected number of proposals to which he responded. In the remainder of this chapter we take up the middle ground and discuss methods for screening and prioritizing alternatives when resources limit the size of the portfolio.

3.2 DYNAMICS OF PROJECT SELECTION

As Fig. 3-1 suggests, project selection can be a very dynamic process. Screening, evaluation, prioritizing, and portfolio analysis decisions may be made at various points, and new ideas may not even go through these steps in sequence. An idea may be shelved or abandoned at any point in time. New information and changed circumstances may call for a previous decision to reject or abandon a project to be reevaluated. For example, efforts to develop lightweight portable computers were given a new impetus with the dramatic improvement in flat-screen display technology. Alternatively, new information or changed circumstances may cause a previously backlogged project to be rejected. The drastic reduction in the price of imported oil in the early 1980s dealt a death blow to the more exotic alternative energy projects, such as coal gasification and shale oil reclamation.

The available budget or labor skill types may constrain the project selection process. If the budget is inadequate to fund a particular project that appears meritorious, it may be necessary to backlog the project. Alternatively, the project may be divided up and certain portions initiated while the remainder is shelved until the economic situation becomes more favorable. Customer complaints, competitive threats, or unique opportunities may result in an urgent need to pursue a particular idea. Depending on the urgency, the project may receive only a cursory screening and evaluation, and may go directly into the portfolio.

Screening, evaluation, prioritizing, and portfolio decisions may be repeated several times over the life cycle of a project in response to emerging technologies and changing environmental, financial, or commercial circumstances. The advent of a new RFP, a change in competitive pressures, or the appearance of a new technology are some factors that may cause management to reevaluate an ongoing project. Moreover, with each advance that is recorded, new technical information will be forthcoming that may influence other efforts and proposed ideas. As current projects near completion, key personnel and equipment may be released so that they can be used on another project, perhaps one that was previously backlogged for lack of appropriate resources.

In general, the evaluation and selection of new product ideas and project proposals is a complex process, consisting of many interrelated decisions. The complexities involve the variety of data that must be collected, and the difficulty of unequivocally measuring and assessing candidate projects on the basis of information derived from these data. Much of the resultant information is subjective and uncertain in nature. Many ideas and proposals exist only as embryonic thoughts and are propelled forward by the sheer force of the sponsor's enthusiasm. But selecting the best idea cannot be done in a political vacuum. The presence of various organizational and behavioral factors tends to politicize the decision-making process. In many cases, the potential costs and benefits of a project play only a small role in the final decision. For example, an extensive 2-year analysis of LANDSAT, an earth-orbiting satellite with advanced resource monitoring capabilities, concluded that the benefits to the user community would fall significantly short of the expected costs associated with operating and maintaining the system over its 10-year lifetime, even under the most optimistic of scenarios (Bard 1984). Nevertheless, pressure from NASA and its congressional allies, who saw LANDSAT as a high-profile nonmilitary application of space technology that might actually return some benefits, persuaded the U.S. Department of Interior to provide funding.

The more sophisticated analytical and behavioral tools that have been developed to aid managers in evaluating projects variously take into account the nonquantitative aspects of the decision. As might be expected, though, the more comprehensive the underlying techniques, the more data and that effort are needed for the analysis.

3.3 CHECKLISTS AND SCORING MODELS

The idea-generation stage of a project, if done properly, will often lead to more proposals than can realistically be pursued. Thus a screening procedure designed to eliminate those proposals that are clearly infeasible or without merit must be established. Compatibility with the organization's objectives and resources is a primary concern. It is also important to keep in mind that when comparing alternatives early on, a wide range of criteria should be introduced in the analysis. The fact that these criteria are often measured on incommensurate scales makes the screening and evaluation much more difficult.

Of the several techniques available to aid in the screening process, perhaps the most commonly used are rating checklists (Augood 1973, Souder 1984). They are appropriate for eliminating the most undesirable proposals from further consideration. Because they require a relatively small amount of information, they can be used when the available data are limited or when only rough estimates have been obtained. Such methods should be viewed as expedient; they do not provide a great deal of depth and should be used with this caveat in mind.

Table 3-1 presents an illustration of a checklist. In constructing a checklist, it is necessary to identify the criteria or set of requirements that will be used in making the decision. In the next step, an (arbitrary) scoring scale is developed to measure how well a project does with respect to each criterion. Words such as "excellent," "good," and so on, may be associated with the numerical values.

In the example displayed in Table 3-1, the criteria include profitability, time to market, development risks, and commercial success. Each candidate is evaluated subjectively and scored using a 3-point scale. The built-in assumption is that each criterion is equally weighted. Total scores are displayed in the rightmost column. Typically, a cutoff point or threshold is specified below which the project is abandoned. Of those exceeding the threshold, the top contenders are held for further analysis while the remainder are back-

TABLE 3-1   EXAMPLE OF A CHECKLIST FOR SCREENING PROJECTS

	Criteria											
	Profitability			Time to market			Development risks			Commercial success			Total score
Score:	3	2	1	3	2	1	3	2	1	3	2	1	
Project A		×		×			×				×		10
Project B		×			×				×			×	6
Project C	×					×	×				×		8

logged or shelved temporarily. Here, if 7 is specified as the threshold, only projects A and C would be pursued.

An alternative means of displaying the information in Table 3-1 is with a multi-dimensional diagram known as a polar graph (Canada and Sullivan 1989), shown in Fig. 3-2. In one sense, this type of representation is more efficient than a table because it allows the analyst quickly to ascertain the presence of dominance. For example, by noting that the performance measure surface of project B is completely within that of project A, we can conclude that B is no better than A on any dimension, and thus can be discarded or backlogged.

Scoring models extend the logic of checklists by assigning a weight to each criterion that signifies the relative importance of one to the other (Baker 1974, Hobbs 1980, Souder and Mandakovic 1986). A weighted score is then computed for each candidate. In deriving the weights, a team approach should be used to head off disagreement after the assessment. One way of accomplishing this is to list all criteria in descending order of importance. Next, assign the least important (last-listed) criterion a value of 10, and assign a numerical weight to each criterion based on how important it is relative to this one. A criterion considered to be twice as important as the least important criterion would be assigned a weight of 20. If team members cannot agree on specific values, sensitivity analysis should be performed.

An example of the use of a scoring model for screening projects associated with the development of new products is shown in Table 3-2. Here eight criteria are to be rated on a numerical scale of 0 to 30, where 0 means poor and 30 means excellent. Because this

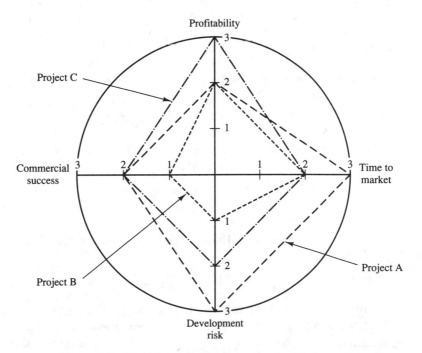

Figure 3-2   Multidimensional diagram for checklist example.

TABLE 3-2   EXAMPLE OF A SCORING MODEL FOR SCREENING PROJECTS

Criteria	Relative weight	Rating Excellent 30	Good 20	Fair 10	Poor 0	Factor score
Marketability	0.20	×				6
Risk	0.20		×			4
Competition	0.15		×			3
Value added	0.15				×	0
Technical opportunities	0.10	×				3
Material availability	0.10			×		1
Patent protection	0.05				×	0
Current products	0.05		×			1
	1.00					18

scale is arbitrary, no significance should be placed on relative values. For convenience, the weights are scaled between 0 and 1. In general, the factor score for project j, call it T_j, is obtained by multiplying the relative weights, w_i for criterion i, by the ratings, s_{ij}, and summing. That is,

$$T_j = \sum_i w_i s_{ij} \tag{3-1}$$

In this example, the project under consideration received a factor score of 18.

A variety of other formulas have been proposed for deriving the relative weights. Three of the simplest are presented below. More elaborate schemes are discussed in Chapter 4.

1. *Uniform or equal weights.* Given N criteria, the weight for each is

$$w_i = \frac{1}{N}$$

2. *Rank sum weights.* If R_i is the rank position of criterion i (with 1 as the highest rank) and there are N criteria, rank sum weights for each criterion may be calculated as

$$w_i = \frac{N - R_i + 1}{\sum_{k=1}^{N} (N - R_k + 1)}$$

3. *Rank reciprocal weights.* These weights may be calculated as

$$w_i = \frac{1/R_i}{\sum_{k=1}^{N} (1/R_k)}$$

The advantage of a scoring model is that it takes into account the trade-offs among the criteria, as defined by the relative weights. The disadvantage is that it lacks precision and relies on an arbitrary scoring system. More sophisticated models should lead to better decisions.

One area where the use of multiattribute scoring models has increased rapidly in the United States over the last few years is in planning studies. Specific applications can be found in regulatory proceedings and in bidding systems used by some utilities to acquire capacity from independent power producers. An environmental scoring form developed by Niagara Mohawk, a New York utility, is depicted in Table 3-3. Note that the procedure for assigning points is specified.

A typical rationale for the interest shown by regulatory commissions is offered by Mintzer (1990), who asserts that the "scaled scoring models are easy to use, facilitate comparisons of environmental impacts measured in different units, and are more comprehensive than economic analysis. In addition, they provide a rank from the lowest to the highest cost on the basis of all economic, environmental and social factors that can be quantified." These are very impressive claims, but unfortunately, many of the applications reported have paid little attention to methodological issues, with the consequence that the validity of the results is highly uncertain.

3.4 BENEFIT-COST ANALYSIS

Within the private sector, evaluating the merits of alternative investment opportunities begins with technical feasibility. The next step involves a comparison at some minimum attractive rate of return (MARR) of the estimated stream of costs over the expected economic life of each project. Engineering studies must be undertaken to establish the fundamental data. The estimated benefits and costs are then compared, usually on a present value basis, using a predetermined discount rate.

In the private sector, the firm generally pays all the costs and receives all the benefits, both quantitative and qualitative. Replacing an outdated piece of equipment is an example where the returns are measurable, while constructing a new company cafeteria illustrates the opposite case. But where the activities of government are concerned, quite a different situation arises. Revenues are received through various forms of taxation and are supposed to be spent "in the public interest." Thus the government pays but receives very few, if any, benefits. This can present all sorts of problems. For one, it means that the intended beneficiaries of a federal project will be very anxious to get the project approved and funded. Such situations may induce otherwise virtuous persons to redefine the standards of acceptable ethical behavior. A second problem concerns the measurement of benefits, which are often widely disbursed. Other difficulties include the selection of an interest rate, and choosing the correct viewpoint from which the analysis should be made. Finally, in the benefit-cost (B/C) analysis, where the B/C ratio is used to rank competing projects, there may be legitimate ambiguity in deciding what goes in the numerator and what goes in the denominator of the ratio.

At first glance it would appear a simple matter of sorting out the consequences into benefits (for the numerator) or costs (for the denominator). This works satisfactorily when applied to projects for a firm or a person. In government projects it may be considerably more difficult to classify the various consequences, as shown in Example 3-1.

TABLE 3-3 ENVIRONMENTAL SCORING FORM USED BY NIAGARA MOHAWK

Environmental attribute	Weight W	0	1	2	3	4	5	Score, W × P
Air emissions								
Sulfur oxides (lb/MWh)	7	>6	4.0–6.0	2.5–3.9	1.5–2.4	0.5–1.4	<0.5	
Nitrogen oxides (lb/MWh)	16	>6	4.0–6.0	2.5–3.9	1.5–2.4	0.5–1.4	<0.1	
Carbon dioxide (lb/MWh)	3	>1,500	1,050–1,500	650–1,049	250–649	100–249	<100	
Particulates (lb/MWh)	1	>0.3	0.2–0.3	0.1–0.199	0.05–0.099	0.01–0.049	<0.01	
Water Effects								
Cooling water flow (annual intake as % of lake volume)	1	80–100	60–79	40–59	20–39	5–19	<5	
Fish protection	1	None		Operational restrictions		Fish protection	No public water used provided	
NY State water quality classification of receiving water	1	A or better	B	C+	C	D	No water use or municipal water/wastewater utilized	

Points, P

Land effects							
	Weight	0.3–0.5	0.2–0.29	0.1–0.19	0.05–0.09	0.01–0.05	<0.01
Acreage required (acres/MW)	1	0.3–0.5	0.2–0.29	0.1–0.19	0.05–0.09	0.01–0.05	<0.01
Terrestrial	1	Unique ecological or historical value		Rural or low-density suburban		Industrial area	No land used
Visual aesthetics	1	Highly visible		Within existing developed area		Not visible from public roads	
Transmission	2	New OH >5 miles	New OH 1–5 miles	New UG >5 miles	New UG 1–5 miles	Use existing facilities	Energy conservation
Noise (L_{eq} − background L_{90})	2	5–10			0–4.9		<0
Solid waste disposal (lb/MWh)	2	>300	200–300	100–199	50–99	10–49	<10
Solid waste as fuel (% of total Btu)	1	0	1–30	31–50	51–80	81–90	91–100
Fuel delivery method	1	New RR spur	Truck and existing RR	New pipeline	Barge	Use existing pipeline	No fuel use
Distance from receptor area (Km)	1	<10	10–39	40–69	70–100	>100	Energy conservation

Total score (210 maximum) =

119

Example 3-1

On a proposed government project, the following consequences have been identified:
- Initial cost of project to be paid by government is $100K.
- Present worth of future maintenance to be paid by government is $40K.
- Present worth of benefits to the public is $300K.
- Present worth of additional public users cost is $60K.

Show the various ways of computing the benefit/cost ratio.

Solution   Putting the benefits in the numerator and all costs in the denominator gives

$$\text{B/C ratio} = \frac{\text{all benefits}}{\text{all costs}} = \frac{300}{100 + 40 + 60} = \frac{300}{200} = 1.5$$

An alternative computation is to consider user costs as disbenefits and to subtract them in the numerator rather than adding them in the denominator:

$$\text{B/C ratio} = \frac{\left(\begin{array}{c}\text{public}\\\text{benefits}\end{array}\right) - \left(\begin{array}{c}\text{public}\\\text{costs}\end{array}\right)}{\text{government costs}} = \frac{300 - 60}{100 + 40} = \frac{240}{140} = 1.7$$

Still another variation would be to consider maintenance costs as disbenefits:

$$\text{B/C ratio} = \frac{300 - 60 - 40}{100} = \frac{200}{100} = 2.0$$

It should be noted that while three different benefit/cost ratios may be computed, the NPV does not change:

$$\text{NPV} = \text{PW of benefits} - \text{PW of costs} = 300 - 60 - 40 - 100 = 100$$

∎

There is no inherently correct way to compute the benefit/cost ratio. Using the notation of Chapter 2, two commonly employed formulations are given below.

1. *Conventional B/C*

$$\text{B/C} = \frac{\text{PW of benefits to user}}{\text{PW of total costs to supplier}} = \frac{\text{PW}[B]}{\text{PW}[\text{CR} + (O + M)]} \qquad (3\text{-}2a)$$

or

$$\text{B/C} = \frac{\text{AW of benefits to user}}{\text{AW of total costs to supplier}} = \frac{B}{\text{CR} + (O + M)} \qquad (3\text{-}2b)$$

where B = annual worth of benefits to user
$\quad$ CR = capital recovery cost (equivalent annual cost of initial investment, considering any salvage value)
$\quad$ O = equivalent uniform annual operating cost
$\quad$ M = equivalent uniform maintenance cost

2. *Modified B/C*

$$B/C = \frac{PW[B - (O + M)]}{PW[CR]}$$

or

$$B/C = \frac{B - (O + M)}{CR}$$

The modified method has become more popular with governmental agencies and departments over the last decade. Although both methods yield the same recommendation when comparing mutually exclusive alternatives, they may yield different rankings for independent investment opportunities. In either case, using PW, AW, or FW should always provide the same results.

Example 3-2 (Single-Project Analysis)

An individual investment opportunity is deemed to be worthwhile if its B/C ratio is greater than or equal to 1.0. Consider the project of installing a new inventory control system with the following data:

Initial cost	$20,000
Project life	5 years
Salvage value	$4,000
Annual savings	$10,000
O&M disbursements	$4,400
MARR	15%

By interpreting annual savings as benefits, the conventional and modified B/C ratios based on annual equivalents are computed as follows:

$$CR = \$20,000(A/P, 15\%, 5) - \$4,000(A/F, 15\%, 5)$$
$$= 20,000(0.2983) - 4,000(0.1483) = \$5,373$$

$$\text{conventional B/C} = \frac{B}{CR + (O + M)} = \frac{\$10,000}{\$5,373 + \$4,400} = 1.02$$

$$\text{modified B/C} = \frac{B - (O + M)}{CR} = \frac{\$10,000 - \$4,400}{\$5,373} = 1.04$$

Because either B/C is greater than 1, the investment is worthwhile. Nevertheless, there is an opportunity cost associated with the investment that may preclude other possibilities. The fact that a B/C > 1 does not necessarily mean that the corresponding project should be pursued.

■

Example 3-3 (Comparing Mutually Exclusive Alternatives)

As was true for rate-of-return calculations, when comparing a set of mutually exclusive alternatives by any B/C method, an incremental approach is preferred. The principles and criterion of choice as explained in Chapter 2 apply equally to B/C methods, the only difference being that each increment of cost (the denominator) must be justified by a B/C ratio ≥ 1.

Consider the data in Table 3-4a associated with the four alternative projects used in Example 2-9 to demonstrate the IRR method. Each is listed in increasing order of investment. The symbol $\Delta(B/C)$ means that the benefit/cost ratio is being computed on the incremental cost. Once again, a MARR of 15% is used. The output data in Table 3-4b confirm the results found using the IRR method. Alternative C would be chosen given that it is the most expensive project for which each increment of cost is justified (by B/C ratio $\geq$ 1).

TABLE 3-4 DATA FOR INCREMENTAL ANALYSIS

(a) Input data

	Project			
	A	B	C	D
Initial cost	$20,000	$30,000	$35,000	$43,000
Useful life	5 years	10 years	5 years	5 years
Salvage value	$4,000	0	$4,000	$5,000
Annual receipts	$10,000	$14,000	$20,000	$18,000
Annual disbursements	$4,400	$8,600	$9,390	$5,250
Net annual receipts—disbursements	$5,600	$5,400	$10,610	$12,750

(b) Results data

	A	A $\rightarrow$ B	A $\rightarrow$ C	C $\rightarrow$ D
Δinvestment	$20,000	$10,000	$15,000	8,000
Δsalvage	4,000	$-4,000$	0	1,000
$\Delta CR = \Delta C$	5,373	605	4,477	2,386
Δ(annual receipts $-$ disbursements) $= \Delta B$	5,600	-200	5,010	2,140
$\Delta(B/C) = \Delta B/\Delta C$	1.04	-0.33	1.12	0.90
Is Δinvestment justified?	Yes	No	Yes	No

■

Benefit-cost studies within the public sector, in particular, may be approached from several points of view. The perspective taken may have a significant impact on the outcome of the analysis. Possible viewpoints include:

1. That of the governmental agency conducting the study
2. That of the local area (e.g., town, municipality)
3. The nation as a whole
4. The targeted industry

Thus it is essential that before proceeding with the study, the analyst have clearly in mind which group is being represented. If the objective is to promote the general welfare of the public, it is necessary to consider the impact of alternative policies on the entire population, not merely on the income and expenditures of a selected group. Practically speaking, however, without regulations the best that can be hoped for is that the broader interests of the community will be taken into account. Most would agree, for example, that without environmental and health regulations and the attendant threat of prosecution, there would be little incentive for firms to treat their waste products before discharging them into local waterways.

The national viewpoint would seem to be the correct one for all federally funded public works projects; however, most such projects provide benefits only to a local area, making it difficult, if not impossible, to trace and evaluate quantitatively the national effects. The following example parallels an actual case history.

Example 3-4

The government wants to decide whether to give a $5,000,000 subsidy to a chemical manufacturer who is interested in opening a new factory in a depressed area. The factory is expected to generate jobs for 200 persons and further stimulate the local economy through commercial ventures and tourist trade. The benefits due to jobs created and improved trade in the area are estimated at $1,000,000 per year. Six percent is considered to be a fair discount rate. The study period is 20 years. Calculate the benefit/cost ratio to determine if the project is worthwhile.

Solution

$$\text{PW of benefits} = \$1,000,000(P/A, 6\%, 20) = \$11,470,000$$

$$\text{B/C ratio} = \frac{11,470,000}{5,000,000} = 2.3$$

Outcome.   The plant was funded based on the foregoing study, but pollution control equipment was not installed. During operations, raw by-products were dumped into the river, causing major environmental problems downstream. Virtually all of the fish died, and the river became a local health hazard. The retrofitting of pollution control equipment sometime later made the entire project uneconomical and the plant eventually closed.

Conclusion.   Because the full costs of the project were not taken into account originally, the results were overly optimistic and misleading. Had the proper viewpoint been established at the outset and all the factors considered, the outcome might not have been so unfortunate.

■

3.4.1 Step-by-Step Approach

To conduct a benefit-cost analysis for an investment project, it is important to complete the following steps.

1. Identify the problem clearly.
2. Explicitly define the set of objectives to be accomplished.
3. Generate alternatives that satisfy the stated objectives.
4. Identify clearly the constraints (e.g., technological, political, legal, social, financial) that exist with the project environment. This step will help narrow the alternatives generated.
5. Determine and list the benefits and costs associated with each alternative. Specify each in monetary terms. If this cannot be done for all factors, this fact should be stated clearly in the final report.

6. Calculate the benefit/cost ratios and other indicators (e.g., present value, rate of return, initial investment required, payback period) for each alternative.

7. Prepare the final report, comparing the results of the evaluation of each alternative examined.

3.4.2 Using the Methodology

As with any decision-making process, the first two steps above are to define the problem and related goals. This may involve identifying a particular problem to be solved (e.g., pollution), or agreeing on a specific program, such as landing an astronaut on the moon. Once this is done, it is necessary to devise a solution that is feasible not only technically and economically, but politically as well.

Implicit in these steps is a twofold selection process: a macro-selection process whereby we choose from among competing opportunities or programs (should more federal funds be expended on space research or pollution cleanup and control?), and a micro-selection process where we strive to find the best of several alternatives (should we build a nuclear- or coal-fired plant?). In the public sector, it has often been claimed that we as a society are ruefully inept in conducting the selection process. Usually, goals and options are chosen in the thick of political debate, and benefits and costs are viewed so narrowly that we end up with projects that are not truly desirable when the needs of society as a whole are considered.

One of the most dramatic examples of such a failure is the East St. Louis public housing project that was torn down only a few years after it was erected. When this project was being planned the main consideration was apparently to maximize the ratio of bricks to dollars; there is no sign that sociological factors, such as the possible wishes and concerns of the potential inhabitants, entered into the analysis. The result was that the project became badly crime-ridden and vandalized, many people refused to live in it, and bankruptcy quickly ensued.

The potential for such mistakes can be greatly minimized if an external as well as internal benefit-cost analysis is made. For a housing project, the planners might examine such external factors as prevailing social conditions (crime rate, poverty levels, family structure, etc.) and the availability of public transportation, schools, and employment. Then the trade-offs must be established that are involved in such internal factors as the type of construction material to use, the size of the rooms versus the number of rooms, cabinetry in the kitchen versus the size of the appliances. The internal and external analyses are interdependent, but most large public-sector engineering projects can be reduced to five or six alternatives.

3.4.3 Classes of Benefits and Costs

Once a set of alternatives has been established, the detailed analysis can begin. The benefits and costs may be broken down into four classes: primary, secondary, external, and intangible. Primary refers to benefits and costs that are a direct result of a particular project. If a corporation manufactures videocassette recorders, the primary costs are in production, and the primary benefits are in profits. In building a canal, the construction costs and the revenues generated from water charges are the primary elements.

"Secondary" benefits and costs are not those that are less important, but rather, the marginal benefits and costs that accrue when an imperfect market mechanism is at work. In such instances, the market prices of a project's final goods and services do not reflect the "true" prices. The use of government funds to build and maintain airports is a good example. There is a hidden cost to society as well as a hidden benefit to the airlines and their more frequent customers. Increased noise pollution and traffic congestion around the airport are illustrative of the costs; benefits can be measured by lower air fares.

External benefits and costs are those that arise when a project produces a spillover effect on someone other than the intended group. Thus a government subsidy to airports produces external benefits by boosting the local economy indirectly. Massive government spending on space has yielded extensive benefits to medical science and the microelectronics industry. Similarly, there are spillover effects of pollution that produce disbenefits in the form of health costs and the loss of recreational facilities.

Intangible benefits and costs are those that are difficult, if not impossible, to measure on a monetary scale. Trademarks and goodwill are intangibles that have always been of concern, whereas the cost of increased urban congestion, or the benefits of advanced automation and robots in the workplace are factors, that engineers have more recently been forced to consider. When intangibles dominate the decision process, the value of multiple-criteria methods such as multiattribute utility theory and the analytic hierarchy process, each discussed in Chapter 4, becomes persuasively self-evident.

After categorizing the benefits and costs in this manner, they should be allocated to the various project time periods in which they are expected to occur. These periods include the planning stage, the implementation stage, the operation stage, and the closeout stage. This distinction is necessary for proper quantitative evaluation. For example, the costs associated with noise, traffic disruption, and hazards of subway construction may occur only in the implementation stage, and must be discounted accordingly.

3.4.4 Shortcomings of the B/C Methodology

Upon completion of the quantitative assessment of the various costs and benefits, the actual desirability of the project can be determined. Use of the benefit/cost ratio to rank or ascertain the best alternative can be deceptive, however, because it disguises the problem of scale. Two projects may have the same ratio, yet involve benefits and costs that differ by millions of dollars. Or one project may have a lower ratio than another and still possess greater benefits. Sometimes, therefore, projects will be selected simply on the basis of whether their benefits exceed their costs. Yet again, scale must be considered, because two projects can obviously have the same net benefit, but one may be far more costly than the other.

As mentioned, another way to evaluate projects is to compare the expected rate of return on investment with the interest rate on an alternative use of the funds. This criterion is implicit in most private-sector decisions but is generally neglected in the public sector, where tangible financial returns are not the sole criterion for investment allocations.

Moreover, there is rarely a consensus on which discount rate should be used. Economists invariably dispute the choice, some arguing for the social rate of time preference while others lean toward the prevailing interest rate. Except when a particular rate is specified by the decision maker, the net present value calculations should be repeated using several values to ascertain sensitivity effects.

The difficulty in agreeing on a discount rate is usually secondary to the problem of determining future costs and benefit streams. Uncertainties in long-term consequences may be large for extended time horizons of more than a few years, although frequently, all alternatives will suffer from a similar fate. Investigating questions of intertemporal equity and methods for dealing with uncertain outcomes are central problems of research, and their logic must be pursued relentlessly. Moreover, all forms of decision making must resolve these questions, whether or not they are dealt with explicitly.

In practice, it is rare that any one criterion will suffice for making a sound decision. Several criteria, as well as their many variations, must be examined in the analysis. The important point, however, is that even if all relevant factors are addressed, the analysis will still possess a high degree of subjectivity, leaving room for both conscious and unacknowledged bias. This leads to the two major shortcomings of benefit-cost analysis.

The first is the need and general failure to evaluate those items that are unquantifiable in monetary terms. The type of question that continually gets raised is: "How do you measure the value of harmony between labor and management?" or "What is the value of pollution-free environment?" The development of indicators other than those that reflect dollar values directly present a considerable challenge to analysts. They must depart from the familiar criteria of economic efficiency as a prime mechanism of evaluation and venture into the unknown areas of social and environmental concerns. Interestingly enough, the nonquantifiable elements bear equally on the governmental, business, and consumer sectors of the economy. In short, these "unmeasurables" may be of utmost significance, as system indicators must be developed to evaluate their impact on the program. It is here where judgment and subjectivity come into play.

The second weakness in the practice of benefit-cost analysis arises from the "judge and jury" characteristic. Invariably, the same government agency that proposes and sponsors a particular project undertakes the analysis. Whether this is done internally or by a subcontractor is not important. Rather, the agency and its contractors will usually display similar attitudes and biases in their approach to a problem. Independent, unbiased assessments are needed if the process is to work correctly and produce believable results.

3.5 COST-EFFECTIVENESS ANALYSIS

When comparing two projects that have the same B/C ratio, the one that costs more will provide greater returns. In some situations, though, there may be a fixed or upper limit on the budget, so a project that is technically feasible may not be economically feasible even if it has a high B/C ratio. Economic barriers to entry are common in many fields, such as automotive or semiconductor manufacturing where the required initial investment may be as high as $1 billion.

In the case where the budget is the limiting factor, we would like to know what is the most effective use of our money. To answer this question, *cost-effectiveness* (C-E) studies are often performed. In these studies, the concern is not so much on the rate of return as on the performance of the resultant system as measured by a composite index which is necessarily subjective in nature. This is because incommensurable and qualitative factors such as development risk, maintainability, and ease of use must all be evaluated collectively.

In general, system effectiveness can be thought of as a measure of the extent to which a system may be expected to achieve a set of specific mission requirements. It is often denoted as a function of the system availability, dependability, and capability.

- *Availability* is defined as a measure of the system condition at the start of a mission. It is a function of the relationship among hardware, personnel, and procedures.
- *Dependability* is defined as a measure of the system condition at one or more points during mission operations.
- *Capability* accounts specifically for the performance spectrum of the system.

The term *effectiveness* possesses a high order of abstraction. Because of its generality, it is subject to many interpretations, which makes it difficult to define precisely. For a product or service, one definition would be the ability to deliver what is called for in the technical specification. Among the terms that are related to (or have been substituted for) *effectiveness* are *value*, *worth*, *benefit*, *utility*, *gain*, and *performance*. Unlike cost, which can be measured in dollars, effectiveness does not possess an intrinsic measure by which it can be uniquely expressed.

Government agencies, in particular, the U.S. Department of Defense, have been the primary exponents of cost-effectiveness analyses. The following eight steps represent a common blueprint for conducting a C-E study.

1. Define the desired goals.
2. Identify the mission requirements.
3. Develop alternative systems.
4. Establish system evaluation criteria.
5. Determine capabilities of alternative systems.
6. Analyze the merits of each.
7. Perform sensitivity analysis.
8. Document results and make recommendations.

A critical step in the procedure is in deciding how the merits of each alternative will be judged. After the evaluation criteria or attributes are established, a mechanism is needed to construct a single measure of performance. Scoring models, such as those described in Section 3.3, are commonly used. Here we assess the relative importance of each system attribute and assign a weight to each. Next, a numerical value, say between 0 and 100, is assigned to represent the effectiveness of each attribute for each system. Once again, these values are subjective ratings but may actually be based on simple mathematical calculations of objective measures, subjective opinion, or engineering judgments. Where an appropriate physical scale exists, the maximum and minimum values can be noted and a straight line between those boundaries can be used to translate outcomes to a scale of 0 to 100. The analyst must ensure that the actual value of the attribute corresponds to the subjective description; for example, $100 \geq$ excellent $\geq 80;\ 80 >$ good ≥ 60.

In many cases it is useful to compare attribute relative values graphically to see if any obvious errors exist in data entry or logic. Figure 3-3 provides a visual comparison of the ratings of each of five attributes for four systems. The corresponding data are displayed in Table 3-5.

At this point in the analysis, two sets of numbers have been developed for each attribute i: the normalized weights, w_i, and the perceived effectiveness assigned to each system j for each attribute i, s_{ij}. To arrive at a composite measure of effectiveness, T_j, for each system j, we could use eq. (3-1). The highest value of T would indicate the system with the best overall performance.

If this system were within budget and none of its attribute values were below a predetermined threshold, it would represent the likely choice. Nevertheless, effectiveness alone does not tell the entire story, and whenever possible, the analysis should be extended to in-

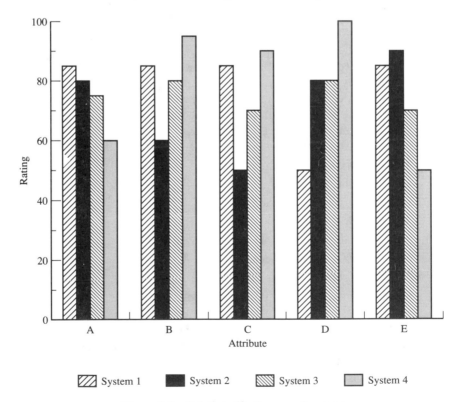

Figure 3-3 Relative effectiveness of systems.

TABLE 3-5 DATA FOR COST-EFFECTIVENESS ANALYSIS

Attribute	Weight	System 1		System 2		System 3		System 4	
		EFF	WT	EFF	WT	EFF	WT	EFF	WT
A. Efficiency	0.32	85	27.2	80	25.6	75	24.0	60	19.2
B. Speed	0.24	85	20.4	60	14.6	80	19.2	95	22.8
C. User friendly	0.24	85	20.4	50	12.0	70	16.8	90	21.6
D. Reliability	0.12	50	6.0	80	9.6	80	9.6	99	22.8
E. Expandability	0.08	85	6.8	90	7.2	70	5.6	50	19.2
Total effectiveness			80.8		68.8		75.2		79.2
Costs			$450K		$250K		$300K		$350K

clude costs as well. In a similar fashion, cost factors can be combined into a single measure to compare with effectiveness. Typically, procurement, installation, and maintenance costs are considered. When the planning horizon extends beyond a year, the effects of time should be included through appropriate discounting. Table 3-5 contains this information.

The final step in an extended version of the C-E methodology compares system effectiveness and costs. A graphical representation may be helpful in this regard. Figure 3-4 plots the two variables for each system. (The unlabeled points represent systems not contained in Table 3-5). The outer envelope denotes the *efficient frontier*. Any system not on

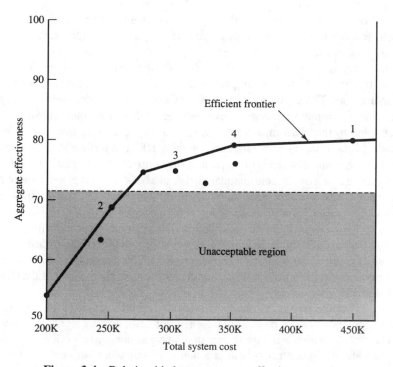

Figure 3-4 Relationship between system effectiveness and cost.

this curve is dominated by one or a combination of two or more systems, implying that it is inferior from both a cost and an effectiveness point of view. Systems falling below the dashed line (predetermined threshold) are arbitrarily deemed unacceptable. Finally, note the relationship between systems 1 and 4. Although system 1 has the highest effectiveness rating, it is only marginally better than system 4. The fact that it is almost 30% more expensive, however, makes its selection problematic, as an incremental analysis would indicate.

3.6 ISSUES RELATED TO RISK

In designing, building, and operating large systems, engineers must face up to such questions as: What can go wrong, and how likely is it to happen? What range of consequences might there be, when, and how could they be averted or mitigated? How much risk should be tolerated or accepted during normal operations, and how can it be measured, reduced, and managed?

Formal risk analysis attempts to pin down, and wherever possible, to quantify the answers to these questions (Bell 1989, Kaplan and Garrick 1981). In new systems it is coming to be accepted as a way of comparing the risks inherent in alternative designs, spotlighting the high-risk portion of a system, and pointing up techniques for attenuating those risks. For older systems, risk analysis conducted after they have been built and operated has often revealed crucial design faults. One such fault cost the lives of 167 workers on the British oil production platform Piper Alpha in the North Sea in July 1988. A simple gas leak in the $3 billion rig led to a devastating explosion. The platform had a vertical structure and risk analysis was not done on the design. Workers' accommodations were on top, above the lower compartments, which housed equipment for separating oil from natural gas. The accommodations were thought to be immune to mishap. But as a post-accident computer simulation revealed, the energy from the explosion in the lower level coupled to the platform's frame. Stress waves were dissipated effectively into the water below, but in short order, reflections at the steel–air interface at the upper levels expanded, weakened, and shattered the structure. In contrast, Norwegian platforms, which are designed using government-mandated risk analysis, are long and horizontal, like aircraft carriers, with workers' accommodations at the opposite end of the structure from the processing facilities and insulated from them by steel doors.

Those working in the field define risk as a combination of the probability of an undesirable event with the magnitude of every foreseeable consequence (damage to property, loss of money, delay in implementation, etc.). The consequences considered can range in seriousness from a mild setback to the catastrophic. Some related definitions are given in Table 3-6.

The first step in risk analysis is to tabulate the various stages or phases of a system's mission and list the risk sensitivities in each phase, be they technical, human, or economic. The time at which a failure occurs may mitigate its consequences. For example, a failure in an air traffic control system at a major airport would disrupt local air traffic far more at week-night rush hour than on a Sunday morning. Similarly, a failure in a chemical

TABLE 3-6   DEFINITIONS RELATED TO RISK

Term	Definition
Failure	Inability of a system, subsystem, or component to perform its required function.
Quality assurance	Probability that a system, subsystem, or component will perform its intended function when tested.
Reliability	Probability that a system, subsystem, or component will perform its intended function for a specified period of time or under normal conditions.
Risk	Combination of the probability of an abnormal event or failure and the consequences of that event or failure to a project's success or system's performance.
Risk assessment	Process and procedures of identifying, characterizing, quantifying, and evaluating risks and their significance.
Risk management	Any techniques used either to minimize the probability of an accident or to mitigate its consequences with, for instance, good engineering design, good operating practices, or preventive maintenance.
Uncertainty	Measure of the limits of knowledge in a technical area, expressed as a distribution of probabilities around a point estimate. The four principal elements of uncertainty are statistical confidence (a measure of sampling accuracy), tolerance (a measure of the relevance of available information to the problem at hand), incompleteness and inaccuracy of the input data, and ambiguity in modeling the problem.

Source: Dougherty and Fragola (1988).

processing plant would be more dangerous if it interfered with an intermediate reaction that produced a toxic chemical than if it occurred at a stage where the by-products were more benign.

Next, for each phase of the mission, the system's operation should be diagrammed and the logical relationships of the components and subsystems during that phase determined. The most useful techniques for the job are failure modes and effects analysis (FMEA), event tree analysis, and fault tree analysis (Henley and Kumamoto 1981). The three complement one another, and when taken together, help engineers identify the hazards of a system and the range of potential consequences. The interactions are particularly important because one piece of equipment might be caused to fail by another's failure to, say, supply fuel or current.

For engineers and managers, the chief purpose of risk analysis—defining the stages of a mission, examining the relationships between system parts, and quantifying failure probabilities—is to highlight any weakness in a design and identify those that contribute most heavily to delays or losses. The process may even suggest ways of minimizing or mitigating risk.

A case in point is the probabilistic risk analysis on the space shuttle's auxiliary power units, completed for NASA in December 1987 by the engineering consulting firm Pickard, Lowe & Garrick, Inc. The auxiliary power units, among other tasks, throttle the orbiter's main engines and operate its wing ailerons. NASA engineers and managers, using qualitative techniques, had formerly judged fuel leaks in the three auxiliary fuel units "unlikely" and the risks acceptable, without fully understanding the magnitude of the risks they accepted, even though a worst-case consequence could be the loss of the vehicle. One of the problems with qualitative assessment is that subjective interpretation of words such as "likely" and "unlikely" allows opportunity for errors in judgment about risk. For example, NASA had applied the word "unlikely" to risks that ranged from 1:250 to 1:20,000.

The probabilistic risk analysis revealed that although the probability of individual leaks was low, there were so many places where leaks could occur that five had in fact occurred in the first 24 shuttle missions. Moreover, in the ninth mission, on Nov. 28, 1983, the escaping fuel self-ignited while the orbiter was hurtling back to earth, and exploded after it had landed.

The probabilistic analysis pinpointed the fact that an explosion was more likely to occur during landing than during launch, when the auxiliary power units are purged with nitrogen to remove combustible atmospheric oxygen. It also suggested several ways of reducing the risk, such as changing the fuels or placing fire barriers between the power units.

3.6.1 Accepting and Managing Risk

Once the risks are determined, managers must decide what levels are acceptable based on economic, political, and technological judgments. The decision can be controversial because it necessarily involves subjective judgments about costs and benefits of the project, the well-being of the organization, and the potential damage or liability.

Naturally, risk is tolerated at a higher level if the payoffs are high or critical to the organization. In the microcomputer industry, for example, where product lifetimes may be no greater than 1 or 2 years, and new products and upgrades are being introduced continually, companies must keep pace with the competition or forfeit market share. Whatever the level of risk finally judged acceptable, it should be compared with and, if necessary, used to adjust the risks calculated to be inherent in the project. The probability of failure may be reduced further by redundant or standby subsystems, or by parallel efforts during development. Also, managers should prepare to counter the consequences of failure or setbacks by devising contingency plans or emergency procedures.

3.6.2 Coping with Uncertainty

Two sources of uncertainty still need to be considered: one intrinsic in probability theory and the other born of all-too-human error. First, the laws of chance exclude the prediction of when and where a particular failure may occur. That remains true even if enough statistical information about the system's operation exists for a reliable estimate of how likely it is to fail. The probability of failure, itself, is surrounded by a band of uncertainty that expands or shrinks depending on how much data is available and how well the system is understood. This statistical level of confidence is usually expressed as a standard deviation about the mean or a related measure. Finally, if the system is so new that few or no data have been recorded for it, and analogous data from similar systems must be used to get a handle on potential risks, there is uncertainty over how well the estimate resembles the actual case.

At the human interface, the challenge is to design a system so that it will not only operate as it should but also leave the operator little room for erroneous judgment. Additional risk can be introduced if a designer cannot anticipate what information an operator may need to digest and interpret under the daily pressures of the job, especially when an emergency starts to develop.

From an operational point of view, poor design can introduce greater risk, sometimes with tragic consequences. After the U.S.S. *Vincennes* on July 3, 1988 mistook Iran Air Flight 655 for an enemy F-14 and shot down the airliner over international waters in the Persian Gulf, Rear Admiral Eugene La Roque blamed the calamity on the bewildering complexity of the Aegis radar system. He is quoted as saying that "we have scientists and engineers capable of devising complicated equipment without any thought of how it will be integrated into a combat situation or that it might be too complex to operate. These machines produce too much information and don't sort the important from the unimportant. There's a disconnection between technical effort and combat use."

All told, human behavior is not nearly as predictable as that of an engineered system. Today there are many techniques for quantifying with fair reliability the probability of slips, lapses, and misperceptions. Still, uncertainty remaining in the prediction of individual behavior contributes to residual risk in all systems and projects.

3.6.3 Risk-Benefit Analysis

Risk-benefit analysis is a generic term for techniques encompassing risk assessment and the inclusive evaluation of risk, costs, and benefits of alternative projects or policies. Like other quantitative procedures, the steps in risk-benefit analysis include specifying objectives and goals for the project options, identifying constraints, defining the scope and limits for the study itself, and developing measures of effectiveness of feasible alternatives. Ideally, these steps should be completed in conjunction with an accountable decision maker, but in many cases this is not possible. It is therefore incumbent upon the analyst to take exceptional care in stating the assumptions and limitations of the each assessment, especially because risk-benefit analysis is frequently controversial.

The principal task of this methodology is to express numerically, insofar as possible, the risks and benefits that are likely to result from project outcomes. Calculating these outcomes may require scientific procedures or simulation models to estimate the likelihood of an accident or mishap, and its probable consequences. Finally, a composite assessment is carried out which aggregates the disparate measures associated with each alternative. The conclusions should incorporate the results of a sensitivity analysis in which each significant assumption or parameter is varied in turn to judge its effect on the aggregated risks, costs, and benefits.

One approach to risk assessment is based on the three primary steps of systems engineering, shown in Fig. 3-5 (Sage and White 1980). These involve the *formulation, analysis*, and *interpretation* of the impacts of alternatives upon the needs, and the institutional and value perspectives of the organization. In risk formulation, we determine or identify the types and scope of the anticipated risks. A variety of systemic approaches, such as the nominal group technique, brainstorming, and the Delphi method are especially useful at this stage (Makridakis and Wheelwright 1978). It is important to identify not only the risk elements but also the elements representing needs, constraints, and alterables associated

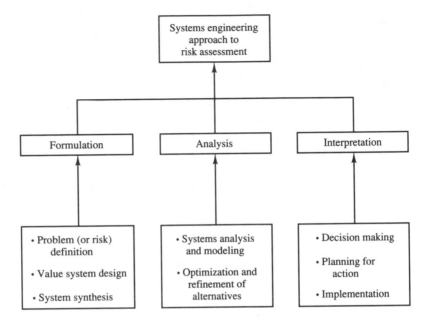

Figure 3-5   Systems engineering approach to risk assessment.

with possible risk reduction with and without technological innovation. This can be done only in accordance with a value system.

In the analysis step, we forecast the failures, mishaps, and other consequences that might accompany the development and implementation of the project. This will include estimation of the probabilities of outcomes and the associated magnitudes. Many methods such, as cross impact analysis, interpretive structural modeling, economic modeling, and mathematical programming, are potentially useful at this step. The inputs are those elements determined during problem formulation.

In the final step, we attempt to give an organizational or political interpretation to the risk impacts. This includes specification of individual and group utilities for the final evaluation. Decision making follows. The economic methods of benefit-cost analysis are most commonly used at this point. Extension to include the results of the risk assessment, however, is not trivial. A principal problem is that risks and benefits may be measured in different units and therefore may not be strictly additive. Rather than trying to convert everything into a single measure, then, it may be better simply to present the risks and net benefits in their respective units or categories. This leaves the decision maker free to impose his or her own values on the results of the analysis.

To aid in interpreting the results, risk–return graphs, similar to the cost-effectiveness graph displayed in Fig. 3-4, can be drawn to highlight the efficient frontier. Risk profiles may also be useful. For example, Fig. 3-6 illustrates a perspective provided by a risk analysis profile. Projects 1 and 2 are most likely to yield lifetime profits of $150,000,000 and

$200,000,000, respectively. So, for some decision makers, project 2 might be considered superior if the benefit/cost ratio was favorable. Nevertheless, it is worth probing the data a bit more. Project 2 has a 0.15 probability of returning zero profits and only a 0.43 probability of achieving the most likely level of $200M. There is also a 0.20 probability that it will yield lower expected profits than will project 1. This is the downside risk if project 2 is chosen. Given these data, a risk-averse person would be inclined to select project 1, which has a big chance (0.83) of realizing a moderate profit of $150M, with very little chance of anything less or greater; that is, project 1 has a small standard deviation. A gambler would lean toward project 2, which has a small chance at a large profit.

The kinds of risk profiles contained in Fig. 3-6 make the consequences of outcomes more visible and thus permit the decision maker consciously to make decisions consistent with his or her attitude toward risk, be it conservative or freewheeling. Generally, the number of data needed to construct a picture such as Fig. 3-6 is small and relatively easy to obtain if a historical database exists. It can be solicited from the engineers and marketing personnel who are familiar with the projects. If no collective experience can be found within the organization, more subjective or arbitrary procedures would be required. A number of software packages are available to help with the construction effort.

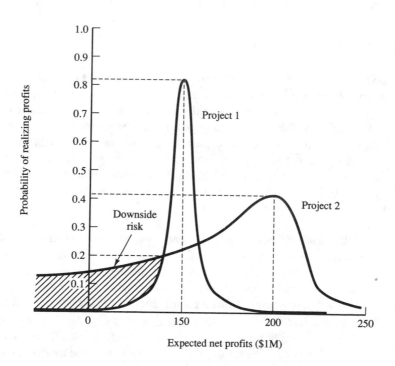

Figure 3-6 Illustration of risk profile. (From W. E. Souder, *Project Selection and Economic Appraisal*, copyright © 1984 by Van Nostrand Reinhold; reprinted by permission of the publisher.)

3.6.4 Limits of Risk Analysis

The ultimate responsibility for project selection and implementation goes beyond any risk assessment and rests squarely on the shoulders of top management. Although formal analysis can point up unexpected vulnerabilities in large complex projects, it remains an academic exercise unless the managers take the results seriously and ensure that the project is managed conscientiously. Safety must be designed into a system from the beginning, and good operating practice is essential to the success of any continuing program of risk management. Controversy still rages, for example, over whether or not the vent-gas scrubber—a key element in the safety system of the Union Carbide pesticide plant in Bhophal, India that exploded in 1984, killing over 3,000 persons—was designed adequately to handle a true emergency. But even if it had been, neither it nor a host of other safety features were maintained in working order.

For risks to be ascertained at all, project managers must agree on the value of assessing them in engineering design. It has often been said that you can degrade the performance of a system by poor quality control, but you cannot enhance a poor design by good quality control. At the point where project managers are responsible for crucial decisions, risk assessment is one more tool that can help them weigh alternatives so that their choices are informed and deliberate rather than isolated, or worse, repetitions of past mistakes.

3.7 DECISION TREES

Decision trees, also known as decision flow networks and decision diagrams, are powerful means of depicting and facilitating the analysis of problems that involve sequential decisions and variable outcomes over time. They have great usefulness in practice because they make it possible to look at a large complicated problem in terms of a series of smaller simple problems, while explicitly considering risk and future consequences.

A decision tree is a graphical method of expressing, in chronological order, the alternative actions that are available to the decision maker and the outcomes determined by chance. In general, they are composed of the following two elements, as shown in Fig. 3-7:

1. *Decision nodes*. At a decision node, usually designated by a square, the decision maker must select one alternative course of action from a finite set of possibilities. Each is drawn as a branch emanating from the right side of the square. When there is a cost associated with an alternative, it is written along the branch. Each alternative branch may result in either a payoff, another decision node, or a chance node.
2. *Chance nodes*. A chance node, designated as a circle, indicates that a random event is expected at this point in the process. That is, one of a finite number of states of nature may occur. The states of nature are shown on the tree as branches to the right of the chance nodes. The corresponding probabilities are similarly written above the branches. The states of nature may be followed by payoffs, decision nodes, or more chance nodes.

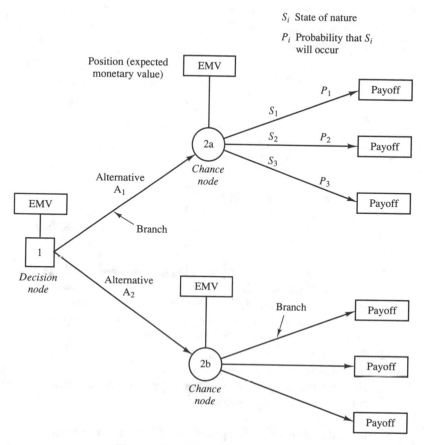

Figure 3-7   Structure of a decision tree.

Constructing a tree.   A tree is started on the left of the page with one or more decision nodes. From these, all possible alternatives are drawn branching out to the right. Then a chance node or second decision node, associated with either subsequent events or decisions, respectively, is added. Each time a chance node is added, the appropriate states of nature with their corresponding probabilities emanate rightward from it. The tree continues to branch from left to right until the final payoffs are reached. The tree shown in Fig. 3-7 represents a single decision with two alternatives, each leading to a chance node with three possible states of nature.

Finding a solution.   To solve a tree, it is customary to divide it into two segments: (1) chance nodes with all their emerging states of nature (Fig. 3-8a), and (2) decision nodes with all their alternatives (Fig. 3-8b). The solution process starts with those segments ending in the final payoffs, at the right side of the tree, and continues to the left, segment by segment, in the reverse order from which it was drawn.

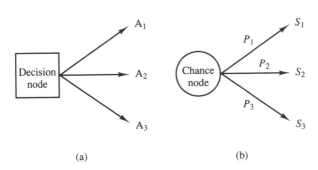

A_j Alternative j
S_i State of nature i
P_i Probability of S_i

(a)   (b)   **Figure 3-8**   Segments of a tree.

1. *Chance node segments.* The expected monetary value (EMV) of all the states of nature emerging from a chance node must be computed (multiply payoffs by probabilities and sum the results). The EMV is then written above the node inside a rectangle (labeled a "position value" in Fig. 3-7). These expected values are considered as payoffs for the branch to the immediate left.

2. *Decision node segments.* At a decision point, the payoffs given (or computed) for each alternative are compared and the best one is selected. All others are discarded. The corresponding branch of a discarded alternative is marked by the symbol || to indicate that the path is suboptimal.

This procedure is based on principles of dynamic programming and is commonly referred to as the "rollback" step. It starts at the endpoints of the tree, where the expected value at each chance node and the optimal value at each decision node are computed. Suboptimal choices at each decision node are dropped, with the rollback continuing until the first node of the tree is reached. The optimal policy is recovered by identifying the choices made at each decision node that maximize the value of the objective function from that point onward.

Example 3-5 (Deterministic Replacement Problem)

The most basic form of a decision tree occurs when each alternative results in a single outcome, that is, when certainty is assumed. The replacement problem defined in Fig. 3-9 for a 9-year planning horizon illustrates this situation. The numbers above the branches represent the returns per year for the specified period should the replacement be made at the corresponding decision point. The numbers below the branches are the costs associated with that decision. For example, at node 3, keeping the machine results in a return of $3K per year for 3 years, and a total cost of $2K.

As can be seen, the decision as to whether or not to replace the old machine with the new machine does not occur just once, but recurs periodically. In other words, if the decision is made to keep the old machine at decision point 1, then later, at decision point 2, a choice

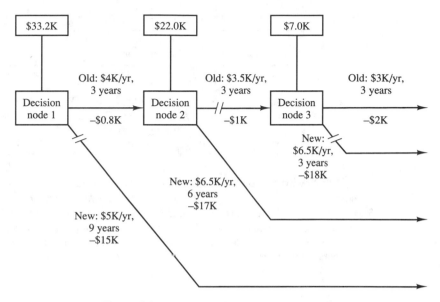

Figure 3-9 Deterministic replacement problem.

again has to be made. Similarly, if the old machine is chosen at decision point 2, a choice has to be made at decision point 3. For each alternative, the cash inflow and duration of the project is shown above the branch, and the cash investment opportunity cost is shown below the branch. At decision point 2, for example, if a new machine is purchased for the remaining 6 years, the net benefits from that point on are (6 years)($6.5K/yr) returns − $17.0K opportunity cost = $22.0K net benefits. Alternatively, if the old machine is kept at decision point 2, we have ($3.5K/yr)(3 years) returns − $1.0K opportunity cost + $7K net benefits associated with decision point 3 = $16.5K net benefits.

For this problem, one is concerned initially with which alternative to choose at decision point 1. But an intelligent choice here should take into account the later alternatives and decisions that stem from it. Hence the correct procedure in analyzing this type of problem is to start at the most distant decision point, determine the best alternative and quantitative result of that alternative, and then roll back to each successive decision point, repeating the procedure until finally, the choice at the initial or present decision point is determined. By this procedure, one can make a present decision that directly takes into account the alternatives and expected decisions of the future.

For simplicity in this example, timing of the monetary outcomes will first be neglected, which means that a dollar has the same value regardless of the year in which it occurs. Table 3-7 displays the necessary computations and implied decisions. Note that the monetary outcome of the best alternative at decision point 3 ($7.0K for the "old") becomes part of the outcome for the old alternative at decision point 2. That is, if the decision at node 2 is to continue to use the current machine rather than replace it, the monetary value associated with this decision equals the EMV at node 3 ($7K) plus the transition benefit from node 2 to 3 ($3.5/yr × 3 years − $1K = $9.5K), or $16.5K. Similarly, the best alternative at decision point 2 ($22.0K for the "new") becomes part of the outcome for the old alternative at decision point 1.

TABLE 3-7   COMPUTATIONAL RESULTS FOR REPLACEMENT PROBLEM IN FIG. 3-9

Decision point	Alternative	Monetary outcome		Choice
3	Old	($3K/yr)(3 years) − $2K	= $7.0K	Old
	New	($6.5K/yr)(3 years) − $18K	= $1.5K	
2	Old	$7K + ($3.5K/yr)(3 years) − $1K	= $16.5K	
	New	($6.5K/yr)(6 years) − $17K	= $22.0K	New
1	Old	$22.0K + ($4K/yr)(3 years) − $0.8K	= $33.2K	Old
	New	($5K/yr)(9 years) − $15K	= $30.0K	

By following the computations in Table 3-7, one can see that the answer is to keep the old now and plan to replace it with the new at the end of 3 years (at decision point 2). But this does not mean that the old machine should necessarily be kept for a full 3 years and then a new machine bought without question at that time. Conditions may change anywhere along the way, necessitating a fresh analysis based on estimates that are reasonable in light of the current conditions.

■

Example 3-6 (Timing Considerations)

For decision tree analyses, which involve working from the most distant decision point to the nearest decision point, the easiest way to take into account the timing of money is to use the present value approach and thus discount all monetary outcomes to the decision points in question. To demonstrate, Table 3-8 gives the computations for the same replacement problem of Fig. 3-9 using an interest rate of 12% per year.

TABLE 3-8   COMPUTATIONS FOR REPLACEMENT PROBLEM WITH 12% INTEREST RATE

Decision point	Alternative	Expected monetary outcome		Choice
3	Old	$3K($P/A$, 12%, 3) − $2K = $3K(2.402) − $2K	= $5.21K	Old
	New	$6.5K($P/A$, 12%, 3) − $18K = $6.5K(2.402) − $18K	= −$2.39K	

2	Old	$3.85K($P/F$, 12%, 3)
		  + $3.5K($P/A$, 12%, 3) − $1K
		  = $3.85K(0.7118) + $3.5K(2.402) − $1K = $10.15K
				Old

	New	$6.5K($P/A$, 12%, 6) − $17K = $6.5K(4.111) − $17K	= $9.72K	

1	Old	$7.79K($P/F$, 12%, 3)
		  + $4K($P/A$, 12%, 3) − $0.8K
		  = $7.79K(.7118) + $4K(2.402) − $0.8K = $14.35K
				Old

	New	$5.0K($P/A$, 12%, 9) − $15K = $5.0K(5.328) − $15K	= $11.64K	

Note from Table 3-8 that when taking into account the effect of timing by calculating present worths at each decision point, the indicated choice is not only to keep the old at decision point 1, but also to keep the old at decision points 2 and 3 as well. This result is not surprising since the high interest rate tends to favor the alternatives with lower initial investments, and it also tends to place less weight on long-run returns. When the interest rate drops to 10%, the solution is the same as for Example 3-5.

■

Example 3-7 (Automation Decision Problem with Random Outcomes)

In this problem the decision maker must decide whether or not to automate a given process. Depending on the technological success of the automation project, the results will turn out to be either poor, fair, or excellent. The net payoffs for possible outcomes (expressed in net present values) are −$90K, $40K, and $300K, respectively. The initially estimated probabilities that each outcome will occur are 0.5, 0.3, and 0.2. Figure 3-10 is a decision tree depicting this simple situation. The calculations for the two alternatives are:

$$\text{Automate: } -\$90K(0.5) \ + \ \$40K(0.3) \ + \ \$300K(0.2) \ = \ \$27K$$
Don't automate: $0

These calculations show that the best choice for the firm is to automate based on an expected NPV of $27K versus $0 if it does nothing. Nevertheless, this may not be a clearcut decision because of the risk of a $90K loss, and because the decision maker might reduce the risk by obtaining further information.

Suppose that it is possible for the decision maker to conduct a technology study at the PW cost of $10K. The study will disclose that the enabling technology is either "shaky," "promising," or "solid." Let us assume that the probabilities of the various possible outcomes, given the technology study findings, are as shown in Fig. 3-11, which is a decision tree for the entire problem. This diagram shows expected future events (outcomes), along with their respective cash flows and probabilities of occurrence. The rectangular blocks represent (decision) points in time at which the decision maker must elect to take one and only one of the paths (alternatives) available. These decisions are normally based on a quantifiable measure, such as money, which has been determined to be the predominant "cost" or "reward" for comparing alternatives. As mentioned, the general approach is to find the action or alternative that will maximize the expected net present value equivalent of future cash flows at each decision point, starting with the furthest decision point(s) and then rolling back until the initial decision point is reached.

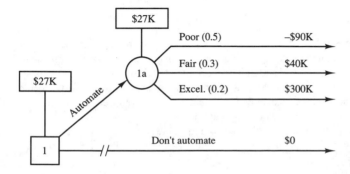

Figure 3-10 Automation problem before consideration of technology study.

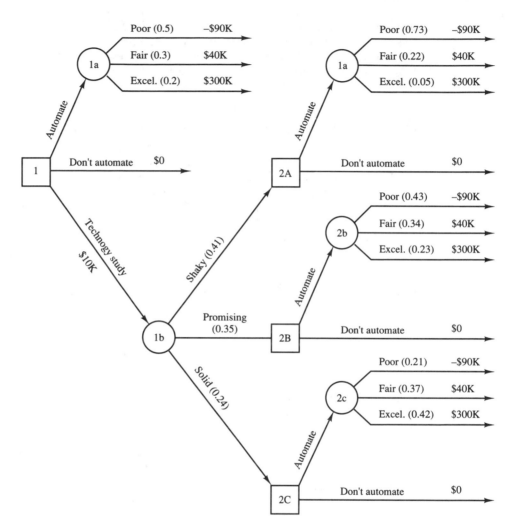

Figure 3-11   Automation problem with technology study taken into consideration.

Once again, the circular (chance) nodes represent points at which uncertain events (outcomes) occur. At a chance node the expected value of all paths leading (from the right) into the node can be calculated as the sum of the anticipated value of each path multiplied by its respective probability. (The probabilities of all paths leading into a node must sum to 1.) As the project progresses through time, the chance nodes are automatically reduced to a single outcome based on the "state of nature" that occurs at that time.

The solution to the problem in Fig. 3-11 is given in Table 3-9. It can be noted that the alternative "technology study" is shown to be best with an expected NPV of $34.62K. (To

TABLE 3-9   EXPECTED NPV CALCULATIONS FOR THE AUTOMATION PROBLEM

Decision point	Alternative	Expected monetary outcome		Choice
2A	Automate	$-\$90K(0.73) + \$40K(0.22)$ $+ \$300K(0.05)$	$= -\$41.9K$	
	Don't automate		$= \$0$	Don't automate
2B	Automate	$-\$90K(0.43) + \$40K(0.34)$ $+ \$300K(0.23)$	$= \$43.9K$	Automate
	Don't automate		$= \$0$	
2C	Automate	$-\$90K(0.21) + \$40K(0.37)$ $+ \$300K(0.42)$	$= \$121.9K$	Automate
	Don't automate		$= \$0$	
1	Automate	(see calculations above)	$= \$27K$	
	Don't automate		$= \$0$	
	Technology study	$\$0(0.41) + \$43.9K(0.35)$ $+ \$121.9K(0.24) - \$10K$	$= \$34.62K$	Technology study

check the solution in Table 3-9, perform the rollback procedure on Fig. 3-11, indicating which branches should be eliminated.)

■

3.7.1 Decision Tree Steps

Now that decision trees (diagrams) have been introduced and the mechanics of using them to arrive at an initial decision have been illustrated, the steps involved can be summarized as follows:

1. Identify the points of decision and alternatives available at each point.
2. Identify the points of uncertainty and the type or range of possible outcomes at each point (layout of decision flow network).
3. Estimate the values needed to conduct the analysis, especially the probabilities of different outcomes and the costs/returns for various outcomes and alternative actions.
4. Analyze the alternatives, starting with the most distant decision point(s) and working back, to choose the best initial decision.

In Example 3-6 we used the expected NPV as the decision criterion. However, if outcomes can be expressed in terms of utility units, it may be appropriate to use the expected utility as the criterion. Alternatively, the decision maker may be willing to express his or her certain monetary equivalent for each chance outcome node and use that as the decision criterion.

Because a decision tree can quickly become discouragingly, if not unmanageably large, it is generally best to start out structuring a problem by considering only major alternatives and outcomes in order to get an initial understanding or feeling for the issues. Then more information on alternatives and outcomes that seems sufficiently important to affect the final decision can be developed. Incremental embellishments can be added until the study is sufficiently complete in view of the nature and importance of the problem, and the time and resources available.

3.7.2 Basic Principles of Diagramming

The proper diagramming of a decision problem is, in itself, generally very useful to the understanding of the problem, as well as being essential to correct subsequent analysis. The placement of decision points and chance nodes from the initial decision point to the base of any later decision point should give an accurate representation of the information that will and will not be available when the decision maker actually has to make the choice associated with the decision point in question. The tree diagram should show the following:

1. All initial or immediate alternatives among which the decision maker wishes to choose.
2. All uncertain outcomes and future alternatives that the decision maker wishes to consider because they may directly affect the consequences of initial alternatives.
3. All uncertain outcomes that the decision maker wishes to consider because they may provide information that can affect his or her future choices among alternatives, and hence, indirectly affect the consequences of initial alternatives.

It should also be noted that the alternatives at any decision point and the outcomes at any payoff node must be:

1. Mutually exclusive; that is, no more than one can possibly be chosen.
2. Collectively exhaustive; that is, when a decision point or payoff node is reached, some course of action must be taken.

Figure 3-11 reflects these points. For example, decision nodes 2A, 2B, and 2C are each reached only after one of the mutually exclusive results of the technology study is known, and each decision node reflects all alternatives to be considered at that point. Further, all possible outcomes to be considered are shown, as evidenced by the fact that the probabilities sum to 1.0 for each chance node.

3.7.3 Use of Statistics to Evaluate the Value of More Information

An alternative that frequently exists in an investment decision is to conduct further research before making a commitment. This may involve such action as gathering more information about the underlying technology, updating an existing analysis of market demand, or investigating anew future operating costs for particular alternatives.

Once this additional information is collected, the concepts of Bayesian statistics provide a means of modifying estimates of probabilities, as well as a means of estimating the economic value of further investigation. To illustrate, consider the one-stage decision situation depicted in Fig. 3-12, in which each alternative has two possible chance outcomes: "high" or "low" demand. It is estimated that each outcome is equally likely to occur, and the monetary result expressed as PW is shown above the arrow for each outcome. Again, the amount of investment for each alternative is written below the respective lines. Based on these amounts, the calculation of the expected monetary outcomes (NPV) is as follows:

Old system: $45M(0.5) + $27.5M(0.5) − $10M = $26.25M
New FMS: $80M(0.5) + $48M(0.5) − $35M = $29.0M

which indicates that the "new FMS" should be selected.

To demonstrate the use of Bayesian statistics, suppose that one is considering the advisability of undertaking a fresh intensive investigation before deciding on the "old system" versus the "new FMS." Suppose also that this new study would cost $2.0M and will predict whether the demand will be high (h) or low (ℓ). To use the Bayesian approach, it is necessary to assess the conditional probabilities that the investigation or "technology study" will yield certain results. These probabilities reflect explicit measures of management's confidence in the ability of the investigation to predict the outcome. Sample assessments are

$$P(h|H) = 0.70, \quad P(h|L) = 0.20, \quad P(\ell|H) = 0.30, \quad \text{and} \quad P(\ell|L) = 0.80,$$

where H and L denote high and low actual demand as opposed to predicted demand. As an explanation, $P(h|H)$ means the probability that the predicted demand is high (h), given that the actual demand will turn out to be high (H).

A formal statement of Bayes' theorem is given in Appendix 3A at the end of this chapter along with a tabular format for ease of computation. Tables 3-10 and 3-11 use this format for revision of probabilities based on the assessment data above and the prior

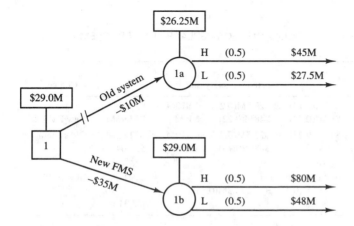

Figure 3-12   One-stage FMS replacement problem.

TABLE 3-10   COMPUTATION OF POSTERIOR PROBABILITIES GIVEN THAT INVESTIGATION-PREDICTED DEMAND IS HIGH (h)

(1)	(2)	(3)	(4) = (2) × (3)	(5) = (4)/Σ(4)
State (actual demand)	Prior probability, P(state)	Confidence assessment, $P(\text{h}\|\text{state})$	Posterior joint probability	Probability, $P(\text{state}\|\text{h})$
H	0.5	0.70	0.35	0.78
L	0.5	0.20	0.10	0.22
			0.45	

TABLE 3-11   COMPUTATION OF POSTERIOR PROBABILITIES GIVEN THAT INVESTIGATION-PREDICTED DEMAND IS LOW (ℓ)

(1)	(2)	(3)	(4) = (2) × (3)	(5) = (4)/Σ(4)
State (actual demand)	Prior probability, P(state)	Confidence assessment, $P(\ell\|\text{state})$	Posterior joint probability	Probability, $P(\text{state}\|\ell)$
H	0.5	0.30	0.15	0.27
L	0.5	0.80	0.40	0.73
			0.55	

probabilities of 0.5 that the demand will be high and 0.5 that the demand will be low [i.e., $P(\text{H}) = P(\text{L}) = 0.5$]. These probabilities are now used to assess the technology study alternative. Figure 3-13 depicts the complete decision tree. Note that demand probabilities are entered on the branches according to whether the investigation indicates high or low demand.

The next step is to calculate the expected outcome for the technology study alternative. This is done by the standard decision tree rollback principle, as shown in Table 3-12.

TABLE 3-12   EXPECTED NPV CALCULATIONS FOR REPLACEMENT PROBLEM IN FIGURE 3-13

Decision point	Alternative	Expected monetary outcome		Choice
2A	Old system	$45M(0.78) + $27.5M(0.22) − $10M	= $31.15M	
	New FMS	$80M(0.78) + $48M(0.22) − $35M	= $37.96M	New FMS
2B	Old system	$45M(0.27) + $27.5M(0.73) − $10M	= $22.23M	Old system
	New FMS	$80M(0.27) + $48M(0.73) − $35M	= $21.64M	
1	Old system	(see calculations above)	= $26.25M	
	New FMS	(see calculations above)	= $29.00M	New FMS
	Technology study	$37.96M(0.45) + $22.23M(0.55) − $2M	= $27.31M	

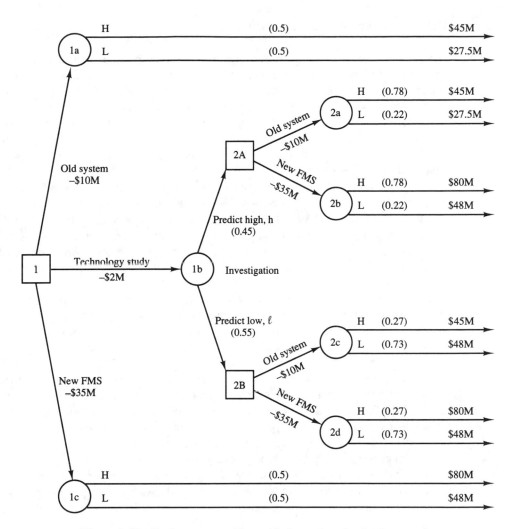

Figure 3-13   Replacement problem with alternative of technology study.

Note that the 0.45 and 0.55 probabilities that the investigation-predicated demand will be high and low, respectively, are obtained from the totals in column (4) of the Bayesian revision calculations depicted in Tables 3-10 and 3-11.

Thus, from Table 3-12, it can be seen that the "new FMS" alternative with an expected NPV of $29.0M is the best course of action by a slight margin. (As an exercise, perform the calculations on Fig. 3-13 and indicate the optimal path.) Although the figures used here do not reflect any advantages to this technology study, the benefit of gathering additional information can potentially be great.

3.7.4 Discussion and Assessment

The unique feature of decision trees is that they allow management to view the logical order of a sequence of decisions. They afford a clear graphical presentation of the various alternative courses of action and their possible consequences. By using decision trees, management can also examine the impact of a series of decisions (over many periods) on the objectives of the organization. Such models reduce abstract thinking to a rational visual pattern of cause and effect. When costs and benefits are associated with each branch and probabilities are estimated for each possible outcome, analysis of the flow network can clarify choices and risks.

On the down side, the methodology has several weaknesses that should not be overlooked. A basic limitation of its representational properties is that only small and relatively simple models can be shown at the level of detail that makes trees so descriptive. Every variable added expands the tree's size multiplicatively. Although this problem can be overcome to some extent by generalizing the diagram, significant information may be lost in doing so. This loss is particularly acute if the problem structure is highly dependent or asymmetric.

Regarding the computational properties of trees, for simple problems in which the endpoints are precalculated or assessed directly, the rollback procedure is very efficient. However, for problems that require a roll-forward procedure, the classic tree-based algorithm has a fundamental drawback: it is essentially an enumeration technique. That is, every path through the tree is traversed in order to solve the problem and generate the full range of outputs. This feature raises the "curse of dimensionality" common to many stochastic models: for every variable added, the computational requirements increase multiplicatively. This implies that the number of chance variables that can be included in the model tends to be small. There is also a strong incentive to simplify the value model, since it is recalculated at the end of each path through the tree.

Nevertheless, the enumeration property of tree-based algorithms in theory can be reduced dramatically by taking advantage of certain structural properties of a problem. Two such properties are referred to as "asymmetry" and "coalescence." For more discussion and some practical aspects of implementation, consult Call and Miller (1990).

TEAM PROJECT

Thermal Transfer Plant

Based on the evaluation of alternatives, TMS management has adopted a plan by which the design and assembly of the rotary combustor will be done at TMS. Most of the manufacturing activity will be subcontracted except for the hydraulic power unit, which TMS decided to build "in-house."

There are three functions involved in charging and rotating the combustor. Two of them, the charging rams and the resistance door, naturally lend themselves to hydraulics. The third, turning the combustor, can be done either electromechanically (by an electric motor and a gearbox) or hydraulically. If the hydraulic method is chosen, there are two alternatives: (1) use a large hydraulic motor as a direct drive; or (2) use a small hydraulic motor with a gearbox. Figure 3-14 contains a schematic.

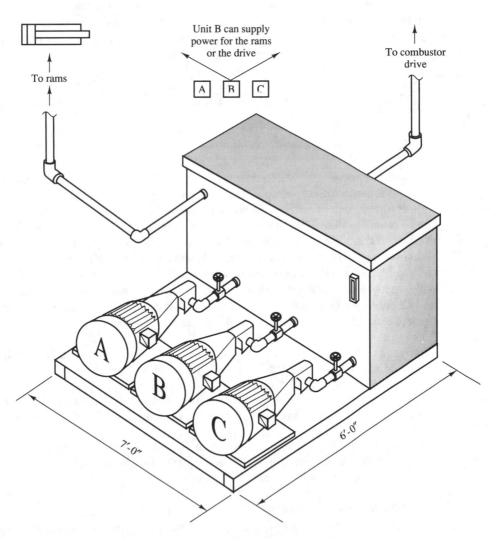

Figure 3-14   Hydraulic power unit.

TMS engineering has produced the following specifications for the hydraulic power unit:

Applicable documents, codes, standards, and requirements.

National Electric Manufacturers Association (NEMA)

American National Standards Institute (ANSI)

Pressure Vessels Code, American Society of Mechanical Engineers (ASME), Section VIII

Hydraulic rams. Two hydraulic cylinders will be provided for the rams. The cylinders will be 8 in. bore × 96 in. stroke. They will operate at 1,500 psi and will have an adjustable extension rate of 2 to 6 ft/min. They will retract in 15 seconds, will operate 180° out of phase, and will retract in the event of a power failure.

Combustor barrel drive. A single-direction, variable-speed drive will be provided for the combustor. The output of this drive will deliver up to 1.6 rpm and 7,500 ft-lb torque.

Resistance door cylinder. This cylinder will be 6 in. bore × 48 in. stroke, and will operate with a constant pressure of 200 psi.

Hydraulic power unit. The hydraulic power unit will be skid mounted and ready for hookup to interfacing equipment. Mounting and lifting brackets will be manufactured as well. Hydraulic pumps will be redundant so that in the event of the failure of one, another can be started to take over its function. Accumulators will be added to retract the rams and close the resistance door in the event of a power failure. The hydraulic fluid is to be E.F. Houghton's Cosmolubric or equivalent. Although system operating pressure is to be 1,500 psi, the plumbing will be designed to withstand 3,000 psi. Water-to-oil heat exchangers will be provided to limit reservoir temperature to 130°C. A method of controlling ram extension speed and combustor rpm within the specifications stated above will be provided. Control concepts may be analog (5 to 20 milliamperes), or digital.

Electrical. Electric motors will be of sufficient horsepower to drive the hydraulic pumps. Motors will operate at 1,200 rpm, 220/440 volts, three phase, 60 hertz.

Solenoids and controls. Solenoids are to be 120 volt, 60 hertz and will have manual overrides. Any analog control function is to respond to a 5- to 20-milliampere signal.

Combustion drive. A single-direction, variable-speed drive will be provided for the combustor. The output of this drive will deliver up to 1.6 rpm and 7,500 ft-lb torque. Three potential alternatives for the combustor drive are:

- Electric motor and gearbox
- Hydraulic motor with gearbox (hydraulic power supplied by hydraulic power unit)
- Hydraulic motor with direct drive (hydraulic power supplied by hydraulic power unit)

Your team assignment is to select the most appropriate drive from these candidates. To do so, develop a scoring model or a decision tree and evaluate each alternative accordingly. State your assumptions clearly regarding technological, economic, and other aspects, and explain the methodology used to support your analysis.

Initial cost estimates available to your team are:

Ram cylinders (two required)	$5,948 each
Resistance door cylinder	$1,505
Hydraulic power unit	$50,000
Low-speed high-torque motor	$22,780
High-speed motor with gearbox	$7,000

DISCUSSION QUESTIONS

1. Where would ideas for new projects and products probably originate in a manufacturing company? What would be the most likely source in an R&D organization such as AT&T Bell Laboratories?

2. Assume that you work in the design department of an aerospace firm and you are given the responsibility of selecting a workstation that will be used by each group in the department. How would you find out what systems are available? What basic information would you try to collect on these systems?

3. How can you extend a polar graph, similar to the one shown in Fig. 3-2, to the case where the criteria are individually weighted?

4. Identify a project that you are planning to pursue either at home or at work. List all the components, decision points, and chance events. What is the measure of success for the project? Assuming that there is more than one measure, how can you reconcile them?

5. If you were evaluating a proposal to upgrade the word processing system used by your organization, what type of information would you be looking for in detail? How would your answer change if you were buying only one or two systems as opposed to a few dozen?

6. What factors in an organization do you think would affect the decision to go ahead with a project, such as automating a production line, other than the benefit/cost ratio?

7. For years before beginning the project to build a tunnel under the English Channel, Great Britain and France debated the pros and cons. Can you speculate on the critical issues that were raised?

8. The project to construct a subway in Washington, D.C. began in the early 1970s with the expectation that it would be fully operational by 1980. A portion of the system opened in 1977, but as of 1993, about 10% remained unfinished. What do you think were the costs, benefits, and risks involved in the original planning? How important was the interest rate used in those calculations? Speculate on who or what was to blame for the lengthy delay in completion.

9. Where does quality fit into the benefit-cost equation? Can you identify some companies or products that compete primarily on the basis of quality rather than price?

10. A software company is undecided on whether it should expand its capacity by using part-time programmers or by hiring more full-time employees. Future demand is the critical factor,

which is not known with certainty but can only be estimated as low, medium, or high. Draw a decision tree for the company's problem. What data are needed?

11. How could benefit-cost analysis be used to help determine the level of subsidy to be paid to the operator of public transportation services in a congested urban area?

12. Why has the Department of Defense been the major exponent of cost-effectiveness analysis? Give your interpretation of what is meant by "diminishing returns" and indicate how it might affect a decision on procuring a military systems versus an office automation system.

13. In which type of projects does risk play a predominant role? What can be done to mitigate the attendant risks? Pick a specific project and discuss.

EXERCISES

3-1. Consider an important decision that you will be faced with in the near future. Construct a scoring model detailing your major criteria and assign weights to each. Indicate which data are known for sure, and which are uncertain. What can be done to reduce the uncertainty?

3-2. Use a checklist and a scoring model to select the best car for a married graduate student with one child. State your assumptions clearly.

3-3. Assume that you have just entered the university and wish to select an area of study.
  (a) Using benefit-cost analysis *only*, what would your decision be?
  (b) How would your decision change if you used cost-effectiveness analysis? Provide the details of your analysis.

3-4. You have just received a job offer in a city 1,000 miles away and must relocate. List all possible ways of moving your household. Use two different analytic techniques for selecting the best approach and compare the results.

3-5. Three new-product ideas have been suggested. These ideas have been rated as follows:

TABLE 3-13

Criteria	Product[a]			Weight (%)
	A	B	C	
Development cost	P	F	VG	10
Sales prospects	VG	E	G	15
Producibility	P	F	G	10
Competitive advantage	E	VG	F	15
Technical risk	P	F	VG	20
Patent protection	F	F	VG	10
Compatibility with strategy	VG	F	F	20
				100

[a]P, poor; F, fair; G, good; VG, very good; E, excellent.

  (a) Using an equal point spread for all five ratings (i.e., P = 1, F = 2, G = 3, VG = 4, E = 5), determine a weighted score for each product idea. What is the ranking of the three products?

(b) Rank the criteria, compute the rank-sum weights, and determine the score for each alternative. Do the same using the rank reciprocal weights.

(c) What are some of the advantages and disadvantages of this method of product selection?

3-6. Suppose that the products from Exercise 3-5 have been rated further as shown in Table 3-14.

TABLE 3-14

	Product		
	A	B	C
Probability of technical success	0.9	0.8	0.7
Probability of commercial success	0.6	0.8	0.9
Annual volume (units)	10,000	8,000	6,000
Profit contribution per unit	$6	$5	$10
Lifetime of product (years)	10	6	12
Total development cost	$50,000	$70,000	$100,000

(a) Compute the undiscounted return on investment for each product idea.

(b) Does this computation change your ranking of the products over that obtained in Exercise 3-5?

3-7. The federal government proposes to construct a multipurpose water project. This project will provide water for irrigation and for municipal uses. In addition, there will be flood control benefits and recreation benefits. The estimated project benefits computed for 10-year periods for the next 50 years are given in Table 3-15.

TABLE 3-15

Purpose	First decade	Second decade	Third decade	Fourth decade	Fifth Decade
Municipal	$ 40,000	$ 50,000	$ 60,000	$ 70,000	$110,000
Irrigation	350,000	370,000	370,000	360,000	350,000
Flood control	150,000	150,000	150,000	150,000	150,000
Recreation	60,000	70,000	80,000	80,000	90,000
	$600,000	$640,000	$660,000	$660,000	$700,000

The annual benefits may be assumed to be one-tenth of the decade benefits. The operation and maintenance cost of the project is estimated to be $15,000 per year. Assume a 50-year analysis period with no net project salvage value.

(a) If an interest rate of 5% is used, and a benefit/cost ratio of unity, what capital expenditure can be justified to build the water project now?

(b) If the interest rate is changed to 8%, how does this change the justified capital expenditure?

3-8. The state is considering the elimination of a railroad grade crossing by building an overpass. The new structure, together with the needed land, would cost $1,800,000. The analysis period is assumed to be 30 years on the theory that either the railroad or the highway above it will be relocated by then. Salvage value of the bridge (actually, the net value of the land on either side of the railroad tracks) 30 years hence is estimated to be $100,000. A 6% interest rate is to be used.

At present, about 1,000 vehicles per day are delayed due to trains at the grade crossing. Trucks represent 40%, and 60% are other vehicles. Time for truck drivers is valued at $18 per hour and for other drivers at $5 per hour. Average time saving per vehicle will be 2 minutes if the overpass is built. No time saving occurs for the railroad.

The installation will save the railroad an annual expense of $48,000 now spent for crossing guards. During the preceding 10-year period, the railroad has paid out $600,000 in settling lawsuits and accident cases related to the grade crossing. The proposed project will entirely eliminate both these expenses. The state estimates that the new overpass will save it about $6,000 per year in expenses due directly to the accidents. The overpass, if built, will belong to the state.

Perform a benefit-cost analysis to answer the question of whether the overpass should be built. If the overpass is built, how much should the railroad be asked to contribute to the state as its share of the $1,800,000 construction cost?

3-9. An existing two-lane highway between two cities is to be converted to a four-lane divided freeway. The distance between them is 10 miles. The average daily traffic (ADT) on the new freeway is forecast to average 20,000 vehicles per day over the next 20 years. Trucks represent 5% of the total traffic. Annual maintenance on the existing highway is $1,500 per lane-mile. The existing accident rate is 4.58 per million vehicle miles (MVM). Three alternative plans of improvement are now under consideration.

Plan A: Improve along the existing development by adding two lanes adjacent to the existing lanes at a cost of $450,000 per mile. It is estimated that this plan will reduce auto travel time by 2 minutes and will reduce truck travel time by 1 minute when compared to the existing highway. The plan A estimated accident rate is 2.50 per MVM. Annual maintenance is estimated to be $1,250 per lane-mile.

Plan B: Improve along the existing alignment with grade improvements at a cost of $650,000 per mile. Plan B would add two additional lanes, and it is estimated that this plan would reduce auto and truck travel time by 3 minutes each compared to the existing facility. The accident rate on this improved road is estimated to be 2.40 per MVM. Annual maintenance is estimated to be $100 per lane-mile.

Plan C: Construct a new freeway on new alignment at a cost of $800,000 per mile. It is estimated that this plan would reduce auto travel time by 5 minutes and truck travel time by 4 minutes compared to the existing highway. Plan C is 0.3 mile longer than A or B. The estimated accident rate for C is 2.30 per MVM. Annual maintenance is estimated to be $100 per lane-mile. Plan C includes the abandonment of the existing highway with no salvage value.

Useful data:

Incremental operating cost	
Autos	6 cents/mile
Trucks	18 cents/mile
Time saving	
Autos	3 cents/minute
Trucks	15 cents/minute
Average accident cost	$1,200

If a 5% interest rate is used, which of the three proposed plans should be adopted?

3-10. A 50-m tunnel must be constructed as part of a new aqueduct system for a city. Two alternatives are being considered. One is to build a full-capacity tunnel now for $500,000. The other alternative is to build a half-capacity tunnel now for $300,000 and then to build a second parallel half-capacity tunnel 20 years hence for $400,000. The cost of repair of the tunnel lining at the end of every 10 years is estimated to be $20,000 for the full-capacity tunnel and $16,000 for each half-capacity tunnel. Determine whether the full-capacity tunnel or the half-capacity tunnel should be constructed now. Solve the problem by benefit-cost ratio analysis, using a 5% interest rate and a 50-year analysis period. There will be no tunnel lining repair at the end of the 50 years.

3-11. Consider the following typical noise levels in decibels (dBA):

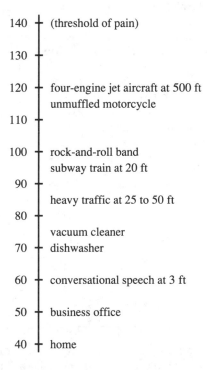

(a) Assume that you are responsible for designing a machine shop. How would you determine an acceptable level of noise? What costs and risks should you weigh?

(b) What would your answer be for the design of a commercial aircraft?

3-12. As of 1994, only a handful of patients have contracted the AIDS (acquired immune deficiency syndrome) virus from health care workers. Many, though, have called for the periodic testing of all health care workers in an effort to protect, or at least, reduce the risks to the public. Identify the costs and benefits associated with such a program. Develop an implementation plan for nationwide testing. How would you go about measuring the costs of the plan? What are the costs and risks of not testing?

3-13. As chief industrial engineer in a manufacturing facility, you are contemplating the replacement of your scheduling and inventory control system with an MRP system. A number of options are available. You can do it all at once and throw out the old system; you can phase in the new system over time; you can run both systems simultaneously, and so on. Identify the costs, benefits, and risks with each approach. Construct a decision tree for the problem.

3-14. The daily demand for a particular type of printed circuit board in an assembly shop can assume one of the following values: 100, 120, or 130 with probabilities 0.2, 0.3, and 0.5. The manager of the shop is thus limiting her alternatives to stocking one of the three levels indicated. If she prepares more boards than are needed in the same day, she must reprocess those remaining at a cost price of 55 cents/board. Assuming that it costs 60 cents to prepare a board for assembly and that each board produces a revenue of $1.05, find the optimal stocking level by using a decision tree model.

3-15. In Exercise 3-14, suppose that the owner wishes to consider her decision problem over a 2-day period. Her alternatives for the second day are determined as follows. If the demand in day 1 is equal to the amount stocked, she will continue to order the same quantity on the second day. Otherwise, if the demand exceeds the amount stocked, she will have the options to order higher levels of stock on the second day. Finally, if day 1's demand is less than the amount stocked, she will have the options to order any of the lower levels of stock for the second day. Express the problem as a decision tree and find the optimal solution using the cost data given in Exercise 3-14.

3-16. Zingtronics Corp. has completed the design of a new graphic-display unit for computer systems and is about to decide on whether it should produce one of the major components internally or subcontract it to another local firm. The advisability of which action to take depends on how the market will respond to the new product. If demand is high, it is worthwhile to make the extra investment for special facilities and equipment needed to product the component internally. For low demand it is preferable to subcontract. The analyst assigned to study the problem has produced the following information on costs (in thousands of dollars) and probability estimates of future demand for the next 5-year period:

	Future demand		
Action	Low	Average	High
Produce	$140	$120	$90
Subcontract	$100	$110	$160
Probability	0.10	0.60	0.30

  (a) Prepare a decision tree that describes the structure of this problem.
  (b) Select the best action based on the initial probability estimates for future demand.
  (c) Determine the expected cost with perfect information (that is, knowing future demand exactly).

3-17. Refer to Exercise 3-16. The management of Zingtronics is planning to hire Dr. Lalith deSilva, an economist and head of a local consulting firm, to prepare an economic forecast for the computer industry. The reliability of her forecasts from previous assignments is provided by the following table of conditional probabilities:

	Future demand		
Economic forecast	Low	Average	High
Optimistic	0.1	0.1	0.5
Normal	0.3	0.7	0.4
Pessimistic	0.6	0.2	0.1
	1.0	1.0	1.0

(a) Select the best action for Zingtronics if Dr. deSilva submits a pessimistic forecast for the computer industry.

(b) Prepare a decision tree diagram for the problem with the use of Dr. deSilva's forecasts.

(c) What is the Bayes' strategy for this problem?

(d) Determine the maximum fee that should be paid for the use of Dr. deSilva's services.

3-18. Allen Konigsberg is an expert in decision support systems and has been hired by a small software engineering firm to help plan their R&D strategy for the next 6 to 12 months. The company wishes to devote up to 3 person-years or roughly $200,000 to R&D projects. Show how Konigsberg can use a decision tree to structure his analysis. State all your assumptions.

3-19. The management of Dream Cruises, Ltd., operating in the Caribbean, has established the need for expanding its fleet capacity and is considering what the best plan for the next 8-year planning period will be. One strategy is to buy a larger 40,000-ton cruise ship now, which would be most profitable if demand is high. Another strategy would be to start with a small 15,000-ton ship now and consider buying another medium 25,000-ton ship 3 years later. The planning department has estimated the probabilities for high and low demand for each period to be 0.6 and 0.4. If the company buys the large ship, the annual profit after taxes for the next 8 years is estimated to be equal to $800,000 if demand is high and $100,000 if it is low. If the company buys the small ship, the annual profits each year will be $300,000 if demand is high and $150,000 if it is low.

After 3 years with the small vessel a decision for new capacity will be reviewed. At this time the firm may decide to expand by adding a 25,000-ton ship or by continuing with the small one. The annual profit after expansion will be $700,000 if demand is high and $120,000 if it is low.

(a) Prepare a decision tree that shows the actions available, the states of nature, and the annual profits.

(b) Calculate the total expected profit for each branch in the decision tree covering 8 years of operation.

(c) Determine the optimum fleet-expansion strategy for Dream Cruises, Ltd.

3-20. Referring to Exercise 3-19, determine the optimal fleet-expansion strategy if projected annual profits are discounted at the rate of 12%.

3-21. *Pipeline Construction Model.* The following exercise is a variation of the classical "machine setup" problem. The installation of an oil pipeline that runs from an oil field to a refinery requires the welding of 1,000 seams. Two alternatives have been specified for performing the welding: (1) use a team of ordinary and apprentice welders (B-team) only, or (2) use a team of master welders (A-team) who check and rework (as necessary) the welds of the B-team. If the the first alternative is chosen, it is estimated from past experience that 5% of the seams will be defective with probability 0.30, 10% will be defective with probability 0.50, or 20% will be defective with probability 0.20. However, if the B-team is followed by the A-team, a defective rate of 1% is almost certain. Material and labor costs are estimated at $400,000 when the B-team is used strictly, whereas these costs rise to $530,000 when the A-team is also brought in. Defective seams result in leaks that must be reworked at a cost of $1,200 per seam, which includes the cost of labor and spilled oil but ignores the cost of environmental damage.

(a) Determine the optimal decision and its expected cost. How might environmental damage be taken into account?

(b) A worker on the pipeline with a Bayesian inclination (from long years of wagering on the ponies) has proposed that management consider x-ray inspections of five randomly selected seams *following* the work of the B-team. Such an inspection would identify defective seams, which would provide management with more information for the decision on whether or not to bring in the A-team. It costs $5,000 to inspect the five seams. Financially, is it worthwhile to carry out the inspection? If so, what decision should be made for each possible result of the inspection?

3-22. A decision is to be made as to whether or not to perform a *complete audit* of an accounts receivable file. Substantial errors in the file can result in a loss of revenue to the company. However, conducting a *complete audit* is expensive. It has been estimated that the average cost of auditing one account is $6. However, if a *complete audit* is conducted, resulting in the true but *unknown* proportion p of the accounts in error being reduced, the loss of revenue may be reduced significantly. Andrew Garland, the audit manager, has the option of first conducting a *partial audit* prior to his decision on the *complete audit*. Using the *prior* probability distribution and *payoffs* given in the table below, develop a single auditing plan based on a *partial audit* of three accounts. (*Note:* Work with opportunity losses.)

Proportion of accounts in error, p	Prior probability of p, $P(p)$	Conditional cost	
		Do not audit	Complete audit
0.05	0.2	$ 1,000	$10,000
0.50	0.7	10,000	10,000
0.95	0.1	29,000	10,000

(a) Develop the opportunity loss matrix—the matrix derived from the payoff matrix (state of mature versus cost) by subtracting from each row entry, the smallest entry in its row.

(b) Structure the problem in the form of a decision tree. Specify all actions, sample outcomes, and events. Indicate opportunity losses and probabilities at all points on the tree. Show all calculations.

(c) Develop the conditional probability matrix, $P(x|p)$.

(d) Develop the joint probability matrix.

(e) Is the single auditing plan better than not conducting a partial audit?

    (1) What is the expected opportunity loss (EOL) with no partial auditing?

    (2) What is the expected value of perfect information (EVPI)? Note that EVPI is the difference between the optimal EMV under perfect information and the optimal EMV under the current uncertainty (before collecting more data).

    (3) What is the expected value of sample information (EVSI), where EVSI = EVPI − EMV? The evaluation of EMV should take into account the results of the partial audit.

    (4) State how you would determine the optimal number of partial audits in a sampling plan.

3-23. A trucking company has decided to replace its existing truck fleet. Supplier A will provide the needed trucks at a cost of $700,000. Supplier B will charge $500,000, but its vehicles may require more maintenance and repair than those from supplier A. The trucking company is also considering modernizing its maintenance and repair facility either by renovation or renovation and expansion. Although expansion is generally more expensive than renovation alone, it enables greater efficiency of repair and therefore reduced annual operating costs of the facility. The estimated costs of renovation alone and renovation and expansion, as well as the ensuing operating costs, depend on the quality of the trucks that are purchased and the extent of the maintenance they require. The trucking company has therefore decided on the following strategy: purchase the trucks now, observe their maintenance requirements for 1 year, then make the decision as to whether to renovate or to renovate and expand. During the 1-year observation period, the company will get additional information about expected maintenance requirements during years 2 through 5.

If the trucks are purchased from supplier A, first-year maintenance costs are expected to be low ($30,000) with a probability of 0.7 or moderate ($40,000) with a probability of 0.3. If they are purchased from supplier B, maintenance costs will be ($30,000) with a probability of 0.3, ($40,000) with a probability of 0.6, or high ($50,000) with a probability of 0.1. The costs of renovation, shown here, depend on the first year's maintenance experience.

One-year maintenance requirements	Renovation costs	Renovation and expansion costs
Low	$150,000	$300,000
Moderate	200,000	500,000
High	300,000	700,000

Expected maintenance costs for years 2 through 5 can best be estimated after observing the maintenance requirements for the first year (Table 3-16). Probabilities of various maintenance levels in years 2 through 5 depend on the types of trucks selected and the maintenance experience during year 1 (Table 3-17). Use decision tree analysis to determine the strategy that minimizes expected costs.

TABLE 3-16

Supplier	First-year maintenance	Renovate Maintenance years 2–5		Renovate and expand Maintenance years 2–5	
		Low	Moderate	Low	Moderate
A	Low	$100,000	$150,000	$40,000	$60,000
	Moderate	100,000	150,000	40,000	60,000
		Moderate	High	Moderate	High
B	Low	$150,000	$200,000	$50,000	$90,000
	Moderate	150,000	200,000	50,000	90,000
	High	250,000	300,000	70,000	100,000

TABLE 3-17

Supplier	First-year maintenance	Maintenance level, years 2–5		
		Low	Moderate	High
A	Low	0.7	0.3	—
	Moderate	0.4	0.6	—
B	Low	—	0.5	0.5
	Moderate	—	0.4	0.6
	High	—	0.3	0.7

REFERENCES

General Models

Augood, D., "A Review of R&D Evaluation Methods," *IEEE Transactions on Engineering Management*, Vol. EM-21, No. 4 (1973), pp. 114–120.

Baker, N. R., "R&D Project Selection Models: An Assessment," *IEEE Transactions on Engineering Management*, Vol. EM-21, No. 4 (1974), pp. 165–171.

Hobbs, B. F., "A Comparison of Weighting Methods in Power Plant Siting," *Decision Science*, Vol. 11, No. 4 (1980), pp. 725–737.

Madey, G. R., and B. V. Dean, "Strategic Planning for Investment in R&D Using Decision Analysis and Mathematical Programming," *IEEE Transactions on Engineering Management*, Vol. EM-32, No. 2 (1986), pp. 84–90.

Mandakovic, T., and W. E. Souder, "An Interactive Decomposable Heuristic for Project Selection," *Research Management*, Vol. 31, No. 10 (1985), pp. 1257–1271.

Mintzer, I., *Environmental Externality Data for Energy Technologies*, Technical Report, Center for Global Change, University of Maryland, College Park, MD, 1990.

Shachter, R. D., "Evaluating Influence Diagrams," *Operations Research*, Vol. 34, No. 6 (1986), pp. 871–882.

Souder, W. E., *Project Selection and Economic Appraisal*, Van Nostrand Reinhold, New York, 1984.

Souder, W. E., and T. Mandakovic, "R&D Project Selection Models," *Research Management*, Vol. 29, No. 4 (1986), pp. 36–42.

Benefit-Cost Analysis

Agogino, A. M, O. Nour-Omid, W. Imaino, and S. S. Wang, "Decision-Analytic Methodology for Cost–Benefit Evaluation of Diagnostic Testers," *IIE Transactions*, Vol. 24, No. 1 (1992), pp. 39–54.

Bard, J. F., "The Costs and Benefits of a Satellite-Based System for Natural Resource Management," *Socio-Economic Planning Science*, Vol. 18, No. 1 (1984), pp. 15–24.

Bordman, S. L., "Improving the Accuracy of Benefit-Cost Analysis," *IEEE Spectrum*, Vol. 10, No. 9 (September 1973), pp. 72–76.

Dicker, P. F., and M. P. Dicker, "Involved in System Evaluation? Use a Multiattribute Analysis Approach to Get the Answer," *Industrial Engineering*, Vol. 23, No. 5 (May 1991), pp. 43–73.

Mishan, E. J., *Economics for Social Decisions: Elements of Cost-Benefit Analysis*, Praeger, New York, 1973.

Newnan, D. G., *Engineering Economic Analysis*, Fourth Edition, Engineering Press, San Jose, CA, 1991.

Sprague, J. C., and J. D. Whittaker, *Economic Analysis for Engineers and Managers*, Prentice Hall, Englewood Cliffs, NJ, 1986.

Walshe, G., and P. Daffern, *Managing Cost Benefit Analysis*, Macmillan Education, London, 1990.

Risk Issues

BELL, T. E., "Special Report on Designing and Operating a Minimum-Risk System," *IEEE Spectrum*, Vol. 26, No. 6 (June 1989).

COMMITTEE ON PUBLIC ENGINEERING POLICY, *Perspectives on Benefit-Risk Decision Making*, National Academy of Engineering, Washington, DC, 1972.

DOUGHERTY, E. M., and J. R. FRAGOLA, *Human Reliability Analysis*, Wiley, New York, 1988.

HENLEY, E. J., and H. KUMAMOTO, *Reliability Engineering and Risk Assessment*, Prentice Hall, Englewood Cliffs, NJ, 1981.

KAPLAN, S., and B. J. GARRICK, "On the Quantitative Definition of Risk," *Risk Analysis*, Vol. 1, No. 1 (1981), pp. 1–23.

LOWRANCE, W. W., *Of Acceptable Risk: Science and the Determination of Safety*, William Kaufmann, Los Altos, CA, 1976.

MAKRIDAKIS, S., and S. C. WHEELWRIGHT, *Forecasting: Methods and Applications*, Wiley, New York, 1978.

SAGE, A. P., and E. B. WHITE, "Methodologies for Risk and Hazard Assessment: A Survey and Status Report," *IEEE Transactions on System, Man, and Cybernetics*, Vol. SMC-10, No. 8 (1980), pp. 425–446.

YATES, J. F. (editor), *Risk Taking Behavior*, Wiley, New York, 1991.

Decision Trees

BAIRD, B. F., *Managerial Decisions under Uncertainty: An Introduction to the Analysis of Decision Making*, Wiley, New York, 1989.

BYRD, J., JR., and L. T. MOORE, *Decision Models for Management*, McGraw-Hill, New York, 1982.

CALL, J. H., and W. A. MILLER, "A Comparison of Approaches and Implementations for Automating Decision Analysis," *Reliability Engineering and System Safety*, Vol. 30 (1990), pp. 115–162.

CANADA, J. R., and W. G. SULLIVAN, *Economic and Multiattribute Evaluation of Advanced Manufacturing Systems*, Prentice Hall, Englewood Cliffs, NJ, 1989.

HOLLOWAY, C. A., *Decision Making under Uncertainty: Models and Choices*, Prentice Hall, Englewood Cliffs, NJ, 1979.

RAIFFA, H., *Decision Analysis: Introductory Lectures on Choices under Uncertainty*, Addison-Wesley, Reading, MA, 1968.

APPENDIX 3A

Bayes' Theorem for Discrete Outcomes

For a given problem, let there be n mutually exclusive, collectively exhaustive possible outcomes $S_1,...,S_i,...,S_n$ whose prior probabilities $P(S_i)$ have been established. If the results of additional study, such as sampling or further investigation, are designated as X, where X is discrete and $P(X) > 0$, Bayes' theorem for the discrete case can be written as

$$P(S_i|X) = \frac{P(X|S_i)P(S_i)}{\sum_{i=1}^{n} P(X|S_i)P(S_i)} \qquad (3A\text{-}1)$$

The posterior probability $P(S_i|X)$ is the probability of outcome S_i given that additional study resulted in X. The probability of X and S_i occurring, $P(X|S_i)P(S_i)$, is the "joint" probability of X and S_i or $P(X, S_i)$. The sum of all the joint probabilities is equal to the probability of X. Therefore, eq. (3A-1) can be written

$$P(S_i|X) = \frac{P(X|S_i)P(S_i)}{P(X)} \qquad (3A\text{-}2)$$

A format for application is presented in Table 3A-1. The columns are as follows:

1. S_i: the potential states of nature.
2. $P(S_i)$: the estimated prior probability of S_i. (*Note:* This column sums to unity.)
3. $P(X|S_i)$: the conditional probability of getting sample or added study results X, given that S_i is the true state.
4. $P(X|S_i)P(S_i)$: the joint probability of getting X and S_i; the summation of this column is $P(X)$, which is the probability that the sample or added study results in outcome X.
5. $P(S_i|X)$: the posterior probability of S_i given that sample outcome resulted in X; numerically, the ith entry is equal to the ith entry of column (4) divided by the sum of the values in column (4). [*Note:* column (5) sums to unity.]

TABLE 3A-1   FORMAT FOR APPLYING BAYES' THEOREM

(1) State	(2) Prior probability	(3) Probability of sample outcome, X	(4) $= (2) \times (3)$ Joint probability	(5) $= (4)/\Sigma(4)$ Posterior probability $P(S_i\|X)$
S_1	$P(S_1)$	$P(X\|S_1)$	$P(X\|S_1)P(S_1)$	$P(X\|S_1)P(S_1)/P(X)$
S_2	$P(S_2)$	$P(X\|S_2)$	$P(X\|S_2)P(S_2)$	$P(X\|S_2)P(S_2)/P(X)$
.	.	.	.	.
.	.	.	.	.
.	.	.	.	.
S_i	$P(S_i)$	$P(X\|S_i)$	$P(X\|S_i)P(S_i)$	$P(X\|S_i)P(S_i)/P(X)$
.	.	.	.	.
.	.	.	.	.
S_n	$P(S_n)$	$P(X\|S_n)$	$P(X\|S_n)P(S_n)$	$P(X\|S_n)P(S_n)/P(X)$
	$\displaystyle\sum_{i=1}^{n} P(S_i) = 1.0$		$\displaystyle\sum_{i=1}^{n} P(X\|S_i)P(S_i) = P(X)$	$\displaystyle\sum_{i=1}^{n} P(S_i\|X) = 1.0$

4

Multiple-Criteria Methods for Evaluation

4.1 INTRODUCTION

It is often the case, particularly in the public sector, that goods and services are either of a collective nature, such as those for defense and space exploration, or else subsidized so that their prevailing market price is an unrealistic measure of the actual cost to the community. In these circumstances an attempt must be made to find a suitable undistorted "price." When the analysis turns to such intangible considerations as safety, health, and the quality of life, it is rarely possible to find a single variable whose direct measurement will provide a valid indicator. Often a surrogate is used. For example, a city's environmental character could be evaluated by means of an index composed of air pollution levels, noise levels, traffic flow rates, and pedestrian densities. Another index might include crime, fire alarms, and suicide rates. At the national level, it is common to cite unemployment percentages, the consumer and producer price indices, the level of the Dow Jones industrial stocks, and the amount of manufacturer inventories, as indicators of general economic well-being. In fact, each of these measures is a composite of a multitude of elements, weighted and summed together in what many would view as an arbitrary manner. A variety of procedures for doing this were presented in Chapter 3. For evaluating large, complex projects, more systematic and rational procedures are required. In this chapter we focus on methods that have been developed to bring greater rigor to the evaluation and selection process.

4.2 FRAMEWORK FOR EVALUATION AND SELECTION

The success of a project depends on a host of factors, the foremost being its ability to meet critical mission requirements. But success also depends on the likelihood that the project will remain within the planned schedule and budget, the technological opportunities it offers beyond the immediate application, and the user's perception regarding its ability to satisfy long-term organization goals. To balance each of these factors, a value model is needed. Such a model offers the decision maker a framework for conducting the underlying trade-offs.

A paradigm for any decision analysis is depicted in Fig. 4-1. In the context of project management, a decision maker must pick the most "preferred" alternative from a finite set of candidates. Here the system model may be as simple as a spreadsheet or as elaborate as a dynamic mathematical simulation. Consideration should be given to the full range of economic, technological, and political aspects of the project. Each alternative, together with the prevailing uncertainties, is fed to the system model and a particular outcome is reported.

When the uncertainties are minimal and the data are reliable, the outcomes will be fairly accurate. When uncertainty dominates, it may not be possible to develop a valid system model. The problems in which decision analysis is most effective lie somewhere between these two extremes. For example, if an advanced energy system is to be developed, certain engineering principles and experience with prototypes should give a good indication of performance. However, some uncertainties will still exist such as the cost of the system in mass production or its reliability in commercial operation.

In the decision analysis paradigm, the outcomes of the system model provide the input to the value model. The output of the latter is a statement of the decision maker's preferences in terms of a rank ordering of the outcomes or as numerical values that indicate strength of preference as well as rank.

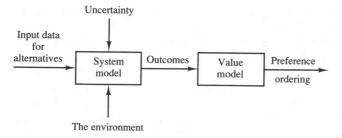

Figure 4-1   Decision analysis paradigm.

4.2.1 Objectives and Attributes[*]

The multicriteria aspect of decision analysis appears because outcomes must be evaluated in terms of several objectives (also called *goals*). These are stated in terms of properties, either desirable or undesirable, that determine the decision maker's preferences for the outcomes. For the design of an automobile, several objectives might be to (1) minimize production costs, (2) minimize fuel consumption, (3) minimize air pollution, and (4) maximize safety. The purpose of the value model is to take the outcomes of the system model, determine the degree to which they satisfy each of the objectives, and then make the necessary trade-offs to arrive at a ranking for the alternatives that correctly express the preferences of the decision maker.

The value model is developed in terms of a hierarchy of objectives and subobjectives, as shown in Fig. 4-2 for an automobile design project. To quantify the model, a unit of measurement must be assigned to the lowest members of the hierarchy. These members are called *attributes*, and may be scaled in any number of ways depending on the evaluation technique employed. In Fig. 4-2, eight attributes are used to quantify the value model. They may be represented by an eight-component vector: $x = (x_1, x_2, x_3, x_4, x_5, x_6, x_7, x_8)$. A specific occurrence of an attribute is called a *state*. An attribute state for the objective "minimize fuel consumption" might be $x_3 = 35$ miles per gallon.

Both theory and practice have shown that the set of attributes should satisfy the following requirements for the value model to be a valid and useful representation of the decision maker's preference structure.

1. *Completeness.* The set of attributes should characterize all the factors to be considered in the decision-making process.

2. *Importance.* Each attribute should represent a significant criterion in the decision-making process, in the sense that it has the potential for affecting the preference ordering of the alternatives under consideration.

3. *Measurability.* Each attribute should be capable of being objectively or subjectively quantified. Technically, this requires that it be possible to establish a utility function (see Chapter 2 for a discussion of utility functions) for the attribute.

4. *Familiarity.* Each attribute should be understandable to the decision maker in the sense that he should be able to identify preferences for different states.

5. *Nonredundancy.* No two attributes should measure the same criterion, a situation that would result in double counting.

6. *Independence.* The value model should be structured so that changes within certain limits in the state of one attribute should not affect the preference ordering for states of another attribute or the preference ordering for gambles over the states of another attribute (more will be said about this later).

---

[*]The word *attribute* is used to describe what is important in a decision problem and is often interchangeable with *objective* and *criterion*. A finer distinction can be made as follows: an *objective* represents direction of improvement or preference for one or more attributes, whereas *criterion* is a standard or rule that guides decision making.

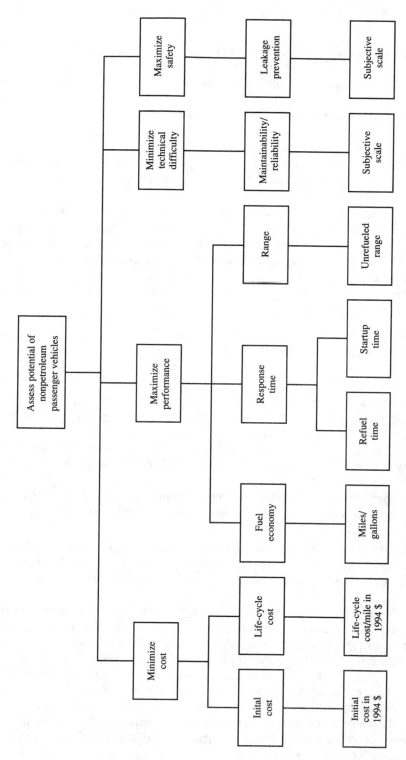

Figure 4-2 Hierarchy of objectives for advanced vehicle systems.

If an attribute does not meet these conditions, it should either be redefined by, say, dividing its range into smaller intervals and introducing "subattributes" corresponding to these intervals, or combined with other attributes until a satisfactory situation is achieved.

4.2.2 Aggregating Objectives into a Value Model

Once attributes have been assigned to all the objectives and attribute states have been determined for all possible outcomes, it is necessary to aggregate the states by constructing a single unit of measurement that will accurately represent the decision maker's preference ordering for the outcomes. This was achieved somewhat arbitrarily in Chapter 3 by specifying weights for each attribute or criterion. A more rigorous and defendable method of doing this is the "willingness to pay" or "pricing out" technique (Keeney and Raiffa 1976). Here, one attribute is singled out as the reference, preferably an attribute measured in dollars, and rates of substitution are determined for the others. Two procedures for operationalizing this concept will now be presented. Complementary techniques have been developed by Graves et al. (1992), Lewandowski and Wirezbicki (1989), and Lotfi et al. (1992), just to name a few.

4.3 MULTIATTRIBUTE UTILITY THEORY

If the set of attributes satisfies the requirements listed above, it is possible to formulate a mathematical function called a *multiattribute utility function* that will assign numbers, called *outcome utilities*, to the each outcome state. In general, the utility $U(x) = U(x_1, x_2, ..., x_N)$, of any combination of outcomes $(x_1, x_2, ..., x_N)$ for N attributes can be expressed as either (1) an additive or (2) a multiplicative function of the individual attribute utility functions $U_1(x_1), U_2(x_2), ..., U_N(x_N)$ provided that each pair of attributes is:

1. Preferentially independent of its complement; that is, the preference order of consequences for any pair of attributes does not depend on the levels at which the other attributes are held.
2. Utility independence of its complement; that is, the conditional preference for lotteries (probabilistic trade-offs) involving only changes in the levels for any pair of attributes does not depend on the levels at which the other attributes are held.

To illustrate condition 1, suppose that four attributes for a given project are profitability, time to market, technical risk, and commercial success. Preferential independence means that if we judge technological risk, for example, to be more important than profitability, this relationship is true regardless of whether the level of profitability is high, low, or somewhere in between; and also regardless of the value of the other attributes.

The second condition, utility independence, means that if we are deciding on the preference ordering (ranking) for probabilistic trade-offs between, for example, technological risk and time to market, this can be done regardless of the value of profitability. An example of preference ordering of probabilistic trade-offs between technological risk and time to market is, for instance: a 25% chance of very low risk and a 70% chance of quick time to market is preferred to, say, a 15% chance of very low risk and a 90% chance of quick time to market.

Before proceeding it is necessary to verify that these two conditions are valid, or more correctly, to test and identify the bounds of their validity. A procedure for doing this is provided by Keeney (1977). The mathematical notation used to describe the model is given below.

x_i = state of the ith attribute

x_i^0 = least preferred state to be considered for the ith attribute

x_i^* = most preferred state to be considered for the ith attribute

x = vector $(x_1, x_2, ..., x_N)$ of attribute states characterizing a specific outcome

x^0 = outcome constructed from the least preferred states of all attributes;
$x^0 = (x_1^0, ..., x_N^0)$

x^* = outcome constructed from the most preferred states of all attributes;
$x^* = (x_1^*, ..., x_N^*)$

$(x_i, \bar{x}_i^0)$ = outcome in which all attributes except the ith attribute are at their least preferred state

$U_i(x_i)$ = utility function associated with the ith attribute

$U(x)$ = utility function associated with the outcome x

k_i = scaling constant for the ith attribute; $k_i = U(x_i^*, \bar{x}_i^0)$

k = master scaling constant

Now, if the two independence conditions hold, $U(x)$ assumes the following multiplicative form:

$$U(x) = \frac{1}{k}\left\{ \prod_{i=1}^{N}[1 + k \cdot k_i U_i(x_i)] - 1 \right\} \qquad (4\text{-}1a)$$

where the master scaling constant k is determined from the equation $1 + k = \prod_i[(1 + k \cdot k_i]$. If $\Sigma_i\, k_i > 1$, then $-1 < k < 0$; if $\Sigma_i\, k_i < 1$, then $k > 0$; if $\Sigma_i\, k_i = 1$, then $k = 0$ and (4-1a) reduces to the additive form:

$$U(x) = \sum_{i=1}^{N} k_i U_i(x_i) \qquad (4\text{-}1b)$$

Because utility is a relative measure, the underlying theory permits the arbitrary assignment of $U_i(x_i^0)$ and $U_i(x_i^*) = 1$; that is, the worst outcome for each attribute is given a utility value of 0, and the best outcome is given a utility value of 1. The shape of the utility function depends on the decision maker's subjective judgment on the relative desirability of possible outcomes. A pointwise approximation of this function can be obtained by asking a series of lottery-type questions such as the following: For attribute i, what certain outcome, x_i, would be equally desirable as realizing the highest outcome with a probability p, and the lowest outcome with a probability of $(1 - p)$? This can be expressed in utility terms using the extreme values x_i^* and x_i^0 as

$$U_i(x_i = ?) = pU(x_i^*) + (1 - p)U_i(x_i^0) = p$$

To construct the curve, p can be varied in fixed increments until either a continuous function can be approximated or enough discrete points have been assessed to give an accurate picture. Alternatively, one could specify the certain outcome x_i over a range of values and ask questions such as: At what p is the certain outcome x_i equally desirable as $pU_i(x_i^*) + (1 - p)U_i(x_i^0)$? Graphically, the assessment of p can be represented as the lottery shown in Fig. 4-3.

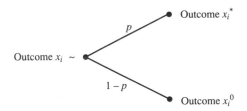

Outcome x_i^*

p

Outcome x_i ~

$1 - p$

Outcome x_i^0

Figure 4-3   Graphical assessment of indifference probability.

Example 4-1

Suppose that we want to estimate a utility function for, say, the relative fuel economy of an automobile under development (attribute 3 in Fig. 4-2). The best achievable might be 80 miles per gallon, and the worst, 20 mpg. These outcomes would give the utility function values of 1 and 0, respectively. For $p = 0.5$ (the 50–50 lottery), the question would be: How many miles per gallon as a "sure thing" would be equivalent to a gamble were there was a 50% chance of realizing 80 mpg and a 50% of realizing 20 mpg? If the answer is, say, 60 mpg, the new utility value would be calculated as:

$$U(x = 60) = 0.5U(x = 80) + 0.5U(x = 20)$$
$$= 0.5(1) + 0.5(0) = 0.5$$

Note that the utility of the certain outcome equals the probability of the best outcome. Figure 4-4 depicts the interview process. A typical utility curve that resulted from the questioning of a representative of a consumer's group is shown in Fig. 4-5 (Feinberg et al. 1985).

                                                    ■

Once utility functions for all attributes have been determined, the next step is to assess the scaling constants, k_i. For both the multiplicative (4-1a) and additive (4-1b) models, $k_i = U(x_i^*, \bar{x}_i^0)$, where $0 \le k_i \le 1$. That is, k_i is the utility value associated with the outcome where attribute i is at its best value, x_i^* and all other attributes are at their worst values, $\bar{x}_i^0$. In assessing the k_i's, the following type of question is usually asked. For what probability p are you indifferent between:

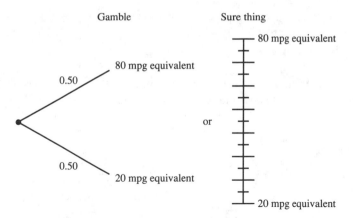

• For which value of the "sure thing" are you indifferent between the "sure thing" and the"gamble"?

    Indifference point _____

• If you knew that all other attributes were at their worst state?

    Indifference point _____

• If you knew that all other attributes were at their best state?

    Indifference point _____

Figure 4-4   Sample interview question for relative fuel economy.

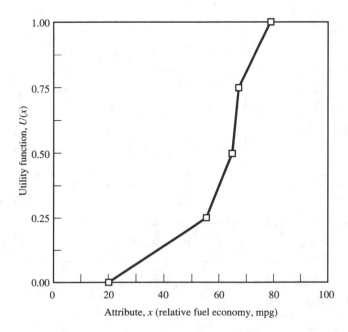

Figure 4-5   Example of utility curve for representative consumer.

1. The lottery giving a p chance at $x^* \equiv (x_1^*,...,x_N^*)$ and a $(1 - p)$ chance at $x^0 \equiv (x_1^0,..., x_N^0)$ versus

2. The consequence $(x_1^0,...,x_{i-1}^0, x_i^*, x_{i+1}^0,..., x_N^0)$.

The interview sheet used for determining the scaling constant associated with relative fuel economy is shown in Fig. 4-6 (the responses to the last two questions give an in-

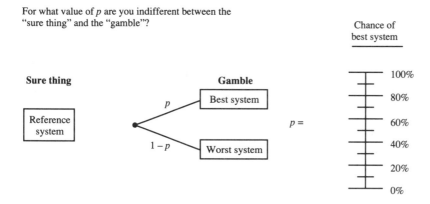

		Relative fuel economy	Initial cost	Life-cycle cost/mile	Maintain-ability	Safety	Refuel time	Unre-fueled range	Maximum startup time
Ref:		80 mpg equivalent							
			$25,000	$1.00/mile	0	0	8 hours	50 miles	600 seconds
Best:		80 mpg equivalent	$5,000	$0.20/mile	10	10	0.17 hour	250 miles	5 seconds
Worst:									
		20 mpg equivalent	$25,000	$1.00/mile	0	0	8 hours	50 miles	600 seconds

Figure 4-6   Sample interview question used to determine scaling constant for the relative fuel economy attribute.

dication of the degree to which the independent conditions hold). The result of the assessment is that, in general, $k_i = p$. Good practice suggests that prior to assessing the scaling constants, the attributes should be ranked in ascending order of importance as they progress from their worst to their best states. Figure 4-7 displays the question sheet that was used for this purpose.

The last step in the evaluation and selection process is to rank the alternatives. This is done by using the multiattribute utility function to calculate outcome utilities for each alternative under consideration. If two or more alternatives appear to be close in rank, their sensitivity to both the scaling constants and utility functions should be examined. The appendix at the end of this chapter contains a more detailed example of the evaluation process.

A final point to make about multiattribute utility theory (MAUT) concerns the possibility that the state of an attribute may be uncertain. "Completion time of a task," "reliability of a subassembly," and "useful life of the system" are some examples of attributes whose states may take on different values with known (or more distressingly, with unknown) probability. In these cases, x_i is really a random variable, so it is more appropriate to compute the *expected* utility of a particular outcome. For the additive model, this can be done with the following equation:

$$E[U(x)] = \sum_{i=1}^{N} \left[k_i \int_{-\infty}^{\infty} U_i(x_i) f_i(x_i) d(x_i) \right] \qquad (4\text{-}2)$$

where $f_i(x_i)$ is the probability density function associated with attribute i, and $E[\cdot]$ is the expectation operator (Keeney and von Winterfeldt 1991). Commercial software (e.g., Smith et al. 1987) is available for helping in the assessment of f_i, as well as the scaling constants k_i and the individual utility functions U_i.

Attribute	Relative fuel economy	Initial cost	Life-cycle cost/mile	Maintain-ability	Safety	Refuel time	Unre-fueled range	Maximum startup time
Best state	80 miles per gallon equivalent	$5,000	$0.20/mile	10	10	0.17 hour (10 min)	250 miles	5 seconds
Worst state	20 miles per gallon equivalent	$25,000	$1.00/mile	0	0	8.0 hours	50 miles	600 seconds (10 min)

Order of importance								

Figure 4-7 Sample interview question used to determine order of importance of attributes.

4.4 ANALYTIC HIERARCHY PROCESS

The analytic hierarchy process (AHP) was developed by Thomas Saaty (1980) to provide a simple but theoretically sound multiple-criteria methodology for evaluating alternatives. Applications can be found in such diverse fields as portfolio selection, transportation planning, manufacturing systems design, and artificial intelligence, just to name a few. The strength of the AHP lies in its ability to structure a complex, multiperson, multiattribute problem hierarchically, and then to investigate each level of the hierarchy separately, combining the results as the analysis progresses. Pairwise comparisons of the factors (which, depending on the context, may be alternatives, attributes, or criteria) are undertaken using a scale indicating the strength with which one factor dominates another with respect to a higher-level factor. This scaling process can then be translated into priority weights or scores for ranking the alternatives.

Like MAUT, the AHP starts with a hierarchy of objectives. The top of the hierarchy provides the analytic focus in terms of a problem statement. At the next level, the major considerations are defined in broad terms. This is usually followed by a listing of the criteria for each of the foregoing considerations. Depending on how much detail is called for in the model, each criterion may then be broken down into individual parameters whose values are either estimated or determined by measurement or experimentation. The bottom level of the hierarchy contains the alternatives or scenarios underlying the problem.

Figure 4-8 shows a three-level hierarchy developed for evaluating five different approaches to assembling the U.S. Space Station while on orbit. The focus of the problem is "Selecting an on-orbit assembly system," and the four major criteria are human productivity, economics, design, and operations. The five alternatives include an astronaut with

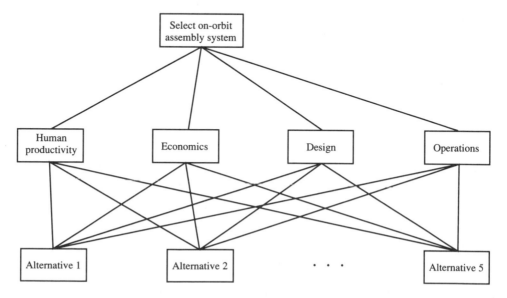

Figure 4-8   Summary three-level hierarchy for selection problem.

tools outside the spacecraft, a dexterous manipulator under human control, a dedicated manipulator under computer control, a teleoperator maneuvering system with a manipulator kit, or a computer-controlled dexterous manipulator with vision and force feedback. In fact, in the full analysis, each of the criteria at level 2 were significantly expanded to capture the detail necessary to make accurate comparisons (Bard 1986). With regard to human productivity, workload, support requirements, crew acceptability, and issues surrounding human–machine interfaces were some of the additional factors included. Figure 4-9 depicts the full portion of the hierarchy used for this criterion.

4.4.1 Determining Local Priorities

Once the hierarchy has been structured, *local priorities* must be established for each factor on a given level with respect to each factor on the level immediately above it. This step is carried out by using pairwise comparisons between the factors to develop the relative weights or priorities. Because the approach is basically qualitative, it is arguably less burdensome to implement from both a data requirement and validation point of view than by using the multiattribute utility approach of Keeney and Raiffa. That is, not all the MAUT independence conditions need be verified nor preference functions derived. Nevertheless, theory requires that the following assumptions, stated in terms of axioms, hold if the methodology is to be valid (Golden et al. 1989).

Axiom 1.   Given any two alternatives (or subcriteria) i and j from the set of alternatives $\mathcal{A}$, the decision maker is able to provide a pairwise comparison a_{ij} of these alternatives under criterion c from the set of criteria $\mathcal{C}$ on a reciprocal ratio scale; that is,

$$a_{ji} = \frac{1}{a_{ij}} \qquad \text{for all } i, j \in \mathcal{A}$$

Axiom 2.   When comparing any two alternatives $i,j \in \mathcal{A}$, the decision maker never judges one to be infinitely better than another under any criterion $c \in \mathcal{C}$; that is, $a_{ij} \neq \infty$ for all $i,j \in \mathcal{A}$.

Axiom 3.   The decision problem can be formulated as a hierarchy.

Axiom 4.   All criteria and alternatives that have an impact on the given decision problem are represented in the hierarchy. That is, all the decision maker's intuition must be represented (or excluded) in the structure in terms of criteria or alternatives.

These axioms above can be used to describe the two basic tasks in the AHP: formulating and solving the problem as a hierarchy (3 and 4), and eliciting judgments in the form of pairwise comparisons (1 and 2). Such judgments represent an articulation of the trade-offs among the conflicting criteria and are often highly subjective in nature. Saaty suggests that a 1:9 ratio scale be used to quantify the decision maker's strength of feeling between

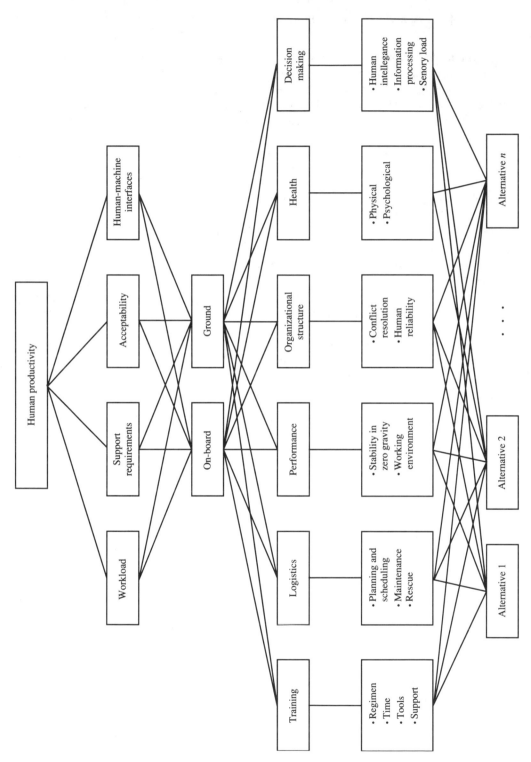

Figure 4-9 Human productivity objective hierarchy.

any two alternatives with respect to a given criterion. An explanation of this scale is presented in Table 4-1. Depending on the context, the word *factors* means either alternatives, attributes, or criteria. We also note that because a ratio scale is being used, the derived weights can be interpreted as the degree to which one alternative is preferred to another. The same cannot be said of the MAUT rankings, which have only a qualitative meaning.

TABLE 4-1   SCALE USED FOR PAIRWISE COMPARISONS

Value	Definition	Explanation
1	Equal importance	Both factors contribute equally to the objective or criterion.
3	Weak importance of one over another	Experience and judgment slightly favor one factor over another.
5	Essential or strong importance	Experience and judgment strongly favor one factor over another.
7	Very strong or demonstrated importance	A factor is favored very strongly over another; its dominance is demonstrated in practice.
9	Absolute importance over another	The evidence favoring one factor is unquestionable.
2,4,6,8	Intermediate values	Used when a compromise is needed.
0	No relationship	The factor does not contribute to the objective.

Example 4-2

To illustrate the nature of the calculations, observe the three-level hierarchy in Fig. 4-8. Table 4-2 contains the input and output data for level 2.

When n factors are being compared, $n(n - 1)/2$ questions are necessary to fill in the matrix A. The elements in the lower triangle are simply the reciprocal of those lying above the diagonal (i.e., $a_{ji} = 1/a_{ij}$, in accordance with Axiom 1) and need not be assessed. In this instance, the entries in the matrix at the center of Table 4-2 are the responses to the 6 ($n = 4$) pairwise questions that were asked. For example, in comparing "human productivity" with "economic" considerations (element a_{12} of the matrix), it was judged that the first "weakly" dominates the second. Note that if the elicited value for this element were $\frac{1}{3}$ instead of 3, the opposite would have been true. Similarly, the value 7 for element a_{34} means that design considerations "very strongly" dominate those associated with operations.

TABLE 4-2   PRIORITY VECTOR FOR MAJOR CRITERIA

Criteria	Criteria 1	2	3	4	Priority weight	Output parameters
1. Human productivity	1	3	3	7	0.521	$\lambda_{max} = 4.121$
2. Economics	0.333	1	1	5	0.205	CI = 0.040
3. Design	0.333	1	1	7	0.227	CR = 0.045
4. Operations	0.143	0.2	0.143	1	0.047	

■

In general, when comparing two factors, the analyst first discerns which factor is more important and then ascertains by how much by asking the decision maker to select a value from the 9-point scale. After the decision maker supplies all the data for the matrix, the following equation is solved to obtain the rankings denoted by w:

$$Aw = \lambda_{max}w \qquad (4\text{-}3)$$

where w is the n-dimensional eigenvector associated with the largest eigenvalue λ_{max} of the comparison matrix, A. The n components of w are then scaled so they sum to 1.

In practice, the priority vector $w = (w_1, w_2, ..., w_n)$ is obtained by raising the matrix A to an arbitrarily large power (16 or greater is usually sufficient). In so doing each element in a given row i converges to the same value, call it v_i. The weights are then computed as follows:

$$w_i = \frac{v_i}{\sum_{k=1}^{n} v_k}, \qquad i = 1, ..., n$$

The value of λ_{max} can be found by solving each row of eq. (4-3) for λ and averaging; that is, let λ_i be the solution to $Aw = \lambda_i w_i$, where A_i is the ith row of A. Then $\lambda_{max} = \frac{1}{n}\sum_{i=1}^{n} \lambda_i$. It should be noted that this procedure works only for the class of positive reciprocal matrices to which A belongs.

4.4.2 Checking for Consistency

Consistency of response or transitivity of preference is checked by ascertaining whether

$$a_{ij} = a_{ik}a_{kj} \qquad \text{for all } i,j,k \qquad (4\text{-}4)$$

In practice, the decision maker is only estimating the "true" elements of A by assigning them values from Table 4-1, so the perfectly consistent case represented by eq. (4-4) is not likely to occur.

Therefore, as an approximation, the elements of A can be thought to satisfy the relationship $a_{ij} = w_i/w_j + \varepsilon_{ij}$, where ε_{ij}, is the error term representing the decision maker's inconsistency in judgment when comparing factor i to factor j. As such, we would no longer expect a_{ij} to equal $a_{ik}a_{kj}$ throughout. Carrying the analysis one step farther, it can be shown that the largest eigenvalue of the matrix A, λ_{max}, satisfies $\lambda_{max} \geq n$, where equality holds for the perfectly consistent case only. This leads to the definition of a consistency index

$$CI = \frac{\lambda_{max} - n}{n - 1} \qquad (4\text{-}5)$$

which can be used to evaluate the quality of the matrix A. To add perspective we compare the CI to the index derived from a completely arbitrary matrix whose entries are randomly chosen. Through simulation, Saaty has obtained the following results:

n	1	2	3	4	5	6	7	8	9	10
RI	0.00	0.00	0.58	0.90	1.12	1.24	1.32	1.41	1.45	1.49

where n represents the dimension of the particular matrix and RI denotes the random index computed from the average of the CI for a large sample of random matrices. It is now possible to define the consistency ratio as

$$CR = \frac{CI}{RI}$$

Experience suggests that the CR should be less than 0.1 if one is to be fully confident of the results. (There is a certain amount of subjectivity in this assertion, much like that associated with interpreting the coefficient of determination in regression analysis.) Fortunately, though, as the number of factors in the model increases, the results become less and less sensitive to the values in any one matrix.

Returning to Table 4-2, it can be seen that the priorities derived for the major considerations were 0.521 for human productivity, 0.205 for economics, 0.227 for design, and 0.047 for operations. These values tend to emphasize the first criterion over the others, probably due to the implicit mandate that the U.S. Space Station must eventually pay for itself. Finally, note that the consistency ratio, 0.045, is well within the acceptable range.

4.4.3 Determining Global Priorities

The next step in the analysis is to develop the priorities for the factors on the third level with respect to those on the second. In our case we compare the five alternatives previously mentioned with each of the major criteria. For the moment, assume that the appropriate data have been elicited and that the calculations for each of the four comparison matrices have been performed, with the results displayed in Table 4-3 (note that each column sums to 1). The first four columns of data represent the local priorities derived from the inputs

TABLE 4-3   LOCAL AND GLOBAL PRIORITIES FOR THE PROBLEM OF SELECTING AN ON-ORBIT ASSEMBLY SYSTEM

	Local priorities				
Alternative[a]	Human productivity (0.521)	Economics (0.205)	Design (0.227)	Operations (0.047)	Global priorities
1	0.066	0.415	0.122	0.389	0.165
2	0.212	0.309	0.224	0.151	0.232
3	0.309	0.059	0.206	0.178	0.228
4	0.170	0.111	0.197	0.105	0.161
5	0.243	0.106	0.251	0.177	0.214

[a]1, Astronaut with tools outside the spacecraft; 2, dexterous manipulator under human control; 3, dedicated manipulator under computer control; 4, teleoperator with manipulator kit; 5, dexterous manipulator with sensory feedback.

supplied by the decision maker. The *global priorities* are obtained by weighting each of these values by the local priorities given in Table 4-2 (and repeated at the top of Table 4-3 for convenience), and summing. The calculations for alternative 1 are as follows: $(0.066)(0.521) + (0.415)(0.205) + (0.122)(0.227) + (0.389)(0.047) = 0.165$. To see how the calculations are performed in general, let:

n_l = number of factors at level l

w_i^l = global weight at level l for factor i

w_{ij}^l = local weight at level l for factor i with respect to factor j at level $l - 1$

The global priorities at level l are obtained from the following equation:

$$w_i^l = \sum_{i=1}^{n_{l-1}} w_{ij}^l w_j^{l-1}$$

Continuing with the example, because there are no more levels left to evaluate, the values shown in the last column of Table 4-3 represent the final priorities for the problem. Thus according to the judgments expressed by this decision maker, alternative 2 turns out to be most preferred.

To complete the analysis, it would be desirable to see how sensitive the results are to changes in judgment and criteria values, that is, to determine how changes in the *A* matrix would affect intralevel and overall priorities and consistencies. This feature is built into *Expert Choice* (Forman et al. 1983), the most popular commercial code for conducting an AHP analysis, and so can be done with little effort. *HIPRE 3+* (Hamalainen and Lauri 1992) also provides this capability. When uncertainty exists in factor values, additional attributes can be defined to account for this randomness (Bard 1992).

In summary, the commonly claimed benefits of the AHP are that:

1. It is simple to understand and use.
2. The construction of the objective hierarchy of criteria, attributes and alternatives facilitates communication of the problem and solution recommendations.
3. It provides a unique means of quantifying judgment and measuring consistency.

4.5 GROUP DECISION MAKING

When more than one person is responsible for making decisions, the issues surrounding group dynamics and consensus building become paramount. Rational procedures must be developed for structuring the problem, soliciting opinions, and making use of the information provided. In general, there are two modes of operation: live sessions and some form of correspondence. In the former, the group takes time to structure its problem, usually weighing all factors and considering all inputs. Still there is a need to trim the structure and eliminate redundancies so that the major effort can be brought to bear on the essential parts of the problem.

With regard to judgments, Saaty (1982) points out that there are four kinds of situations.

1. People are completely antagonistic to the process and do not wish to participate in a constructive way, as they may feel the outcome would dilute their own influence.
2. The participants wish to cooperate to arrive at a rational decision, and in so doing, wish to determine every judgment by agreement and consensus.
3. The group is willing to have their individual judgments synthesized after some debate.
4. The group consists of experts each of whom knows his mind exactly and does not wish to interact. They are willing to accept an outcome but are not willing to compromise on their judgments.

After the session in which the substance is hammered out, the group may be willing to revise their structure and judgments by conducting additional sessions, or by correspondence using questionnaires.

The second alternative is to do the entire process by correspondence without organized meetings. The question here is how to solicit opinions and interact most effectively. The Delphi method is one particular approach for doing this that has gained strong adherents.

In the last few years, several researchers (e.g., DeSanctis and Gallupe 1987, Huber and McDaniel 1986) have pointed to the following trends in decision making:

1. Organizational decisions are much more technically and politically complex and require frequent meetings attended by a wide range of people.
2. Decisions must be reached quickly, usually with greater participation of low-level or staff personnel than in the past.
3. There is an increasing focus on the development of computer-based systems that support the formulation and solution of unstructured decision problems by a group [i.e., a group decision support system or (GDSS)].

In what follows, we highlight some of the important considerations in the group decision-making process.

4.5.1 Group Composition

The inherent complexity and uncertainty surrounding an organization's major activities usually necessitates the participation of many persons in the decision-making process. In some cases, the composition of the group is fixed (such as the board of directors advising the chief executive officer of a corporation), whereas in others, it is necessary to select a mix of members (such as choosing a panel to investigate the *Challenger* disaster). The latter selection process requires specifying the number of experts, nonexperts, staff personnel, and upper-level managers to participate, as well as choosing the appropriate people.

This process can be difficult and time consuming for many reasons. First, participants who are considered "experts" are likely to be troublesome. They may have strong ideas on the appropriate course of action and may not be easily swayed in their assessments. Second, decision makers who are considered "powerful" members of the organization might refuse to participate. These members are aware that their level of control and influence might be diminished in a group setting. They fear that the social and interactive nature of the group process might dilute their power and ability to direct policy within the organization (Saaty 1989). However, if powerful persons actively participate, they are likely to dominate the process. In contrast, results generated by a group that consists solely of "low-level" managers with little power may not be useful. The danger in all this is that powerful managers will implement their preferred solutions without taking into account the opinions and observations of others.

One way of dealing with the "power differential" problem is to assemble a group of participants who have equal responsibility and stature within the organization. Collectively, these people can be treated as a decision-making "subgroup" that could help formulate and solve a part of the problem with which they are most knowledgeable. They could also contribute to discussions that involve higher or lower levels of management. This can be viewed as a sort of "shared" decision-making responsibility in which high-level management cooperates with subordinates. In practice, high-level management often depends on low-level employees to gather the appropriate information on which to base their decisions.

4.5.2 Running the Decision-Making Session

After the group has been chosen, the members should begin preparing for the decision-making session by formalizing their agenda, structuring the allowable interactions between participants, and clearly defining the purpose of the session in advance. They can seek answers to several questions (such as the ones listed below) that are designed to establish the operating ground rules.

- Is the purpose of the session simply to improve the group's understanding of the problem, or is the purpose to reach a final solution?
- Are the participants committed to generating and implementing a final solution?
- What is the best way to combine the judgments of the participants on various issues in order to produce a united course of action?

Often we model decision problems as if the people we are dealing with know their minds and can give answers inspired by a clear or telling experience. But this is seldom the case. People have a habitual domain. They are conditioned, biased, but also learning and adaptive. Instead of trying to cajole or coerce them prematurely, they must be given the opportunity to learn and solidify their ideas. After much experimentation and trial and error, something useful may emerge. If you hurry, all you get is a hurried answer, no matter how scientific you try to be. People must be given an adequate chance to understand their own minds before they can be expected to commit themselves. Persons with different assumptions and different backgrounds, though, may never be on the same wavelength, and will change their minds later if they are forced to agree. Moreover, interpersonal comparisons should only be undertaken with the utmost of care. Peer pressure, concealed and distorted preferences, and the inequalities of power all conspire to prejudice the group decision-making process.

4.5.3 Implementing the Results

After the final results have been generated, the group should evaluate the effort and cost of implementing the highest-priority outcome. It must be determined whether it is likely that the participants and their constituencies will cooperate in the implementation phase of the effort. To be useful, the decision-making process must be acceptable to the participants and they must be willing to abide by the outcome. Finally, it is important for the group to view whichever GDSS was used, not as a tool for isolated, one-time applications, but rather as a process that has ongoing validity and usefulness to an organization.

4.5.4 Group Decision Support Systems (GDSS)

A GDSS aims to improve the process of group decision making by removing common communications barriers, providing techniques for structuring decision analysis, and systematically directing the pattern, timing, and content of the discussion. The more sophisticated the GDSS technology, the more dramatic is the intervention into the group's natural (unsupported) environment. Of course, more dramatic intervention does not necessarily lead to better decisions, but its appropriate design and use can produce the desired effects.

Communications technologies available within a GDSS include electronic messaging, local- and wide-area networks, teleconferencing, and store-and-forward facilities. Computer technologies include multiuser operating systems, fourth-generation languages, databases, data analysis methodologies, and so on. Decision support technologies include agenda setting, decision modeling methods (such as decision trees, risk assessment, forecasting techniques, the AHP, and MAUT), and rules for directing discussion.

Concerning the information-exchange aspect of group decision making, DeSanctis and Gallupe (1987) have proposed three levels of support. Level 1 GDSSs provide technical features aimed at removing communications barriers, such as large screens for instantaneous display of ideas, voting solicitation and compilation, anonymous input of ideas and preferences, and electronic message exchange between members. Level 1 features are found in meeting rooms normally referred to as "computer-supported conference rooms" or "electric board rooms."

Level 2 GDSSs provide decision modeling and group decision techniques designed to reduce the uncertainty and "noise" that occur in the group decision process. The result is an enhanced GDSS, as opposed to a level 1 system, which is a communications medium only. A level 2 GDSS might provide automated planning tools or other aids found in individual DSSs for group members to work on and view simultaneously, again using a large common screen. Modeling tools to support analyses that ordinarily are performed in a qualitative fashion, such as social judgment formation, risk assessment, and multiattribute utility methods, can be introduced to the group via a level 2 GDSS. In addition, group structuring techniques found in the organizational development literature can be administered efficiently.

Level 3 GDSSs are characterized by machine-induced group communication patterns and can include expert advice in the selecting and arranging of rules to be applied during a meeting. As an example, Hiltz and Turoff (1985) have experimented with automating the Delphi method and the nominal group technique. But to date, very little research has been done with such high-level systems.

In summary, the objective of GDSSs is to discover and present new possibilities and approaches to problems. They do this by facilitating the exchange of information among the group. Message transfer can be hastened and smoothed by removing barriers (level 1); systematic techniques can be used in the decision process (level 2); and rules for controlling pattern, timing, and content of information exchange can be imposed on the group (level 3). The higher the level of the GDSS, the more sophisticated the technology and the more dramatic the intervention compared to the natural decision process. Table 4-4 highlights the major tasks of a decision-related meeting, the main activities, the corresponding level of GDSS, and the possible support features.

TABLE 4-4   EXAMPLE GDSS FEATURES TO SUPPORT SIX TASK TYPES

Task purpose	Task type	GDSS level	Possible support features
General	Planning	Level 1	Large screen display, graphical aids
		Level 2	Planning tools (e.g., PERT); risk assessment, subjective probability estimation for alternative plans
	Creativity	Level 1	Anonymous input of ideas, pooling and display of ideas; search facilities to identify common ideas, eliminate duplicates
		Level 2	Brainstorming; nominal group technique
Choose	Objective	Level 1	Data access and display; synthesis and display of rationales for choices
		Level 2	Aids to finding the correct answer (e.g., forecasting models, multiattribute utility models)
		Level 3	Rule-based discussion emphasizing thorough explanation of logic
	Preference	Level 1	Preference weighting and ranking with various schemes for determining the most favored alternative; voting schemes
		Level 2	Social judgment models; automated Delphi method
		Level 3	Rule-based discussion emphasizing equal time to present opinion
Negotiate	Cognitive conflict	Level 1	Summary and display of members' opinions
		Level 2	Using social judgment analysis, each member's judgments are analyzed by the system and then used as feedback to the individual member or the group
		Level 3	Automatic mediation; automate Roberts' rules
	Mixed-motive	Level 1	Voting solicitation and summary
		Level 2	Stakeholder analysis
		Level 3	Rule base for controlling opinion expression; automatic mediation; automate parliamentary procedure

TEAM PROJECT

Thermal Transfer Plant

TMS management is considering the following aspects in selecting a hydraulic power unit for the rotary combustor:

- Size
- Weight
- Power consumption
- Required maintenance
- Noise
- Cost
- Reliability

The power unit provides power to operate three components of the system: feed rams, resistance door, and combustor. Three design alternatives are available:

1. Electric motor on a gearbox
2. Low-speed, high-torque hydraulic motor with direct drive
3. High-speed, low-torque hydraulic motor on a gearbox

Initial data include:

	Electromechanical	Low speed, high torque	High speed, low torque
Delivery	90–120 days	1–6 weeks	90–120 days
Overall efficiency	96%	94%	88%
Useful life	20 years	25 years	25 years
Noise level	85 dB	78 dB	100 dB

Using the criteria above as guidance, develop a MAUT and an AHP model for evaluating the three alternatives. It will be necessary to collect data or make assumptions about the values of all the attributes. For one of the models, perform the analysis with the help of a computer program, and give your recommendation. Be sure to justify and document your results, basing part of your recommendation on a sensitivity analysis.

DISCUSSION QUESTIONS

1. How might you measure the benefits associated with space exploration or the Superconducting Super Collider? Can you put a dollar value on these benefits? What are the real costs and opportunity costs of these types of projects?

2. Identify an advanced technology project that you feel should be undertaken, such as high-definition TV or coal gasification. Who should be responsible for funding the project? the government? industry? a consortium? What are the major attributes or criteria associated with the project?

3. What type of technical background, if any, do you think is needed to understand MAUT? the AHP?

4. You have just completed a MAUT evaluation of a number of data communications systems under consideration by your company. How would you present the results to upper management? Assuming that they know nothing about the technique, how much background would you give them? How would your answer differ if the AHP were used instead?

5. What do you think are the strengths and weaknesses of the AHP and MAUT?

6. How would you go about constructing an objective hierarchy? Who should be consulted? Identify a project from your personal experience or observations, and construct such a hierarchy.

7. When performing an evaluation using any multiple-criteria method, from whose perspective should the analysis be undertaken? Would the answer differ if it were a public rather than a private project?

8. What experiences have you had with group decision making? What difficulties do you see arising when trying to perform a multiple-criteria analysis with many interested parties involved? How might these difficulties be overcome, or at least mitigated?

9. Are benefit-cost analysis and multiple-criteria analysis mutually exclusive techniques? In which circumstances is either most appropriate?

10. You just inherited a large sum of money and would like to develop an investment strategy for it. Use the AHP to fashion such a strategy. Construct an objective hierarchy listing all criteria and subcriteria, and principal alternatives. What data are needed to perform the evaluation? How would you go about obtaining the data?

11. From a practical point of view, how would you verify the independence assumptions associated with MAUT?

12. Are the axioms underlying the AHP reasonable and unambiguous? In which circumstances do you think one or more of them could be relaxed?

13. Both the AHP and MAUT are value models that facilitate making trade-offs between incommensurable criteria. Can you come up with your own value model or procedure for doing this?

14. In conducting a group study using a multiple-criteria method, you reach a point where two of the participants cannot agree on a particular response. What course of action would you take to placate the parties and avoid further delay?

15. For which type of projects or problems might MAUT be more amenable than the AHP? Similarly, when is the AHP more appropriate than MAUT?

EXERCISES

4-1. Assume that you work for a company that designs and fabricates VLSI chips. You have been given the job of selecting a new computer-aided design software package for the engineering group.

(a) Develop a MAUT model to assist in the selection process.

(b) Develop an AHP model to assist in the selection process.

In both cases, begin by enumerating the major criteria and the associated subcriteria. Explain your assumptions. Who are the possible decision makers? How do you think the outcome of the analysis would change with each of these decision makers?

4-2. Develop a flowchart detailing input, output, and processes for a software package that supports:

(a) MAUT applications

(b) AHP applications

4-3. Using MAUT and the AHP, perform an analysis to select a graduate program. Explain your assumptions and indicate which technique you feel is most appropriate for this application.

4-4. You are the vice-president of planning for Zingtronics, a small-scale manufacturer of IBM-compatible personal computers and peripherals based in Silicon Valley. Business is growing and the company would like to open a second facility. Three options are being considered: (1) a second plant in Silicon Valley, (2) a new plant in Mexico as a Maquiladora, and (3) a new plant in Singapore. Most of the workforce will be low-skilled assembly and machine operators, but training in the use of computers and information systems will be required. It is also desirable to set up a small design group of engineers for new product and process development. Of course, each option has its pros and cons. For example, Silicon Valley has a high-skill labor pool but is a very expensive place to do business. Singapore offers the same level of worker skills at lower cost but is distant from the market and headquarters. Mexico is the least expensive place to set up a business, due to favorable tax laws and cheap labor, but has a less educated workforce. Develop two objective hierarchies, one for costs and one for benefits, that can be used to investigate the location problem. Use the AHP to rank the three alternatives on both hierarchies, and then compute the benefit/cost ratios of each. According to your analysis, which alternative is best?

4-5. Referring to Exercise 4-4, combine the two hierarchies into one so that there are no more than eight subobjectives at the bottom level. Define either a quantitative or a qualitative scale for each of these subobjectives and construct a utility function for each. Use MAUT to evaluate and rank the three alternatives.

4-6. Use the criteria below to construct a two-level objective hierarchy (major criteria with one set of subcriteria under each) to help evaluate political candidates. Consider as alternatives the major candidates running in the last U.S. presidential election, and use the AHP to make your choice.

Criteria for choosing a national political candidate:

- *Charisma:* personal leadership qualities inspiring enthusiasm and support
- *Glamor:* charm, allure, personal attractiveness; associations with other attractive people
- *Experience:* past officeholding relevant to the position sought; preparation for the position
- *Economic policy:* coherence and clarity of a national economic policy
- *Ability to manage international relations:* coherence and clarity of foreign policy plus ability to deal with foreign leaders
- *Personal integrity:* quality of moral standards, trustworthiness
- *Past performance:* quality of role fulfillment—independent of what the role was—in previous public offices; public record
- *Honesty:* lawfulness in public life, law-abidingness

4-7. Louise Ciccone, head of industrial engineering for a medium-sized metalworking shop, wants to move the CNC machines from their present location to a new area. Three distinct alternatives are under consideration. After inspecting each alternative and determining which factors reflect significant differences among the three, Louise has decided on five independent attributes to evaluate the candidates. In *descending order of importance*, they are:

A. Distance traveled from one machine to the next (more distance is worse)
B. Stability of foundation [strong (excellent) to weak (poor)]
C. Access to loading and unloading [close (excellent) to far (poor)]
D. Cost of moving the machines
E. Storage capacity

(*Note:* Once the machines have been moved, operational costs are independent of the area chosen and hence are the same for each area.) The data associated with these factors for the three alternatives are shown in Table 4.5. Using the multiattribute utility methodology, determine which alternative is best. For at least one attribute, state all the probabilistic trade-off (lottery-type) questions that must be asked together with answers to obtain at least four utility values between the "best" and "worst" outcomes so that the preference curve can be plotted. For the other attributes you may make shortcut approximations by determining if each is concave or convex, upward or downward, and then sketching an appropriate graph for each. Next, ask questions to determine the scaling constants k_i, and compute the scores for the three alternatives. [*Note:* If you follow the recommended procedure for deriving the scaling constants, probably $\Sigma_i\, k_i \neq 1.0$, so you should use the multiplicative model (4-1a). After comparing alternatives by that model, "normalize" the scaling constants so $\Sigma_i\, k_i \neq 1.0$, and then compare the alternatives using the additive model (4-1b). (It is not theoretically correct to normalize the k_i values to enable use of the additive model.) How much difference does use of the "correct" model make?]

TABLE 4-5

Attribute	Area I	Area II	Area III	Ideal	Standard	Worst
		Alternative				
A	500 ft	300 ft	75 ft	0 ft	300 ft	1,000 ft
B	Good	Very good	Good	Excellent	Good	Poor
C	Excellent	Very good	Good	Excellent	Good	Poor
D	$7,500	$3,000	$8,500	$0	$5,000	$10,000
E	60,000 ft^2	85,000 ft^2	25,000 ft^2	10,000 ft^2	25,000 ft^2	150,000 ft^2

4-8. Starting with the environmental scoring model in Table 3-3, construct an objectives hierarchy that can be used to evaluate capital development and expansion projects being considered by an electric utility company.

4-9. The six major objectives listed below are used by the British Columbia Hydro and Power Authority to evaluate new projects. Use this list to construct an objectives hierarchy by providing subobjectives and their respective attributes where appropriate. Also, estimate the "worst" and "best" levels for all the factors at the lowest level of the hierarchy.

1. Maximize the contribution to economic development
2. Act consistently with the public's environmental values
3. Minimize detrimental health and safety impacts

4. Promote equitable business arrangements
5. Maximize quality of service
6. Be recognized as public service oriented

4-10. (a) Use the three weighting techniques in Section 3.3 to make a selection of one of the three automobiles for which some data are given in Table 4-6. State your assumptions regarding miles driven each year; life of the automobile (how long *you* would keep it); market (resale) value at end of life, interest cost, price of fuel, cost of annual maintenance, attribute weights, and other subjectively based determinations.

(b) Repeat the analysis using MAUT; that is, construct utility functions and scaling functions for each attribute, and determine the overall utility of each alternative. Does your answer agree with the one obtained in part (a)? Explain why they should (or should not) agree.

TABLE 4-6

	Alternative		
Attribute	Domestic	European	Japanese
Price	$8,100	$12,600	$10,300
Gas mileage	25 mpg	30 mpg	35 mpg
Type of fuel	Gasoline	Diesel	Gasoline
Aesthetic appeal	5 out of 10	7 out of 10	9 out of 10
Passengers	4	6	4
Performance on road	Fair	Very good	Very good
Ease of servicing	Excellent	Very good	Good
Stereo system	Poor	Good	Excellent
Headroom	Excellent	Very good	Poor
Storage space	Very good	Excellent	Poor

4-11. An aspiration level for a criterion or attribute is a level at which the decision maker is satisfied. For example, we would all like our investment portfolio to provide an annual rate of return of 50% or higher, but most of us would happily settle for a return of 10% above the Dow Jones. Develop an interactive multicriteria methodology that is based on aspiration levels of the criteria. Construct a flowchart for the logic and computations. Use your methodology to select one of the alternatives in Exercise 4-10.

REFERENCES

Multiattribute Utility Theory

BARD, J. F., and A. FEINBERG, "A Two-Phase Methodology for Technology Selection and System Design," *IEEE Transactions on Engineering Management*, Vol. EM-36, No. 1 (1989), pp. 28–36.

BELL, D. E., R. L. KEENEY, and H. RAIFFA (editors), *Conflicting Objectives in Decisions*, Wiley, New York, 1977.

DYER, J. S., and R. F. MILES, JR., "An Actual Application of Collective Choice Theory to the Selection of Trajectories for the Mariner Jupiter/Saturn 1977 Project," *Operations Research*, Vol. 24 (1976), pp. 220–244.

EDWARDS, W., "How to Use Multiattribute Utility Measurement for Social Decisionmaking," *IEEE Transactions on Systems, Man, and Cybernetics*, Vol. SMC-7, No. 5 (1977), pp. 326–340.

FEINBERG, A., R. F. MILES, JR., and J. H. SMITH, *Advanced Vehicle Preference Analysis for Five-Passenger Vehicles with Unrefueled Ranges of 100, 150, and 250 Miles*, JPL D-2225, Jet Propulsion Laboratory, Pasadena, CA, March 1985.

KEEFER, D. L., "Allocation Planning for R&D with Uncertainty and Multiple Objectives," *IEEE Transactions on Engineering Management*, Vol. EM-25, No. 1 (1978), pp. 8–14.

KEENEY, R. L., "The Art of Assessing Multiattribute Utility Functions," *Organizational Behavior and Human Performance*, Vol. 19 (1977), pp. 267–310.

KEENEY, R. L., and H. RAIFFA, *Decisions with Multiple Objectives: Preference and Value Tradeoffs*, Wiley, New York, 1976.

KEENEY, R. L., and D. VON WINTERFELDT, "Eliciting Probabilities from Experts in Complex Technical Problems," *IEEE Transactions on Engineering Management*, Vol. 38, No. 3 (1991), pp. 191–201.

MEHREZ, A., and Z. SINUARY-STERN, "Resource Allocation to the Interrelated Risky Projects Using Multiattribute Utility Function," *Management Science*, Vol. 29, No. 4 (1983), pp. 430–439.

SHOEMAKER, P. J. H., and C. C. WAID, "An Experimental Comparison of Different Approaches to Determining Weights in Additive Utility Models," *Management Science*, Vol. 28, No. 2 (1982), pp. 182–196.

SMITH, J. H., A. FEINBERG, and T. LEE, *SmartEdge: A Decision Support System for Expert Knowledge Acquisition and Evaluation*, Haviland-Lee, Northridge, CA, 1987.

Analytic Hierarchy Process

BARD, J. F., "Evaluating Space Station Applications of Automation and Robotics," *IEEE Transactions on Engineering Management*, Vol. EM-33, No. 2 (1986), pp. 102–111.

BARD, J. F., and S. F. SOUSK, "A Tradeoff Analysis for Rough Terrain Cargo Handlers Using the AHP: An Example of Group Decision Making," *IEEE Transactions on Engineering Management*, Vol. 37, No. 3 (1990), pp. 222–227.

BELTON, V., and T. GEAR, "On a Shortcoming of Saaty's Method of Analytic Hierarchies," *Omega*, Vol. 11 (1984), pp. 228–230.

FORMAN, E. H., T. L. SAATY, M. A. SELLY, and R. WALDRON, *Expert Choice*, Decision Support Software, McLean, VA, 1983.

GOLDEN, B. L., E. A. WASIL, and P. T. HARKER (editors), *The Analytic Hierarchy Process: Applications and Studies*, Springer-Verlag, Berlin, 1989.

HAMALAINEN, R. P., and H. LAURI, *HIPRE 3 + Decision Support Software*, Systems Analysis Laboratory, Helsinki University of Technology, Helsinki, Finland, 1992.

LIBERTORE, M. J., "An Extension of the Analytic Hierarchy Process for Industrial R&D Project Selection and Resource Allocation," *IEEE Transactions on Engineering Management*, Vol. EM-34, No. 1 (1987), pp. 12–18.

SAATY, T. L., *The Analytic Hierarchy Process*, McGraw-Hill, New York, 1980.

SAATY, T. L., "Axiomatic Foundations of the Analytic Hierarchy Process," *Management Science*, Vol. 32, No. 7 (1986), pp. 841–855.

SHTUB, A., and E. M. DAR-EL, "A Methodology for the Selection of Assembly Systems," *International Journal of Production Research*, Vol. 27, No. 1 (1989), pp. 175–186.

Group Decision Making

ACZEL, J., and C. ALSINA, "Synthesizing Judgments: A Functional Equation Approach," *Mathematical Modelling*, Vol. 9 (1987), pp. 311–320.

DESANCTIS, G., and GALLUPE, R. B., "A Foundation for the Study of Group Decision Support Systems," *Management Science*, Vol. 33, No. 5 (1987), pp. 589–609.

FRANZ, L. S., G. R. REEVES, and J. J. GONZALEZ, "Group Decision Processes: MOLP Procedures Facilitating Group and Individual Decision Orientations," *Computers & Operations Research*, Vol. 19, No. 7 (1992), pp. 695–706.

HILTZ, S. R., and M. TUROFF, "Structuring Computer-Mediated Communication Systems to Avoid Information Overload," *Communications of the ACM*, Vol. 28, No. 7 (1985), pp. 680–689.

HUBER, G., and R. MCDANIEL, "Decision-Making Paradigm of Organizational Design," *Management Science*, Vol. 32, No. 5 (1986), pp. 572–589.

POOLE, M. S., M. HOLMES, and G. DESANCTIS, "Conflict Management in a Computer-Supported Meeting Environment," *Management Science*, Vol. 37, No. 8 (1991), pp. 926–953.

SAATY, T. L., *Decision Making for Leaders*, Lifetime Learning Publications, Belmont, CA, 1982.

SAATY, T.L., "Group Decision Making and the AHP," in B. L. Golden, E. A. Wasil, and P. T. Harken (editors), *The Analytic Hierarchy Process: Applications and Studies*, Springer-Verlag, Berlin, pp. 59–67, 1989.

SHAW, M. E., *Group Dynamics: The Psychology of Small Group Behavior*, Third Edition, McGraw-Hill, New York, 1981.

Comparison of Methods

BARD, J. F., "A Comparison of the Analytic Hierarchy Process with Multiattribute Utility Theory: A Case Study," *IIE Transactions*, Vol. 24, No. 5 (1992), pp. 111–121.

BELTON, V., "A Comparison of the Analytic Hierarchy Process and a Simple Multi-attribute Value Function," *European Journal of Operational Research*, Vol. 26 (1986), pp. 7–21.

KAMENENTZKY, R. D., "The Relationship between the AHP and the Additive Value Function," *Decision Science*, Vol. 13 (1982), pp. 702–713.

Additional MCDM Techniques

GRAVES, S. B., J. L. RINGUEST, and J. F. BARD, "Recent Developments in Screening Methods for Nondominated Solutions in Multiobjective Optimization," *Computers & Operations Research*, Vol. 19, No. 7 (1992), pp. 683–694.

LEWANDOWSKI, A., and A. P. WIREZBICKI (editors), *Aspiration Decision Support Systems*, Vol. 331, Lecture Notes in Economics and Mathematical Systems, Springer-Verlag, Berlin, 1989.

LOTFI, V., T. J. STEWART, and S. ZIONTS, "An Aspiration-Level Interactive Model for Multiple Criteria Decision Making," *Computers & Operations Research*, Vol. 19, No. 7 (1992), pp. 671–681.

APPENDIX 4A

Comparison of MAUT with the AHP: Case Study*

In this appendix we present a case study in which the analytic hierarchy process (AHP) and multiattribute utility theory (MAUT) are used to evaluate and select the next generation of rough-terrain cargo handlers for the U.S. Army. Three alternatives are identified and ultimately ranked using the two methodologies. A major purpose of this study is to demonstrate the strengths and weaknesses of each methodology and to characterize the conditions under which one might be more appropriate than the other.

The evaluation team consisted of five program managers and engineers from the Belvoir Research, Development & Engineering Center. The objective hierarchy used for both techniques contained 12 attributes. In general, the AHP was found to be more accessible and conducive to consensus building. Once the attributes were defined, the decision makers had little difficulty in furnishing the necessary data and discussing the intermediate results. The same could not be said for the MAUT analysis. The need to juggle 12 attributes at a time produced a considerable amount of frustration among the participants. In addition, the lottery questions posed during the data collection phase had an unsettling effect that was never satisfactorily resolved.

4A.1 INTRODUCTION AND BACKGROUND

In an ongoing effort to reduce risk and to boost the productivity of material handling crews, the Army is investigating the use of robotics to perform many of the dangerous and labor-intensive functions normally undertaken by enlisted personnel. To this end, a number of programs are currently under way at several government facilities. These include the development of a universal self-deployable cargo handler (USDCH) at Belvoir Research, Development, & Engineering Center (Belvoir 1987b), the testing of a field material handling robot (FMR) at the Human Engineering Laboratory, and the prototyping of an advanced robotic manipulator system (ARMS) at the Defense Advanced Research Projects Agency (more details are given in Sievers and Gordon 1986, Sousk et al. 1988).

In each of these efforts, technological risk, time, and cost ultimately intervene to limit the scope and performance of the final product. But to what extent and in what manner? To

*The material presented in this appendix has been excerpted from Bard (1992).

192

answer these questions, a model is needed that is capable of explicitly addressing the conflicts that arise among system and organizational goals. Such a model must also be able to deal with the subjective nature of the decision-making process. The two approaches examined, the AHP and MAUT, each offer an analytic framework in which the decision maker can conduct trade-offs among incommensurate criteria without having to rely on a single measure of performance.

4A.2 CARGO HANDLING PROBLEM

Although the Army is generally viewed as a fighting force, the bulk of its activity involves the movement of massive amounts of material and supplies in the field. This is achieved with a massive secondary labor force whose risk exposure is comparable to those engaged in direct combat. From an operational point of view, cargo must be handled in all types of climates, regions, and environments. At present, this is accomplished by three different sized rough terrain forklifts with maximum lifting capacities of 4,000, 6,000, and 10,000 lbs each. These vehicles are similar in design and performance to those used by industry, and at best, can reach speeds of 20 mph. For the most part, this means that the fleet is not self-deployable (i.e., it cannot keep pace with the convoy on most surfaces). As a consequence, additional transportation resources are required for relocation between job sites. This restriction severely limits the unit's maneuverability and hence its survivability on the battlefield.

A second problem relates to the safety of the crew. Although protective gear is available for the operator, his or her effectiveness is severely hampered by its use. Heat exhaustion, vision impairment, and the requirement for frequent changes are the problems cited most commonly. Logistics units thus lack the ability to provide continuous support in extreme conditions.

4A.2.1 System Objectives

To overcome these deficiencies as well as to improve crew productivity, a heavy-duty cargo-handling forklift is needed. This vehicle should be capable of operating in rough terrain and traveling over paved roads at speeds in excess of 40 mph. To permit operations in extreme conditions, internal cooling (microcooling) should be provided for the protective gear worn by the operator. As technology progresses, it is desirable that the basic functions be executable without human intervention, implying some degree of autonomy.

At a minimum, then, the vehicle should be:

- Able to substitute for the existing 4,000, 6,000, and 10,000 (4K, 6K, 10K)-lb forklifts while maintaining current material handling capabilities
- Capable of unaided movement (self-deployability) between job sites at convoy speeds in excess of 40 mph
- Capable of determining if cargo is contaminated by nuclear, biological, or chemical agents

- Capable of handling cargo in all climates and under all contamination conditions
- Transportable by C-130 and C-141B aircraft
- Operable in the near term as a human–machine system expandable to full autonomy
- Capable of robotic cargo engagement
- Operable remotely from up to 1 mile away

4A.2.2 Possibility of Commercial Procurement

A market survey of commercial forklift manufacturers, including those currently under contract for the 4K-, 6K-, and 10K-lb vehicles, indicates little opportunity for a suitable off-the-shelf buy. With Army needs constituting less than 15% of the overall market, lengthy procurement cycles and uneven demand work to dampen corporate interest. In the commercial environment, the use of rough-terrain forklifts is limited to construction and logging operations; highway travel and teleoperations have no real applications. Therefore, few, if any, incentives exist for the industry to undertake the research and development (R&D) effort implied by the design requirements to build a prototype vehicle.

4A.2.3 Alternative Approaches

To satisfy the system objectives, then, the existing fleet must either be replaced outright or substantially overhauled. However, given the low priority of logistics relative to combat needs, a full-scale R&D program is not a realistic option. A more likely approach involves either an improvement in the existing system, a modification of a commercial system, or the adaptation of available technology to meet specific requirements. Each of these approaches occasions a different level of risk, cost, and performance that must be evaluated and compared before a final decision can be made. This is the subject of the remainder of the appendix, but first, the leading alternatives are defined.

Taking into account mission objectives and the fact that the Army has functioned with the existing system up until now, the following alternatives have been identified. This set represents a consensus of the program managers and engineers at Belvoir, and the customer at the Quartermaster School.

1. *Baseline:* the existing system, comprising the 4K-, 6K-, and 10K-lb rough-terrain forklifts augmented with the new 6K-lb variable-reach vehicle
2. *Upgraded system:* baseline upgraded to be self-deployable
3. *USDCH:* teleoperable, robotic-assisted universal self-deployable cargo handler with microcooling for the protective gear and the potential for full autonomy

The new 6K-lb variable-reach (telescoping boom) forklift is scheduled to be introduced into the fleet in early 1990. Its performance characteristics, along with those of the USDCH, have been discussed by Belvoir (1987a,b). Figure 4A-1 depicts a schematic of the robotic-assisted cargo handler. Note that the field material handling robot and the advanced robotic manipulator system have been omitted from the list above. At this juncture,

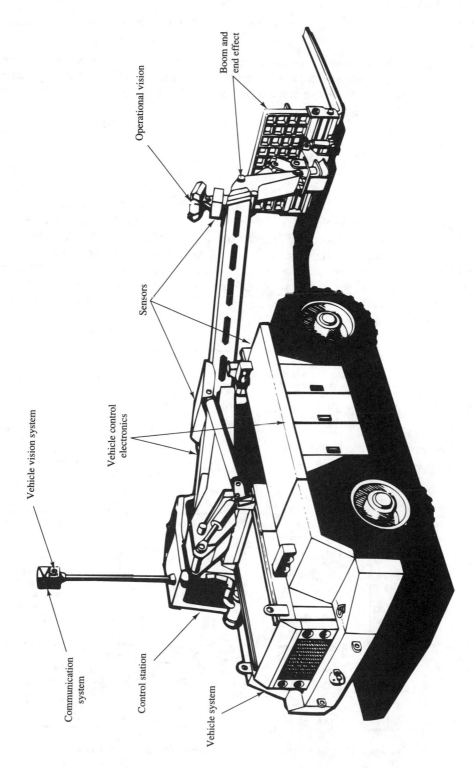

Figure 4A-1 Universal self-deployable cargo handler.

Operational vision

Boom and
end effect

Sensors

Vehicle vision system

Vehicle control
electronics

Communication
system

Control station

Vehicle system

195

the primary interest in these systems centers on their robotic capabilities rather than on their virtues as cargo handlers. In fact, almost none of the operational deficiencies mentioned previously would be overcome by either the FMR or ARMS. Consequently, each was dismissed from further consideration.

4A.3 ANALYTIC HIERARCHY PROCESS

The first step in any multiobjective methodology is to identify the principal criteria to be used in the evaluation. These should be expressed in fairly general terms and be well understood by the study participants. For our problem, the following four criteria were identified: performance, risk, cost, and program objectives. The next step is to add definition by associating a subset of attributes (subcriteria) with each of the above. Figure 4A-2 depicts the resultant objective hierarchy. Risk, for example, has been assigned the following attributes: system integration, technical performance, cost overrun, and schedule overrun. The alternatives are arrayed at the bottom level of the diagram. The connecting lines indicate points of comparison.

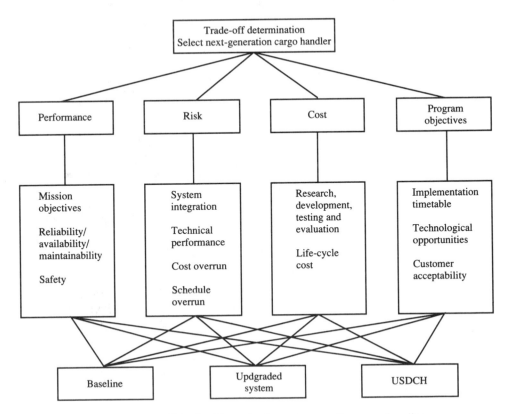

Figure 4A-2   Objective hierarchy for next-generation cargo handler.

In constructing the objective hierarchy, consideration must be given to the level of detail appropriate for the analysis. This is often dictated by the present stage of the development cycle, the amount of data available on each alternative, and the relative importance of prospective criteria and attributes. For example, if human productivity were a major concern, as it is in the space program, a fifth criterion might have been included at the second level. The inclusion or exclusion of a particular attribute depends on the degree to which its value differs among the alternatives. Although transportability and survivability are important design considerations, all candidates for the cargo handling mission are expected equally to satisfy basic requirements with respect to these attributes. Consequently, it is not necessary to incorporate them in the model.

To avoid too much detail, aggregation is recommended. This permits overly specific factors to be taken into account implicitly by including them in the attribute definitions. For example, "life-cycle cost" could have been further decomposed into unit purchase price, operations and maintenance costs, spare parts, personnel and training, and so on, but at the expense of overtaxing the current database and cost accounting system. As a result, these factors were left undifferentiated. Similar reasoning applies to the attribute "reliability/availability/maintainability."

4A.3.1 Definition of Attributes

Each of the attributes displayed at level 3 in Fig. 4A-2 is described in more detail below. These descriptions, in the form of instructions, were used by the analyst to elicit responses from the decision makers during the data collection phase of the study.

Performance

1. *Mission objectives.* Compare the alternatives on the basis of how close they come to satisfying mission objectives and requirements. Consideration should be given to such factors as lifting capacity, deployability, productivity improvement, and operation in a nuclear–biological–chemical (NBC) environment.

2. *Reliability, availability, and maintainability (RAM).* Using military standards for reliability, availability, and maintainability, compare the alternatives relative to the likelihood that each will meet these standards. If possible, take into account mean time between failures, mean time to repair, and the most probable failure modes.

3. *Safety.* Compare the alternatives on the basis of how well they protect the crew in all climatic conditions and in an NBC environment. Consider the probable degree of hazard exposure, the vehicle response under various driving conditions, and the ability of the crew to work effectively for extended periods.

Risk

4. *System integration.* Compare the effort required to achieve full system integration for the alternatives, taking into account the degree of upgrading and reengineering associated with each.

5. *Technical performance.* Considering the performance goals of each system, evaluate the relative likelihood that these goals will be met within the current constraints of the program. Take into account the Army's experience with similar systems, and the state of commercially available technologies.

6. *Cost overrun.* Based on the maturity of the technology and the funding histories of similar programs, compare the alternatives as to whether one is more likely to go over budget than the other.

7. *Schedule overrun.* Based on the maturity of the technology and the development histories of similar programs, compare the alternatives as to whether one is more likely than the other to result in a schedule overrun.

Cost

8. *Research, development, testing, and evaluation (RDT&E).* Compare the alternatives from the standpoint of which is likely to have the least cost impact during its development cycle. Consideration should be given to each phase of the program prior to implementation.

9. *Life-cycle cost.* Compare the total cost of buying, operating, maintaining, and supporting each alternative over its expected lifetime. Exclude RDT&E, but take into account personnel needs, training, and the degree of standardization achieved by each system.

Program Objectives

10. *Implementation timetable.* Compare the alternatives with respect to their individual schedules for implementation. Consider the effect that the respective timetables will have on military readiness.

11. *Technological opportunities.* Compare the alternatives on the basis of what new technologies might result from their development, as well as the likelihood that new applications will be found in other areas. Consideration should be given to the prospect of spin-offs, potential benefits, and the development of long-term knowledge.

12. *Customer Acceptability.* Compare the alternatives from both the user representative's and operator's points of view. Take into account the degree to which each alternative satisfies basic objectives, as well as the potential for growth, risk reduction, and the adaptation of new technologies. Also consider secondary or potential uses, operator comfort, and program politics.

4A.3.2 AHP Computations

To illustrate the nature of the calculations, observe Fig. 4A-3, which depicts a three-level hierarchy—an abbreviated version of Fig. 4A-2 used in the analysis. Table 4A-1 contains the input and output data for level 2. Recall that when n factors are being compared, $n(n - 1)/2$ questions are necessary to fill in the matrix. The elements in the lower triangle (omitted

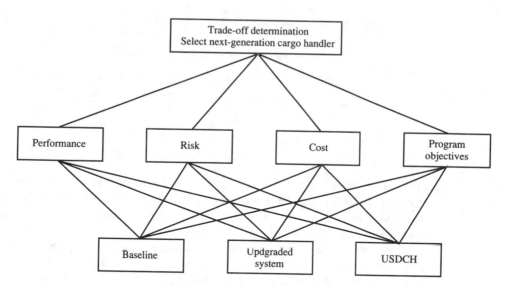

Figure 4A-3   Abbreviated version of the objective hierarchy.

here) are simply the reciprocal of those lying above the diagonal; that is, $a_{ji} = 1/a_{ij}$. The entries in the matrix at the center of Table 4A-1 are the responses to the 6 ($n = 4$) pairwise questions that were asked. These responses were drawn from the 9-point scale shown in Table 4-1. For example, in comparing "performance" with "risk" (element a_{12} of the matrix), it was judged that the first "strongly" dominated the second. Note that if the elicited value for this element were $\frac{1}{5}$ instead of 5, the opposite would have been true. From Table 4A-1 it can be seen that the priorities derived for the major criteria were 0.517 for performance, 0.059 for risk, 0.306 for cost, and 0.118 for program objectives. Also note that the consistency ratio (0.097) is a bit high but still within the acceptable range.

The next step in the analysis is to develop the priorities for the factors on the third level with respect to those on the second. In our case, we compare the three alternatives against the major criteria. For the moment, assume that the appropriate data have been elicited and that the calculations have been performed for each of the four comparison matrices, giving the results displayed in Table 4A-2. The first four columns of data are the

TABLE 4A-1   PRIORITY VECTOR FOR MAJOR CRITERIA

Criteria	Criteria 1	2	3	4	Priority weights	Output parameters
1. Performance	1	5	3	4	0.517	$\lambda_{max} = 4.262$
2. Risk		1	$\frac{1}{6}$	$\frac{1}{3}$	0.059	CR $= 0.097$
3. Cost			1	4	0.306	
4. Program objectives				1	0.118	

TABLE 4A-2 LOCAL AND GLOBAL PRIORITIES

	Local priorities				
Alternative	Performance (0.517)	Risk (0.059)	Cost (0.306)	Program objectives (0.118)	Global priorities
Baseline	0.142	0.704	0.384	0.133	0.248
Upgrade	0.167	0.229	0.317	0.162	0.216
USDCH	0.691	0.067	0.299	0.705	0.536

local priorities derived from the inputs supplied by the decision maker; note that each column sums to 1. The global priorities are found by respectively multiplying these values by the higher-level local priorities given in Table 4A-1 (and repeated at the top of Table 4A-2 for convenience), and then summing. Because there are no more levels left to evaluate, the values contained in the last column of Table 4A-2 represent the final priorities for the problem. Thus, according to the judgments expressed by this decision maker, alternative 3 turns out to be most preferred. Finally, it should be noted that other schemes are available for determining attribute weights (e.g., see Shoemaker and Waid 1982).

4A.3.3 Data Collection and Results for AHP

In the formative stages of the study, two questions quickly arose: (1) "Who should provide the responses?" and (2) "Whose point of view should be represented?" With regard to the first, it was felt that the credibility of the results depended on having a broad spectrum of opinion and expertise as input. Subsequently, five people from Belvoir's Logistics Equipment Directorate with an average of 15 years' experience in systems design, R&D program management, and government procurement practices were assembled to form the evaluation team. After some discussion it was agreed that the responses should reflect the position of the material developer—the US Army Materiel Command. Other candidates included the Army as a whole, the customer, and the mechanical equipment division at Belvoir.

At the first meeting, the group was introduced to the AHP methodology and examined the objective hierarchy developed previously by the analyst. Eventually, a consensus grew around the attribute definitions, and each member began to assign values to the individual matrix elements. A bottom-up approach was found to work best. Here the alternatives are first compared with respect to each attribute; next, a comparison is made among the attributes with respect to the criteria; and finally, the four criteria at level 2 are compared among themselves. After the data sheets had been filled out for each criterion, individual responses were read aloud to ascertain the level of agreement. In light of the ensuing discussion, the participants were asked to revise their entries to better reflect their renewed understanding of the issues. This phase of the study took about 6 hours and was done in two sessions over a 5-day period.

As with the Delphi procedure, the challenge was to come as close to a consensus as possible without coercing any of the team members. Unfortunately, this proved more difficult than expected, due to the speculative nature of much of the attribute data. In practice, many researchers have found that uniformity within a group can rarely be achieved without stretching the limits of persuasion (Shaw 1981). Biases, insecurities, and stubbornness often develop their own constituencies. And although none of these factors was openly present at the meetings, organizational and program concerns were clearly seen to influence individual judgments.

In the extreme, when there is no possibility of reconciling conflicting perceptions, it is best to stratify responses along party lines. In our case, sufficient agreement emerged to permit the averaging of results without obscuring honest differences of opinion. Table 4A-3 highlights individual preferences for the level 2 criteria and for the problem as a whole. The numbers in parentheses represent the local weights computed for the four criteria: performance, risk, cost, and program objectives. Global weights and rankings are given in the last two columns.

Table 4A-4 summarizes the computations for each decision maker and presents two collective measures of comparison: (1) the arithmetic mean and (2) the geometric mean. (Issues surrounding the synthesis of judgments is discussed by Aczel and Alsina 1987.) The latter is obtained by a geometric averaging of the group's individual responses at each point of comparison to form a composite matrix, followed by calculation of the eigenvectors in the usual manner. As can be seen, both methods give virtually identical results and rankings. The strongest preference is shown for the USDCH, closely followed by

TABLE 4A-3   COMPARISON OF RESPONSES USING THE AHP

		Local results								Global results	
		Performance		Risk		Cost		Program obj.			
Respondent	Alternative	Weight	Rank	Weight	Rank	Weight	Rank	Weight	Rank	Weight	Rank
1		(0.517)		(0.059)		(0.306)		(0.118)			
	Baseline	0.142	3	0.704	1	0.384	1	0.133	3	0.248	2
	Upgrade	0.167	2	0.229	2	0.317	2	0.162	2	0.216	3
	USDCH	0.691	1	0.067	3	0.229	3	0.705	1	0.536	1
2		(0.553)		(0.218)		(0.147)		(0.082)			
	Baseline	0.144	3	0.497	1	0.432	1	0.202	3	0.268	3
	Upgrade	0.213	2	0.398	2	0.383	2	0.269	2	0.282	2
	USDCH	0.643	1	0.105	3	0.185	3	0.529	1	0.450	1
3		(0.458)		(0.240)		(0.185)		(0.117)			
	Baseline	0.252	3	0.677	1	0.467	1	0.350	2	0.405	1
	Upgrade	0.273	2	0.249	2	0.375	2	0.371	1	0.298	2
	USDCH	0.474	1	0.074	3	0.158	3	0.280	3	0.297	3
4		(0.359)		(0.315)		(0.210)		(0.116)			
	Baseline	0.214	3	0.666	1	0.602	1	0.529	1	0.474	1
	Upgrade	0.263	2	0.266	2	0.313	2	0.313	2	0.280	2
	USDCH	0.524	1	0.068	3	0.085	3	0.158	3	0.246	3
5		(0.469)		(0.252)		(0.194)		(0.085)			
	Baseline	0.184	3	0.655	1	0.565	1	0.176	3	0.376	2
	Upgrade	0.227	2	0.274	2	0.285	2	0.178	2	0.246	3
	USDCH	0.589	1	0.071	3	0.150	3	0.646	1	0.378	1

TABLE 4A-4 SUMMARY OF RESULTS FOR THE AHP ANALYSIS

| | Respondent | | | | | | | | | | Arithmetic mean | | Geometric mean | |
| | 1 | | 2 | | 3 | | 4 | | 5 | | | | | |
Alternative	Weight	Rank	Weight	Rank	Weight	Rank	Weight	Rank	Weight	Rank	Weight	Rank	Weight	Rank
Baseline	0.248	2	0.268	3	0.405	1	0.474	1	0.376	2	0.354	2	0.358	2
Upgrade	0.216	3	0.282	2	0.298	2	0.280	2	0.246	3	0.265	3	0.258	3
USDCH	0.536	1	0.450	1	0.297	3	0.246	3	0.378	1	0.381	1	0.383	1

the baseline. The upgraded system is a distant third. All computations were done on an IBM-PC-AT with a code written by J. F. Bard.

4A.3.4 Discussion of AHP Process and Results

The output in Tables 4A-3 and 4A-4 represents the final judgments of the participants and was obtained only after holding two additional meetings to discuss intermediate results. All persons were given the opportunity to examine the priority weights calculated from their initial responses and to assess the reasonableness of the rankings. When their results seemed counterintuitive, they were encouraged to reevaluate their input data, determine the source of the inconsistency, and make the appropriate changes. The debate taking place during these sessions proved to be extremely helpful in clarifying attribute definitions and surfacing misunderstandings. In a few instances, well-reasoned arguments persuaded some people to reverse their position completely on a particular issue. This was more apt to occur when the advocate was viewed as an expert and was able to furnish the supporting data. Ordinarily, one- or two-point revisions were the rule and had no noticeable effect on the outcome.

Looking at the data in Table 4A-3, a great deal of consistency can be seen across the group. In all but one instance, performance is given the highest priority, followed by risk, cost, and program objectives. For the first three criteria, each alternative has the same ordinal ranking; the only differences arise in the case of program objectives. Nevertheless, the real conflict is reflected in the magnitude of the weights. Although some variation is inevitable, it is frustrating to observe the results for "cost." In particular, there is little agreement concerning the extent to which personnel and transportation resource reductions accompanying the USDCH will be offset by increased operations and maintenance expenses; or how these factors will affect life-cycle costs. The third and fourth decision makers were more skeptical than the first two, and hence showed a greater preference for the baseline.

The results for "risk" also inform a divergence of opinion. Respondent 1 was most forthright in acknowledging its presence in the USDCH program by assigning it an extremely low weight (0.067) relative to the baseline (0.704). The effect of this assignment was minimal, though, due to the fact that he judged risk to be considerably less important than the other three criteria. Compare his corresponding weight (0.059) with those derived for respondents 2 through 5 (0.218, 0.240, 0.315, 0.252). From the data in Table 4A-3 it can be seen that the latter four decision makers all viewed risk as the second-most-important criterion. This observation was corroborated indirectly in the utility analysis.

4A.4 MULTIATTRIBUTE UTILITY THEORY

MAUT is a methodology for providing information to the decision maker for comparing and selecting among complex alternatives when uncertainty is present. It calls similarly for the construction of an objective hierarchy as depicted in Fig. 4A-2, but addresses only the bottom two levels.

4A.4.1 Data Collection and Results for MAUT

After agreeing on the attributes, the next step in model development is to determine the scaling constants, k_i, and the attribute utility functions, U_i. This is done through a series of questions designed to probe each decision maker's risk attitude over the range of permissible outcomes. Before the interviews can be conducted, though, upper and lower bounds on attribute values must be specified. Table 4A-5 lists the values elicited from respondent 1 for the 12 attributes. Notice that seven of these are measured on a qualitative (ordinal) scale, the meanings of which were made precise at the first group session. Table 4A-6 defines the range of scores for the "mission objectives" attribute and is typical of the 10-point scales used in the analysis.

To determine the scaling constants, the decision maker must specify an indifference probability, p, related to the best (x^*) and worst (x^0) values of the attribute states. The following scenario is posed:

1. Let attribute i be at its best value and the remaining attributes be at their worst values. Call this situation the "reference."
2. Assume that a "gamble" is available such that the "best outcome" occurs with probability p, and the "worst outcome" occurs with probability $1 - p$. If you can achieve the "reference" for sure, for what value of p are you indifferent between the "sure thing" and the "gamble"?

TABLE 4A-5 ATTRIBUTE DATA FOR DECISION MAKER 1

No.	Attribute	Scale	A1	A2	A3	Range	Order of importance[b]	Scaling constant
Performance								
1	Mission objectives	Ordinal	4	4	8	4–8	1	0.176
2	RAM	Ordinal	6	4	3	3–6	11	0.044
3	Safety	Ordinal	4	4	10	4–10	2	0.162
Risk								
4	System integration	Ordinal	9	7	3	3–7	8	0.059
5	Technical performance	Ordinal	9	7	3	3–9	9	0.059
6	Cost overrun	$M	0	1	5	0–5	12	0.044
7	Schedule overrun	Years	0	2	4	0–4	7	0.059
Cost								
8	RDT&E	$M	0	6	13	0–13	6	0.059
9	Life-cycle cost	$B	3.0	2.8	2.5	2.5–3.0	4	0.088
Program objectives								
10	Timetable	Years	2	6	8	2–8	10	0.044
11	Technical opportunity	Ordinal	1	2	7	1–7	5	0.074
12	Customer acceptability	Ordinal	1	3	9	1–9	3	0.132

The "Value" columns A1, A2, A3 fall under a spanning header "Value[a]".

[a] A1, baseline; A2, upgraded system; A3, USDCH.

[b] Order of importance for the given *range* of attribute values.

TABLE 4A-6   SCALE USED FOR "MISSION OBJECTIVES" ATTRIBUTE

Value	Explanation
10	All mission objectives are satisfied or exceeded, and some additional capabilities are provided. The design is expected to lead to significant improvements in human productivity and military readiness.
8	All basic mission objectives are met and some improvement in productivity is expected. The design readily permits the incorporation of new technologies when they become available.
6	Minor shortcomings in system performance are evident, but the overall mission objectives will not be compromised. Some improvement in operator efficiency is expected.
4	Not all performance levels are high enough to meet basic mission objectives. However, no more than one major objective (such as self-deployability or microcooling) is compromised and no threat exists to military readiness.
2	An inability to meet one or more major mission objectives exists. With the current design, it is not economically feasible to bring overall performance up to standards.
0	Significant shortcomings exist with respect to the mission objectives. Implementation or continued use could seriously jeopardize military readiness.

The resultant scaling constants for each of the five decision makers are displayed in Table 4A-7 along with the corresponding AHP weights. The former have been normalized to sum to 1 to facilitate the comparison and to permit the use of the additive model (4-1b). At a superficial level, the group showed a remarkable degree of consistency from one set of responses to the next. (Theoretical speaking, the AHP weights and the MAUT scaling constants measure different phenomena, and hence cannot be given the same interpretation; see Kamenentzky 1982.) In almost all cases, mission objectives, safety, technical performance, and life-cycle cost emerged as the dominant concerns. A look at individual values shows some discrepancies, but rankings and orders of magnitude are quite similar.

The procedure used to assess the utility functions is nearly identical to that used for the scaling constants. Not surprisingly, the respondents evidenced a slight risk aversion for the attribute ranges considered. A commercial software package called SmartEdge was used for data input and analysis (Smith et al. 1987). Further explanation of the methodology is given by Bard and Feinberg (1989) and Feinberg et al. (1985).

The computational results for the utility analysis are displayed in Table 4A-8 and are seen to parallel closely those for the AHP. Only decision makers 3 and 5 partially reversed themselves but without consequence; the others maintained the same ordinal rankings. Note again that it would be inappropriate to compare the final AHP priority weights with the final utility values obtained for each alternative (see Belton 1986). The former are measured on a ratio scale and have relative meaning; the latter simply indicate the order of preference.

An examination of the last four columns of Tables 4A-4 and 4A-8 shows that the two methods give the same general results. Here the geometric mean, also known as the Nash bargaining rule, is computed from the five entries in the table. In making comparisons, only the rankings (and not their relative values) should be taken into account.

TABLE 4A-7   COMPARISON OF AHP WEIGHTS AND MAUT SCALING CONSTANTS FOR THE FIVE DECISION MAKERS

		Respondent									
		1		2		3		4		5	
No.	Attribute	AHP	MAUT	AHP	MAUT	AHP	MAUT	AHP	MAUT	AHP	MAUT
Performance											
1	Mission objectives	0.324	0.176	0.341	0.287	0.245	0.199	0.215	0.171	0.293	0.222
2	RAM	0.048	0.044	0.047	0.031	0.092	0.081	0.072	0.105	0.064	0.033
3	Safety	0.145	0.162	0.164	0.144	0.092	0.103	0.072	0.075	0.112	0.098
Risk											
4	System Integration	0.006	0.059	0.080	0.061	0.061	0.016	0.021	0.013	0.018	0.031
5	Technical Performance	0.018	0.059	0.080	0.085	0.141	0.093	0.203	0.225	0.155	0.182
6	Cost overrun	0.018	0.044	0.037	0.074	0.025	0.097	0.058	0.076	0.048	0.018
7	Schedule overrun	0.018	0.059	0.023	0.023	0.013	0.016	0.033	0.047	0.031	0.018
Cost											
8	RDT&E	0.038	0.059	0.018	0.025	0.023	0.038	0.023	0.013	0.024	0.046
9	Life-cycle cost	0.268	0.088	0.129	0.111	0.162	0.191	0.187	0.170	0.170	0.138
Program objectives											
10	Timetable	0.012	0.044	0.027	0.025	0.066	0.094	0.079	0.032	0.015	0.018
11	Technical opportunity	0.030	0.074	0.027	0.057	0.017	0.021	0.010	0.044	0.045	0.092
12	Acceptability	0.075	0.132	0.027	0.077	0.033	0.051	0.027	0.029	0.025	0.104

TABLE 4A-8   SUMMARY OF RESULTS FOR MAUT ANALYSIS

	Respondent										Arithmetic mean		Geometric mean	
	1		2		3		4		5					
Alternative	Weight	Rank	Weight	Rank	Weight	Rank	Weight	Rank	Weight	Rank	Weight	Rank	Weight	Rank
Baseline	0.302	2	0.299	3	0.481	1	0.539	1	0.482	1	0.421	2	0.408	2
Upgrade	0.273	3	0.328	2	0.261	3	0.426	2	0.378	3	0.333	3	0.327	3
USDCH	0.595	1	0.567	1	0.337	2	0.273	3	0.432	2	0.441	1	0.422	1

4A.4.2 Discussion of MAUT Process and Results

The interview sessions in which the scaling constants and utility functions were assessed took about a half hour each and were conducted individually while the analyst and decision maker were seated at a terminal. Three difficulties arose immediately. The first related to the probabilistic nature of the questions. None of the respondents could make sense out of the relationship between the posed lotteries and the overall evaluation process. Repeated coaxing was necessary to get them to concentrate on the gambles and to give a deliberate response.

In this regard, it might have been possible to develop more perspective by employing a probabilistic rather than a deterministic utility model. This would have required the attribute outcomes to be treated as random variables (which, in fact, they are) and for probability distributions to be elicited for each. It was felt, however, that this additional burden would have strained the patience and understanding of the group without producing credible results. It was difficult enough to collect the basic attribute data on each alternative without having to estimate probability distributions.

The second issue centered on the assessment of the scaling constants. Here the decision makers were asked to balance best and worst outcomes for 12 attributes at a time. This turned out to be nearly impossible to do with any degree of accuracy and created a considerable amount of tension. The problem was compounded by the fact that in most instances the group believed that a low score on any one of the principal attributes, such as mission objectives or safety, would kill the program. This produced an unflagging reluctance to accept the sure thing unless the gamble was extremely unfavorable. Because most people are unable to deal intelligently with low-probability events, this called into question, at least in our minds, the validity of the accompanying results.

The third concern relates to the use of ordinal scales to gauge attribute outcomes. Although time and cost have a common frame of reference, ordinal scales generally defy intuition. This was the case here. None of the respondents felt comfortable with this part of the interview, even when they were willing to accept the overall methodology.

4A.5 ADDITIONAL OBSERVATIONS

The level of abstraction surrounding the use of MAUT strongly suggests that the AHP is more acceptable to decision makers who lack familiarity with either method. For problems characterized by a large number of attributes, most of whose outcomes can only be measured on a subjective scale, the AHP once again seems best. When the data are more quantifiable, the major attributes few, and the alternatives well understood, MAUT may be the better choice.

This is not to say that the AHP does not have its drawbacks. The most serious relates to the definition and use of the 9-point ratio scale. At some point in the analysis, each of the decision makers found it difficult to reconcile the fact that by expressing a "weak" preference for one alternative over another they were saying that they preferred it by a factor of 3:1. Although this might have seemed reasonable in some instances, in others they felt that a score of 2 was equivalent to showing a "strong" preference. Perhaps this problem could be alleviated by the use of a logarithmic scale.

From the standpoint of consensus building, the AHP methodology provides an accessible data format and a logical means of synthesizing judgment. The consequences of individual responses are easily traced through the computations and can quickly be revised when the situation warrants. In contrast, the MAUT methodology hides the implications of the input data until the final calculations. This makes intermediate discussions difficult because no single point of focus exists. Sensitivity analysis offers a partial solution to this problem but in a backward manner that undercuts its theoretical rigor.

As a final observation we note that the enthusiasm and degree of urgency that the participants brought to the study varied directly with their involvement in the program. Those with vested interests were eager to grasp the methodologies and were quick to respond to requests for data. The remainder viewed each new request as a frustrating and unnecessary ordeal that was best dealt with through passive resistance.

4A.6 CONCLUSIONS FOR THE CASE STUDY

The collective results of the analysis indicated that the group had a modest preference for the USDCH over the baseline. The trade-off between risk and performance for the upgraded system did not appear favorable enough to make it a serious contender for the cargo handling mission. We therefore recommended that work continue on the development of the basic USDCH technologies, including self-deployability and robotic cargo engagement, to demonstrate the underlying principles. If more supportive data are needed, the place to start would be with a full-scale investigation of life-cycle costs, and some of the more quantifiable performance measures such as reliability. The effort required to gather these statistics would be considerable, though, and does not seem justified in light of the overall findings.

In summary, the group felt that the idea of imposing new technologies on an existing system would probably increase its life-cycle cost without achieving the desired capabilities. The extensive improvements in performance ultimately sought could best be realized through a structured research and development program that fully exploited technological advances and innovative thinking in design. Such an approach would significantly reduce risk while permitting full systems integration. In fact, this is the approach now being pursued.

ADDITIONAL REFERENCES

BARD, J. F., "A Comparison of the Analytic Hierarchy Process with Multiattribute Utility Theory: A Case Study," *IIE Transactions*, Vol. 24, No. 5 (1992), pp. 111–121.

BELVOIR RD&E CENTER, *Test and Evaluation Master Plan for the Variable Reach Rough Terrain Forklift*, U.S. Army Troop Support Command, Logistics Equipment Directorate, Fort Belvoir, VA, 1987a.

BELVOIR RD&E CENTER, *Universal Self-Deployable Cargo Handler*, Contract DAAK-70-87-C-0052, U.S. Army Troop Support Command, Fort Belvoir, VA, Sept. 25, 1987b.

SIEVERS, R. H., and B. A. GORDON, *Applications of Automation Technology to Field Material Handling*, SAIC-86/1987, Science Applications International Corporation, McLean, VA, December 1986.

SOUSK, S. F., H. L. KELLER, and M. C. LOCKE, *Science and Technology for Cargo Handling in the Unstructured Field Environment*, U.S. Army Belvoir RD&E Center, Logistics Equipment Directorate, Fort Belvoir, VA, 1988.

5

Structuring the Project: Organizational Structure and Work Breakdown Structure

5.1 Introduction
5.2 Organizational Structures
5.3 Organizational Structure of Projects
5.4 Work Breakdown Structure
5.5 Combining the Organizational and Work Breakdown Structures
5.6 Management of Human Resources in Projects

5.1 INTRODUCTION

Project management deals with the one-time effort to achieve a specific goal within a given set of resource and budget constraints. It is most common for companies to use a project organization when the work content is too large to be accomplished by a single person within the scheduled time frame. The fundamentals of project management involve the allocation of work to the participating units at the planning stage, and the continuous integration of output through the execution stage. How the efforts of the participants are coordinated to accomplish their assigned tasks, and how the final assembly of their work is achieved on time and within budget, is as much an art as it is a science.

The breakdown of work, the allocation of specific tasks to individuals and subcontractors, the management and control during execution, and finally, the integration of the parts into the prespecified whole is the focus of project management. Adequate technical skills and the availability of resources are necessary but rarely sufficient to guarantee project success. There is a need for coordinated teamwork and leadership—the essence of sound project management.

Three types of "structures" are involved in the overall process: (1) the organizational structure of each unit participating in the project (the client, the prime contractor, subcontractors, and perhaps one or more government agencies); (2) the organizational structure of the project itself, which specifies the relationship between the organizations and people doing the work; and (3) the work breakdown structure of the project, that is, the way the work content is divided into small, manageable parcels that can be allocated to the participating units.

In general, organizations set up management structures to facilitate the achievement of their overall mission, as defined in both strategic and tactical terms. In so doing, compromise is needed to balance short-term objectives with long-term goals. As a practical matter, the project manager has very little say in the final design of the organization or in any restructuring that might occur from time to time. Organizations may be involved in many activities and cannot be expected to reorient themselves with each new project. However, both the project organizational structure and the work breakdown structure should be designed to achieve the project's objectives, and therefore should be directly under project management control. The thoughtful design and implementation of these structures is critical, due to their effect on project success.

The design of a project organizational structure is among the early tasks of the project manager. In performing this task, issues of authority, responsibility, and communications should be addressed. The project organizational structure should fit the nature of the project, the nature of the participating organizations, and the environment in which the project will be performed. For example, the transportation of U.S. forces in Operation Desert Shield in 1991 required a project organization capable of coordinating logistical activities across three continents (North America, Europe, and the Arab Peninsula). The authority to decide which forces to transport, when and by what means, as well as the channels through which such decisions were communicated, had to be defined by the project organizational structure. The participating parties were many, including all branches of the U.S. armed services, NATO, and various members of the Arab League. To facilitate coordination among these parties, a well-structured project organization with clear definitions of authority, responsibility, and communication channels was needed.

The work content of the project can usually be structured in a variety of ways. For example, if the project is aimed at developing a new passenger airplane (such as the Boeing 777), the work content can be structured around the main systems, including the body, wings, engines, avionics, and controls. Alternatively, it can be broken down by the technology used; that is, materials, electronic systems, electrical power systems, hydraulic systems, and so on. The actual decision is usually based on the project work content and on the design of the project organizational structure, since elements in the organizational structure perform the required tasks on the components of the work breakdown structure.

To reach the most advantageous organizational arrangement and work assignments, the project manager needs to know what types of structures are common, their strengths and weaknesses, and under what conditions each structure performs best.

5.2 ORGANIZATIONAL STRUCTURES

Projects are performed by organizations using human, capital, and other resources to achieve a specific goal. Many projects cut across organizational lines. To understand the organizational structure of a project, the structure of the participating organizations must first be addressed.

Theorists have devised various ways of partitioning an organization into subunits to improve efficiency and to decentralize authority, responsibility, and accountability. The mechanism through which this is accomplished is called *departmentalization*. In all cases, the objective is to arrive at an orderly arrangement of the interdependent components. Departmentalization is integral to the delegation process.

1. *Functional*. The organizational units are based on distinct common specialties, such as manufacturing, engineering, and finance.
2. *Product*. Distinct units are organized around, and given responsibility for, a major product or product line.
3. *Customer*. Organizational units are formed to deal explicitly with a single customer group, such as the Department of Defense.
4. *Territorial*. Management and staff are located in units defined along geographical lines, such as a southern United States sales zone.
5. *Process*. Human and other resources are based on a flow of work, such as an oil refinery.

Thus organizations may be structured in many different ways, based on functional similarity, types of processes used, product characteristics, customers served, and territorial considerations.

Functional organization.   Perhaps the most widespread organizational structure found in industry is designed around the functions performed by each organizational unit. This structure derives from the assumption that each unit should specialize in a specific functional area and perform all the tasks that require its expertise. Common functional organizational units are engineering, manufacturing, information systems, and marketing. The engineering department is responsible for such activities as product and process design, while marketing is responsible for advertising, sales, and so on. The division of labor is based on the function performed, not on the specific process or product. Figure 5-1 depicts a typical functional structure.

When the similarity of processes is used as a basis for the organizational structure, departments such as metal cutting, painting, and coating are formed in industry, and departments such as word processing and financial management are formed in the service sector. When similar processes are performed by the same organizational elements, capital investment is minimized and expertise is built through repetition within the particular group.

The problems that arise in the functional organization revolve around the fact that there is no strong central authority that worries about each project individually. Major decisions relating to resource allocation and budgets are seldom based on what is best for a particular project, but rather, on how they affect the strongest functional unit. In addition, considerable time is spent in evaluating alternative courses of action since any project decision requires coordination and approval of all the functional groups in addition to upper management. Finally, there is no single point of interface to the customer.

Despite these limitations, the functional organization offers the most fundamental and stabilizing arrangement for large concerns, be they electronic manufacturers, brokerage

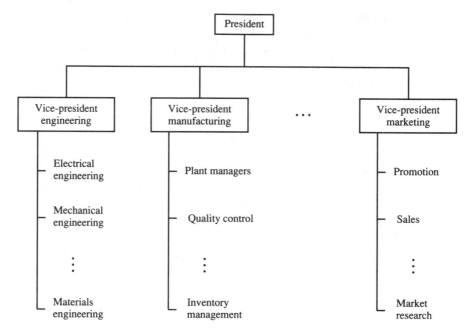

Figure 5-1   Portion of typical functional organization.

houses, or research laboratories. Meredith and Mantel (1989) and Thamhain (1984) summarize its advantages and disadvantages as follows:

Advantages

Efficient use of collective experience and facilities

Institutional framework for planning and control

All activities receive benefits from most advanced technology

Allocates resources in anticipation of future business

Effective use of production elements

Career continuity and growth for personnel

Well suited for mass production of items

Disadvantages

No central project authority

Little or no project planning and reporting

Weak interface with customer

Poor horizontal communications across functions

Difficult to integrate multidisciplinary tasks

Tendency of decisions to favor strongest functional group

Project organization.   A second organizational structure is based on assigning projects to each organizational unit. Here the various functions, such as engineering and finance, are performed within each unit. This type of structure results in duplication of resources, as similar activities and processes are performed by different elements of the organization on different projects. A second disadvantage, due to the limited life span of projects, is that work assignments and reporting hierarchies are subject to continuous change. This can have a detrimental effect on career paths and professional growth.

Figure 5-2 depicts the project-oriented organizational structure. As can be seen, functional units are duplicated across projects. These units are coordinated by the corresponding central functional unit, but the degree of coordination may vary sharply. The higher the level of coordination, the closer the organizational structure is to the pure functionally oriented structure. Low levels of coordination represent organizational structures closer to the project-oriented structure. For example, consider an organization that has to select a new CAD/CAM (computer-aided design/computer-aided manufacturing) system.

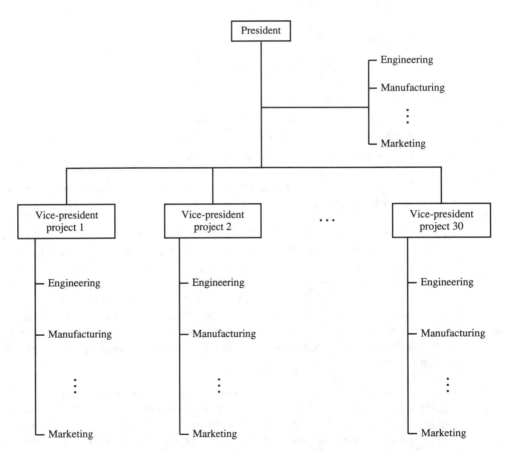

Figure 5-2   Project-oriented organizational structure.

In a functional organization, the engineering department might have the responsibility of selecting the most appropriate system. In a project-oriented organization, each engineering group will select the system that fits its needs best. If, however, it is desirable to achieve commonality and have all engineering groups use the same system, the central engineering department will have to solicit input from the various groups, and based on this input, make a decision that balances the concerns of each. The characteristics of a fully "projectized" organization are highlighted below.

Advantages

Strong control by a single project authority

Rapid reaction time

Encourages performance, schedule, and cost tradeoffs

Personnel loyal to a single project

Interfaces well with outside units

Good interface with customer

Disadvantages

Inefficient use of resources

Does not develop technology with an eye on the future

Does not prepare for future business

Less opportunity for technical interchange among projects

Minimal career continuity for project personnel

Difficulty in balancing workloads as projects phase in and out

In addition to the functional organization and project organization, the following structures may be appropriate.

Product organization.   In a mass production environment where large volumes are common, such as in consumer electronics or chemical processing, the organizational structure may be based on the similarity among products. An organization specializing in domestic appliances, for example, may have a refrigerator division, washing machine division, and small appliances division. This structure facilitates the use of common resources, marketing channels, and subassemblies for similar products. By exploiting commonality it is possible for mixed model lines and group technology cells handling a family of similar products to achieve performance that rivals the efficiency of dedicated facilities designed for a unique product.

Customer organization.   Some organizations have a few large customers. This is frequently the case in the defense industry, where contractors typically deal with one branch of the service. By structuring the contractor's organization around its principal client, it is much easier to establish good working relationships. In many such organizations, as exemplified by consulting and law firms, there is a tendency to hire veteran employees from the customer's organization to smooth communications and exploit personal friendships.

Territorial organization.   Organizational structures can be based on territorial considerations, too. Service organizations that have to be located close to the customer tend to be structured along geographical lines. The same applies to marketing organizations that need to keep close contacts with the specific segment of the market they serve. More recently, with the push toward reduced inventories and just-in-time delivery, large manufacturers are encouraging their suppliers to set up plants or warehouses in the neighborhood of the main facility.

5.2.1 Project–Functional Interface and Matrix Organization

Projects are essentially horizontal, while the functional organization, as exemplified by the traditional organization chart, is vertical. The basic dichotomy between the two can be better understood by comparing the types of questions project and functional managers ask. Table 5-1, suggested by Cleland and King (1983), highlights the differences.

A project can be viewed as a small business within a larger enterprise whose ultimate goal is to go out of business when all tasks are completed. At any point in time, the enterprise has a stream of projects flowing through it in various phases of completion. The challenge faced by upper management is to juggle budgets, resources, and schedules to keep the stream of projects flowing smoothly.

Matrix organization.   A hybrid structure known as the matrix organization provides a sound basis for balancing the use of human resources and skills as people are shifted from one project to another. The matrix organization can be viewed as a project organization superimposed on a functional organization with well-defined interfaces between project teams and functional elements. In the matrix organization, duplication of functional units is eliminated by assigning specific resources of each functional unit to each project. Figure 5-3 depicts an organization that is performing several projects concurrently. Each project has a manager who must secure the required resources from the functional groups. Technical support, for example, is obtained from the engineering department and sales

TABLE 5-1 CONCERNS OF PROJECT AND FUNCTIONAL MANAGERS

Project manager	Functional manager
What is to be done?	How will the task be done?
When will the task be done?	Where will the task be done?
What is the importance of the task?	Who will do the task?
How much money is available to do the task?	How well has the functional input been integrated into the project?
How well has the total project been done?	

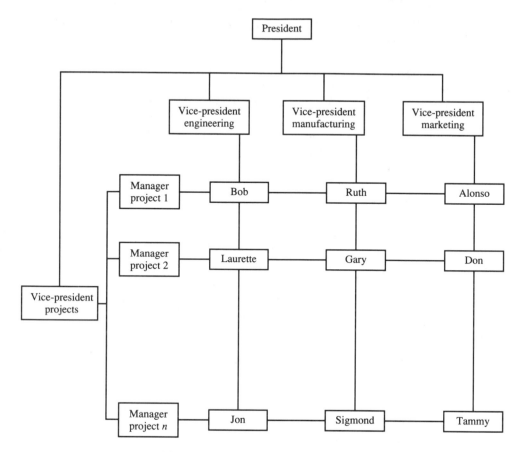

Figure 5-3 Typical matrix organization.

estimates are provided by marketing. The project manager's request for support is handled by the appropriate functional manager, who assigns resources based on availability and the project's relative level of need. The two must act as partners to coordinate operations and the use of resources. It is the project manager, though, who is ultimately responsible for the success or failure of the project. Important benefits of the matrix organization are:

1. *Better utilization of resources.* Because the functional manager assigns resources to all projects, he or she can allocate resources in the most efficient manner. The limited life span of projects does not reduce utilization of resources since they can be reassigned as the need arises.

2. *State-of-the-art technology.* The knowledge gained from various projects is accumulated at the functional level. The most sophisticated projects are sources of new technology and skills that can be transferred to other projects and activities performed by the organization.

3. *Adaptation to changing environment.* The matrix organization can adapt to changing conditions, including the arrival of new competition in the market, the termination of existing projects, and the realignment of suppliers and subcontractors. The functional skeleton is not affected by such changes, and resources can be reallocated and rescheduled as needed. No loss of knowledge is experienced when projects terminate since the experts are kept within the functional units.

The matrix organization benefits from having focused effort in both the functional and project dimensions. However, this advantage may be offset by several potential difficulties:

1. *Authority.* Whereas the resources are under the control of the functional manager in the long run, it is the project manager who assigns them work on a day-to-day basis. In a matrix organization, this can lead to a conflict of interest and to a "dual boss" phenomenon.
2. *Technical knowledge.* The project manager is not likely to be an expert in all technical aspects of a project. The manager has to rely on the functional experts and the functional managers for their inputs but, once again, must take responsibility for the overall outcome.
3. *Communications.* Workers have to report to their functional manager and to the project manager to whom they are assigned. Double reporting and simultaneous horizontal/vertical communication channels are difficult to develop, manage, and maintain.
4. *Goals.* The project manager tends to see the short-term objectives of the project most clearly, whereas the functional manager typically focuses on the longer-term goals, such as accumulation of knowledge and the acquisition and efficient use of resources. These different perspectives frequently conflict and create friction within an organization.

The design and operation of a matrix organization is a complicated, time-consuming task. A well-conceived and well-managed structure is necessary if the impact of the problems listed above is to be minimized.

On the plus side, a matrix organization is well suited to apply "management by objectives" (MBO) techniques (Carroll and Tosi 1973), which gained popularity in the early 1970s. The basic elements of an effective MBO program include:

1. Precise goal setting and planning by top-level management
2. Organizational commitment to the approach
3. Mutual goal setting
4. Frequent performance reviews
5. Some degree of freedom in developing means
   for the achievement of the objectives

In general, each project and functional unit has a set of objectives that must be balanced against a set of mutually agreed upon performance measures. This balance depends on the weight given to each objective and is an important determinant is selecting the organizational structure. For example, if the successful completion of projects on time and within budget is considered most important, the matrix organization will be more project oriented. In the case where functional goals are emphasized, the matrix organization can be designed to be functionally oriented.

The orientation of a matrix organization can be measured to some degree by the percentage of workers who are fully committed to a single project. If this number is 100%, the organization has a project-oriented structure. If none are fully committed, the organization has a functional structure. A range of matrix organizations can be defined between these two extremes, as depicted in Fig. 5-4. In the figure, functional organizations are located on the left-hand side and project-oriented organizations on the right. Those in between are hybrids of varying degree. An organizational structure that is based on one part-time person managing each project while everyone else is a member of a functional unit represents a very weak matrix structure with a strong functional orientation. On the other hand, if the common arrangement is project teams with only a few shared experts among them, we have a matrix organization with a strong project structure orientation.

In summary, the principal advantages and disadvantages of the matrix organization are:

Advantages

Combines strengths of functional and project-oriented organizations

Good interface with outside contacts

Effective multidisciplinary task integration

Effective use of production resources

Effective project control

Career continuity and professional growth

Perpetuates technology

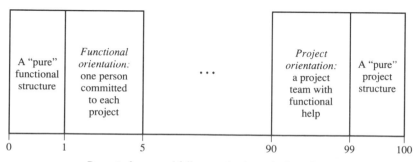

Percent of personnel fully committed to a single project

Figure 5-4   Range of matrix organizations.

Disadvantages

Dual accountability of personnel

Conflicts between project and functional managers

Profit-and-loss accountability difficult

5.2.2 Criteria for Selecting an Organizational Structure

The decision to adopt a specific organizational structure is based on several criteria, as discussed below.

1. *Technology.* A functional organization and a process-oriented organization have one focal point for each type of technology. The knowledge gained in all operations, projects, and products is accumulated at that focal point and is available to the entire organization. Furthermore, experts in different areas can be used efficiently since they, too, are a resource available to the entire organization.

2. *Finance and accounting.* These functions are easier to perform in a functional organization, where the budgeting process is controlled by one organizational element capable of understanding the "whole picture." Such an entity is in the best position to develop a budget that integrates the organizational goals within individual project objectives.

3. *Communications.* The functional organization has clear lines of communication that follow the organizational structure. Instructions flow from the top down, while progress reports are directed over the same channels from the bottom up. The functional organization provides a clear definition of responsibility and authority, and thus minimizes ambiguity in communications.

Product-, process-, or project-oriented structures have vertical as well as horizontal communication lines. In many cases, communications between units responsible for the same function on different projects, processes, or product lines might not be well defined. The organizational structure itself is subject to frequent changes as new projects or products are introduced, existing projects are terminated, or obsolete lines are discontinued. These changes affect the flow of information and cause communications problems.

4. *Responsibility to a project/product.* The product- or project-oriented organization removes any ambiguity over who has responsibility for each product manufactured or project performed. The project manager has complete control over all resources allocated to the project along with the authority to use those resources as he or she sees fit. The one-to-one relationship between an organizational element and a project or product eliminates the need for coordination of effort and communication across organizational units, and thus makes management easier and more efficient.

5. *Coordination.* As mentioned, the project/product-oriented structure reduces the need for coordination of activities related to the project or product; however, more coordination is required between organizational units performing the same function on different products.

6. *Customer relations.* The project/product-oriented organization provides the customer with a single point of contact. Any need for service, documentation, or support can be handled by the same organizational unit. Accordingly, this structure supports better communications and frequently better service for the customer compared to the functional structure. Its performance closely approximates that of a pure customer-oriented organizational structure.

This partial list demonstrates the fact that no single structure is optimal for all organizations in all situations. Therefore, each organization must analyze its own operations and select the structure that best fits its needs, be it functional, process oriented, customer oriented, project/product oriented, or a combination thereof.

5.2.3 Linear Responsibility Chart

An important tool for the design and implementation of a project oriented organization is the *linear responsibility chart* (LRC). The LRC (also known as the *matrix responsibility chart* or *responsibility interface matrix*) summarizes the relationships between project participants and their responsibilities in each element of a project. An element can be a specific activity, an authorization to perform an activity, a decision, or a report. The columns of the LRC represent managers of organizational elements; the rows correspond to project elements performed by the organization. Each cell denotes an activity and the organizational element to which it is assigned. The level of participation of the organizational element in the project element is defined within each cell.

By reading down a column of the LRC, one can get a picture of the responsibilities of each manager; reading across a row gives an indication of which department is responsible for the work. Thus each cell in the LRC corresponds to a work package, as explained later in the chapter. An example of a LRC is shown in Table 5-2. The notation used in the table is defined as follows:

A   *Approval.* Approves the work package.

P   *Primary responsibility.* Indicates who is responsible for accomplishing the work package.

TABLE 5-2 EXAMPLE OF AN LRC

Activity	Engineering	Manufacturing	Contracts	Project manager	Marketing	Quality assurance
Respond to RFP	I	I	O,A	P	B	A
Negotiating contract	I,N	I,N	I,R	P		A
Preliminary design	P	A	R	O,B		A
Detailed design	P	A	R	O		A
Execution	R	P		O,B		R
Testing	I	I		O,B		P
Delivery	N	N	P	A	N	A

R *Review*. Reviews output of the work package. For example, the legal department reviews a proposal bid package submitted by the team leader.

N *Notification*. Notified of the output of the work package. As a result of this notification, the person makes a judgment as to whether or not any action should be taken.

O *Output*. Receives the output of the work package and integrates it into the work being accomplished. For example, the contract administrator receives a copy of the engineering change orders so that the effects of changes on the terms and conditions of the contract can be determined.

I *Input*. Provides input to the work package. For example, a "bid/no bid" decision on a contract cannot be made by a company unless inputs are received from the manufacturing manager, financial manager, contract administrator, and the marketing manager.

B *Initiation*. Initiates the work package. For example, new product development is the responsibility of the R&D manager, but the process generally is initiated with a request from the marketing manager.

If A, R, and B are not separately identified, P is assumed to include them. The LRC in Table 5-2 corresponds to a single project. Similar matrices can be constructed for each project in the portfolio.

The LRC depicts authority, responsibility, and communication channels in a clear, concise way. Taken as a whole, it is a blueprint of the activity and information flows that occur at the interfaces of an organization. As Cleland (1990) points out, once the LRC is developed, it can be sorted for each organizational position, first by the P's, then by the I's, and so on. When the manager reviews the sorted work packages associated with the unit, he immediately sees a list of activities for which he has direct responsibility and others in which he plays a supportive role.

The LRC is a robust tool and conveys as much information as several pages of job descriptions and policy documents. It provides a means for all participants in a project to view their responsibilities and agree upon their assignments. It shows the extent or type of authority exercised by each participant in performing an activity in which two or more parties have overlapping involvement, and it clarifies command relationships that may otherwise be ambiguous when persons share work. By using the LRC and by defining authority, responsibility, and communications within its framework, the relationship between the organizational units and the work content of the project is made clear.

5.3 ORGANIZATIONAL STRUCTURE OF PROJECTS

The organizational structure of a project should be designed as early as possible in the project's life cycle. A clear definition of communication channels, responsibilities, and the authority of each participating unit is a key element affecting project success. The most appropriate structure depends on the nature of the project, on the environment in which the work is performed, and on the structure of the participating organizations. In most projects it is not enough to adopt the organizational structure of the prime contractor. At a minimum, both the client and contractor organizations must be considered. The client organization usually initiates the project by defining its specific needs, while the contractor is responsible for developing the plan to satisfy those needs. The two may be members of the same organization (e.g., an engineering department that develops a new product "for" the marketing department), or they may be unrelated (e.g., a contractor for the National Aeronautics and Space Administration). In either case the relationship between these organizations is defined by the project organizational structure. This definition should specify the responsibility of each party—the client's responsibility to supply information or components for the project, such as government-furnished equipment, and the contractor's responsibility to perform certain tasks, to supply progress reports, to consult periodically with the client, and so on.

5.3.1 Factors in Selecting a Structure

The primary factors that should be taken into consideration when selecting an organizational structure for managing projects are as follows:

1. *Number of projects and their relative importance.* Most organizations are involved in projects. Common examples are the installation of a new management information system, the introduction of an organizational change, or the cultivation of a new market. If an organization is dealing with projects only infrequently, a functional structure supported by ad hoc project coordinators may be best. As the number of projects increases and their relative importance (measured by the budget of all projects as a percentage of the organizational budget, or any other method) increases, the organizational structure should adapt by moving to a matrix structure with a stronger project orientation.

2. *Level of uncertainty in projects.* Projects may be subject to different levels of uncertainty that affect cost, schedule, and performance. To handle uncertainty, a feedback control system is used to detect deviations from original plans, and to detect trends that might lead to future deviations. It is easier to achieve tight control and to react faster to the effects of uncertainty when each project manager controls all the resources used in the project and gets all the information regarding actual performance directly from those who are actively involved. Therefore, a project-oriented structure is preferred when high levels of uncertainty are presented.

3. *Type of technology used.* When a project is based on a number of different technologies, and the effort required in each area does not justify continuous effort throughout the project life cycle, the matrix organization is preferred. If each project concentrates on a technology that is mastered by one functional area, the functional organization and project coordinators of each functional area are the best choice. When projects are based on several technologies and the work content in each area is sufficient to employ at least one full-time person, a project-oriented structure is preferred.

Research and development projects where new technologies or processes are developed are subject to high levels of uncertainty regarding task completion times, the likelihood of a contemplated breakthrough, or simply the chances that the project's components can be integrated successfully. Therefore, a project-oriented structure may be best for them.

4. *Project complexity.* High complexity that requires very good coordination among the project team is best handled in a project-oriented structure. Here communication is most rapid and unobstructed. Low-complexity projects can be handled effectively in a functional organization or a matrix arrangement with a functional orientation.

5. *Duration of projects.* Short projects do not justify a special organization and are best handled within the matrix organization. Long projects that span many months or years justify a project-oriented structure.

6. *Resources used by projects.* When common resources are shared by two or more projects, the matrix arrangement with a functional orientation tends to be best. This is the case when expensive resources are used or when each project does not need a fully devoted unit of a resource. If the number of common resources among projects is small, the project-oriented structure is preferred.

7. *Overhead cost.* By sharing facilities and services among projects, the overhead cost of each project is reduced. A matrix organization should be preferred if an effort to reduce overhead cost is required.

8. *Data requirements.* If many projects have to share the same databases and the information generated by projects should be made available immediately to organizational elements not directly involved in these projects, an organizational structure with a functional orientation is preferred.

In addition to these factors, the organizational structures of the client and the contractor also must be taken into account. If both have a functional orientation, direct communication between similar functions in the two organizations might be most appropriate. If both are project/product oriented, an arrangement that supports direct communication links between project managers in their respective organizations would be most efficient.

The situation is complicated when the contractor and the client do not have similar organizational structures, or when there are several participating units. If the contractor is functionally oriented, the client project manager may have to deal simultaneously with many departments as well as a host of subcontractors, government agencies, and private consultants. In such cases the project organization should be designed to anticipate and moderate conflicts that arise in scheduling activities competing for resources, setting priorities for different phases of the work, assigning workers to tasks, and managing technological changes.

5.3.2 Project Manager

The organizational structure selected for a project is a primary determinant of its success. Once this structure is defined, the selection of personnel for management positions within this structure must be considered. Important attributes of project managers were mentioned briefly in the introduction (Section 1.4.2). These attributes are now discussed in more detail.

Leadership.   The most essential attribute of a project manager is leadership. The project manager has to lead the project team through each phase of its life cycle, dealing swiftly and conclusively with any number of problems as they arise along the way. This is made all the more difficult given that the project manager usually lacks full control and authority over the participants. His or her ability to guide the project team smoothly from one stage to the next depends on the person's stature, temperament, skills of persuasion, and the degree of commitment, self-confidence, and technical knowledge. A manager who possesses these characteristics in some measure is more likely to be successful even when his or her formal authority is limited.

Interpersonal skills.   The project manager (as any manager) has to achieve a given set of goals through other people. The manager must deal with his or her own superiors, the members of the project team, the functional managers, and perhaps, an array of clients. In addition, the manager frequently must interact with representatives from other organizations, including subcontractors, laboratories, and government agencies. To achieve the goals of the project, the ability to develop and maintain good personal relationships with all parties is crucial.

Communication skills.   The interaction between groups involved in a project and the project manager takes place through a combination of verbal and written communications. The project manager must be kept abreast of progress and be able to transmit directions in a succinct and unambiguous manner. By building reliable communication channels and by using the best channel for each application, the project manager can achieve a fast, accurate response from the team with some degree of confidence that his or her directions will be carried out correctly. The more up-to-date and comprehensive the information, the smoother the implementation route will be.

Decision-making skills.   The project manager has to recognize problems and establish the procedures by which they will be documented and addressed. Once the source and nature of a problem are identified, the manager must evaluate alternative solutions, select the best corrective action, and see that it is implemented. These are the fundamental steps in project control.

In some instances, the project manager gets involved early enough to participate in discussions regarding the organizational structure of the project and the selection of technological solutions. An understanding of the basic technical issues gives the project manager the credibility needed to influence resource allocation, budget, and schedule decisions before they are finalized. His or her input on these matters in the initial stages increases the probability that the project will get started in the right direction.

Negotiation and conflict resolution.   Many of the problems the project manager faces do not have a "best solution": for example, when a conflict of interest exists between the project manager and the client over a contract issue contingent on various interpretations. There are many sources of conflicts:

- *Scheduling:* disagreements that develop around the timing, sequencing, duration of projects, and feasibility of schedule for project-related tasks or activities
- *Managerial and administrative procedures:* disagreements that develop over how the project will be managed: the definition of reporting relationships and responsibilities, interface relationships, project scope, work design, plans of execution, negotiated work agreements with other groups, and procedures for administrative support
- *Communication:* disagreements resulting from poor information flow among staff or between senior management and technical staff, including such topics as misunderstanding of project-related goals, the strategic mission of the organization and the flow of communication from technical staff to senior management
- *Goal or priority:* disagreements arising from lack of goals or poorly defined project goals, including disagreements regarding the project mission and related tasks, differing views of project participants over the importance of activities and tasks, or the shifting of priorities by superiors/customers
- *Resource allocation:* disagreements resulting from the competition for resources (e.g., personnel, materials, facilities, and equipment) among project members or across teams, or from lack of resources or downsizing of organizations
- *Reward structure/performance appraisal or measurement:* disagreements that originate from differences in understanding the reward structure and from the insufficient match between the project team approach and the performance appraisal system
- *Personality and interpersonal relations:* disagreements that focus on interpersonal differences rather than on "technical" issues; includes conflicts that are ego-centered, personality differences, or conflicts caused by prejudice or stereotyping
- *Costs:* disagreements that arise from the lack of cost control authority with either the project management or functional group. Disagreements related to the allocation of funds
- *Technical, opinion:* disagreements that arise, particularly in technology-oriented projects, over technical issues, performance specifications, technical trade-offs, and the means to achieve performance
- *Politics:* disagreements that center on issues of territorial power (not-invented-here attitudes) or hidden agendas
- *Poor input or direction from leaders:* disagreements that arise from a need for clarification from upper management on project-related goals and the strategic mission of the organization

- *Ambiguous roles/structure:* disagreements, especially in the matrix structure, where two or more persons or sections have related or overlapping assignments or roles
- *Unresolved prior conflict:* disagreements stemming from prior unresolved conflicts

Thamhain and Wilemon (1975) have observed empirically that conflict intensity and its source vary over the life cycle of a project. Figure 5-5 illustrates this relationship. The project manager must have the negotiation skills required to minimize the occurrence of disputes and to resolve them satisfactorily when they arise.

Trade-off analysis skills.   Because most projects have multidimensional goals (performance, schedule, budget, etc.) that are incommensurate, the project manager often has to perform trade-off analyses to reach a compromise solution. Questions like, "Should the project be delayed if extra time is required to achieve the performance levels specified?" or "Should more resources be acquired at the risk of a cost overrun to reduce a schedule delay?" are common and must be resolved by trading off one objective for another.

In addition to these skills and attributes, a successful project manager will embody good organizational skills, the ability to manage time effectively, a degree of openmindedness, and loyalty to his or her charge. The correct selection of the project organizational

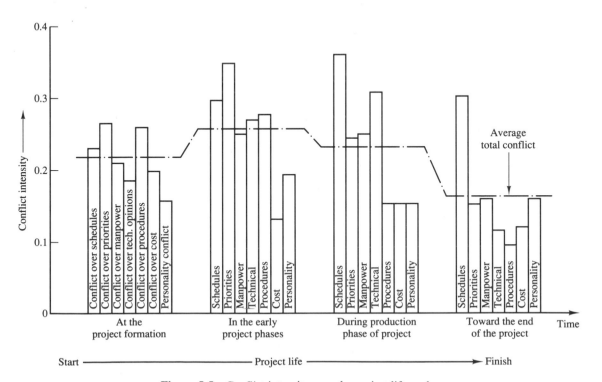

Figure 5-5   Conflict intensity over the project life cycle.

structure and a good choice for project manager are two important decisions that are made early in the life cycle of a project and have a lasting impact.

When a project organizational structure is selected and managers are appointed, the responsibility of each manager within this structure can be defined with respect to each project task. As mentioned, the LRC can be used to depict the respective roles of the participants in the project.

5.4 WORK BREAKDOWN STRUCTURE

The blueprint for the execution of a project is a plan that defines the objectives, deliverables, and specifications for each task to be accomplished. In addition, a schedule, a budget, the required resources, and most important, an indication of individual responsibility must be provided. Tasks may be characterized by their length, work content, level of technology, and cost. Some tasks are complex and expensive, spanning months or years, whereas others are short and present little technical difficulty. It is often convenient to break the longer tasks down into subtasks of shorter duration. Once this is done, it is possible to arrange all tasks and subtasks in a network resembling a directed tree with a single root node. The resultant figure, known as the project work breakdown structure (WBS), is an important aid in planning and managing the project. The construction of an actual WBS is situation dependent. However, some guidelines relating to the design of military systems are presented in MIL-STD-881A, where the WBS is defined as follows: "A work breakdown structure is a product-oriented family tree composed of hardware, services and data which result from project engineering efforts during the development and production of a defense material item, and, which completely defines the project/program. A WBS displays and defines the product(s) to be developed or produced and relates the elements of work to be accomplished to each other and to the end product."

The WBS is a schematic presentation of the disaggregation–integration process by which the project manager plans to execute the project. This process, once again, is the heart of project management. The work content of a project has to be divided into tasks that can be assigned and performed by one of the participating organizational units. If such tasks cannot be defined, the project plan is not feasible. The definition of a task at the lowest level of the WBS should include the following elements.

1. *Objectives.* A statement is made of what is to be achieved by performing this task. The objectives may include tangible accomplishments such as the successful production of a part or a successful integration of a system. Nontangible objectives are also possible, such as learning a new computer language.

2. *Deliverables.* Some deliverables may be part of the hardware and software used in the project, like a pump in the hydraulic system of a new airplane. Other deliverables might include a report that documents the findings of an economic analysis, or a recommendation made after evaluating a number of scenarios with a computer model.

3. *Schedule.* For each task at least two milestones should be specified: its planned start time and its planned finish time. The elapsed time between these two dates is not

necessarily the estimated duration of the task since some leeway may be built into the plan.

4. *Budget.* A time-phased budget should be prepared for each task. In so doing, projected outlays should be synchronized with the planned schedule and estimated cost of the respective task.

5. *Performance measures.* The successful completion of a task has to be judged by predefined performance measures. These measures are used during project execution to compare actual and planned performance in order to establish project control. A total quality management approach to project management integrates cost, schedule, and quality into a common performance measure called the earned value. These concepts are explained in Chapter 6.

6. *Responsibility.* The organizational unit responsible for on-budget and on-schedule performance of each task has to be defined. This is done by associating a lower-level element in the WBS (a task) with an organizational unit in the project organizational breakdown structure (OBS). The entity that consists of a task to be performed by an organizational unit for a given schedule and budget is called a *work package.* The role of work packages is elaborated in Section 5.5.

The division of the project work content into work packages should reflect the way in which the project will be executed. If, for example, a university initiates a project to develop a new MBA program, the development of a specific course for the program can be defined as a task, and the organizational unit responsible for that course (a professor) can be associated with the task to form a work package. There are, however, different ways to decompose the work content of this project. One way is to divide the entire project directly into work packages. If there are 30 courses required in the program and each course is developed by one professor, there will be 30 work packages in the WBS. This is illustrated in Fig. 5-6. This structure can be represented by the following coding scheme:

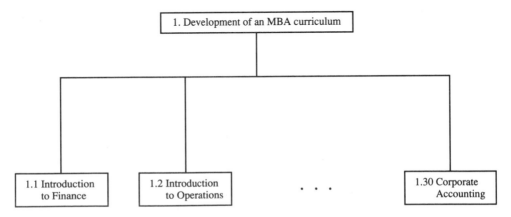

Figure 5-6 Two-level WBS.

1. Development of an MBA program curriculum
   1.1 Introduction to Finance
   1.2 Introduction to Operations
    .
    .
    .

   1.30 Corporate Accounting

Alternatively, the project manager may decide to disaggregate the work content by functional area and have each such area divide the work content further into specific courses assigned to professors. This situation is illustrated in Fig. 5-7. Using an expanded coding scheme, the WBS in this case might take the following form:

1. Development of an MBA program curriculum
   1.1 Development of courses in finance
      1.1.1 Introduction to Finance
      1.1.2 Financial Management
     .
     .
     .

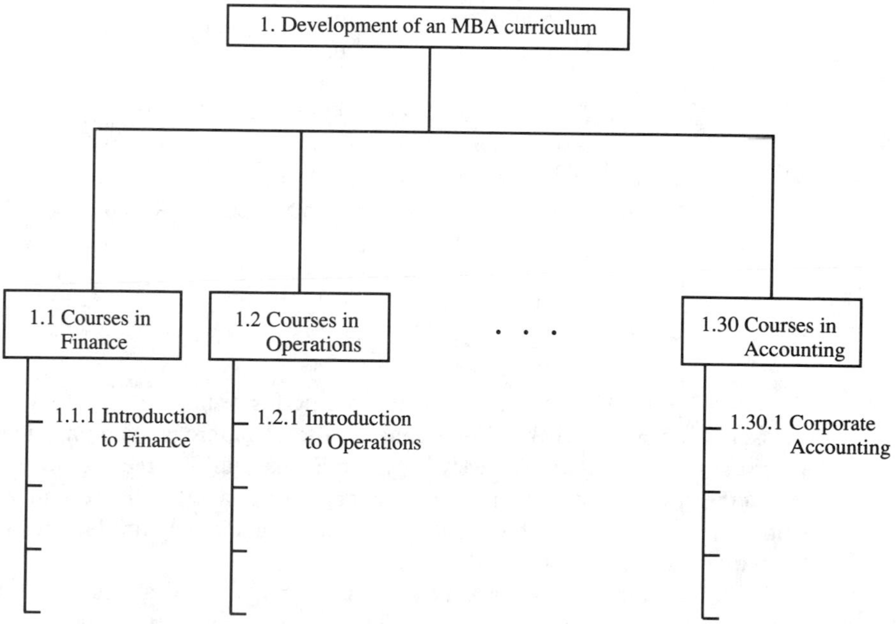

Figure 5-7   Three-level WBS.

1.2 Development of courses in operations
  1.2.1 Introduction to Operations
  1.2.2 Practice of Operations Management
    .
    .
    .

1.6 Development of courses in accounting
  1.6.1 Fundamentals of Accounting
    .
    .
    .

  1.6.4 Corporate Accounting

A third option that the project manager might consider is to divide the work content according to the year in the program in which the course is taught and then divide it again by functional areas. This WBS is illustrated in Fig. 5-8 and might take the following form:

1. Development of an MBA curriculum
  1.1 First-year courses
    1.1.1 Development of courses in finance
      1.1.1.1 Introduction to Finance
      .
      .
      .

  1.2 Second-year courses
    1.2.1 Development of courses in finance
      1.2.1.1 Financial Management
      .
      .
      .

    1.2.6 Development of courses in accounting
      1.2.6.1 Management Information Systems in Accounting
      .
      .
      .

    1.2.6.4 Corporate Accounting

For all three WBSs, the same 30 tasks are performed at the lowest level by the same professors. However, each WBS represents a different approach to organizing the project. The first structure is "flat." There are only two levels and from the organizational point of view all the professors report directly to the project manager. In the second WBS, there is one intermediate level—the functional committee—and in the third instance there are two intermediate levels.

As a second example consider the construction of a new assembly line for an existing product. To capitalize on experience and minimize risk, the design may be identical to that of the existing facilities; alternatively, a new design that exploits more advanced

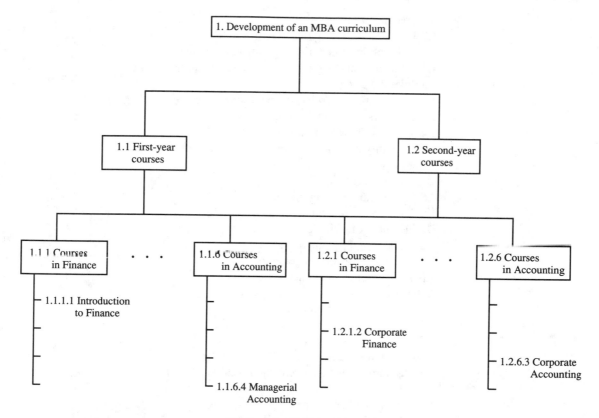

Figure 5-8 Four-level WBS.

technology may be sought. In the latter case, the WBS might include automated material handling equipment, an updated process design, and the development of production planning and control systems. One possible WBS follows:

1. New assembly line
    1.1 Process design
        1.1.1 Develop a list of assembly operations
        1.1.2 Estimate assembly time for each operation
        1.1.3 Assignment of operations to workstations
        1.1.4 Design of equipment required at each station
    1.2 Capacity planning
        1.2.1 Forecast of future demand
        1.2.2 Estimates of required assembly rates
        1.2.3 Design of equipment required at each station
        1.2.4 Estimate of labor requirements

1.3 Material handling
    1.3.1 Design of line layout
    1.3.2 Selection of material handling equipment
    1.3.3 Integration design for the material handling system
1.4 Facilities planning
    1.4.1 Determination of space requirements
    1.4.2 Analysis of energy requirements
    1.4.3 Temperature and humidity analysis
    1.4.4 Facility and integration design for the whole line
1.5 Purchasing
    1.5.1 Equipment
    1.5.2 Material handling system
    1.5.3 Assembly machines
1.6 Development of training programs
    1.6.1 For assembly line operators
    1.6.2 For quality control personnel
    1.6.3 For foremen and managers
1.7 Actual training
    1.7.1 Assembly line operators
    1.7.2 Quality control
    1.7.3 Foremen and managers
1.8 Installation and integration
    1.8.1 Shipment of equipment and machines
    1.8.2 Installations
    1.8.3 Testing of components
    1.8.4 Integration and testing of line
    1.8.5 Operations
1.9 Management of project
    1.9.1 Design and planning
    1.9.2 Implementation monitoring and control

The decision on how to disaggregate the work content of a project is entwined with the decision on how to structure the project organization. In making these decisions, the project manager not only establishes how the work content will be decomposed and then integrated, but also lays the foundation for the project planning and control systems.

Most projects can be structured in several ways. The choice depends on any number of factors, such as the duration of the project, the work content of the project, levels of uncertainty, the organizational structure, resource availability, and management style, so there is no one "correct" way. Nevertheless, the WBS selected should be complete in the sense that it captures all the work to be performed during the project. It should be detailed in the sense that at its lowest level, executable tasks with specific schedules, budgets, and objectives are specified. And it should be accurate in the sense that it represents the way management envisions first decomposing the work content and then integrating the completed tasks into a unified whole.

The OBS should also be complete, detailed, and accurate: complete in the sense that it should depict all the organizational units that will participate in the project; detailed in the sense that each organizational unit is represented down to the level where the work is actually being performed; and accurate in the sense that it reflects the true lines of authority, responsibility, and communication.

The linear responsibility chart integrates the two structures by assigning lower-level WBS elements to lower-level OBS elements. This can be done only if both the WBS and OBS are accurate and comprehensive, as explained above.

The design of the work breakdown structure is guided by the complexity of the project, its goals, and the constraints imposed on it. A complex project requires detailed design and several levels of management control. Therefore, the accompanying WBS should have several levels, with the most detailed breakdown at the bottom. The goals of the project may influence the WBS as well. Risky projects that require close monitoring of costs and schedules should also be supported by a detailed WBS.

5.5 COMBINING THE ORGANIZATIONAL AND WORK BREAKDOWN STRUCTURES

The two structures—the organizational structure and the work breakdown structure—form the basis for project planning, management, and control. The hierarchical nature of these structures provides for a *roll-up* mechanism whereby the information gathered and processed at any level can be aggregated and rolled up to its predecessor. This mechanism supports the layers of management typically found in modern organizations. At the intersection of the lowest levels of these structures, building blocks are formed. These building blocks are work packages in which a specific lowest-level organizational unit is assigned a specific task on the lowest-level WBS. At the work package level, the task to be performed is subdivided into specific activities each defined by its work content, timetable, budgeted cost, and expected results or output.

In operational terms, the work package is the smallest unit used by the project manager for planning and control. Further disaggregation is undertaken by the person charged with getting the work done (say, the group leader), who converts the work package into a set of basic tasks and activities. For example, "Introduction to Operations" is a work package in the project outlined in Fig. 5-6. Let's assume that the corresponding execution responsibilities have been assigned to an operations management instructor. To complete the assignment properly, the instructor must divide the work package into tasks and activities. These might include collecting syllabi from institutions that offer a similar course, establishing a list of possible topics, deciding what material to cover on each topic, developing a detailed bibliography, evaluating case studies, generating exercises and discussion questions, and so on.

The person responsible for a work package is responsible for the detailed scheduling, budgeting, and resource planning of its constituent tasks. The development of the OBS–WBS relationship is the first step in the integration of these functions. By planning, controlling, and managing the execution of a project at the work package level, lines of responsibility are clear and the effect of each decision made on each element of the project can be traced to any level of the OBS or WBS. The persons responsible for work packages are those dealing with the actual performance of the work. The relationship between project managers, client representatives, and functional management is defined by the organizational structure and the responsibility matrix.

The work breakdown structure, the organizational breakdown structure, and their integration into work packages along with the linear responsibility chart are the cornerstones of project management. They provide the framework for developing and integrating the tools needed for scheduling, budgeting, technological management, and control.

5.6 MANAGEMENT OF HUMAN RESOURCES IN PROJECTS

Of the many types of resources used in projects (people, equipment, machinery, data, capital), human resources are the most difficult to manage. Unlike other resources, human beings seek motivation, satisfaction, and security and need an appropriate climate and culture to achieve high performance. The problem becomes even more complicated in a project environment since successful completion of the project depends on team effort. Working groups or teams are the common organizational units within which individual efforts are coordinated to achieve a common goal. A team is well integrated when information flows smoothly, trust exists among its members, each person knows his or her role in the project, morale is high, and the desire is for a high level of achievement.

5.6.1 Project Team

In the project environment where workers from many disciplines join to perform multifunctional tasks, the importance of teamwork is paramount. The issues center on how to build a team, how to manage it, and what kind of leadership is most appropriate for a project team. The objective of team building is to transform a collection of individuals with different objectives and experiences into a well-integrated group in which the objectives of each person promote the goals of the group. The limited life of projects and the frequent need to cross the functional organizational lines make team building a complicated task.

Members of a new project team may come from a variety of organizational units or may be newly hired personnel. To build an efficient team, personal and organizational uncertainty and ambiguity must be reduced to a minimum. This is done by clearly defining, as early as possible, the project, its goals, its organizational structure (organizational chart), and the procedures and policies that will be followed during execution.

Each person joining the project must be given a job description that defines reporting relationships, responsibilities, and duties. Task responsibilities must also be defined. Each person needs to know the tasks in which he or she is expected to participate and in what capacity. The linear responsibility chart is a useful tool for defining individual tasks and responsibilities. Once the roles of all team members have been established, they should be introduced to each other properly and their functions explained. Continuous efforts on the part of the project manager are required to keep the team organized and highly motivated.

An ongoing effort is also required to detect any problems and to ensure that appropriate correction measures are taken.

Physical proximity can help in building team spirit. If possible, team members should be located in the same office area to facilitate communication and understanding. Open channels of communication should be established and maintained, especially with regard to the project objectives and the role of each team member, in the effort to achieve those objectives.

The roles of team members tend to change over time as the project moves from one phase of its life cycle to the next. Because confusion and uncertainty cause conflict and inefficiency, the project manager should frequently update team members regarding their roles in the project. Furthermore, the manager should detect any morale or image problems as early as possible in an effort to identify and eliminate the cause of such problems. For example, the appearance of cliques or isolated members should serve as a signal that the team is not being managed properly.

The project manager should also help in reducing anxieties and uncertainty related to "life after the project." When a project reaches its final stages, the project manager should discuss the future role in the organization of each team member and prepare a plan that ensures a smooth transition to that new role. By providing a stable environment and a clear project goal, the managers can help team members focus on the job at hand.

A recommended practice for management is to conduct regular team meetings throughout the life cycle of the project, but perhaps more frequently in the early phases where uncertainty is highest. In a team meeting, plans, problems, operating procedures, and policies should be discussed and explained. By identifying future problems and preparing an agreed-upon plan, the probability of success is increased and the probability of conflicts is reduced or eliminated altogether.

5.6.2 Encouraging Creativity and Innovation

The one-time nature of projects requires solutions to problems that have not been dealt with in the past. The ability to apply past solutions to present problems may be limited. The human ability to innovate and create new ideas should thus be exploited properly by the project manager and maintained at top performance by encouraging team members to think, create, and innovate.

To flourish, creativity and innovation require the appropriate climate. The various ways and means by which management has tried to establish the proper conditions have been well documented in literature and include quality circles, suggestion boxes, and rewards for new ideas that are implemented. Sherman (1984) interviewed key executives in eight leading U.S. companies to study the techniques used to encourage innovation. Following are some of his findings:

Organizational level

- The search for new ideas is part of the organizational strategy. Continuous effort is encouraged and supported at all levels.

- Innovation is seen as a means for long-term survival.
- Small teams of people from different functions are used frequently.
- New organizational models such as quality circles, product development teams, and decentralized management are tested frequently.

Individual level

- Creative and innovative team members are rewarded.
- Fear that the status quo will lead to disaster is a common motivator for individual innovation.
- The importance of product quality, market leadership, and innovation is stressed repeatedly and thus is well known to employees.

These findings are in line with the work of Drucker (1985), who suggested that innovation and creativity can be encouraged and properly managed. To enhance innovation, Drucker proposed a systematic process that starts by analyzing the sources of new opportunities in the market: namely, users' needs and expectations. The analysis uses such techniques as *quality function deployment* and the *house of quality* (Hauser and Clausing 1988).

Once a need is identified, a focused effort is required to fulfill that need. Such an effort is based on knowledge, ingenuity, free communication, and well-coordinated hard work. The entire process should aim at a solution that will be the standard and trend setter for that industry. Techniques that support individual creativity and innovation are usually designed to organize the process of thinking and include:

1. A list of questions regarding the problem or the status quo
2. Influence diagrams that relate elements of a problem to each other
3. Models that represent a real problem in a simplified way, such as physical models, mathematical programs, and simulation models

A project manager can enhance innovation by selecting team members who are experts in their technical fields with a good record as problem solvers and innovators in past projects. The potential of individuals to innovate is further enhanced by teamwork and the application of proper techniques, such as brainstorming (Osborn 1957) and the Delphi method.

Brainstorming is used as a tool for developing ideas by groups of individuals headed by a session chairman. The session starts by the chairman presenting a clear definition of the problem at hand. Group members are invited to present ideas, subscribing to the following rules:

- Criticism of an idea is barred absolutely.
- Modification of an idea, or its combination with another idea, is encouraged.
- Quantity of ideas is sought.
- Unusual, remote, or wild ideas are encouraged.

A major function of the chairman is to stimulate the session with new ideas or direction. A typical session lasts up to an hour and is brought to an end at the onset of fatigue.

The Delphi technique is used to structure intuitive thinking. It was developed by the RAND Corporation (Bright 1968) as a tool for the systematic collection of expert opinions from a group of experts. Unlike brainstorming, the members of the group need not be in the same physical location. Each member gets a description of the problem and submits a response. These responses are collected and fed back anonymously to the group members. Each person then considers whether he or she wants to modify earlier views or contribute more information. Iterations continue until there is convergence to some form of consensus.

Turoff (1973) implemented the Delphi method on a time-shared computer with terminals available to all participants. His pioneering work is the basis of many computer-based systems used today for brainstorming and Delphi-like sessions.

In addition to these two approaches, a number of other techniques are available to support creativity and innovation by groups. For a comprehensive review, see Warfield et al. (1975). As a final example, we mention the *nominal group technique*, which works as follows:

1. A problem or topic is given and each team member is asked to prepare a list of ideas that might lead to a solution.
2. Participants present their ideas to the group, one at a time, taking turns. The team leader records the ideas until all lists are exhausted.
3. The ideas are presented for clarification. Team members can comment on or clarify each of the ideas on the list.
4. Participants are asked to rank the ideas on the list.
5. The group discusses the ranked ideas and ways to expand or implement them.

5.6.3 Leadership, Authority, and Responsibility

Leadership is the ability to influence a group and to direct its activities. Jago (1982) states that leadership is the use of noncoercive influence to direct the activities of members of a group to accomplish its collective goals. Tannenbaum and Schmidt (1958) suggested that there is a continuum of leadership behavior between two extremes: boss-centered leadership (level 1) and subordinate (or people)-centered leadership (level 7):

1. A boss-centered leader who makes a decision and announces it
2. A leader who makes decisions but "sells" the decisions to the group
3. A leader who presents ideas and invites questions
4. A leader who presents a tentative decision that is subject to change
5. A leader who presents a problem, gets suggestions, and makes a decision
6. A leader who presents a problem, defines limits, and asks the group to make a decision

7. A subordinate-oriented leader who sets limits and lets the group members function within these limits

By adopting the right leadership style, the project manager can stimulate teamwork and creativity among project team members. By adopting a people-oriented leadership style, the manager can increase the generation and presentation of new ideas, promote communication, and increase individual job satisfaction. However, if this style is adopted, the project manager must feel that his or her authority and leadership are unquestioned.

The responsibility of a project manager is typically to execute the project in such a way that the prespecified deliverables will be ready within the time and budget planned. This responsibility must come with the proper level of legal authority, implying that leadership and authority are related: a manager cannot be a leader unless he or she has authority. Authority is the power to command or direct other people. There are two sources of authority: legal authority and voluntarily accepted authority. *Legal authority* is based on the organizational structure and a person's organizational position. It is delegated from the owners of the organization to the various managerial levels and is usually contained in a document. *Voluntarily accepted authority* is based on personal knowledge, interpersonal skills, or a person's experience, which enable him or her to exercise influence over and above their legal authority. The project manager should have well-defined legal authority in the organization and over the project. However, a good project manager will seek voluntarily accepted authority from the team members and organizations involved in the project, on the basis of his or her personal skills.

The importance of legal authority is most pronounced in a matrix organization, where the need to work with functional managers and to utilize resources that "belong" to functional units can trigger conflicts. Reduction of these conflicts depends on the formal authority definition, as well as on the ability of both the project manager and the functional manager to be flexible.

5.6.4 Ethical and Legal Aspects of Project Management

The legal authority of a project manager and his or her role as a leader require proper understanding of the legal and ethical aspects of project management. Ireland et al. (1982) developed the following "Code of Ethics" to address these points:

Preamble: Project managers, in the pursuit of their profession, affect the quality of life for all people in our society. Therefore, it is vital that project managers conduct their work in an ethical manner to earn and maintain the confidence of team members, colleagues, employees, clients and the public.

Article I: Project managers shall maintain high standards of personal and professional conduct.

    a. Accept responsibility for their actions.
    b. Undertake projects and accept responsibility only if qualified by training or experience, or after full disclosure to their employers or clients of pertinent qualifications.

c. Maintain their professional skills at the state-of-the-art and recognize the importance of continued personal development and education.

d. Advance the integrity and prestige of the profession by practicing in a dignified manner.

e. Support this code and encourage colleagues and co-workers to act in accordance with this code.

f. Support the professional society by actively participating and encouraging colleagues and co-workers to participate.

g. Obey the laws of the country in which work is being performed.

Article II: Project managers shall, in their work:

a. Provide the necessary project leadership to promote maximum productivity while striving to minimize costs.

b. Apply state-of-the-art project management tools and techniques to ensure schedules are met and the project is appropriately planned and coordinated.

c. Treat fairly all project team members, colleagues and co-workers, regardless of race, religion, sex, age or national origin.

d. Protect project team members from physical and mental harm.

e. Provide suitable working conditions and opportunities for project team members.

f. Seek, accept and offer honest criticism of work, and properly credit the contribution of others.

g. Assist project team members, colleagues and co-workers in their professional development.

Article III: Project managers shall, in their relations with employers and clients:

a. Act as faithful agents or trustees for their employers or clients in professional or business matters.

b. Keep information on the business affairs or technical processes of an employer or client in confidence while employed, and later, until such information is properly released.

c. Inform their employers, clients, professional societies or public agencies of which they are members or to which they may make any presentations, of any circumstances that could lead to a conflict of interest.

d. Neither give nor accept, directly or indirectly, any gift, payment or service of more than nominal value to or from those having business relationships with their employers or clients.

e. Be honest and realistic in reporting project cost, schedule and performance.

Article IV: Project managers shall, in fulfilling their responsibilities to the community:

a. Protect the safety, health and welfare of the public and speak out against abuses in those areas affecting the public interest.

b. Seek to extend public knowledge and appreciation of the project management profession and its achievements.

This code establishes guidelines for the ethical responsibilities of the project manager. Similar guidelines are required for the manager's legal responsibilities. These, however, depend on the specific organizations involved, the contract, and the laws of the country where the project is performed. The following legal aspects are common to most projects:

- Contractual issues regarding clients, suppliers, and subcontractors
- Government laws and regulations
- Labor relations legislation

As a rule of thumb, whenever the project manager is not sure of the legal aspects of a decision or a situation, he or she should consult the legal staff of the organization.

Legalities are very important when an organization contracts to carry out a project or parts of a project for a customer or when an organization uses subcontractors. A large variety of contract types exist, commonly classified into fixed-cost and cost-reimbursable contracts. Among the first class, two major subclasses can be identified: (1) firm fixed price (FFP) contracts, and (2) fixed price incentive fee (FPIF) contracts. Under FFP contracts, the contractor assumes full responsibility for cost, schedule, and technical aspects of the project. This type of contract is suitable only when the levels of uncertainty are low, technical specifications are well defined, and schedule and cost estimates are subject to minimal errors. The FPIF contract is designed to encourage performance above a preset target level. Thus if the project is completed ahead of schedule or under cost, an incentive is paid to the contractor. In some FPIF contracts, a penalty is also specified in case of cost overruns or late deliveries. By specifying a target that can be achieved with high probability, the risk that the contractor takes is minimized, while the incentive motivates the contractor to try and do better than the specified target.

Cost-reimbursable contracts are also classified into two major types: (1) cost plus fixed fee (CPFF), and (2) cost plus incentive fee (CPIF) contracts. The former are designed for projects in which most of the risk associated with cost overrun is borne by the customer. This type of contract is appropriate when it is impossible to estimate costs accurately, as, for example, in R&D projects. On top of the actual cost of performing the work, an agreed-upon fee is paid to the contractor. CPIF contracts are designed to guarantee a minimum profit to the contractor while motivating the contractor to achieve superior cost, schedule, and technical performance. This is done by paying an incentive for performance higher than expected and tying the level of incentive to the performance level.

Within the four types of contracts, there are many variations. The proper contract for a specific project depends on the levels of risk involved, the ability of each party to assume part of the risk, and the relative negotiating power of the participants. Although the legal staff is usually responsible for contractual arrangements, the project manager has to execute the contract, so his or her ability to establish good working relationships with the client, suppliers, and subcontractors within the framework of the contract is extremely important.

In addition to the contracts, the project manager should be familiar with government laws and regulations in areas such as labor relations, safety, environmental issues, patents, and trade regulations. Whenever a question arises, the project manager should consult the legal staff. It is highly recommended that the project manager be trained in the basics of labor relations legislation, as an important part of his or her work is to manage the human resources.

Each country has its own labor relations legislation, and managers of international projects must not assume that these regulations are the same or even similar from one country to the next. Typically, these regulations have to do with minimum wages, benefits, work conditions, equal employment opportunity, the employment of the handicapped, and occupational safety and health.

To summarize, the management of human resources is probably the most difficult aspect of project management. It requires the ability to create a project team, to manage it, to encourage creativity and innovation without being threatened, and to deal with human resources in and out of the organization. The project manager can learn some of these skills, but a majority of them only come with experience, common sense, and inherent leadership qualities.

TEAM PROJECT

Thermal Transfer Plant

At the last TMS board meeting, approval was given to develop a new area of business: recycling and waste management. Because your supporting analysis was the determining factor, your team has been asked to develop an organizational structure for TMS that will integrate this new area with its current business. You are also required to develop a detailed OBS and WBS for a project aimed at designing and assembling a prototype rotary combustor, for which only the power unit will be manufactured in-house; other parts will be purchased or subcontracted. In developing the OBS and WBS for the project, clearly identify the corresponding hierarchies and show who has responsibility at each level.

In your report explain your objectives and the criteria used in reaching a decision. Show why the selected structure is superior to the alternatives considered, and explain how this structure relates to the TMS organization as a whole. Your report will be submitted to TMS management for review. Be prepared to present the major points to your management and to defend your recommendations.

DISCUSSION QUESTIONS

1. Describe the organizational structure of your school or company. What difficulties have you encountered working within this structure?
2. Explain how a matrix organization can perform a project for a functional organization. What are the difficulties, contact points, and communication channels?
3. In the matrix management structure, the functional expert on a project has two bosses. What considerations in a well-run organization reduce the potential for conflict?
4. Write a job description for a project manager in a matrix organization. Assume that only the project manager is employed full time by the project.
5. How does the work breakdown structure affect the selection of the organizational breakdown structure of a project?
6. Under what conditions can a functional manager act as a project manager?
7. Develop a list of advantages and disadvantages of the following structures.
   (a) Product organization
   (b) Customer organization
   (c) Territorial organization
8. What kind of OBS is used in the company or organization to which you belong? What are the limitations that you have perceived?
9. What are the activities and steps involved in developing an LRC?
10. Describe the "team building" inherent in the development of an LRC. How is team building accomplished on large projects? How does this relate to development of the LRC?
11. Discuss the applicability of the nominal group technique, the Delphi method, and brainstorming to the process of scheduling and budgeting a project.
12. Compare the advantages and disadvantages of the four types of contracts discussed in this chapter.
13. Of the types of leadership discussed, which is most appropriate for a high-risk project?

EXERCISES

5-1. Develop an organizational structure for a project performed in your school (e.g., the development of a new degree program). Explain your assumptions and objectives.
5-2. You are in charge of designing and building a new solar heater. Develop the organizational breakdown structure and the work breakdown structure. Explain the relationship between the two.
5-3. Develop an organizational breakdown structure for an emergency health care unit in a hospital. How should this unit be related to the other departments in the hospital?
5-4. Develop a work breakdown structure for a construction project.
5-5. Consider the development of a new electric car by an auto manufacturer and a manufacturer of high-capacity batteries.
   (a) Develop an appropriate four-level WBS.
   (b) Develop the OBS.
   (c) Use work packages to relate the WBS elements to the OBS.

5-6. Suggest three approaches (OBS–WBS combinations) for the development of a new undergraduate program in electrical engineering.

5-7. Develop an LRC for a project done for a client having a functional organization by a contractor having a customer-oriented organization.
    (a) Describe the project and its WBS.
    (b) Describe the OBS of the client and the contractor.

5-8. You are the president of a startup company specializing in computer peripherals such as optical backup units, tape drives, signature verifications systems, and data transfer devices. Construct two OBSs and discuss the advantages and disadvantages of each.

5-9. List two activities that you have recently performed with two or more other people. Explain the role of each participant using an OBS, a WBS, and an LRC.

5-10. Give an example of an organization with an ineffective or cumbersome structure. Explain the problems with the current structure and how these problems could be solved.

5-11. You have been awarded the contract to set up a new restaurant in an existing building at a local university (in other words, there is no need for external construction). The WBS for the project, as developed by the planning team, is presented in Fig. 5-9. Using this WBS, carry out the following exercises.

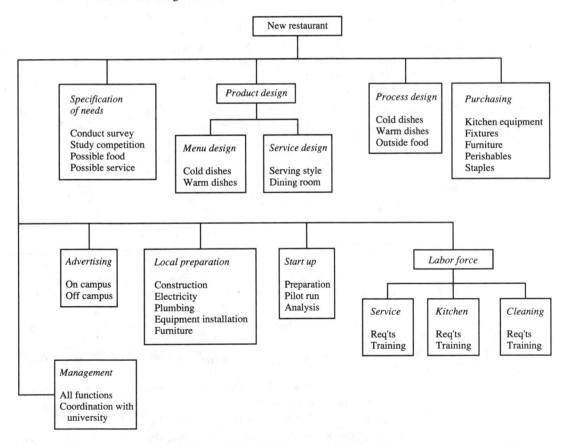

Figure 5-9   WBS for new restaurant.

(a) Develop a coding system for the project.

(b) Identify other types of projects that could use this coding system. For which types of projects would it be inappropriate? Explain.

(c) If you wish to use a more general coding system that deals with construction, what would be the differences between the latter and the more specific coding system developed in part (a)?

5-12. You have been offered a contract to undertake the restaurant project in Exercise 5-11 at several campuses belonging to the same university.

(a) Suggest an OBS for these projects.

(b) Generate three work packages and assign them to the appropriate organizations.

(c) Identify some areas that will require coordination among the organizations included in the OBS to assure that the three work packages will be completed properly.

(d) Construct a linear responsibility chart for coordinating the work among the various functions that are to be carried out.

5-13. For the restaurant project in Exercise 5-11:

(a) Develop another WBS, making sure that it includes the same work packages that are shown in the original WBS in Fig. 5-9.

(b) Generate additional work packages for the project and add them to the new WBS.

5-14. You have been assigned the task of developing a network representation of the project in Exercise 5-11 (network construction is taken up in much greater detail in Chapter 7).

(a) Design the network for the WBS in Fig. 5-9. In so doing, each work package in the WBS should correspond to a node in the network, and each arc should indicate a precedence relation. Include in your diagram a dummy start node and a dummy end node.

(b) Extend your network by including several activities for each work package.

5-15. Prepare a Delphi session for selecting the best project manager for a given project.

5-16. Develop a set of guidelines for project managers in international projects dealing with legal and ethical issues.

REFERENCES

Organizational Structures

ANDERSON, C. C., and M. M. K. FLEMING, "Management Control in an Engineering Matrix Organization: A Project Engineer's Perspective," *Industrial Management*, Vol. 32, No. 2 (1990), pp. 8–13.

CARROLL, S. J., JR., and H. L. TOSI, JR., *Management by Objectives: Applications and Research*, Macmillan, New York, 1973.

CHAMBERS, G. J., "The Individual in a Matrix Organization," *Project Management Journal*, Vol. 20, No. 4 (1989), pp. 37–42, 50.

DIMARCO, N., J. R. GOODSON, and H. F. HOUSER, "Situational Leadership in a Project/Matrix Environment," *Project Management Journal*, Vol. 20, No. 1 (1989), pp. 11–18.

GALBRAITH, J. R., *Organization Design*, Addison-Wesley, Reading, MA, 1977.

HELLRIEGEL, D., and J. W. SLOCUM, JR., "Preferred Organizational Designs and Problem Solving Styles: Interesting Companions," *Human Systems Management*, Vol. 1 (1980), pp. 151–158.

KINGDON, D. R., *Matrix Organization: Managing Information Technologies*, Tavistock, London, 1973.

LIKERT, R., and J. G. LIKERT, *New Ways of Managing Conflict*, McGraw-Hill, New York, 1976.

LORSCH, J. W., and P. R. LAWRENCE, *Studies in Organization Design*, Irwin Dorsey, Chicago, 1970.

McCOLLUM, J. K., and J. D. SHERMAN, "The Effects of Matrix Organization Size and Number of Project Assignments on Performance," *IEEE Transactions on Engineering Management*, Vol. 38, No. 1 (1991), pp. 75–78.

MINTZBERG, H., *The Structure of Organizations*, Prentice Hall, Englewood Cliffs, NJ, 1979.

TAKAHASHI, N., "Sequential Analysis of Organization Design: A Model and a Case of Japanese Firms," *European Journal of Operational Research*, Vol. 36, No. 3 (1988), pp. 297–310.

Project Organization

ANDERSON, J., *Organizing for Large Project Management: The Client's Needs*, McKinsey, Chicago, 1978.

BAKER, B. N., D. C. MURPHY, and D. FISHER, "Factors Affecting Project Success," in D. I. CLELAND and W. R. KING (editors), *Project Management Handbook*, Van Nostrand Reinhold, New York, 1983, pp. 669–685.

BARKER, J., D. TJOSVOLD, and I. R. ANDREWS, "Conflict Approaches of Effective and Ineffective Project Managers: A Field Study in a Matrix Organization," *Journal of Management Studies*, Vol. 25, No. 2 (1988), pp. 167–178

CABLE, D., and J. R. ADAMS, *Organizing for Project Management*, Project Management Institute, Pittsburgh, PA, 1982.

CLELAND, D. I., *Project Management: Strategic Design and Implementation*, TAB Books, Blue Ridge Summit, PA, 1990.

CLELAND, D. I., and W. R. KING, *Systems Analysis and Project Management*, Third Edition, McGraw-Hill, New York, 1983.

DINSMORE, C., "Human Factors," in *Project Management*, Revised Edition, American Management Association, New York, 1990.

FRAME, J. D., *Managing Projects in Organizations*, Jossey-Bass, San Francisco, 1987.

HOUSE, R. S., *The Human Side of Project Management*, Addison-Wesley, Reading, MA, 1988.

LARSON, E. W., and D. H. GOBELI, "Matrix Management: Contradictions and Insights," *California Management Review*, Vol. 29, No. 4 (1987), pp. 126–138.

MEREDITH, J. R., and S. J. MANTEL, JR., *Project Management: A Managerial Approach*, Second Edition, Wiley, New York, 1989.

THAMHAIN, H. J., *Engineering Program Management*, Wiley, New York, 1984.

THAMHAIN, H. J., and D. L. WILEMON, "Conflict Management in Project Life Cycles," *Sloan Management Review*, Vol. 16, No. 3 (1975), pp. 31–50.

YOUKER, R., "Organizational Alternatives for Project Management," *Project Management Quarterly*, Vol. 8, No. 1 (1977), pp. 18–24.

Work Breakdown Structure

LAVOLD, G. D., "Developing and Using the Work Breakdown Structure," in D. I. CLELAND and W. R. KING (editors), *Project Management Handbook*, Van Nostrand Reinhold, New York, 1983, pp. 303–323.

McCann, J. E., and T. N. Gilmore, "Diagnosing Organizational Decision Making through Responsibility Charting," *Sloan Management Review*, Vol. 24 (1983), pp. 3–15.

MIL-STD-881, *A Work Breakdown Structure for Defense Military Items*, U.S. Department of Defense, Washington, DC, 1975.

Stewart, R. D., and A. L. Stewart, *Proposal Preparation*, Wiley, Toronto, Ontario, Canada, 1984.

Human Resources

Bright, J. R., *Technological Forecasting for Industry and Government*, Prentice Hall, Englewood Cliffs, NJ, 1968.

Drucker, P., "The Discipline of Hard Work," *Harvard Business Review*, Vol. 62 (May–June 1985), pp. 67–72.

Hauser, J. R., and D. Clausing, "The House of Quality," *Harvard Business Review*, Vol. 66, No. 3 (1988), pp. 62–73.

Ireland, L. R., W. J. Pike, and J. L. Schrock, "Ethics for Project Managers," *Proceedings of the 1982 PMI Seminar Symposium of Project Management*, Toronto, Ontario, Canada.

Jago, A. G., "Leadership: Perspectives in Theory and Research," *Management Science*, Vol. 28, No. 3 (1982), pp. 315–336.

Osborn, A., *Applied Imagination*, Scribner, New York, 1957.

Sherman, P. S., "Eight Big Masters of Innovation," *Fortune* (October 15, 1984), pp. 66–81.

Tannenbaum, R., and W. H. Schmidt, "How to Choose a Leadership Pattern," *Harvard Business Review*, Vol. 36, No. 2 (1958), pp. 95–101.

Turoff, M., "Human Communication via Data Networks," *Computer Decisions* (January 1973).

Warfield, J. N., H. Geschka, and R. Hamilton, *Methods of Idea Management*, Battelle Institute and Academy of Contemporary Problems, Columbus, OH, 1975.

6

Technological Aspect: Configuration Selection, Management, and Control

6.1 TECHNOLOGICAL, FUNCTIONAL, QUALITY, AND RISK CONSIDERATIONS

In its February 1989 report on management of technology, the special task force of the National Research Council lists "project management" as one of eight primary areas in which U.S. industry must improve if it is to remain competitive. The report views project management as a tool for integrating the technological aspects of projects with scheduling, budgeting, and resource concerns. The more efficiently this can be done, the easier it will be to develop and implement new technologies. In previous chapters, the problems associated with project selection and project structuring were addressed. In this chapter we focus on the technological aspects of project management and the various components of the decision-making process.

6.1.1 Competition and Technology

Technology has a competitive impact in two primary ways. First, it provides a market advantage through differentiation of value added; and second, it provides a cost advantage through improved overall system economies. To use technology effectively, companies must be explicit about its role. This requires answering four elementary questions: (1) What is the basis of competition in our industry? (2) To compete, which technologies must we master? (3) How competitive are we in these areas? (4) What is our technology strategy? In embryonic and growth industries, technology frequently drives the strategy, while in more mature fields, technology must be an enabling resource for manufacturing, marketing, and customer service. The U.S. excels at technology-driven innovation that creates whole new enterprises. By contrast, Japan excels at incremental advances in existing products and processes.

6.1.2 Product, Process, and Support Design

The technological aspects of projects are related to the design, manufacturing, and support of the product or system for which the project exists. Design activities begin with an analysis of the client's or organization's needs, which are translated into required methods of operation. Once approved by the client or upper management, these requirements are transformed into functional and technical specifications. The last link in the chain is detailed product, process, and support design. Product design centers on the structure and shape of the product. Performance, cost, and quality goals must all be defined. Process design deals with the preparation of a series of plans for manufacture, integration, testing, and quality control. In the case of an item to be manufactured, this means selecting the processes and equipment to be used during production, setting up the part routings, defining the information flows, and assuring that adequate testing procedures are put into place.

Support design is responsible for selecting the hardware and software that will be used to track and monitor activities once the system becomes operational. This means developing databases, defining report formats, and specifying communication protocols for the exchange of data. A second support function concerns the preparation of manuals for operators and maintenance personnel. Related issues center on the design of maintenance facilities and equipment, and the development of policies for inventory management. Both process design and support design include the design of training for those who manufacture, test, operate, and maintain the system.

Design efforts are present in many nonengineering projects. Such efforts are required to transform needs into the blueprint of the final product. For example, consider the design of a new insurance policy or a change in the structure of an organization. In the first case, new needs may be detected by the marketing department: for example, a need to provide insurance for pilots of ultralight airplanes. The designer of this new policy should consider the risk involved in flying ultralights and the cost and probability of occurrence associated with each risk. In addition to the risk to the pilot due to accidents, damage to the ultralight or to a third party needs to be considered. The designer of this new policy has to decide what options should be available to the customer and how the different options should be combined with each other.

In the second case, new products, changes in the business environment, and new technologies may generate a need to change the structure of an organization. For example, if a new product is very successful in a traditional organization and the business associated with this product becomes a major part of the total organizational business, a special division may be needed to manufacture, market, and support this product. The designer of the new organizational structure should consider questions related to the size of the new division, its mission, and its relationship with the existing parts of the organization.

In some projects, the design effort represents the most important component of the work. An example would be an architect who is designing a new building, or a team of communication experts designing a satellite relay network. Usually, design is the basis for production or implementation, depending on the context. In many situations, the design effort may consume only a small portion of the assigned budget and resources. Nevertheless, decisions made in the conceptual design and advanced development phases are likely to have a significant effect on the total budget, schedule, resource requirements, and overall success of the project.

Management of the design effort, from identifying a specific need to implementation of the end product, is the core of the technological aspect of project management. The fact that design takes place in the early stages of most projects does not imply that technological management efforts cease once the blueprints are drawn. Changes in design are notoriously common throughout the life cycle of a project and have to be managed carefully.

The project manager has to consider four major factors when dealing with the management of technology: quality, cost, risk, and performance, the latter being measured by the functional attributes of the system. The tools for assessing each of these factors in the initial stages of a project were discussed in Chapter 2, Engineering Economic Analysis; Chapter 3, Project Screening and Selection; and Chapter 4, Multiple-Criteria Methods for Evaluation. In Chapter 10 we discuss life-cycle costing and show how (design) decisions made early in the project affect the total life-cycle cost. To underscore the importance of good tools, we cite a National Science Foundation study which has shown that over 70% of the life-cycle cost of a product is defined at the conceptual and preliminary design stages. Information and decision support systems play a dynamic role in these stages by focusing management's efforts on the technology and providing feedback to the design team in the form of assessment data.

Our aim here is to indicate how the techniques discussed previously can be used throughout the life cycle of a project to manage its technological aspects. Frequently, the design is subject to change due to newly identified needs, changing business conditions, and evolution of the technology. Therefore, the management of the design (or technological management) is a continuing process that ends only with delivery of the finished product to the customer and perhaps, not even then. Manufacturer warranties and an insistent desire for product improvement in some markets may keep a project alive well after delivery of the product(s).

6.1.3 Importance of Time

In the global market, successful companies will be those that learn to make and deliver goods and services faster than do their competitors. These "turbo marketers," as Kotler and Stonich (1991) term them, will have a distinct advantage in markets where customers value time compression enough to pay a premium or to increase purchases. Moreover, in certain high-tech areas such as semiconductor manufacturing and telecommunications, where performance is increasing and price is dropping almost daily, survival depends on the rapid introduction of new technologies.

Once a company has examined the demand for its product, it can begin to reduce cycle time. In general, three principles can be applied: (1) reorganize the work, (2) organize and reward to encourage time compression, and (3) pursue cycle time reduction aggressively. Although the implementation effort and cost required to reduce cycle time will be substantial, the payoff can be great. To create a sustainable advantage, companies must couple the so-called "soft" aspects of management with programs aimed at achieving measurable time-based results.

In this regard, a recent trend in technology management is to perform all major components of design concurrently. This approach, aptly known as *concurrent engineering*, is based on the concept that the parallel execution of the major design components will shorten project life cycles and thus reduce the time to market for new products. In an era of time-based competition where the shelf life of some high-tech items may be as short as 6 months, this can make the difference between mere survival and material profits.

Studies by the consulting firm McKinsey & Co. have shown repeatedly that being a few months late to market is even worse than having a 30% development cost overrun. Figure 6-1 points up the difference in revenue when a product is on time or late. The model underlying the graph assumes that there are three phases in the product's commercial life: a growth phase (where sales increase at a fixed rate regardless of entry time), a stagnation phase (where sales level off), and a decline phase (where sales decrease to zero). The figure shows that a delay causes a significant decline in revenue. Suppose that a market has a 6-month growth period followed by a year of stagnation and a decline to zero sales in the succeeding 8 months. Then, being late to market by 3 months would reduce revenues by 36%. Thus a delay of one-eighth of the product lifetime reduces income by over one-third. Such a loss can be especially severe since the largest profits are usually realized during the growth phase.

The application of concurrent engineering principles to technology management requires thoughtful planning and oversight. There is a clear need to inform the product

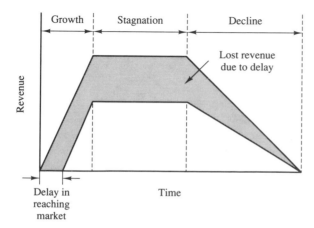

Figure 6-1   Lost revenue due to delay in reaching market.

engineers, process engineers, and support specialists of current status of the design and to keep them updated on all change requests. Due to the rapid propagation of errors, the risk of cost and schedule overruns, faltering performance, and team bickering is greatly increased when control is lax. An error in one aspect of the design may have repercussions that quickly extend to all functional areas.

In the following sections we explore the issues surrounding concurrent engineering and describe the risk aspects of technological management. We also discuss several approaches for managing design information and introduce the concept of total quality management for projects.

6.2 CONCURRENT ENGINEERING AND TIME-BASED COMPETITION

The ability to design and produce high-quality products that satisfy a real need at a competitive price was, for many years, almost a sure guarantee for commercial success. With the explosion of electronic and information technology, a new factor—time—has become a critical element in the equation. The ability to reduce the time required to develop new products and bring them to market is considered by many the next industrial battleground. For example, the Boeing 777 transport design took a year and half less than its predecessor the 767, permitting the company to introduce it in time to stave off much of the competition from the European Airbus. Similarly, John Deere's success in trimming development time for new products by 60% has enabled it to maintain its position as world leader in farm equipment in the face of a growing challenge from the Japanese. This was done using the concurrent engineering (CE) approach to support time-based competition.

The basic idea of concurrent engineering is to use project scheduling and resource management techniques in the design process. These techniques, discussed in Chapters 7 and 9, have always been common to the production phase but are now recognized as vital to all life-cycle phases of a project from start to finish. In a CE environment, teams of experts from different disciplines work together to ensure that the design progresses smoothly and that all the participants share the same, most recent information.

The concurrent engineering approach replaces the conventional sequential engineering approach, in which new product development starts by one organizational unit (e.g., marketing), which lays out product specifications based on customer needs. These specifications are used by engineering to come up with a product design, which in turn serves as the basis for manufacturing engineering to develop the production processes and flows. Only when this last step is approved does support design begin.

Sequential engineering takes longer because all the design activities are strictly ordered. Furthermore, cycles are quite common throughout the project's life cycle. This is the case when product specifications prepared by marketing cannot be met by the available technology (unless, perhaps, a risky or costly R&D effort is undertaken). Similarly, manufacturing engineering may not be able to translate product design into process design, due to technological difficulties or the absence of adequate support (e.g., it may not be economically practical to develop test equipment for a product that has not been designed with testing in mind). In each of these examples, primary activities have to be repeated, creating cycles, and thus lengthening the entire design process, making it that much more expensive.

Put succinctly, concurrent engineering depends upon designing, developing, testing, and building prototype parts and subsystems concurrently, not serially, while at the same time designing and developing the equipment to fabricate the new product or system. This does not necessarily mean that all tasks are performed in parallel, but rather, that the team members from the various departments make their contribution in parallel. A prime objective of CE is to shorten the time from conception to market (or deployment in the case of government or military systems), so as to be more competitive or responsive to evolving needs.

The basis of CE is teamwork, parallel operations, information sharing, and constant communication among the team members. To be most effective, the group should be multidisciplinary, comprised of one or more representatives from each functional area of the organization. The watchword is cooperation. After a century of labor–management confrontation and sequestering employees in job categories, hierarchies, and functional departments, many manufacturers are now seeking teamwork, dialogue, and barrier bashing, if only for survival. By performing product, process, and support design in parallel there is a much greater likelihood that misunderstandings and problems of incompatibility will be averted over the project's life cycle. By reducing the length of the design process, overhead and management costs are reduced proportionally, while the elimination of design cycles reduces direct costs as well. From a marketing point of view, a shorter design process results in the ability to introduce new models more frequently and to target specific models to specific groups of customers. This strategy leads to higher market share and consequently, higher sales and profits.

The implementation of CE is based on shared databases, good management of design information (this is the subject of configuration management, which is discussed later in this chapter), and computerized design tools such as CAD/CAM (computer-aided design/computer-aided manufacturing). It is also important to understand that CE is risky and that without proper technological management the results can be calamitous. The two most prominent risks are:

1. *Organizational risks.* The attempt to cross the lines of functional organizations and to introduce changes into the design process is usually met with resistance. One way to overcome this resistance, often manifested by inertia and subtle acts of neglect, is to form, prior to the implementation of CE, project teams made up of people from the various functional areas. In addition, an educational effort aimed at teaching the advantages and the logic of CE can create a positive atmosphere for this new approach.

2. *Technological risks.* The simultaneous effort of product, process, and support design should be well coordinated to ensure that the information used by all the designers is current and correct. The risks associated with a failure to manage this design information in the CE environment is much higher than in sequential engineering, where it is possible to freeze product design once process design starts, and to freeze process design once support design starts.

Companies thinking about using concurrent engineering techniques for the first time should consider projects that have the following characteristics:

1. The project is of a developmental (novel applications of known technology) or applications (routine applications of known technology) type.

2. The team has experience with the technology.

3. The team has received training in total quality management and has had the opportunity to apply the concepts in its work.

4. The scale of the project falls somewhere in the range of 5 to 35 full-time staff members for a period of 3 to 30 months.

5. The goal is a product or family of products with clearly defined features and functions.

6. Success is not dependent on invention or significant innovation.

6.2.1 Time Management

One of the goals of concurrent engineering is to reduce the time it takes to develop and market new products. Before we can say that a reduction has been achieved, we must have some idea of what the current standards are and what controls them. This is not as clear cut as it sounds because few projects proceed smoothly without interference from outside forces. Also, most companies modify their goals as work progresses, making it that much more difficult to measure project length.

Every industry and its constituent firms are in continuous flux, but they are all limited in their flexibility to achieve change. A number of inhibiting factors combine to create a rhythm or tempo in a company that is very difficult to break. Table 6-1 lists some of these factors for manufacturing companies, although each may not be universally applicable at all times. Thoughtful engineering managers develop a feeling for the important factors in their business and how these affect their operations. As far as possible, they quantify them. This provides a baseline against which improvement can be measured. It is clear that many time-sensitive decisions have an impact on the successful operation of a business and that focusing on only one or two factors to the exclusion of the others is rarely optimal. Concurrent engineering is a business activity, not just an engineering activity. Market success is a function of a firm's ability to improve all of its key tempo factors by integrating current engineering decisions with business decisions. Important issues are:

■ *Multifunctional teams.* Many people have written about time management for individuals. Concurrent engineering requires time management for organizations. The prin-

TABLE 6-1   FACTORS AFFECTING THE TEMPO OF MANUFACTURING FIRMS

Technology life	Market forces
Product lifetime costs	Product life
Product development cycle	Process development cycle
Market development cycle	Economic cycle
Workforce hiring/training	Capital/loan acquisition
Long-lead items	Access to limited resources
Manufacturing capacity planning	Competitive product introductions

ciples are the same, but their implementations are somewhat different. Two notorious time wasters are senior people doing junior work and everyone repeating the same tasks. These are both addressed by forming a multifunctional project team from the appropriate departments and carefully assigning responsibilities to the members. Not everyone is needed full time on every team, but the organizing plan should indicate where to get resources when needed on a part-time basis. All team members, whether or not active, should be kept informed of progress so that they do not have to waste time catching up when called into play. Examples of persons falling in this category are patent attorneys, illustrators, and technical specialists needed for tricky problems.

The participation of staff from all major functions—marketing, development, manufacturing, finance, and so on—from day 1 makes a direct contribution to the reduction of duplicate effort. The marketing person can immediately comment on the desirability of some feature before the development person has spent time on it. Similarly, the development staff can get immediate feedback from manufacturing on the producibility of a particular design.

■ *Tools.* The team organization will lose effectiveness if the members are not provided with appropriate tools; today, this usually means access to personal computers, applications software, and system support for CAD, CAE, CIM, and other computer-aided disciplines. Team members must also be trained in the effective use of the tools. Just as the craftsman can do better work than the journeyman, so the trained specialist with a detailed understanding of the capabilities of computer-aided systems can do better work than an untrained person using the same tools.

■ *Team empowerment.* The team organization will also lose effectiveness if there are unnecessary delays in decision making. This is another major source of time wasting. The solution is the well-known empowerment approach, which gives the team the authority to make the great majority of the decisions. This is not to say that upper management should not be informed of those decisions on a regular basis, but that the natural tendency to second guess those doing the work should be restricted. The initial program plan should include some major review points when upper management can influence the course of the project. These meetings should not be determined by the calendar but rather by progress. The same principle is true of meetings among team members. Setting them every Tuesday at 8:00 A.M. usually leads people to spend all day Monday preparing for Tuesday and all day Wednesday responding to Tuesday. Have frequent team meetings but schedule them at short notice to deal with issues as they arise. To use the project scheduling terminology, team meetings are activities, not events. Many companies have difficulty implementing the empowerment requirement because it encroaches on established lines of authority. This is one area where concurrent engineering can actually increase risks, not reduce them, unless special attention is paid to it.

If there is an important role for upper management to play during the course of day-to-day activities, it is in assigning access to limited resources. If two or more teams need access to a special piece of equipment, say, for production trials, there has to be a responsive mechanism in place to set priorities. Again, the initial project plan must cover this situation.

■ *Use of design authorities.* Another approach to facilitate decision making is to appoint design authorities in various areas. For example, there could be a technology design authority, a product design authority, a process design authority, and an equipment design authority. The authorities must be legitimate experts in their fields. They do not necessarily do the design work and may not, in fact, be full-time members of the team. Their role is to help the project manager make the final decision when two or more conflicting approaches have been recommended. The design authority should not be called in until the competing approaches have been documented in equivalent detail. He or she is a last resort to help resolve the sticky issues. By having them available and identified in the plan with their role clearly spelled out, it is possible to facilitate decision making even in complex situations. Nevertheless, the ultimate decision maker is the project manager. The design authorities are consultants who are called on only to evaluate competing solutions and offer their expertise.

■ *Quality.* A major time waster is repeating work because of poor quality. This is where the principles of total quality management come into play. Every person on the team must realize that they have "customers," many of whom are fellow team members, while some are outsiders who will use their products. Developing procedures and habits that focus on delivering satisfaction to the customers, both internal and external, goes a long way in reducing the need to correct or redo work. Obviously, careful selection of team members also goes a long way in assuring high-quality results. Here is where the best interests of a CE team can conflict with the best interests of individuals. Unless the company implementing the procedures takes special steps to prevent it, working on a CE team can limit growth opportunities for individuals and even eliminate career paths. The project manager wants to be assured of high-quality work in all areas and will tend to select people who have already demonstrated their ability to deliver. The problem can be especially acute for junior staff members who have demonstrated their skills in one area but are not given a chance to expand into other areas because they are continually asked to work on projects that require their known skills.

■ *Bureaucracy.* The final time waster of note is lengthy administrative and bureaucratic procedures. Eberhardt Rechtin, a former vice-president of engineering at Hewlett-Packard, once said that an approval takes $2n$ days, where n is the number of levels of approval needed. The obvious solution to this problem is to empower the project team in advance with all the necessary approval authority. Again, this means that the initial project plan must be prepared very carefully. Another approach to shortening the time required for administration is to provide the team leader with the authority to eliminate competitive bidding procedures on certain development items involving known vendors. Other bureaucratic red tape should also be eliminated, although this makes sense even in the absence of concurrent engineering. Many companies assign a full-time administrator/facilitator to CE teams to relieve the project manager of the burden of handling all the details.

■ *External participation.* The best users of concurrent engineering also extend the concept of the project team to involve key vendors and customers. The customers can help minimize the time required to define and specify the product, they can facilitate product acceptance procedures, and they can reduce project risk by either ordering early or at least indicating through a letter of intent what their purchases may be. Vendors can be extraordinarily helpful members of the team by providing technical support for the application of their products and materials, and by providing preferential access to scarce resources. In return, they get some indication of likely sales. If a company uses formal vendor certification procedures, they should extend them to "certifying" selected key vendors as participants in concurrent engineering development programs.

6.2.2 Guideposts for Success

On the surface, the idea of multifunctional product design and development teams along with self-managing work groups seems unassailable. Nevertheless, a critical issue remains: How do you keep such teams from becoming committees with plenty of talk but little action? Tom Peters (1991), a well-known management consultant, has postulated the following guideposts to help organizations implement the team concept.

1. *Set goals/deadlines/or key subsystem tests.* Committees deliberate. Project teams *do.* Successful project teams are characterized by a clear goal, although the exact path is left unclear to induce creativity. Also, three to six strict due dates for subsystem technical and market tests/experiments are set and adhered to religiously.

2. *Insist on 100% assignment to the team.* Members must be obsessed by the project. Forget "one-fifth obsession": key function members must be assigned full-time for the project's duration.

3. *Place key functions on board from the outset.* Members from sales, distribution, marketing, finance, purchasing, operations/manufacturing, and design/engineering should be part of the project team from day 1. Legal, personnel, and others should provide full-time members for part of the project.

4. *Give members authority to commit their function.* With few exceptions, each member should be able to commit resources from his or her function to project goals and deadlines without second-guessing from higher-ups. Top management must establish and enforce this rule from the start. If commitments from members' primary functional areas (e.g., engineering) are conditional, you've got a committee.

5. *Keep team-member destiny in the hands of the project leader.* For consulting firms such as Booz, Allen & Hamilton and McKinsey & Co., life is a series of projects. The team leader might be from San Francisco or Sydney, Australia; either way, his or her evaluation of team members' performance will make or break a career. In general, then, the project boss rather than the functional boss should evaluate team members. Otherwise, the project concept falls flat.

6. *Make careers a string of projects.* A career in a "project-minded company" is viewed as a string of multifunction tasks. How one does on these determines career prospects.

7. *Live together.* Project teams should be sequestered away from headquarters as much as possible. Team camaraderie and commitment depend to a surprising extent on "hanging out" together, isolated from one's normal set of functional colleagues.

8. *Remember the social element.* Spirit is important: "We're in it together." "Mission impossible." High spirits are not accidental. The challenge of the task per se is central. Beyond that, the successful team leader facilitates what psychologist call "bonding." This can take the form of "signing up" ceremonies upon joining the team, frequent (at least monthly) milestone celebrations, and humorous awards for successes and setbacks alike.

9. *Allow outsiders in.* The product development team notion is incomplete unless outsiders participate. Principal vendors, distributors, and "lead" (future test-site) customers should be full-time members. Outsiders not only contribute directly, but also add authenticity and enhance the sense of distinctiveness and task commitment.

10. *Construct self-contained systems.* At the risk of duplicating equipment and support, the engaged team should have its own workstations, local area network, database, and so on. This is necessary to foster an "its-up-to-us-and-we've-got-the-wherewithal" environment. However, the additional risk created by too much isolation must be balanced with the need for self-sufficiency. Problems may arise when it comes time to integrate the project with the rest of the firm.

11. *Permit the teams to pick their own leader.* A champion blessed by management gets things under way. But successful project teams usually *select* and *alter* their own leaders as circumstances warrant. It is expected that leadership will shift over the course of the project, as one role and then another dominates a particular stage (engineering first, then manufacturing, distribution later).

12. *Honor project leadership skills.* No less than a wholesale reorientation of the firm is called for away from "vertical" (functional specialists dominate) and toward "horizontal" (cross-functional teams are the norm). In this environment, horizontal project leadership becomes the most cherished skill in the firm, rewarded by dollars and promotions. Good team skills, for junior members, are also valued and rewarded.

Few experts or practitioners disagree: To halve product development time and constantly improve quality and service, all companies must destroy the walls between functions and commit to perpetual "horizontal" improvement projects. But to become project-focused requires more than appointing teams. Teams and task forces have often ended up adding to, rather than subtracting from, bureaucracy. Care must be taken to avoid the project-turned-committee trap.

6.2.3 Industrial Experience

The success stories of CE implementation in companies such as Boeing, Hewlett-Packard, Raytheon, and John Deere have been amply documented and reported at professional conferences such as those sponsored by the Computer-Aided Acquisition and Logistics Support Society, the Society for Computer-Aided Engineering, the Institute of Industrial

Engineers, and the Society of Manufacturing Engineers. We cite a few examples briefly to gain a better appreciation of the issues.

For *Cadillac*, a 1990 winner of the Malcolm Baldrige National Quality Award, concurrent engineering has come to mean a new culture and a new way of designing and building its extraordinary complex product—luxury cars. Engineers, designers, and assemblers are now members of vehicle, vehicle-systems, and product (parts) teams that work in close coordination rather than belonging to separate, isolated functional areas as before. Assembly line workers, dealers, repair shop managers, and customers provide insight to engineers involved in all stages of design. To inspire cultural change, Cadillac created a position of champion of simultaneous engineering (a role that combines keeping the process on track, preaching to the believers, and motivating the recalcitrant) and sent 1,400 employees to seminars on total quality management. They also established an "Assembly Line Effectiveness Center" where production workers rub shoulders with engineers, critiquing prototypes for manufacturability.

John Deere's Industrial Equipment Division in Moline, Illinois, has had two concurrent engineering efforts. The first, begun in 1984, failed because management retained the traditional manufacturing departments. Designers and process engineers who were assigned to task groups remained loyal to the interests of their disciplines rather than to the overall enterprise. In 1988 the division reorganized. Staff members now report to product teams and answer to team leaders, not to functional department heads. Early in the design stage, teams create a product definition document that describes the product precisely, sets deadlines, and lays out the manufacturing plan. Products no longer change as departments work on them. The result has been gradual improvements in manufacturing processes. There are now fewer experimental designs, and it is possible to produce prototypes in the production environment. The advantage of this is that in addition to checking for flaws in the prototypes themselves, engineers can simultaneously perfect the manufacturing process.

The third example comes from *Federal-Mogul*, a precision parts manufacturer in Southfield, Michigan. The first Federal-Mogul unit to adopt concurrent engineering was its troubled oil-seal business. Other units quickly followed. Success in the oil-seal business, where products are simple but must meet exacting standards, requires rapid turnaround on bids and prototypes and strong customer service. By providing estimates to customers in minutes instead of weeks and producing sample seals in 20 working days instead of 20 weeks, market share has soared. Federal-Mogul accomplished this by adopting a cross-functional product team approach to manufacturing, encouraging consensus building and empowerment, and introducing new information technologies. Key applications include networks allowing all plants to share CAD drawings and machine tools, a scheduling system that notifies appropriate team members automatically when a new order comes in, an engineering data management system, and an on-line database of past orders.

6.2.4 Unresolved Issues

From a technical point of view, what makes concurrent engineering possible are the recent advances in hardware and software, database systems, electronic communications, and the various components of computer-integrated manufacturing. At the first International

Workshop on Concurrent Engineering Design sponsored by the National Science Foundation (Hsu et al. 1990) four themes emerged from the discussions: models, tools, training, and culture. Participants identified measurement issues and trade-offs that will inform future models of new product development. They concluded that tools must focus on expanded CAD/CAM/CAPP capabilities with strong interfaces. Training is needed for multiple job stations, in the impact of design on downstream tasks, and in teamwork and individual responsibility. Corporate culture, and how to change it, must be better understood. Important aspects of culture to be clarified include incentives and performance, myths that inhibit an organization's progress, and the management of change.

One of the primary roles of CE is in identifying the interdependencies and constraints that exist over the life cycle of a product, and ensuring that the design team is aware of them. Nevertheless, care must be taken in the early stages to avoid overwhelming the design team with constraints and in stifling their creativity for the sake of simplicity. A truly creative design that satisfies customer requirements in a superior manner may justify the expense of relaxing some of the development and process guidelines.

Although a basic tenet of CE is that input to the design process should come from all life-cycle stages, there is much ambiguity about how to achieve this. At exactly what point in the CE process should discussion of assembly, sequences, tolerances, and support requirements be introduced? Also, trade-offs abound. For example, consolidation of parts is desirable, yet too much consolidation implies costly and inefficient procurement and inventorying. A balance must be struck between meeting the customer's specifications, designing for manufacturability, and life-cycle cost. This means that cost information should be available to the design teams, not just the accounting department, throughout the project.

6.3 RISK MANAGEMENT

Risk is a major factor in technological management. Whenever the design process or the design itself deviates from current procedures and established techniques, technological risks are introduced. These risks can be related to the product design, to the process design, or to the design of the support system, and can vary widely in magnitude. For example, in product design a low-level risk might be one associated with the modification of an existing subassembly. A moderate-level risk would concern the design of a new product based on currently used technologies and parts (integration risks); a third, even higher level of risk is related to the use of new materials, such as ceramics, in a product that was previously fabricated out of conventional metal alloys.

The development of the first transistor was a high-risk project involving a completely new technology. SONY's work on the first radio transistor was also a high-risk project because this technology was being implemented in a new product—the portable radio. However, development of subsequent models of the transistor radio represented much lower risks, as both the technology and the basic product were known.

The probability of success (or the risk of failure) should be estimated and monitored throughout the life cycle of a project. The selection of projects for implementation, the evaluation of alternative designs for a specific project, the decision to adopt or to reject proposed design changes, and the implementation of such changes measurably affect outcomes. Risk management is therefore part of the project manager's responsibilities. The scope of associated activities includes risk assessment, risk analysis, risk reduction, and risk handling.

Risk can be defined as a function of the probability of an undesirable event and by the severity of the consequences of that event. In general, high risk corresponds to a strongly adverse event that has a high probability of occurrence, while low risk corresponds to a low probability of occurrence and low severity. Moderate levels of risk correspond to combinations of probabilities and consequences that fall between these extremes.

Consider, for example, the risk associated with flying. The possible outcome of an accident is death, but due to high safety standards in the airline industry the probability of accidents is very low, and therefore the overall risk is low. This is evidenced by the cost of traveler insurance policies. When the probability of an undesirable event is higher (e.g., with parachuting or hang gliding), insurance policies that cover the same outcome (death) are much more expensive. Thus the level of risk depends on both the severity of an adverse outcome and its probability of occurrence.

It is possible to develop a scale for assessing the level of risk based on these two measures. Each risk, however, may affect one or more of the dimensions of project success. Thus a project may face a schedule risk related to the event of delays, a cost risk associated with the event of a budget overrun, one or more performance risks accompanying the failure to achieve technical/operational goals, and a program risk related to the success of the project as perceived by the participating parties.

Risks are caused by several factors.

■ *Technology.* The rapid pace with which technology (e.g., information systems and integrated circuits) is expanding may make a new product obsolete the day the first unit rolls off the production line. To avoid this risk, design engineers prefer to use the latest technologies available, which frequently are immature and unproven. This increases the risk of technological failure. Simple lack of experience heightens the chances that the project will be saddled with unforeseen problems. The trade-off between well-proven technologies with lower performance levels and new, unproven technologies requires detailed risk analysis. When NASA decided to build a new space shuttle in 1987 to replace the *Challenger*, which exploded on launch, it opted for a design that was nearly identical to the original. Rather than exploit recent advances in microelectronics, expert systems, and robotics, 20-year-old technology was used to avoid additional risks.

■ *Complexity and integration.* The adoption of well-known technologies for a project reduces the risk of component failure but may do little to mitigate the risk of integration failure. Modern, complex systems are based on the integration of parts and subsystems, the compatibility of software modules, and integration of hardware and software. The interfaces between components of a system are a source of integration problems and risks. For example, problems related to RFI (radio-frequency interference) or EMI (electromagnetic interference) should be considered in the design of electrical devises. Parts of the same system may affect each other in an undesired and unexpected fashion. Complex interfaces within a system, between systems, and between systems and humans are sources of risks that need management attention throughout the project.

■ *Changes.* Virtually all projects are subject to design changes throughout their life cycle. A reassessment of needs, revitalized competition, and emerging technologies are some of the factors that may call the original design into question. Design changes are risky, as each change may have a different effect on the system or its components. As a result, the risks of integration may go up sharply. A control system that evaluates each proposed change and its possible consequences is required. This system should provide information on approved changes to the design engineers in an effort to reduce the risk of integration failure. The same system should provide updated design information to manufacturing and quality control so that the product is manufactured and tested according to the most recent configuration.

■ *Supportability.* Good design and workmanship are not enough to guarantee a successful project. The ultimate test of success is customer or end-user satisfaction. To achieve this goal, the design should be based on the customer's needs, and the product should conform to the design. At the conclusion of the project, the product or system delivered should be operational (i.e., all the support required for maintenance and operations should be available). In the case of the rough-terrain cargo handler, for example, this includes trained personnel, transportation, storage and maintenance facilities, spare parts, and manuals. To prepare for worst-case contingencies, the design effort should cover the risks of a system delivered without adequate logistics support.

The multiple sources of risks and the different aspects of a project that are subject to failure or delay make risk management a demanding, time-consuming activity. Because technological risks affect performance as well as cost and schedule, we shall discuss the relationship between technological management and the effect these risks have on overall project success.

To demonstrate the problems faced by management during the various project phases, consider an organization that has decided to initiate a project aimed at automating its production planning and control system. Among the large variety of available options, the organization focuses on two alternatives: (1) purchasing the most suitable system off the shelf and modifying it according to its individual needs, or (2) developing a system that will support all of the specific production planning policies and procedures currently in use. In this example, the first alternative represents a project of relatively low development risk; however, the benefits may be minimal. This is because most off-the-shelf software packages have limited flexibility and can only rarely be made 100% compatible with the existing work environment. The second alternative offers a higher chance of achieving the technological and functional goals but involves a significant software development effort. As such, development and integration risks, and consequently the risk of schedule delays and budget overruns, are higher.

To perform a trade-off analysis between the two alternatives, the techniques presented in earlier chapters can be implemented. This should be done in such a way that all parties feel involved in arriving at the final choice and are satisfied with the decision-making process. Consensus building is the key. Achieving a high level of satisfaction depends on the process used in selecting and implementing the alternative. When management, potential users, and future operators of the new system select the alternative and define its specific configuration, the probability that the project will be successful is greatly increased.

In addition to the economic, scheduling, and cost aspects that have to be analyzed, risk analysis is part of the selection process. Risk analysis starts with the identification of all possible events that might have a negative impact on the project. In the example above, typical negative events for the first alternative are an inability to modify the software to accommodate a given need, and an inability to integrate the package with existing management information systems and databases. Negative events for the second alternative include unexpected difficulties in integrating the modules of the new software package, and excessive CPU time requirements that slow down information processing and retrieval.

In the next stage of the analysis, the severity of each event is estimated and the level of risk (based on the severity of the event and the probability of occurrence) is calculated. The events are then ranked, with those exhibiting the highest risks placed at the top of the list. Next, the source of high-risk events is investigated. Technological risks are usually generated by one or more of the following factors:

- Unproven technology
- System complexity
- Integration requirements
- Physical or chemical properties
- Modeling assumptions
- Interfacing with other systems
- Interfacing with operators, service personnel, and so on
- Operating environment

Major sources of risks require special attention. The risk management plan starts by identifying each of these sources, their magnitude, their relation to the various design stages, and their possible effects on cost, schedule, quality, and performance. The next step is to look for modifications or alternatives that would permit risk reduction. Continuing with the example above, the thoughtful selection of a computer language or operating system may reduce some of the integration risks. If management decides to develop a new software package, contingency plans that cut expenses and development time at the cost of lower performance should be prepared. These plans are used in case the undesired event takes place. By preparing contingency plans in advance, time is saved when the anticipated problem surfaces.

Special techniques for the analysis of risk include fault tree analysis, event tree analysis, synergistic contingency evaluation and review technique, and reliability analysis evaluation and review. The specifics are explained in books dealing with risk analysis, such as that of Cooper and Chapman (1987), and in technical articles such as that by

Rasmussen (1981). The analysis of risk is based on experience gathered in past projects, expert opinion, and physical or mathematical models. If the project manager does not have the technical expertise to perform the task, he or she should call upon those in the organization who are more qualified. In some cases it may be appropriate to contract an outside consultant to undertake the assessment. By initiating a risk management activity at the outset of the project, unnecessary risks can be avoided while those deemed necessary can be minimized or transformed.

6.4 CONFIGURATION SELECTION

In large, technologically sophisticated projects, selection of the best design by a direct comparison among the functional efficiencies of the alternatives is difficult and sometimes impossible, due to technological uncertainties, the absence of a single agreed-upon objective, the size of the system, and the system's complexity. In such projects it might not be appropriate to make a decision based solely on the cost of development and manufacturing. Once the system is put into place, its operations and maintenance costs may be significant enough, even after discounting over the useful life, to warrant consideration when the original decision is being made.

Cost-effectiveness and benefit-cost analyses are intended to assist in the selection of the most appropriate design alternative for system development or system modification projects. These techniques are supported by a variety of models used to estimate the functional efficiency, the risk, and the life-cycle cost of each technological alternative.

The selection process may be driven by the available budget or by the functional requirements. In the first case, the available budget for the project is viewed as a binding constraint and an effort is made to design a system with the best possible capabilities without exceeding the budget. This is known as the *design-to-cost* approach. In the second case, the design effort is aimed at minimizing the ratio between the cost of the system and its effectiveness. This is known as the *cost-effectiveness* approach. In either case there is a need to define and estimate the value of some performance measures for cost and effectiveness. Both approaches are used in the process of configuration selection. This process takes place before and during the detailed design phase, when the exact configuration of the system and each of its components is selected.

The techniques discussed earlier in this book for project selection are used for configuration selection as well. Checklists and scoring models, benefit-cost analysis, cost-effectiveness analysis, and multiple-criteria methods all have a role. In the configuration selection process, each alternative design (configuration) is analyzed by a life-cycle cost model as explained in Chapter 10 and is evaluated with respect to its expected performance. Performance measures are project dependent (they would be different for the development of a new car and for the construction of a new building), but some are common to many systems.

1. *Operational or functional capability.* This is a measure of the system's ability to perform tasks and satisfy the market or customer's needs. For example, the range of an electric passenger vehicle, its payload, and its speed are possible measures of operational or functional capabilities. In the software selection example, the ability to perform all planning and control activities required within acceptable time standards is an operational performance measure.

2. *Timeliness.* This measure relates to the point of time when the system is available to perform its mission (i.e., the successful completion of acceptance tests and the start of regular operations). In the software selection example, an off-the-shelf system can perform better with respect to this measure.

3. *Quality.* The measure of quality relates to the degree that the system's design reflects the market or customer needs, and to the degree that the product or service meets its design specifications. Therefore, the quality of an alternative refers to the system's components, the integration of those components, and the compatibility of the proposed system with the environment in which it will interact. Quality is defined in specific terms for military systems (known as MIL-STD or military standards); for nonmilitary systems such as planes, boats, buildings, and computers, a host of national and international standards exist. The *Institute of Electrical and Electronics Engineers* is in the forefront of setting standards for electrical equipment and devices. If adequate standards are not available, desired quality levels should be specified for both the operational (functional) and technical (design and workmanship) aspects of the system. The relationship between these aspects through QFD (quality function deployment) is explained later. Returning to the software selection example, in addition to specific needs and requirements, a set of standards and testing procedures has been established for software quality and reliability. The *Software Engineering Institute* based at Carnegie–Mellon University has taken the lead in this effort.

4. *Reliability.* This measure is a function of how well the system performs for a given period of time under specific operating conditions. The two factors used in the calculations are the mean time between failures (MTBF) and the mean time to repair (MTTR) the system. Reliability can then be defined as follows:

$$\text{reliability} = \frac{\text{MTBF}}{\text{MTBF} + \text{MTTR}} \times 100\%$$

There is a correlation between reliability and quality, as high quality of design, workmanship, and integration usually leads to a high level of reliability. However, reliability also depends on the type of technology used and the operating environment. In the software selection example, the presence of bugs and the average time required to repair them affect the reliability of the system.

5. *Compatibility.* This measure corresponds to the system's ability to operate in harmony with existing or future systems. For example, a new management information system has a higher degree of compatibility if it can use existing databases. Electronic systems are said to be compatible if they can operate without interference from the electromagnetic radiation put out by other systems in the same vicinity. In the software selection example, the ability to import and export data from other information systems and databases in the desired format is an important measure of compatibility.

6. *Adaptability.* This measure evaluates a system's ability to operate in conditions other than those initially specified. For example, a communication system designed for ground use would be considered a highly adaptable system if it could be used in high-altitude supersonic aircraft without losing any of its functionality. Systems with high adaptabilities are preferred when future operating conditions are difficult to forecast. In the software selection example, a highly adaptable software package is one that can run on different computer types under a variety of operating systems in addition to the computer and operating system specified.

7. *Life span.* This measure has a direct impact on both cost and effectiveness. Due to learning and efforts at continued process improvement, systems with a longer life span tend to improve over time. This eliminates the need for frequent capital investments and hence reduces total life-cycle cost.

8. *Simplicity.* The process of learning a new system while it is being introduced into an organization depends on its simplicity. A system that is easy to maintain and operate is usually accepted faster and creates fewer difficulties for the user. Furthermore, complicated systems may not be maintained and exercised adequately, especially during startup or periods of change, when there is high turnover in the organization. In the software selection example, a software package that is simple to operate and maintain is one that is developed according to software engineering standards regarding modularity, documentation, and so on.

9. *Safety.* The methods by which a system will be operated and maintained should be considered in the advanced development phase. Safety precautions should be introduced and evaluated to minimize the risk of accidents. As with quality, designing a safe system from the start can provide significant benefits over the long run.

10. *Commonality.* A high level of commonality with other systems in the organization should be a driving force in the design of a system. Commonality has many facets, such as common parts and subsystems, input sources, communication channels, databases, and equipment for troubleshooting and maintenance. Many airlines insist that all aircraft they buy within a particular class, regardless of manufacturer, have the same engines. In a similar vein, the U.S. Department of Defense developed the computer language Ada in the late 1970s and now requires that all programs commissioned by any of its branches be written in Ada.

11. *Maintainability.* Providing adequate maintenance for a system is important. There is a question, though, as to how much preventive maintenance should be scheduled. When a system is out of service, it cannot perform its assigned tasks. This loss in operational time must be weighed against the probability of system failure and the need for unscheduled maintenance, which in turn, reduces the system's overall effectiveness. Higher levels of maintainability lead to better labor utilization and lower personnel training costs. Part of maintainability is testability—the ability to detect a system failure and pinpoint its source in a timely manner. Higher levels of maintainability and testability contribute to the effectiveness of a system. In the software selection example, a well-documented source code as well as clearly defined interfaces between the modules of the software package help in the detection and correction of bugs.

12. *Friendliness.*This performance measure quantifies the effort and time required to learn how to operate and maintain a system. A friendly system requires less time and skill to learn, and hence reduces both direct and indirect labor costs. In the software selection example, the use of menus, on-line help, and pointing devices such as a mouse can increase the friendliness of the software package.

The measures listed above form a general, nonexhaustive list of indicators that should be taken into account when evaluating a system from a technological point of view. By combining them with specific project objectives related to budget and schedule, they provide a framework for selecting the design configuration and foreshadow the capabilities of the final system.

For a particular project, each of these measures should be subdivided until the desired level of detail is reached. For example, compatibility might be broken down by hardware, software, operations and maintenance personnel, training requirements, and logistics support. Software then might be decomposed into databases, controls, interface protocols, and applications. Quantifying each element in the resultant hierarchy for each alternative is the first step in the analysis. The selection process itself can be supported by scoring models, MAUT-based models, or the AHP, as discussed in Chapters 3 and 4.

The cost of each alternative must also be evaluated. The notion of life-cycle cost is widely used for this purpose. The life-cycle cost of a system is defined as its total cost from the start of the conceptual design phase until it completes active service. Related methodologies and techniques are discussed in Chapter 10.

Along with a benefit-cost analysis, a risk analysis of each alternative design should be conducted. Risk analysis includes the following steps:

- Identification of risk drivers
- Estimation of probabilities of undesired outcomes
- Evaluation of the impact of each undesired outcome (on cost, schedule, quality, and operational and technological capabilities)
- Elimination and reduction of risks
- Preparation of contingency plans

The procedures used for selecting the best design alternative can also be adopted for managing configuration changes. This is discussed presently, but first we offer some guidelines for system definition. The selection process is complete when the specifications of the proposed system are robust enough to at least answer the following questions:

Technological specifications

- *Operational/functional:* What tasks should the system perform, and what performance levels are expected?
- *Timeliness:* When should the system be operational?

- *Quality:* Which standards are applicable? Which customer needs are to be supported by the system, and to what extent?
- *Reliability:* What are the expected MTBF and MTTR in the environment in which the system has to operate?
- *Compatibility:* With which other systems must the system contemplated operate in harmony? What interfaces are required?
- *Adaptability:* Under what environmental conditions is the system designed to operate, be maintained, and stored? Under what conditions is the system required to operate, be maintained, and stored?
- *Life span:* For how long is the system expected to be in service?
- *Simplicity:* What level of training is required to operate and to maintain the system?
- *Safety:* What safety standards are applicable to the system?
- *Commonality:* What level of commonality is required with each existing or planned system?
- *Maintainability:* What logistics support is required: spare parts, training, technical manuals, test equipment, and so on?
- *Friendliness:* What features should be included in the system to enhance its friendliness?

Life-cycle cost

- What are the estimated costs of design, manufacture, operation, maintenance, and phaseout for the system?
- What is the expected timing of each cost component?

Risk assessment

- What are the major risk drivers?
- What are the probabilities of undesired outcomes?
- What is the expected impact of each undesired outcome?
- What are the plans to handle undesired outcomes?

The selected design alternative defines the technological aspects of the project. Based on the specifications, estimates of cost and schedule are made and the proposed project is either approved or rejected. Project approval is a management decision that may affect the entire organization. When several projects are being considered, the final choice is based on strategic and tactical considerations including:

General considerations

- Organizational goals
- Current or pending projects
- Existing and future products and markets

- Introduction of new technologies
- Image of the organization
- Organizational growth

Research and development

- The availability of required technology
- Future use of new technologies developed or acquired for the project
- Development risks
- Opportunity to acquire new technologies and new knowledge
- Availability of resources required
- Future use of new resources acquired for the project

Logistics and production

- Project's need for logistics support
- Future use of investment in logistics support
- Project's production resource requirements
- Availability of production resources needed
- Effect on utilization of existing resources
- Need for new facilities
- Future use of facilities required for the project

Marketing

- Potential markets
- Estimate of future sales or business
- Availability of marketing resources
- Effect on existing products markets

Finance

- Project net present value
- Project rate of return
- Project payback period
- Project budgetary risks
- Project cash flow

This partial list, together with any specific considerations unique to the organization, underlies the selection process. Once again, actual decision making can be based on multi-criteria techniques or any of the other methods discussed for evaluating and selecting alternatives.

6.5 CONFIGURATION MANAGEMENT

Configuration management (CM) concentrates on the management of technology by identifying and controlling the functional and physical design characteristics of a system, and its support documentation. The medium of implementation is a set of tools designed to provide accurate information on what is to be built, what is currently being built, and what has been built in the past. The mission of CM is to support concurrent engineering and to assist management in evaluating and controlling proposed technological changes. Through quality assurance activities CM ensures the integrity of the design and engineering documentation, and supports production, operation, and maintenance of the system.

In configuration management, a baseline is established in each phase of the system's life cycle with well-defined procedures for handling proposed deviations. The initial baseline, known as the *functional* (or *program requirements*) *baseline*, is prepared in the first phase of the life cycle—the conceptual design phase. This baseline contains technical data regarding functional characteristics, demonstration tests, interface and integration characteristics, and design constraints imposed by operational, environmental, and other considerations. Approval is subject to a preliminary design review (PDR).

The advanced development phase produces the second baseline, the *allocated* (or *design requirements*) *baseline*. This document contains performance specifications guiding the development of subsystems and components, including characteristics derived from the system's design. Laboratory or computer simulation may be used to demonstrate achievement of functional characteristics, interface requirements, and design constraints. This baseline is subject to a critical design review (CDR).

The *product* (or *product configuration*) *baseline* is last and includes information on the system as built, including results of acceptance tests for a prototype, supporting literature, operation and maintenance manuals, and part lists. Acceptance is subject to a physical configuration audit (PCA). In addition to these three baselines, other baselines and additional design reviews are frequently needed when complicated systems are involved. Examples are a baseline that defines the initial design and a baseline that defines the detailed design of the system. The transition from one baseline to the next is controlled by design reviews.

The configuration management system ensures smooth transition and provides updated information on the configuration of the system and all pending change requests at all times. To function properly, it should perform the tasks discussed in the following subsections.

6.5.1 Configuration Identification

Configuration identification sets the foundation of the CM system. It starts with the selection of configuration items (CI), both software and hardware, that have one or more of the following characteristics:

- End-use function
- New or modified design

- Technical risk or technical complexity
- Many interfaces with other items
- High rate of future design changes expected
- Logistic criticality

The selection of configuration items is a critical task of systems engineering. Too few configuration items will not provide adequate management control, and too many may overload the system, sparking a waste of time and money.

Next, a coding system is adopted and configuration identification numbers are assigned. These numbers are designed to assist in providing the following information on each configuration item or a lower-level item:

1. Technical requirements which form the basis for detail design. These are provided by *specification numbers.*
2. Identification of the equipment designed and built to the applicable specification. These are provided by *equipment numbers.*
3. Technical descriptions for the equipment and its lower-level items. These are provided by *drawing and part numbers.*
4. Description of the sequence of manufacturing the equipment and its lower-level items. These are provided by equipment and item *serialization numbers.*
5. Change documents. These are provided by *change identification numbers.*
6. Sources of manufacture at all levels. These are provided by *manufacturer's code identification numbers.*

As an example, consider configuration item 123. For this CI, the following identification numbers are defined:

Specification number	SPEC 123
Equipment number	CI 123
Drawing number	123A
Serial number	123 SN5
Manufacturer number	00375
Change identification numbers:	
Engineering change request	ECR 123 N 005
Engineering order	EO 123 N 005

To control the allocation of numbers, they should be assigned from a single point and a standard procedure established to prevent errors in identification.

6.5.2 Configuration Change Control

Configuration change control involves the development of procedures that govern three steps:

1. *Preparation of a change request.* This step requires that a formal change request be prepared and submitted. The initiation of a change can be internal (the project team) or external (the customer, a subcontractor, or a supplier). The change request specifies the reason for the modification and forewarns management of increases in cost, schedule, and risk, as well as changes in quality, contractual arrangements, and system performance. Each change request is assigned an identification number and is evaluated after input is received from all organizational units affected. The principal aim is to collect the relevant data on each proposed change and to assess its expected impact.

A typical change request form will include the following information:

- Change request number _____
- Originator _____
- Date issued _____
- Contract or project number _____
- Configuration items affected by the change _____
- Type of changes: temporary _____   permanent _____
- Description of change _____
- Justification for change _____
- From serial number _____ through serial number _____
- Priority _____

Effect on:	Cost _____
	Schedule _____
	Resource requirements _____
	Operational aspects _____
	Timeliness _____
	Quality _____
	Reliability _____
	Compatibility _____
	Life span _____
	Simplicity _____
	Safety _____
	Commonality _____
	Maintainability _____
	Friendliness _____
Remarks:	Engineering _____
	Marketing _____
	Manufacturing _____
	Logistics support _____
	Configuration management _____
	Other organizational units _____
CCB decision:	Accept _____ Reject_____ More information needed _____
	Acceptance date _____
	Rejection date _____

2. *Evaluation of a change request.* A team of experts representing the different organizational functions is responsible for the evaluation of change requests. This team, known as a change control board (CCB) or configuration management board, evaluates each proposed change based on its effect on the form, fit and function of the system, logistics (manuals, training, support equipment, spare parts, etc.), and project cost and schedule. This review leads to a decision to approve or reject the change request, or to reconsider it after more data are collected.

Changes are classified as either permanent or temporary. A temporary change might be needed for test programs or debugging software. Approval can usually be obtained in a short time compared to a request for a permanent change.

All information regarding each proposed change is accumulated and analyzed by the change control board, which also functions as a central repository for historical records. The CCB decision is based on cost-effectiveness and risk analysis in which the need for the change and its expected benefits are weighed against implementation and project life cycle costs, its impact on project quality and schedule, and the expected risks associated with implementation.

3. *Management of the implementation of approved changes.* Approved changes are integrated into the design. This is accomplished by preparing and distributing a change approval form or an engineering change order to all parties involved, including engineering, manufacturing, quality control, and quality assurance.

The CCB is responsible for the pivotal task of conducting a comprehensive impact analysis of each change proposed. A well-functioning change control system assures tight control of the technological aspects of a project. In addition, it provides accurate configuration records for the smooth, coordinated implementation of changes and effective logistics support during the life cycle of the system.

6.5.3 Configuration Status Accounting

This task provides for the updated recording of:

- Current configuration identification, including all baselines and configuration items
- Historical baselines and the registration of approved changes
- Register and status of all pending change requests
- Status of implementation of approved changes

Configuration status accounting provides the link between different baselines of the system. It is the tool that supports the CCB in its analysis of new change requests. The effect of these changes on the current baseline must be evaluated and their relationship to all pending change requests must be determined before a decision can be taken.

6.5.4 Review and Audits

This configuration management task provides both the contractor and the customer with the assurance that test plans demonstrate the required performance, and that test results prove conformance to requirements. Functional configuration audit includes a review of development test plans and test results, as well as a list of required tests not performed, deviations from the plan, and waivers. In this task, the relationship between quality assurance and configuration management is established. Configuration management provides the baselines and a record of incorporated and outstanding changes. Quality assurance first checks the configuration documentation to gauge requirements; then it verifies that the system conforms to the approved configuration.

Configuration management is a tool designed to help the project team know what they are developing, producing, testing, and delivering so that the appropriate support and maintenance can be given to the product throughout its life cycle. It specifies the procedures and information required for the project to be carried out in the most cost-effective manner.

6.6 TECHNOLOGICAL MANAGEMENT OF PROJECTS

The technological aspect of project management is the link between the engineering, science, and management disciplines. The understanding of these links and the connection between product design, process design, support design, cost, schedule, resources, and project success are the main contributions that the project manager makes to the technological aspects of a project.

Technological management starts when a need is identified for a new product, service, or system. The first step is to assess the feasibility of providing that need. Favorable results from a feasibility study and a positive economic evaluation of the proposed project trigger the next step—the definition of specifications and design characteristics that satisfy the needs in the best possible way. As an example, if the development of an electric passenger vehicle is being considered, questions related to the required driving range on a single charge, the payload required, and the maximum speed are translated into design attributes such as vehicle size, battery type, and power consumption. These attributes are then translated into the blueprint of the finished product through a detailed design. In the car example, questions as to the types of materials that should be used and the best manufacturing process are answered by teams of engineers. If a concurrent engineering approach is adopted, issues such as logistics support throughout the life cycle of the product are studied simultaneously.

The organizational aspect should be addressed at the outset of the project during the conceptual design phase. The structure of the CCB is a critical decision. Functional areas such as engineering and quality assurance should be included along with the various representatives of the project. The role that each CCB member plays and the procedures adopted for change request evaluations are central to the technological management process.

A time aspect is present because each phase of the project ends with an updated, more complete, and more detailed configuration baseline. The baselines and the corresponding design reviews—the functional baseline and system design review, the allocated baseline and preliminary design review, the product baseline and critical design review, and any additional baselines and design reviews—are key milestones in the project life cycle. Well-prepared reviews and a supporting configuration management plan provide the necessary foundation. In addition, many products and systems require ongoing technological management, due to a continuing stream of engineering change requests.

Finally, the cost aspect of technological decisions and the analysis of risk should be integrated within technological management. Each design decision has an impact on the project's life-cycle cost and on its probability of success. The ability to integrate all the different factors affected by technological decisions, and consequently to produce the best design based on all these factors, is the essence of technological management.

6.7 TOTAL QUALITY MANAGEMENT IN PROJECTS

The industrial world is in the midst of a quality revolution brought on by the Japanese. The introduction of the just-in-time philosophy supported by kanban for production and inventory control, continuous process improvement on the shop floor, and the general goal of zero part and product defects has allowed Japanese firms to capture the bulk of the consumer electronics market, a large share of the semiconductor market, and a troublesome proportion of the U.S. automobile market. This success, which has come in less than two decades, can be attributed to a knack for squeezing a few more percentage points of performance out of a system or process after logic and economics indicate that diminishing returns have long set in. But such an explanation is too glib. At the heart of Japanese manufacturing is an emphasis on education and training, a cross-functional workforce, teamwork, and a commitment to excellence—some of the basic components of total quality management (TQM).

6.7.1 What Is Total Quality Management?

TQM is a system that combines quality control techniques and organizational models developed over the last 40 years in both the United States and Japan. It is a logical evolution of management by objectives, strategic planning, quality circles, quality assurance, and many other systems. TQM offers a structural approach to creating organization-wide participation in planning and implementing a continuous improvement process that exceeds the expectations of the customer or client. It is built on the assumption that 90% of our problems are process related rather than employee related.

TQM has three major components:

1. *Breakthrough planning, sometimes called Hoshin planning* (see Fig. 6-2 for the full planning process). The purpose is to:

- Clarify a vision of where the organization wants to go in the next 5 or 10 years.
- Identify goals and objectives that move the organization toward its vision.
- Identify critical processes that must deliver the goods and services provided to clients in a way that exceeds their expectations.
- Select a few (no more than four) breakthrough items that can help the organization reach its vision quickly.

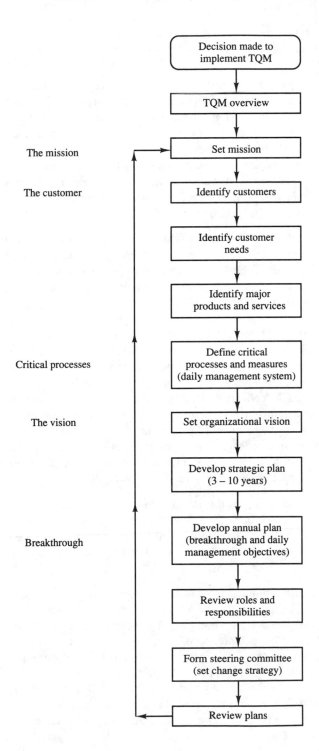

The mission

The customer

Critical processes

The vision

Breakthrough

Figure 6-2   Basic components of TQM.

- Communicate the vision and the methods to reach all employees.
- Provide a structure for monitoring progress toward the vision.

Vision planning allows the organization to identify and focus on key areas. The ultimate objective is to facilitate full development of the methods and pathways by which "breakthroughs"—major, fundamental quality improvements—can be implemented. All vertical levels of the organization then participate in the planning, development, and deployment of the strategic breakthroughs identified (see Fig. 6-2).

Breakthrough planning is an evolution of strategic planning that adds value statements, reminding us that "how we got here" is just as important as where we are going. A related concept is *breakthrough thinking*, which has recently been formalized by Nadler and Hibino (1990). Its major components are outlined in Table 6-2.

2. *Daily management.* This system shows people what they must do personally and what they must measure and control to keep the organization running smoothly. It helps them define and understand the processes they use in producing goods and services to meet customer needs and expectations. Once these processes are understood, individuals and departments can continuously improve them, then standardize the methods to ensure that gains are maintained.

This continuous improvement is achieved by problem-solving teams who engage in identifying customer problems, finding solutions, and then providing ongoing control of the improved process. Use of several basic quality control tests and statistical methods helps people to manage with facts, not opinions, and to solve the real problems, not just symptoms. Problem solving requires the collective efforts of everyone in an organization, working in study teams following a multistep procedure. This includes customer surveys, identification and selection of issues, establishment of performance measures, data collection, modeling and analysis, action, and feedback. Daily management is the most revolutionary of the three components of TQM. It empowers employees at all levels of the organization and focuses management improvement efforts on process problems.

3. *Cross-functional management.* This is the integration of team activities across divisions/departments to achieve organizational goals. It is the vehicle for breaking down department/divisional barriers. Through cross-functional management, top-level managers can ensure that all groups in the organization—technicians, engineers, staff, and so on—are working together with a commonality of purpose.

Concurrent engineering and configuration management establish the framework in which TQM can be implemented. The impacts can be felt throughout the life cycle of the project. In the detailed design and production phases TQM is geared toward customer needs and preventing future defects. After completion of the production process, TQM is aimed at detecting defects and more important, eliminating their sources. The third aspect of TQM is to achieve control by trend analysis and preventive actions.

The TQM approach is based on the philosophy that quality should be designed into the product and process, and that defects should be avoided at almost all cost. This is because defect detection by itself is expensive and prone to error. Once a defect is created, it

TABLE 6-2   BREAKTHROUGH THINKING METHODOLOGY

1. *Assemble the problem-solving team.*
   - Brainstorm major functions that could be improved.
   - Use consensus to select one function on which to work.
   - Size the work group between 5 and 8.
   - Select people who will contribute to the success of implementation.
   - Include one "outsider"—someone unconnected to the issue—and a customer.

2. *Develop the function statement.*
   - Discuss the uniqueness of this particular issue.
   - Expand the function statement.
   - Select the largest function that can be addressed with the resources available.
   - Review the chosen function and function expansion with decision makers.

3. *Establish measures to gauge improvement.*
   - Select ways to measure your improvement progress.
   - Use existing data and systems whenever possible.
   - Establish goals for improvement.

4. *Brainstorm general solutions.*
   - Define regularly occurring conditions or assumptions.
   - Brainstorm solutions for the chosen function. Ask the question, "How can we achieve our selected function?"
   - Ask, "What would be the solution after that?" Focus on an ideal future.
   - Organize solutions into major alternatives.
   - Add limited details to each solution: who, when, where, how, how much.
   - Review solutions with decision makers.

5. *Select the ideal solution.*
   - Narrow the solutions to those two or three that best address the function.
   - Add further details to these solutions, including plans for irregularities.
   - Analyze the remaining solutions. Which solution provides the best answer, taking into account the entire system and any irregularities?
   - Select the solution to implement.
   - Establish a plan of actions to achieve implementation.
   - Review solution and plan with decision makers.

6. *Implement the plan of action.*
   - Individuals or small teams working within their own areas of responsibility further develop the details of implementation.
   - Further analysis or tests may be required as the detailed plan is implemented.
   - The team leader acts as the project manager to coordinate diverse activities.

7. *Monitor progress and repeat process.*
   - Manage at your own level of responsibility. Do not micromanage implementation.
   - At regular intervals, ask each participant what barriers exist to the success of the implementation.
   - Exercise control at the group rather than at the individual level.

might be a nightmare to find and remove it. In electronic assembly, for example, an oft-cited rule of thumb is that the cost of finding a defective component goes up by a factor of 10 for each level of assembly: device, board, system, and field installation. Furthermore, even if the defect is found, correction is not only expensive and time consuming but is likely to reduce the quality of the product. A reworked part will often not measure up to the same standards of one manufactured properly the first time.

6.7.2 Cost of Quality

The modeling and analysis of trade-offs is at the very foundation of decision making. In manufacturing, one of the trade-offs most enshrined in our thinking has been the perceived relation between quality and cost. Vaughn (1990) points out that when people go to buy a product, they usually want it to be perfect at a minimum price. However, defect-free products are impossible to attain in any production environment at any cost.

The traditional view holds that near approaches to zero-defect operations are too expensive for manufacturers of most products to attempt to reach. When the manufacturer must work to such high quality standards, production costs require that the price charged be outrageously high. As a consequence, a balance is struck between cost and quality, as shown in Fig. 6-3. This frequently leads to the distribution of goods and services that fail to meet customer expectations.

What drives this trade-off according to this traditional view? Vaughn explains that efforts to reduce rework, repairs, warranty costs, and liability losses generate increasing costs associated with the time, materials, engineering, and overhead required to achieve these ends. From an economic perspective, the point at which the marginal cost of improving quality 1 unit is just equal to the marginal loss due to poor quality is the point at which the optimum is achieved. Even Joe Juran, the guru who has long been at the forefront of the quality improvement movement, encouraged this kind of thinking (Juran 1988). In his optimization model, he shows how failure costs decline until they are overtaken by the increasing costs of appraisal and prevention. At this point, total quality costs begin to rise.

Why, then, have the Japanese been so successful in increasing quality while simultaneously bringing down production costs? As Cole (1992) asks, "Have they abolished the laws of economics?" Hardly, but they have made us realize that the point at which total quality costs start to rise again as failure costs are driven down has shifted sharply to the right in Fig. 6-3. Moreover, continuous improvement makes perfect economic sense as long as the search follows a minimum cost path. In terms of organizational dynamics, it has long been observed that quality achievements tend to regress over time. If one is not

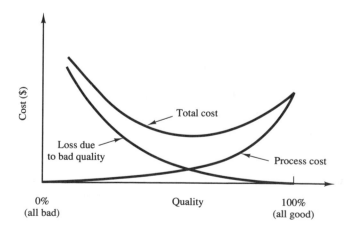

Figure 6-3   Relationship between quality and cost.

going forward, one is going backward. Thus to press for continuous improvement at least helps ensure that no ground is lost.

According to Cole, the Japanese achievement has been sixfold. First, Japanese managers realized that the traditional calculations dramatically underestimated the costs of poor quality. Typically, such calculations ignored the customers who were lost or who had never bought the product. A declining reputation among customers and the effects of negative word-of-mouth publicity were never considered, partly because they were difficult to quantify. There is every reason to believe that these effects are substantial. Japanese managers recognize these costs. They stress the fragility of their reputations with their customers and the importance of winning the customer's trust. They find it entirely appropriate to spend generously for this purpose. In short, once one recognizes the high costs associated with poor quality, one sees that it is economically rational to invest more in quality improvement.

Second, the traditional approach vastly underestimated the payback that a corporatewide quality improvement culture yields in terms of worker motivation and a broad array of performance indicators. A 1991 study undertaken by the U.S. Government Accounting Office (GAO) of the 1988 and 1989 Baldrige Award finalists revealed that companies who adopted TQM practices achieved better employee relations, higher productivity, greater customer satisfaction, increased market share, and improved profitability. GAO calculated that, on average, these measures increased 4.5% per year from the mid-1980s on. Other measures, related to employee turnover, product reliability, number of employee suggestions, on-time delivery, order processing time, number of defects, production lead times, customer complaints, and inventory turnover, improved at even greater rates. Although these findings are still preliminary (no control groups were included and some data were missing), they are suggestive of the broad impact that a quality initiative can have.

Third, the Japanese pursuit of quality is accompanied by intense pressures to minimize the attending costs. In the United States, some of the quality zealots have made the mistake of separating the two and regarding support for any quality initiative as a kind of litmus test of enlightened management. This misses the point altogether and substitutes an unthinking repetition of the quality mantra for real understanding. The widespread mobilization of production workers, equipped with elementary but powerful statistical problem-solving methods, to improve quality is a concrete manifestation of this low-cost approach. Such employees are a lot cheaper than design engineers and there are a lot more of them to work on an endless supply of problems. The incremental costs of their involvement in quality improvement are extremely modest (Schneiderman 1986).

Fourth, preventing problems at the source has become the preferred approach to improving quality. The Japanese gradually recognized that the costs of poor quality could be reduced more effectively by moving their efforts "upstream." In practice, this means concentrating on the process of new product development. This approach dramatically reduces appraisal costs and has the beneficial side effect of eliminating white-collar rework (e.g., downstream engineering changes).

Fifth, Japanese managers came to see quality improvement not as a matter of adding product attributes (which inevitably add costs) but as a matter of improving the quality of all business processes. By doing things right the first time, massive amounts of rework could be avoided and costs could actually be reduced. Those involved in the business processes were trained and given responsibility for improving them. Proceeding along these lines, Japanese firms actually eliminated a good deal of the traditional quality–cost trade-off.

Finally, the traditional trade-off model assumed that what the customer wanted in quality and was willing to pay for did not change over time. In fact, what Japanese manufacturers discovered was that by achieving the highest quality standards, they could charge a premium for their products, and in so doing, they educated consumers to demand higher and higher quality. As the company that changed customers' tastes, they were then in a unique market position to satisfy those new tastes. This in turn could be translated into higher prices or a greater share of the market.

Cole's analysis has a resonance that is being felt by all players in the global marketplace. New attitudes toward quality improvement and the results that have been achieved have made the traditional quality–cost trade-off model obsolete. The Japanese have not abolished the economics of quality, but they have changed the way we approach, conceive of, and measure the relevant variables.

6.7.3 Captains of Quality

The TQM approach was developed over the last 40 years by many researchers and practitioners. Four of the most prominent are W. Edwards Deming, Joseph Juran, Philip Crosby, and Masaaki Imai. Their message is basically the same:

- Commit to quality improvement throughout the organization.
- Attack the process, not the employees.
- Strip down the process to find and eliminate problems that diminish quality.
- Identify your customers and satisfy their requirements.
- Instill teamwork and create an atmosphere for innovation and permanent quality improvement.

The leitmotif is worker enablement and empowerment; that is, train the workers and give them responsibility. In the remainder of this section we highlight the main points made by each of these pioneers and mention how they can be applied to project management. Like any discipline, a unique vocabulary has grown around TQM. Appendix 6A contains a glossary of the principal terms.

Deming approach.   Deming, originally a physicist with a Ph.D. from Yale, after many years in industry came to believe: "Improve quality and you automatically improve productivity. You capture the market with lower prices and better quality. You stay in business and you produce jobs. It's so simple." In his work he stresses statistical process control, statistical quality control, and a 14-point plan for managers that emphasizes the human element. The philosophy is to treat people as intelligent human beings who want to do a good job. Although statistical control methods are difficult to implement in a project environment due to the one-of-a-kind nature of projects, the 14 points are readily adoptable.

Deming was the American who took his message to Japan in 1950 after being shown the door by most major U.S. corporations. It was a time when U.S. firms dominated international markets; there was virtually no competition from abroad, so as long as the product worked, the concern for quality was minimal. Deming was instrumental in changing this attitude and in turning Japanese industry into an economic world power. His 14 principles for achieving competitiveness through quality are as follows (Deming 1986):

1. *Create constancy of purpose toward improvement of products and services; emphasize long-term needs rather than short-term objectives.* This principle should guide management throughout the life cycle of a project. In early stages of project selection, long-term goals should be emphasized. The acquisition of new knowledge and the ability to master new technologies are leading considerations. Furthermore, the specific configuration selected for a project should support these long-term objectives as well. The required constancy can be achieved only by a learning process that promotes improvement from one project to the next. Whatever learned during project execution is as important as the final results and deliverance. Top management has to facilitate the diffusion of knowledge throughout the organization and the transfer of technology between different projects. This calls for an investment in resources and the development of procedures to support these activities.

2. *Adopt a new philosophy.* The philosophy that productivity and cost are the most important performance measures should be modified based on the recognition that improved quality can reduce cost and improve productivity while increasing customer satisfaction. Thus, quality is the most important performance measure. Defects are unacceptable and problem-solving tools should be used at all levels of the organization to eliminate their sources.

3. *Cease dependence on mass inspection.* Quality should be built into the product design, the process design, and the support design. The responsibility for quality should not be with quality control and quality assurance but rather with all members of the organization regardless of function and level. Advanced presentation analysis should be used to support group problem solving so that improved process capability will be maintained.

4. *Reduce the number of vendors; don't select vendors on the basis of cost.* The decision on project structure and selecting subcontractors should be based on quality considerations; i.e., employ a limited number of high-quality subcontractors with whom long-term relationships predicated on loyalty and trust can be established. Quality should be the predominant factor in choosing a subcontractor, rather than price tag alone.

5. *Search for problems and improve constantly.* Successful project management is based on good design and good planning, and a dogged determination to remain on course. Thus, an ongoing effort to identify and solve problems is the key to smooth implementation. The earlier a problem is detected, the easier it is to correct. Control systems should be established with this adage in mind. One way of doing this is with trend analysis in areas such as cost, schedule, and quality, as discussed in Chapter 11.

6. *Institute on-the-job training to make better use of human resources.* Instituting training in technological and managerial fields supports continuous improvement in the process. Training employees in new technologies developed in one project enhances the likelihood of success for future projects. It is an investment that will pay for itself many times over. Promulgating the philosophy of built-in quality enables the whole organization to move in the direction of quality improvement. Training in managerial techniques used for planning, scheduling, budgeting, and control is important for the project manager and the project team. Input from the various team members on all problems relevant to their expertise should be continually sought, and the reasons behind all decisions should be made explicit.

7. *Improve supervision.* The role of supervision is leadership aimed at helping people do a better job. Since recognition and reinforcement are critical to good job performance, supervisors should be trained to support this continued improvement process by providing leadership, example, and training to their teams. Managers should become coaches rather than feudal lords.

8. *Drive out fear.* Encourage open communication. Open communication lines and the ability to report problems without fearing the consequences are essential to the ongoing improvement process. Employees are the first to know about problems in their specific areas of responsibility. Problems can be resolved early in the project life cycle by encouraging open discussion. Furthermore, employees usually know their part of the project better than anybody else. By encouraging them to initiate change in the product, process, or support design when they see fit, a continuous improvement is likely to take place. Management should institute an "open door" policy and a "How can I help you in doing your job?" approach to promote communication and to effectively use its most important resource—people.

9. *Break down barriers and promote communication among the different organizations participating in the project.* By eliminating communication barriers between functional areas, departments, and subcontractors, concurrent engineering can be implemented. All the participants in the project should be viewed as a team with a common goal. The organizational breakdown structure should clearly define the formal communication channels within the team. However, informal communication between the members of the project should also be encouraged. Each organization participating in a project should learn to view the other organizations as its customers (or suppliers), striving to understand their needs in performing their tasks. By adopting the customer-provider point of view, integration between the various elements of the WBS assigned to different organizations in the OBS greatly increases the likelihood that it will be smooth and error free.

10. *Eliminate slogans, posters, and targets for the workforce that ask for a new level of productivity without appropriate methods and solutions.* The focus should be on the process as well as the outcome. An effort to improve the process of design and implementation will result in higher levels of achievement. Management should help employees develop better ways of performing their tasks; i.e., provide leadership in problem identification and problem-solving methodologies.

11. *Use work standards (quotas) carefully.* Work standards used in a project environment can be dangerous, especially when they depend on environmental factors external to the project. Standards and quotas are important in the planning process (they can be used for time and cost estimates as explained in Chapters 7 and 8), but should be carefully used as a foundation for performance evaluations. When standard time or cost goals are not achieved, the source of the problem should be determined. It is rarely a good idea to apply sanctions solely on the basis of cost or schedule overruns, as the cause may be outside the worker's control.

12. *Remove barriers that eliminate the worker's pride of workmanship. The responsibility of supervisors must be changed from sheer numbers to quality.* Employees should be permitted to evaluate their own work and to take pride in it. This means the abolishment of annual or merit rating and of management by objectives. By assigning the responsibility of better quality to the employees who perform the work, a link is established between their satisfaction and the improvement process. This link is necessary to promote improvement in quality.

13. *Institute an education and training program to teach workers new skills as new technologies are developed or assimilated by the organization.* Technologies developed or acquired for one project can serve the whole organization in future projects if proper training regimens are established. In addition to on-the-job training, a training program should be instituted to transfer knowledge between different parts of the organization.

14. *Everyone in the organization should team up in the quality improvement process.* This process should not be an isolated effort of the quality control or quality assurance departments. Everyone in the organization should be involved in the transformation to quality. Too often, advice and opinions of low-level staff are either not sought or ignored. Too many times, managers act as if they know the answer to every problem. Top management must set the example in implementing a TQM program by insisting that the basic principles be adopted by each unit in the organization.

Juran approach.   Juran (1988) believes management must establish top-level plans for annual improvement and encourage projects as a means to achieve this end. The underlying philosophy seems to appeal to boss-type managers, since it gives them a strong sense of control. Juran asserts that poor planning by management results in poor quality. His approach for improving quality, known as the *Juran trilogy*, is to (1) plan, (2) control, and (3) improve. More specifically:

■ *Quality planning.* In preparing to meet organizational goals, the end result should be a process that is capable of meeting those goals under operating conditions. Quality planning might include identifying internal and external customers, determining customer needs, developing a product or service that responds to those needs, establishing goals that meet the needs of customers and suppliers at a minimum cost, and proving that the process is capable of meeting quality goals under operating conditions. A necessary step is for managers to engage cross-functional teams and openly supply data to team members so that they may work together with unity of purpose.

■ *Quality control.* At the heart of this process is the collection and analysis of data for the purpose of determining how best to meet project goals under normal operating conditions. One may have to decide on control subjects, units of measurement, standards of performance, and degrees of conformance. To measure the difference between the actual performance before and after the process or system has been modified, the data should be statistically significant and the processes or system should be in statistical control. Task forces working on various problems need to establish baseline data so that they can determine if the implemented recommendations are responsible for the observed improvements.

■ *Quality improvement.* This process is concerned with breaking through to a new level of performance. The end result is that the particular process or system is obviously at a higher level of quality in delivering either a product or a service.

Juran's approach, like those of his colleagues, stresses the involvement of employees in all phases of a project. The philosophy and procedures require that managers listen to employees and help them rank the processes and systems that require improvement. This can be done with the help of any of the techniques described in Chapters 3 and 4.

Crosby approach.   Crosby's philosophy seems to appeal to the human resources type of manager. He enforces the belief that quality is a universal goal and that management must provide the leadership to compel an enterprise in which quality is never compromised. Crosby defines quality as conformance to requirements, and asserts that the mechanism for attaining quality is prevention (i.e., the first mission is appraisal). He encourages a performance standard of zero defects and says that a casualness toward quality is the price of nonconformance—doing something over rather than doing it right the first time. He believes that managers should be facilitators and should be considered as such by employees, rather than as punishment sent from the heavens.

Like Deming, Crosby (1984) has 14 steps for quality improvement. They are:

1. Management commitment
2. Quality improvement teams
3. Measurement
4. Cost of quality
5. Quality awareness
6. Corrective action
7. Zero-defect planning
8. Employee education
9. Zero-defect day
10. Goal setting
11. Error-cause removal
12. Recognition
13. Quality councils
14. Doing it over again

Imai approach.   Imai (1986) supports the continuous improvement process, where people are encouraged to focus on the environment in which they work rather than on the results. He believes that by continually improving processes and systems, the end result will be a better product or service. This has become known as the "P" or *process approach*, rather than the "R" or *results approach* of Frederick Taylor, a pioneer in work measurement and the father of industrial management. The process approach is also known as the *Kaizen approach*.

In the "R" approach, management examines the anticipated result(s), usually specified by a management-by-objectives plan, and then rates the performance of the individual(s). A person's performance is influenced by reward and punishment; that is, the use of "carrot and stick" motivation. In the "P" approach, management supports individual and team efforts to improve the processes and systems leading to the end result.

The effects of the continuous improvement, or Kaizen approach, can be elusive because they are long-term and often undramatic. Change is gradual and consistent. The approach involves everyone, with the group effort focused on processes and systems rather than one person's performance evaluation. And while the monetary investment is low, a great deal of management support is required to maintain the momentum of the group. The Kaizen approach is people oriented.

6.7.4 Quality Function Deployment

Each of the strategies put forth by the TQM gurus encourages a systematic effort to achieve excellence by the entire organization. Excellence or quality has many aspects that can be grouped into two major categories: (1) the quality of the design, and (2) the quality of conformance. To facilitate a discussion of these two categories, the following definition of quality is used: A quality product is one that meets or exceeds customer expectations. Thus the design quality is the degree to which product, process, and support design meets or exceeds customer expectations, and the quality of conformance is the degree to which the product, service, or system delivered meets the design specifications.

Clearly, a quality design is the translation of customer needs and expectations into the blueprints of the product, process, and support system. An important technique that accompanies quality design and concurrent engineering is *quality function deployment* (QFD), introduced by Yoji Akao. QFD is based on using interdisciplinary teams. The members of the teams study the market (customers) to determine the required characteristics of the product or system. These characteristics are classified into customer attributes and are listed in order of their relative importance to the customer.

The ranked attributes, also called the "WHATS," are the input to a second step in which the team members translate the attributes into technical specifications, or "HOWS." Thus an attribute such as "a tape recorder that is easy to carry around" can be translated into physical dimensions and weight that can be used to guide product development. This example, of course, led to Sony's Walkman. The joint effort by the team members promotes concurrent engineering while ensuring better communication and easier integration of the basic functions.

A matrix called the *quality chart* is used in the QFD process. The rows of the quality chart list in hierarchical order the attributes (the "WHATS"), while the design characteristics (the "HOWS") are similarly listed across the columns. Each cell in the resulting matrix corresponds to a lower-level attribute intersection with a lower-level design characteristic.

Entries indicate the correlation between the corresponding attribute and design character-istic. From the matrix the team members can infer the relative importance of the attributes along with their correlated design characteristics and the degree of correlation. Based on this information, a weight, w_i, is calculated for each design characteristic, i. This weight is the sum of all attribute weights, a_j, multiplied by the corresponding correlation, c_{ij}, be-tween the specific design characteristic and the particular attribute. The formula for calcu-lating w_i is

$$w_i = \sum_j a_j c_{ij}$$

QFD is a powerful tool that helps the concurrent engineering team focus on the de-sign characteristics that influence the attributes viewed as most important by customers. The weights derived as part of the process serve this purpose directly. To illustrate the ideas behind QFD, consider a project aimed at designing a new cross-country bicycle. By using market research, the project team can identify the most important attributes of this product for its potential customers. Suppose that the four top-ranking attributes were found to be durability, convenience, speed, and cost. Next, the team considers the three major components of the new bicycle: the frame, the gears, and the wheels. Table 6-3 il-lustrates the relationship between the attributes required by the customers and the design characteristics.

TABLE 6-3   QUALITY CHART FOR NEW BICYCLE DESIGN

			Design characteristics[a]					
			1. *Frame*		2. *Gears*		3. *Wheels*	
Attribute		Weight, a_j	1.1 Material	1.2 Design	2.1 Material	2.2 Design	3.1 Material	3.2 Design
1. *Durability*	1.1 Corrosion	2	H	L	H	M	H	M
	1.2 Impact	1	H	H	H	H	H	H
	1.3 Pressure	3	H	H	H	M	H	H
	1.4 Wear	2	M	L	H	H	H	M
2. *Convenience*	2.1 Carrying	3	H	M	L	L	H	H
	2.2 Riding	3	M	M	M	H	H	H
	2.3 Maintenance	2	L	M	H	H	H	H
3. *Speed*	3.1 Flat surface	1	M	H	M	H	L	H
	3.2 Up hill	3	M	H	M	H	L	H
	3.3 Down hill	2	M	H	M	H	L	H
4. *Cost*	4.1 Purchase	2	H	H	M	H	H	M
	4.2 Maintenance	2	M	M	H	H	H	M
	4.3 Salvage value	1	H	H	M	H	H	M

[a]H, high correlation; M, medium correlation; L, low correlation.

Now, assuming that the correlations used are H = 0.9, M = 0.5, and L = 0.3, the weight of, say, the frame material (w_1) is

$$w_1 = \sum_{j=1}^{13} a_j c_{1j} = 2 \times 0.9 + 1 \times 0.9 + 3 \times 0.9 + 2 \times 0.5 + 3 \times 0.9$$
$$+ 3 \times 0.5 + 2 \times 0.3 + 1 \times 0.5 + 3 \times 0.4 + 2 \times 0.5$$
$$+ 2 \times 0.9 + 2 \times 0.5 + 1 \times 0.9 = 17.9$$

In Table 6-3, only two levels of attributes and design characteristics are presented. Lower levels such as the dimensions and shape of the frame and the size of each gear in the transmission can be added if more detail is deemed necessary. Additional information frequently found in the quality chart are the relative importance of each attribute, target value of design characteristics, information about similar products available in the market, and the correlation between design characteristics.

In general terms, QFD uses four "houses" to integrate the informational needs of marketing, engineering, R&D, manufacturing, and management. It is best known by the first house, the house of quality, shown conceptually in Fig. 6-4. For new-product development, the team begins by obtaining the "voice of the customer" in the form of 200 to 300 detailed customer needs such as (on-screen programming) "menu appears on the TV screen with easy-to-read instructions." These customer needs are grouped hierarchically into a relatively few primary needs (to establish the strategic position), 20 to 30 secondary needs (to design the basic product and its marketing), and 150 to 250 tertiary needs (to provide specific design direction to engineers). Customer perceptions of competitive products provide goals and opportunities for new products. The importance of customer needs establishes design priorities.

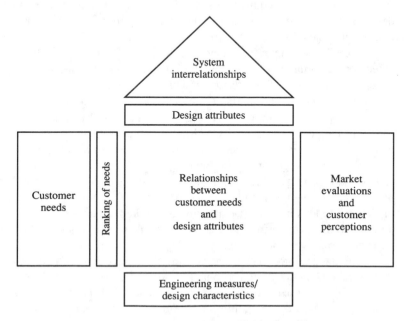

Figure 6-4   House of quality.

The relationship matrix translates customer needs, the language of marketing, into engineering language. Engineering design attributes, such as an automatic shutoff time delay, provide the means to satisfy customer needs. Performance measures of the design attributes (seconds of delay, etc.) establish competitor capabilities. Finally, the "roof matrix" (upper triangle in Fig. 6-4) quantifies the physical interrelations among the design attributes—instructions must be succinct and correlate with the design.

The house of quality encourages cooperation and communication among functions by requiring input from marketing (the customer's voice) and engineering (engineering measures and the roof matrix) and agreement on interrelationships. The entire team should participate with all members understanding and accepting these inputs and relationships. Once the house of quality is complete, the other "houses" link design attributes to part characteristics, part characteristics to manufacturing processes, and manufacturing processes to the production line. A complete set of QFD houses represents the customer's needs through every product development function. Further discussion on these ideas can be found in the work of Hauser and Clausing (1988).

By integrating product design, process design, and support design in this fashion, producibility and inspectability are built into the product and the process by which it is manufactured. Producibility is achieved when the product design has been verified through prototypes and qualification testing. Issues related to good process design include the use of standard equipment and tooling for fabrication, assembly, and test. Inspectability is achieved when all possible defects that can be created due to design errors or manufacturing problems are detectable by those performing the actual work.

A major goal of TQM is defect prevention. To achieve this end, design should be started only after the requirements are clearly understood. Product, process, and support design should be integrated so that manufacturing technology is compatible with product complexity, and all training requirements are identified and performed prior to the production phase. The configuration management system provides the project manager with updated configuration and engineering information needed as references for quality control. When applying TQM, the detection of a defect is not only a trigger for rework, but it initiates a study aimed at eliminating future defects, that is, a study of the process and product design as well as the processes and methods used in manufacturing that might be the source of the problem. Again, TQM tries to eliminate the source of defects so that defect detection and rework do not become the normal mode of operation.

When a project consists of building several identical units in series, product trend analysis is used to avoid repeating mistakes. This is done by monitoring the performance of consecutive units and studying related trends. When the trend is toward higher performance due to learning, no special action is required. If, however, deterioration (or simply no improvement) in performance is observed, the source of such a trend should be identified and corrective action taken.

The integration of TQM with concurrent engineering and configuration management greatly facilitates the design of quality into new products and their manufacturing processes. This minimizes dependence on inspection and the need for costly rework. High quality is achieved by doing things right from the beginning, not by removing defects that should not have been there in the first place.

6.7.5 Quality Assurance Plan

There are many definitions of quality, such as "meeting or exceeding customer requirements" or "fitness for use," but these can be vague and even troublesome when it comes time for action. In the conceptual design phase of a system, quality is often undervalued, due to the difficulty we have in quantifying it, or by incorrectly equating it with time and money, as discussed in Section 6.7.2. But even if these two problems are remedied, there is a third that catches most people unaware—the lack of planning. To rephrase a point made above, quality cannot be added to a system upon completion; it must be built in.

The vehicle for doing this is the quality assurance (QA) plan. This is a before-the-fact document that states the rules that will be followed during project execution. Of course, wherever there is a plan, there must be a way to verify that it is being carried out correctly. This is the function of the quality assurance review, which is an after-the-fact checkpoint. The QA plan and the QA review provide a means for close monitoring of a project, in terms of both meeting requirements and conformance to standards.

The necessity for quality planning is much the same as the necessity for any planning activity. The underlying rationale and expected benefits are outlined below.

1. A plan is needed to make something happen. Conversely, without a plan, objectives are likely to be shortchanged. For example, during development, if all levels of testing (parts, subassemblies, integration, system, acceptance) are not mentioned explicitly, one of them could easily be overlooked. Or, if no procedure has been defined for updating all documentation after a change is made, blueprints and manuals may become dangerously out of date.

2. A plan is needed to prevent corners from being cut. The idea here is an extension of what was said above. For example, there may be an implicit requirement to conduct a walk-through on test plans. But with a project behind schedule, the temptation may be great to skip over this step and thus lose the benefit of peer criticism on the testing procedures.

3. A plan is a statement of procedures. It describes how quality will be examined and measured. If prototypes are to be subjected to quality inspection, for example, the people involved should know beforehand what is going to be examined and measured so that there are no surprises after the fact. It is easier to play the game when one knows all the rules.

4. A plan states the amounts of time and money required. Thus quality is less susceptible to cuts when it is planned for explicitly. It becomes a stated requirement of the system being developed for which resources must be allocated.

5. A plan becomes a yardstick for measuring improvement. After quality planning becomes an integral step in the system development life cycle, the degree to which expectations are met can be weighed against the amount of planning undertaken. This comparison should reveal whether an increase in overall quality planning is required or whether it is simply necessary to shift the emphasis of observation and measurement from one phase of the system life cycle to another.

6. A QA plan tends to generate uniform quality. Differing levels of experience, ability, style of work, and even attitude can cause variations in quality levels within the same company or department. But if quality plans are mandatory and are produced according to standard guidelines, the variations in quality should diminish.

7. Finally, quality plans encourage attention to standards. The problem is not that standards do not exist, but simply that they frequently fall into disuse. People rarely read manuals from cover to cover, but if at the start of a project they were told what the relevant standards were and exactly where they could be found, the chances of the standards being applied correctly would increase.

Plans must be tailored. No single quality plan will suit all project environments and circumstances. The *IEEE Standard for Software Quality Assurance Plans* is a comprehensive document that can provide guidance for software development. For two-party contractual arrangements, the International Standards Organization (ISO) has propounded a series of standards known as *ISO 9000*. Coverage includes the selection and use of equipment, the development, installation, and servicing of facilities, final inspection and test, and general management responsibilities. ISO 9000 certification is critical for companies that wish to compete in the European marketplace. Additional references for military system standards are given in the reference list at the end of the chapter.

A customized QA plan can be developed easily by asking some fundamental "what–who–how" questions: What has to be accomplished? Who has the responsibility? How are the tasks to be done? Consolidating answers to these questions into a concise, practical format yields a simple, yet effective plan for assuring a smooth, problem-free transition from one phase of a project to the next. A common format is the QA matrix, which arrays standards for each task against the three headings "what," "who," "how." It is similar to the linear responsibility chart discussed in Chapter 5.

Generally, the QA matrix is developed by the QA team if one exists, or alternatively, by the project manager. But regardless of who prepares it, agreement must be obtained from all parties mentioned before work begins. Consensus carries with it a number of automatic benefits.

- It verifies responsibility and identifies the type of involvement for each of the participants.
- It presents a complete picture of responsibility, such that one party can see the involvement of other responsible parties.
- It is a forewarning of required standards knowledge.
- It is an explicit sign of acceptance of shared responsibility for deliverable quality.

A QA matrix can be developed at several different levels: deliverables, project phase, or the complete system life cycle. In each case, the elements remain the same.

6.7.6 Implementing TQM

In TQM, how a process is implemented is as important as what the process includes. Many organizations still labor under the remnants of a departmentalized Taylor approach in which some employees plan improvements, others carry out the work, and still others inspect to see if procedures and results are correct. In TQM, all employees, every day, commit to improving the quality of their service so that customer needs are not only met but exceeded. In the language of projects, TQM is a set of activities not an event. This is illustrated in Fig. 6-5, which indicates that the first step is awareness of a problem. Next goals arc set, a strategy is formulated, and resources are deployed. If the commitment is there, improved performance should follow. But this is only the beginning. New circumstances lead to a heightened awareness and a continuous repetition of the process.

How to realize this model, though? The TQM literature is bristling with techniques, prescriptions, admonitions, and anecdotes. Nevertheless, little attention is devoted to how firms have implemented TQM, the hurdles they encountered, and how they adapted the underlying principles to their existing cultures. Furthermore, the lack of agreement among the TQM gurus produces contradictions and inconsistent prescriptions that are puzzling to would-be users. Deming says "eliminate slogans," while Crosby invokes the slogan of "zero defects." Deming says "drive out fear," while Juran argues that "fear can bring out the best in people." Deming's process starts at the top and works down while Juran starts with middle management and works in both directions.

A nonprofit TQM research company in Massachusetts called GOAL/QPC found that six implementation models are currently being used.

1. *TQM element approach.* This approach, used in the early 1980s, employs elements of quality improvement programs such as quality circles, statistical process control, and quality functional deployment rather than full implementation of TQM.

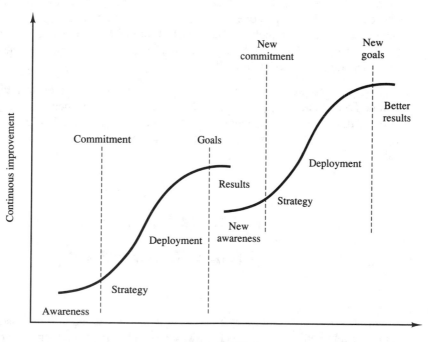

Figure 6-5   TQM journey.

2. *Guru approach.* This approach uses writings of a guru such as Deming, Juran, and Crosby as a benchmark to determine what the organization lacks, then adopts the guru's systems to make changes. Use of Juran's trilogy is an example.

3. *Japanese model approach.* Organizations using this method begin by studying the Japanese "Deming prizewinners" as a way to develop a master plan that fits their specific circumstances. This approach was used by Florida Power and Light, which was the first American company to win the prize.

4. *Industrial company model approach.* In this approach, people visit a company using TQM, identify its successes, and integrate this information with their own ideas to create a customized approach. This method was used in the late 1980s by many of the Baldrige winners.

5. *Hoshin planning approach.* This approach, developed by a Japanese firm, Bridgestone, was used successfully by Hewlett-Packard. It is a 10-step procedure that focuses on successful planning, deployment, execution, and monthly diagnosis. Its key elements are continuously improved planning and implementation processes; participation and coordination at all levels and departments in the planning, development, and deployment of strategic breakthroughs; fact-driven decision making; and the development of capability-driven goals and action plans that cascade through the organization.

6. *Baldrige award criteria approach.* In this model an organization uses criteria for the Malcolm Baldrige National Quality Award to identify areas of improvement. The criteria cover seven key components of TQM. The program is managed by the U.S. Department of Commerce through the National Institute of Standards and Technology and administered by the American Society for Quality Control. Information and guidelines are published annually (see NIST 1993).

All of these approaches, as well as derivatives, can work well with the proper motivation. To cite one example, Oregon State University developed a strategy most closely associated with the Hoshin planning model and then adopted the Baldrige award criteria to lay out a 5-year plan. Here, as in all cases, top management commitment was critical for success.

TEAM PROJECT

Thermal Transfer Plant

The approved rotary combustor project is now in the detailed design phase. Recently, the chief operating officer (COO) at TMS was exposed to the following three concepts: time-based competition, total quality management, and configuration management. Since a task force under his supervision is now examining the potential benefits and risks of the rotary combustor project, you have been asked to explain how these three concepts will be implemented in the project's production phase to maximize its probability of success.

Specifically, the COO would like to know how the configuration will be managed, what level of concurrent engineering will be implemented, and which aspects of and by what means TQM will be designed into the project. Your analysis is part of the detailed design phase, and the COO is expecting a thoroughly documented report that at least answers the following questions:

1. What forms will be used for change requests?
2. Who should sit on the CCB?
3. How can continuous improvement be encouraged? Be specific.

Submit a detailed report that can be implemented within TMS's current organizational structure. Discuss the costs and benefits of introducing each of these ideas into the project based on your assumptions and analysis.

DISCUSSION QUESTIONS

1. What are the design aspects of writing a term paper?
2. Describe a design process you are familiar with that is performed sequentially. Explain how concurrent engineering could be implemented (if possible) in this case. If in your opinion, concurrent engineering is not possible, explain why.
3. Give an example in which concurrent engineering cannot be applied.
4. What is the difference between the communication needs in sequential engineering and concurrent engineering?
5. What is the relationship between configuration management and TQM?
6. In what ways does concurrent engineering affect TQM?
7. What are the similarities and differences between Deming's 14 points and Crosby's 14 points?
8. What does worker empowerment mean at a university from the point of view of the student, staff, faculty, and administration?
9. In a factory, what does worker empowerment mean from the point of view of an assembly line worker, first-line supervisor, design engineer, and middle-level manager?
10. The Kaizen approach of Imai stresses gradual, long-term improvement. In what situation, or under what conditions, might this approach not work very well? Under what conditions is this approach best?
11. Contrast Juran's approach with the Kaizen approach. Identify situations in which either would be more or less appropriate.
12. The idea of training workers and giving them responsibility for improving the processes with which they are involved seems unassailable. Why then has it taken U.S. management so long to embrace this idea?

13. Henry Ford invented mass production. In so doing, he perfected the assembly line concept, where each worker does only one or a handful of jobs and is given little other responsibility. This worked well for 70 years; however, it is now apparent that an increasing number of U.S. companies cannot produce a high-quality product by sticking to the assembly line model. What has changed?

14. U.S. manufacturers spend about 80% of their R&D budgets on new technology, while their Japanese counterparts spend about 80% on process improvement. What do you think have been the positive and negative impacts of these allocations on product quality? What, in your opinion, is the best division of the R&D budget? Your answer should be industry specific.

15. Giving the factory worker responsibility for improving the manufacturing process, inspecting intermediate products, and troubleshooting operations sounds like a good idea, but why might a worker resist such empowerment?

16. Discuss the problem of assigning weights and estimating correlations in QFD. Suggest a way to solve this problem.

17. Discuss the risk involved in the project "buying a used car." Develop a risk management plan for this project.

18. One of the requirements for graduation in engineering is the successful completion of a design project. Discuss the criteria and the logic a student should use in selecting a project.

19. What are the risks associated with the project in Question 18?

EXERCISES

6-1. Prepare a risk management plan for the project of finding a job after graduation.

6-2. Select a project that you are familiar with and explain the most important factors affecting the configuration selection decisions of this project.

6-3. Prepare a configuration identification system for the project you have selected in Exercise 6-2.

6-4. Prepare a form for a configuration change request for the project selected in Exercise 6-2.

6-5. Write a job description for the configuration manager of a project.

6-6. Develop a flow diagram for the data handling and data processing required for configuration management, including:
   1. Definition of files
   2. Sources of data
   3. Data processing requirement
   4. Required output

6-7. Assume that you are an instructor in either an engineering or a business college. Interpret the meaning of and indicate how you would apply each of Deming's 14 points to a typical class that you teach.

6-8. Do the same as in Exercise 6-7 for Crosby's 14 points.

6-9. TQM principles are only now being adopted by universities. Develop a plan for the administration in your college for implementing Juran's approach.

6-10. Do the same as in Exercise 6-9 for the chairman of an academic department.

6-11. A basic tenet of TQM is worker empowerment. Develop a reward system for motivating factory workers to do their jobs more conscientiously and to take on more responsibility.

6-12. How would the reward system developed in Exercise 6-11 be different for (a) first-line supervisors; (b) middle managers?

6-13. In implementing TQM, one of the first steps is to propose a "vision," specify a "mission," and detail the "guiding principles." The vision is an idealized statement of what the organization would like to be, the mission is a statement of what business we are in, and the guiding principles are concrete statements that form the basis of an action plan. The latter include "goals" that tell us what outcomes we should work toward and "objectives" that articulate how we are going to get where we choose to go. For your company, agency, or academic department, write down a vision, a mission, and a list of guiding principles.

6-14. With respect to a university:
(a) Define the customers, both internal and external.
(b) List the critical processes, such as teaching, research, and hiring, and indicate how each can be measured.
(c) List the products and services.
(d) Construct a matrix of customers versus services and indicate in each cell the relationship between the two (i.e., strong, moderate, or questionable).
(e) Construct a similar matrix for products and services identified in part (c) versus critical processes identified in part (b).

6-15. Do the same as in Exercise 6-14 for the following organizations.
(a) Manufacturing company
(b) Government agency
(c) Retail store

6-16. Use QFD to analyze the project "developing a new course in project management."

6-17. List the major risks of a military operation such as Desert Storm, the United Nations' effort to oust Iraq from Kuwait in 1991. Outline a risk management plan for such projects.

REFERENCES

Concurrent Engineering

ALLEN, C. W., R. E. KING, and D. A. SKIVER (editors), *Simultaneous Engineering*, Society of Manufacturing Engineers, Dearborn, MI, 1990.

ALTER, A. E., "Concurrent Affairs," *CIO*, Vol. 5, No. 3 (1991), pp. 31–41.

CLELAND, D. I., "Product Design Teams: The Simultaneous Engineering Perspective," *Project Management Journal*, Vol. 22, No. 4 (1991), pp. 5–10.

HSU, J. P., J. S. GERVAIS, and F. Y. PHILLIPS, *International Workshop on Concurrent Engineering Design*, Final Report, National Science Foundation, Washington, DC, September 1991.

O'GRADY, P., and J. S. OH, "A Review of Approaches to Design for Assembly," *Concurrent Engineering* (May–June 1991), pp. 5–11.

PETERS, T., "Get Innovative or Get Dead," *California Management Review*, Vol. 33, No. 2 (1991), pp. 9–23.

REDDY, R., T. R. WOOD, and K. J. CLEETUS, "The DARPA Initiative: Encouraging New Industrial Practices," *IEEE Spectrum*, Vol. 28, No. 7 (1991), pp. 26–30.

SHINA, S. G., *Concurrent Engineering and DFM for Electronic Products*, Van Nostrand Reinhold, New York, 1991.

TURINO, J. L., *Managing Concurrent Engineering: Buying Time to Market*, Van Nostrand Reinhold, New York, 1992.

VESEY, J. T., "The New Competitors: They Think in Terms of 'Speed-to-Market,'" *The Executive*, Vol. 5, No. 2 (1991), pp. 23–33.

Configuration Selection

BELL, D. E., R. L. KEENEY, and H. RAIFFA, *Conflicting Objectives in Decisions*, Wiley, New York, 1977.

BLANCHARD, S. B., *Design and Manage to Life Cycle Cost*, M/A Press, Portland, OR, 1978.

BLANCHARD, B. S., and W. J. FABRYCKY, *Systems Engineering and Analysis*, 2/E, Prentice Hall, Englewood Cliffs, NJ, 1990.

CANADA, J. R., and W. G. SULLIVAN, *Economics and Multiattribute Evaluation of Advanced Manufacturing Systems*, Prentice Hall, Englewood Cliffs, NJ, 1989.

Design to Cost, Directive 5000.28, U.S. Department of Defense, Washington, DC, 1975.

ENGLISH, J. M. (editor), *Cost Effectiveness: The Economic Evaluation of Engineering Systems*, Wiley, New York, 1968.

OSTWALD, P. F., *Cost Estimating*, Second Edition, Prentice Hall, Englewood Cliffs, NJ, 1974.

SAATY, T. L., *The Analytical Hierarchy Process*, McGraw-Hill, New York, 1980.

Configuration Management

EGGERMAN, W. V., *Configuration Management Handbook*, TAB Books, Blue Ridge Summit, PA, 1990.

BERSOFF, E. H., and A. M. DAVIS, "Impacts of Life Cycle Models on Software Configuration Management," *Communications of the ACM*, Vol. 34, No. 8 (1991), pp. 104–118.

MORRIS, M. L., "Configuration Management: Getting the Facts Right," *Inform*, Vol. 5, No. 5 (1991), pp. 52–56, 66.

STEVENS, C. A., and K. WRIGHT, "Managing Change with Configuration Management," *National Productivity Review*, Vol. 10, No. 4 (1991), pp. 509–518.

SWEETMAN, S. L., "Utilizing Expert Systems to Improve the Configuration Management Process," *Project Management Journal*, Vol. 21, No. 1 (1990), pp. 5–12.

Standards

DOD-STD-480A, *Engineering Changes, Deviations and Waivers*, U.S. Department of Defense, Washington, DC, 1978.

MIL-STD-482A, *Configuration Status Accounting, Data Elements, and Related Features*, U.S. Department of Defense, Washington, DC, 1974.

MIL-STD-483, *Configuration Management for Systems, Equipment, Munitions and Computer Programs*, U.S. Department of Defense, Washington, DC, 1985.

Management of Technology

BABCOCK, D. L., *Managing Engineering and Technology: An Introduction to Management for Engineers*, Institute of Industrial Engineers, Norcross, GA, 1991.

BETZ, F., *Managing Technology*, Prentice Hall, Englewood Cliffs, NJ, 1987.

FLEMING, S. C., "Using Technology for Competitive Advantage," *Research-Technology Management*, Vol. 34, No. 5 (1991), pp. 38–41.

KOTLER, P., and P. J. STONICH, "Turbo Marketing through Time Compression," *Journal of Business Strategy*, Vol. 12, No. 5 (1991), pp. 24–29.

MILLER, D. B., *Managing Professionals in Research and Development: A Guide for Improving Productivity and Organizational Effectiveness*, Jossey-Bass, San Francisco, 1986.

National Research Council, *Management of Technology: The Hidden Competitive Advantage*, U.S. Department of Commerce, National Institute of Standards and Technology, Springfield, VA, 1987.

Risk Management

COOPER, D., and C. CHAPMAN, *Risk Analysis for Large Projects: Models, Methods, and Cases*, Wiley, New York, 1987.

GROSE, V. L., *Managing Risk: Systematic Loss Prevention for Executives*, Prentice Hall, Englewood Cliffs, NJ, 1987.

LICHTENBERG, S., "Alternatives to Conventional Project Management," *International Journal of Project Management*, Vol. 1, No. 2 (1983), pp. 101–102.

RASMUSSEN, N. C., "The Application of Probabilistic Risk Assessment Techniques to Energy Technologies," *Annual Review of Energy*, Vol. 6 (1981), pp. 123–138.

Total Quality Management

CLAUSING, D., and B. H. SIMPSON, "Quality by Design," *Quality Process*, Vol. 23, No. 1 (1990), pp. 41–44.

COLE, R., "The Quality Revolution," *Production and Operations Management*, Vol. 1, No. 1 (1992), pp. 118–120.

CROSBY, P., *Quality without Tears: The Art of Hassle-Free Management*, McGraw-Hill, New York, 1984.

DEMING, W. E., *Out of the Crisis*, MIT Center for Advanced Engineering, Cambridge, MA, 1986.

EVANS, J. R., and W. M. LINDSAY, *The Management and Control of Quality*, West, St. Paul, MN, 1989.

FEIGENBAUM, A. V., *Total Quality Control: Engineering and Management*, McGraw-Hill, New York, 1961.

FEIGENBAUM, A. V., *Total Quality Control*, Third Edition, McGraw-Hill, New York, 1986.

GARVIN, D. A., *Managing Quality*, Free Press, New York, 1988.

General Accounting Office, *Management Practices: U.S. Companies Improve Performance through Quality Efforts*, U.S. Government Printing Office, Washington, DC, 1991.

HAAVIND, R., *Road to the Baldrige Award*, Butterworth Heinemann, Stoneham, MA, 1992.

IMAI, M., *Kaizen: The Key to Japan's Competitive Success*, Productivity Press, Cambridge, MA, 1986.

JURAN, J., *Juran's Quality Control Handbook*, Fourth Edition, McGraw-Hill, New York, 1988.

NADLER, G., and S. HIBINO, *Breakthrough Thinking*, Prima Publishing & Communications, St. Martins Press, New York, 1990.

NIST, *Award Criteria: Malcolm Baldrige National Quality Award*, U.S. Department of Commerce, National Institute of Standards and Technology, Gaithersburg, MD, 1993.

SCHNEIDERMAN, A., "Optimum Quality Costs and Zero Defects: Are They Contradictory Concepts?" *Quality Progress*, Vol. 19 (1986), pp. 28–31.

VAUGHN, R., *Quality Assurance*, Iowa State University Press, Ames, IA, 1990.

WALTON, M., *The Deming Management Method*, Perigee Books, Putnam Publishing Group, New York, 1986.

Quality Function Deployment

AKAO, Y., *Quality Deployment: A Series of Articles*, translated by Glen Mazur, GOAL/QPC Methuen, MA, 1987.

AKAO, Y. (editor), "Quality Function Deployment," in *Integrating Customer Requirements into Product Design*, Productivity Press, Cambridge, MA, 1990.

CHANG, C.-H., "Quality Function Deployment (QFD) Processes in an Integrated Quality Information Systems," *Computers & Industrial Engineering*, Vol. 17, No. 1–4 (1989), pp. 311–316.

COOPER, R. G., "New Product Strategies: What Distinguishes Top Performers?" *Journal of Product Innovation Management*, Vol. 2 (1984), pp. 151–164.

COOPER, R. G., and E. J. KLEINSCHMIDT, "An Investigation into the New Product Process: Steps, Deficiencies, and Impact," *Journal of Product Innovation Management*, Vol. 5 (1987), pp. 71–85.

GRIFFEN, A., and J. R. HAUSER, "Patterns of Communication among Marketing, Engineering and Manufacturing: A Comparison between Two New Product Teams," *Management Science*, Vol. 38, No. 3 (1992), pp. 360–373.

HAUSER, J. R., and D. CLAUSING, "The House of Quality," *Harvard Business Review*, Vol. 66, No. 3 (1988), pp. 62–73.

KING, B., *Better Designs in Half the Time: Implementing Quality Function Deployment (QFD) in America*, GOAL/QPC, Methuen, MA, 1987.

KOGURE, M., and Y. AKAO, "Quality Function Deployment and CWQC," *Quality Process*, Vol. 16, No. 10 (1983), pp. 25–29.

MADDUX, G. A., R. W. AMOS and A. R. WYSKIDA, "Organizations Can Apply Quality Function Deployment as Strategic Planning Tool," *Industrial Engineering*, Vol. 23, No. 9 (1991), pp. 33–37.

SULLIVAN, L. P., "Quality Function Deployment," *Quality Progress*, Vol. 19 (1986), pp. 39–50.

APPENDIX 6A

Glossary of TQM Terms

- *Breakthrough items—innovative improvement:* activities that are specifically geared to generate breakthroughs in systems and procedures that can result in improvements in quality.
- *Continuous improvement:* operating philosophy within an organization by which all employees seek to make systems and processes better on an ongoing basis.
- *Cross-functional teams:* project teams made up of representatives from a variety of functions with the charter to resolve specific quality problems or to spearhead a specific breakthrough activity. Participants can be both management and nonmanagement personnel.
- *Customer relationship management:* organization's utilization and responsiveness to external customer data for the enhancement of TQM efforts.
- *Data management:* systems and processes by which data are generated, analyzed, recorded, and retained in the organization.
- *Employee involvement:* involvement of individual, nonsupervisory employees in the improvement of the systems and processes of the organization.
- *Executive (top) management:* group of senior executives responsible for operation of the organization. Logically, this group includes the chief executive officer (CEO) and all functional department managers who report directly to the CEO. In organizations that do not have a formal hierarchy, this would apply to those people who make the policy, administration, and operational decisions for the organization.
- *Hoshin planning:* process used to ensure that the directions, goals, and objectives of the organization are rooted in customer requirements and are rationally developed, well defined, clearly communicated, monitored for compliance, and adapted based on system feedback so that customer demands are satisfied within reasonable cost.
- *Internal customer/supplier relationships:* interaction among persons who are linked by a functional process or system.
- *Numbers in pictures:* data analysis tools such as time-series graphs, scattergrams, Pareto diagram, and histogram, used in the process of identifying the causes of the issue/problem.
- *Process operations:* regular, recurring job processes that directly affect the quality of the organization's product or service.

- *Quality assurance:* system and procedures by which an organization ensures the quality performance of its products and services.

- *Quality function deployment:* system that identifies, through charting and analysis, the needs of customers and communicates those needs throughout the organization in an effort to exceed its competition in satisfying customer requirements.

- *Quality of worklife:* physical environment and the work atmosphere in which the organization conducts its operations.

- *Recognition/rewards:* formal and informal procedures designed to reinforce activities and behaviors that improve the total quality of an organization.

- *Supplier quality:* relationship between an organization and the suppliers of its materials, components, or supporting services.

- *TQM momentum:* totality and forward movement of the organization toward achieving TQM.

- *Trends in improvement:* historical and current process and system improvement trends of the organization in its implementation of TQM.

7

Project Scheduling

7.1 INTRODUCTION

Project scheduling deals with the planning of timetables and the establishment of dates during which various resources, such as equipment and personnel, will perform the activities required to complete the project. Schedules are the cornerstone of the planning and control system, and because of their importance, are often written into the contract by the customer.

The scheduling activity integrates information on several aspects of the project, including the estimated duration of activities, the technological precedence relations among activities, constraints imposed by the availability of resources and the budget, and if applicable, due-date requirements. This information is processed into an acceptable schedule by appropriate models, most often of the network type. The aim is to answer the following questions:

1. If each of the activities goes according to plan, when will the project be completed?
2. Which tasks are most critical to ensure the timely completion of the project?
3. Which tasks can be delayed, if necessary, without delaying project completion, and by how much?
4. More specifically, at what times should each activity begin and end?
5. At any given time during the project, what is the range of dollars that should have been spent?
6. Is it worthwhile to incur extra costs to accelerate some of the activities?

The first four questions relate to time, which is the chief concern of this chapter; the latter two deal with the possibility of trading off time for money and are taken up in Chapter 8.

The schedule itself can be presented in several ways, such as a timetable or a Gantt chart, which is essentially a bar chart that shows the relationship of activities over time. Different schedules can be prepared for the various participants in the project. A functional manager may be interested in a schedule of tasks performed by members of the group. The project manager may need a detailed schedule for each WBS element and a master schedule for the entire project. The vice president of finance may need a combined schedule for all projects that are under way in the organization to plan cash flows and capital requirements. Each person involved in the project may need a schedule with all the activities in which he or she is involved.

Schedules provide an essential communications and coordination link between the individuals and organizations participating in the project. They facilitate the coordination of effort among people coming from different organizations working on different elements of the WBS in different locations at different time periods. By developing a schedule the project manager is *planning* the project. By authorizing work according to the scheduled start of each task, he or she triggers execution of the project; and by comparing the actual execution dates of tasks with the scheduled dates, he or she *monitors* the project. When actual performance deviates from the plan to such an extent that corrective action must be taken, the project manager is exercising *control*.

Although schedules come in many forms and levels of detail, they should all relate to the master program schedule, which gives a time-phased picture of the principal activities and highlights the major milestones associated with the project. For large programs a modular approach that reduces the prospects of getting bogged down in the excess detail that necessarily accompanies work assignments is recommended. To implement this approach, the schedule should be partitioned according to its functions or phases and then disaggregated to reflect the various work packages. For example, consider the work breakdown structure shown in Fig. 7-1 for the development of a microcomputer. One possible modular array of project schedules is depicted in Fig. 7-2. The details of each module would have to be worked out by the individual project leaders and then integrated by the project manager to gain the full perspective.

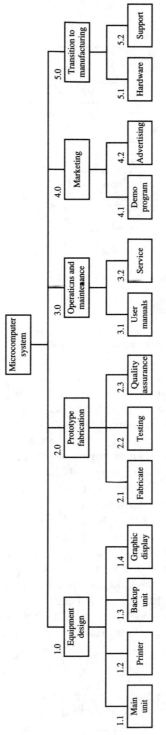

Figure 7-1 Work breakdown structure for a microcomputer.

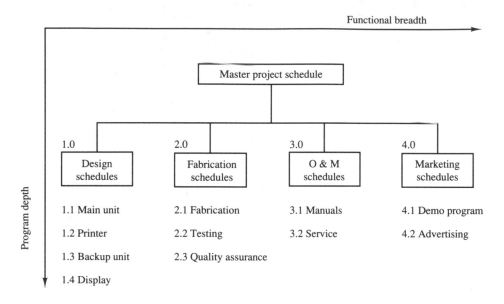

Figure 7-2 Modular array of project schedules. (From H. J. Thamhain, *Engineering Program Management*, copyright © 1984 by John Wiley & Sons, Inc.; reprinted by permission of the publisher.)

Schedules are working tools for program planning, evaluation, and control. They are developed in many iterations with project team members and continuing feedback from the client. The reality of changing circumstances requires that they remain dynamic throughout the project life cycle. Every project has unique management requirements. When preparing the schedule, it is important that the dates and time allotments for the work packages be in precise agreement with those set forth in the master schedule. These times are control points for the project manager. It is his or her responsibility to insist on and maintain consistency, but the actual scheduling of tasks and work packages is usually done by those responsible for their accomplishment—after the project manager has approved the due dates. This procedure ensures that the final schedule reflects the interdependencies among all the tasks and participating units, and that it is consistent with available resources and upper management expectations.

It is worth noting that the most comprehensive schedule is not necessarily best in all situations. In fact, too much detail can impede communications and divert attention from critical activities. Nevertheless, the quality of a schedule has a major effect on the success of the project and frequently affects other projects that compete for the same resources.

7.1.1 Key Milestones

A place to begin the development of any schedule is to define the major milestones for the work to be accomplished. For ease of viewing, it is often convenient to array this information on a time line depicting events and their due dates. Once agreed upon, the resultant milestone chart becomes the skeleton for the master schedule and its disaggregated components. A key milestone is defined as an important event in the project life cycle, and may include, for instance, the fabrication of a prototype, the start of a new phase, a status review, a test, or the first shipment. Ideally, the completion of these milestones should be easily verifiable. But in reality, this may not be the case. Design, testing, and review tend to run together. There is always a desire to do a bit more work to correct superficial flaws or to extract a marginal improvement in performance. This blurs the demarcation points and makes project control that much more difficult.

Key milestones should be defined for all major phases of the project prior to startup. Care must be taken to arrive at an appropriate level of detail. If the milestones are spread too far apart, continuity problems in tracking and control can arise. On the other hand, too many milestones can result in unnecessary busywork, over control, confusion, and increased overhead costs. As a guideline for long-term projects, four key milestones per year seem to be sufficient for tracking without overburdening the system.

The project office, in close cooperation with the customer and the participating organizations, typically has the responsibility for defining key milestones. Selecting the right type and number is critical. Every key milestone should represent a checkpoint for a collection of activities at the completion of a major project phase. Some examples with well-defined boundaries include:

- Project kickoff
- Requirements analysis complete
- Preliminary design review
- Critical design review
- Prototype fabricated
- Integration and testing completed
- Quality assurance review
- Start volume production
- Marketing program defined
- First shipment
- Customer acceptance test complete

7.1.2 Network Techniques

The basic approach to all project scheduling is to form an actual or implied network that graphically portrays the relationships between the tasks and milestones in the project. Several techniques evolved in the late 1950s for organizing and representing this basic information. Best known today are PERT (program evaluation and review technique) and CPM (critical path method). PERT was developed by Booz, Allen & Hamilton in conjunction with the U.S. Navy in 1958 as a tool for coordinating the activities of over 11,000 contractors involved with the Polaris missile program. CPM was the result of a joint effort by

DuPont and the UNIVAC division of Remington Rand to develop a procedure for scheduling maintenance shutdowns in chemical processing plants. The major difference between the two is that CPM assumes that activity times are deterministic, while PERT views the time to complete an activity as a random variable that can be characterized by an optimistic, a pessimistic, and a mostly likely estimate of its durations. Over the years a host of variants has arisen, mainly to address specific aspects of the tracking and control problem, such as budget fluctuations, complex intertask dependencies, and the multitude of uncertainties found in the R&D environment.

PERT/CPM is based on a diagram that represents the entire project as a network of arrows and nodes. The two most popular approaches are either to place the activities on the arrows (AOA) and have the nodes signify milestones, or to place activities on the nodes (AON) and let the arrows show precedence relations among activities. A precedence relation states that, for example, activity X must be completed before activity Y can begin, or that X and Y must end at the same time. It allows tasks that must precede or follow other tasks to be clearly identified, in time as well as function. The resulting diagram can be used to identify potential scheduling difficulties, to estimate the time needed to finish the entire project, and to improve coordination among the participants.

To apply PERT/CPM, a thorough understanding of the project's requirements and structure is needed. The effort spent in identifying activity relationships and constraints yields valuable insights. In particular, four questions must be answered to begin the modeling process:

1. What are the chief project activities?
2. What are the sequencing requirements or constraints for these activities?
3. Which activities can be conducted simultaneously?
4. What are the estimated time requirements for each activity?

PERT/CPM networks are an integral component of project management and have been shown to provide the following benefits (Clark and Fujimoto 1989, Meredith and Mantel 1989):

- They furnish a consistent framework for planning, scheduling, monitoring, and controlling projects.
- They illustrate the interdependencies of all tasks, work packages, and work units.
- They aid in setting up the proper communications channels between participating organizations and points of authority.
- They can be used to estimate the expected project completion dates as well as the probability that the project will be completed by a specific date.
- They identify so-called critical activities which, if delayed, will delay the completion of the entire project.
- They also identify activities with slack that can be delayed for specific periods of time without penalty, or from which resources may temporarily be borrowed without negative consequences.

- They determine the dates on which tasks may be started, or must be started, if the project is to stay on schedule.
- They illustrate which tasks must be coordinated to avoid resource or timing conflicts.
- They also indicate which tasks may be run, or must be run, in parallel to achieve the predetermined completion date.

As we will see, PERT and CPM are easy to understand and use. And while computerized versions are available for both small and large projects, manual calculation is quite suitable for many everyday situations. Unfortunately, though, some managers have placed too much reliance on these techniques at the expense of good management practice (Vazsonyi 1970). For example, when activities are scheduled for a designated time slot, there is a tendency to meet the schedule at all costs. This may divert resources from other activities and cause much more serious problems downstream, the effects of which may not be felt until a near-catastrophe has set in. If tests are shortened or eliminated as a result of time pressure, design flaws may be discovered much later in the project. As a consequence, a project that appeared to be under control is suddenly several months behind schedule and substantially over budget. When this happens, it is convenient to blame PERT/CPM even though the real cause is poor management.

In the remainder of this chapter we discuss and illustrate the techniques used to estimate activity durations, to construct PERT/CPM networks, and to develop the project schedules. The focus is on the timing of activities. Issues related to budget and resource constraints as they affect the project's schedule are taken up in Chapters 8 and 9.

7.2 ESTIMATING THE DURATION OF PROJECT ACTIVITIES

A project is composed of a set of tasks. Each task is performed by one organizational unit on one WBS element. Most tasks can be broken down into activities. Each activity is characterized by its technological specifications, drawings, list of required materials, quality control requirements, and so on. The technological processes selected for each activity affect the resources required, the materials needed, and the timetable. For example, to move a heavy piece of equipment from one point to another, resources such as a crane and a tractor-trailer might be called for, as well as qualified operators. The time required to perform the activity may also be regarded as a resource. If the piece of equipment is mounted on a special fixture prior to moving, the required resources and the performance time might be affected. Thus the schedule of the project as well as its cost and resources requirements are a function of the technological decisions.

Some activities cannot be performed unless certain activities are completed beforehand. For example, if the piece of equipment to be moved is very large, it might be necessary to disassemble it or at least remove a few of its parts before loading it onto the truck. Thus the "moving" task has to be broken down into activities with precedence relations among them.

The process of dividing a task into activities, and activities into subactivities, should be performed carefully to strike a proper balance between size and duration. The following guidelines are recommended:

1. The length of each activity should be approximately in the range of 0.5 to 2% of the length of the project. Thus if the project takes about 1 year, each activity should be between a day and a week.

2. Critical activities that fall below this range should be included. For example, a critical design review that is scheduled to last 2 days on a 3-year project should be included in the activity list because of its pivotal importance.

3. If the number of activities is very large (say, above 250), the project should be divided into subprojects, perhaps by functional area as suggested in Section 7.1, and individual schedules developed for each. Schedules with too many activities quickly become unwieldy and are difficult to monitor and control.

We start our discussion with techniques commonly used to estimate the length of activities. We then describe the effects that precedence relations among activities have on the overall schedule.

Two approaches are used for estimating the length of an activity: the deterministic approach and the stochastic approach. The deterministic approach ignores uncertainty and thus results in a point estimate. The stochastic approach addresses the probabilistic elements in a project by estimating both the expected duration of each activity and its corresponding variance. Although tasks are subject to random forces and other uncertainties, the majority of project managers prefer the deterministic approach because of its simplicity and ease of understanding. A corollary benefit is that it yields satisfactory results in most situations.

7.2.1 Stochastic Approach

Only in rare instances is the exact duration of a planned activity known in advance. Therefore, to gain an understanding of how long it will take to perform the activity, it is logical to analyze past data and to construct a frequency distribution of related activity durations. An example of such a distribution is illustrated in Fig. 7-3. From the plot we observe that previously, the activity under consideration was performed 40 times, requiring anywhere from 10 to 70 hours. We also see that in 3 of the 40 observations the actual duration was 45 hours and that the most frequent duration was 35 hours. That is, in some 8 of the 40 repetitions, the actual duration was 35 hours.

The information in Fig. 7-3 can be summarized by two measures: the first is associated with the center of the distribution (commonly used measures are the mean, the mode, and the median), and the second is related to the spread of the distribution (commonly used measures are the variance, the standard deviation, and the interquartile range). The mean of the distribution in Fig. 7-3 is 35.25, its mode is 35, and its median is also 35. The standard deviation is 13.3 and the variance is 176.89.

When working with empirical data, it is often desirable to fit the data with a continuous distribution that can be represented mathematically in closed form. This approach

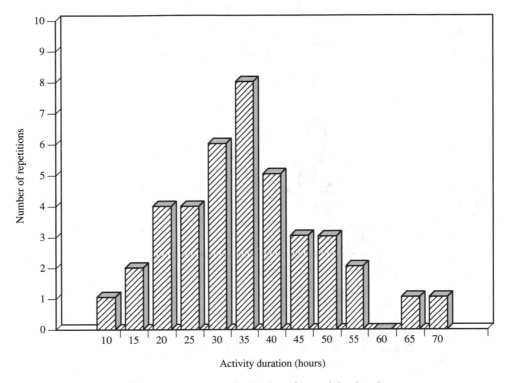

Figure 7-3   Frequency distribution of an activity duration.

facilitates the analysis. Figure 7-4 shows the superposition of a normal distribution with the parameters $\mu = 35.25$ and $\sigma = 13.3$ on the original data.

While the normal distribution is symmetrical and easy to work with, the distribution of activity durations is likely to be skewed. Furthermore, the normal distribution has a long left-hand tail while actual performance time cannot be negative. A better model of the distribution of activity lengths has proven to be the beta distribution, which is illustrated in Fig. 7-5.

A visual comparison between Figs. 7-4 and 7-5 reveals that the beta distribution provides a closer fit to the frequency data depicted in Fig. 7-3. The left-hand tail of the beta distribution does not cross the zero duration point, nor is it necessarily symmetric. Nevertheless, in practice a statistical test (such as the chi-square goodness-of-fit test; Bain and Engelhardt 1987) must be used to determine whether a theoretical distribution is a valid representation of the actual data.

In project scheduling, probabilistic considerations are incorporated by assuming that the time estimate for each activity can be derived from three different values:

a = optimistic time, which will be required if execution goes extremely well
m = most likely time, which will be required if execution is normal
b = pessimistic time, which will be required if everything goes badly

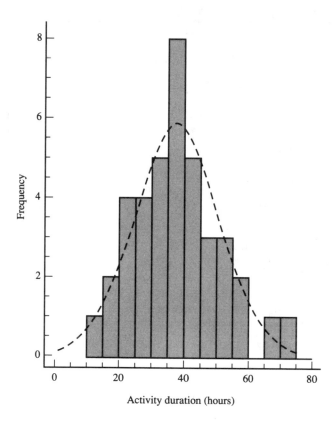

Figure 7-4   Normal distribution fitted to the data.

Statistically speaking, a and b are estimates of the lower and upper bounds of the frequency distribution, respectively. If the activity is repeated a large number of times, only in about 0.5% of the cases would the duration fall below the optimistic estimate, a, or above the pessimistic estimate, b. The most likely time, m, is an estimate of the mode (the highest point) of the distribution. It need not coincide with the midpoint $(a + b)/2$ but may occur on either side.

To convert m, a, and b into estimates of the expected value $\bar{d}$ and variance (v) of the elapsed time required by the activity, two assumptions are made. The first is that the standard deviation, s (square root of the variance), equals one-sixth the range of possible outcomes; that is,

$$s = \frac{b - a}{6} \tag{7-1}$$

The rationale for this assumption is that the tails of many probability distributions (such as the normal distribution) are considered to lie about 3 standard deviations from the mean, implying a spread of about 6 standard deviations between tails. In industry, statistical quality control charts are constructed so that the spread between the upper and lower control limits is approximately 6 standard deviations (6σ). If the underlying distribution is normal,

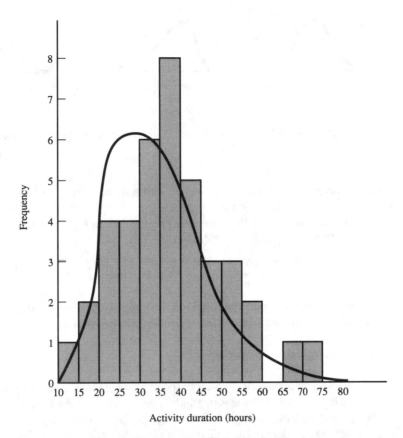

Figure 7-5   Beta distribution fitted to the data.

the probability is 0.9973 that $\bar{d}$ falls within $b - a$. In any case, according to Chebyshev's inequality, there is at least an 89% chance that the duration will fall within this range (see, e.g., Bain and Engelhardt 1987).

The second assumption concerns the form of the distribution and is needed to estimate the expected value, $\bar{d}$. In this regard, the definition of the three time estimates above provide an intuitive justification that the duration of an activity may follow a beta distribution with its unimodal point occurring at m and its endpoints at a and b. Figure 7-6 shows the three cases of the beta distribution: (a) symmetric, (b) skewed to the right, and (c) skewed to the left. The expected value of the activity duration is given by

$$\bar{d} = \frac{1}{3}\left[2m + \frac{1}{2}(a + b) \right] = \frac{a + 4m + b}{6} \qquad (7\text{-}2)$$

Notice that $\bar{d}$ is a weighted average of the mode m and the midpoint $(a + b)/2$, where the former is given twice as much weight as the latter. Although the assumption of the beta distribution is an arbitrary one and its validity has been challenged from the start (Grubbs

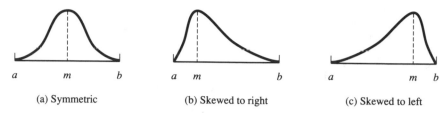

(a) Symmetric   (b) Skewed to right   (c) Skewed to left

Figure 7-6   Three cases of the beta distribution: (a) symmetric; (b) skewed to the right; (c) skewed to the left.

1962), it serves the purpose of locating $\bar{d}$ with respect to m, a, and b in what seems to be a reasonable way (Hillier and Lieberman 1986).

The following calculations are based on the data in Fig. 7-3 from which we observe that $a = 10$, $b = 70$, and $m = 35$.

$$\bar{d} = \frac{10 + (4)(35) + 70}{6} = 36.6 \quad \text{and} \quad s = \frac{70 - 10}{6} = 10$$

Thus assuming that the beta distribution is appropriate, the expected time to perform the activity is 36.6 hours with an estimated standard deviation of 10 hours.

7.2.2 Deterministic Approach

When past data for an activity similar to the one under consideration are available and the variability in performance time is negligible, the duration of the activity may be estimated by its mean; that is, the average time it took to perform the activity in the past. A problem arises when no past data exist. This problem is common in organizations that do not have an adequate information system to collect and store past data, and in R&D projects where an activity is performed for the first time. To deal with this situation, three techniques are available: the modular technique, the benchmark job technique, and the parametric technique. Each is discussed below.

7.2.3 Modular Technique

This technique is based on decomposing each activity into subactivities (or modules), estimating the performance time of each module, and then totaling the results to get an approximate performance time for the activity. As an example, consider a project to install a new flexible manufacturing system (FMS). A training program for employees has to be developed as part of the project. The associated task can be broken down into the following activities:

1. Definition of goals for the training program
2. Study of the potential participants in the program and their qualifications
3. Detailed analysis of the FMS and its operation

4. Definition of required topics to be covered

5. Preparation of a syllabus for each topic

6. Preparation of handouts, transparencies, and so on

7. Evaluation of the proposed program (a pilot study)

8. Improvements and modifications

If possible, the time required to perform each activity is estimated directly. If not, the activity is broken into modules and the time to perform each module is estimated based on past experience. Although the new training task may not be wholly identical to previous tasks undertaken by the company, the modules themselves should be common to many training programs, so historical data may be available.

7.2.4 Benchmark Job Technique

This technique is best suited for projects containing many repetitions of some standard activities. The extent to which it is used depends on the performing organization's diligence in maintaining a database of the most common activities along with estimates of their duration and resource requirements.

To see how this technique is used, consider an organization that specializes in construction projects. To estimate the time required to install an electrical system in a new building, the time required to install each component of the system would be multiplied by the number of components of that type in the new building. If, for example, the installation of an electrical outlet takes on the average 10 minutes, and there are 80 outlets in the new building, a total of $80 \times 10 = 800$ minutes is required for this type of component. After performing similar calculations for each component type or job, the total time to install the electrical system would be determined by summing the resultant times.

The benchmark job technique is most appropriate when a project is composed of a set of basic elements whose execution time is additive. If the nature of the work does not support the additivity assumption, another method—the parametric technique—should be used.

7.2.5 Parametric Technique

This technique is based on cause–effect analysis. The first step is to identify the independent variables. For example, in digging a tunnel an independent variable might be the length of the tunnel. If it takes on average 20 hours to dig 1 ft, the time to dig a tunnel of length l can be estimated by $T(l) = 20 \times l$, where time is considered the dependent variable and the length of the tunnel is considered the independent variable.

When the relationship between the dependent variable and the independent variable is known exactly, as it is in many physical systems, one can plot a response curve in two dimensions. Figure 7-7 depicts two examples of length versus time: line (a) represents a linear relationship between the independent and dependent variables, and line (b) a nonlinear one. In general, if the dependent variable, Y, is believed to be a linear function of the

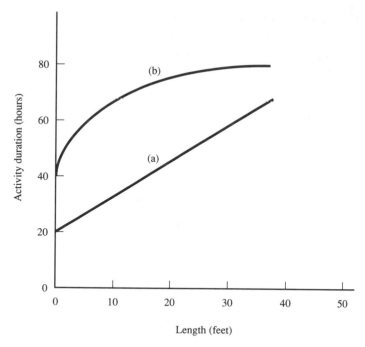

Figure 7-7   Two examples of activity duration as a function of length.

independent variable, X, regression analysis can be used to estimate the parameters of the line $Y = b_0 + b_1X$. Otherwise, either a transformation is performed on one or both of the variables to establish a linear relationship and then regression analysis applied, or a non-linear curve-fitting technique is used.

In the simple case, we have n pairs of sample observations on X and Y which can be represented on a scatter diagram as in Fig. 7-8. Since the line $Y = b_0 + b_1X$ is unknown, we hypothesize that

$$Y_i = b_0 + b_1X_i + u_i, \qquad i = 1,...,n$$

$$E(u_i) = 0 \qquad\qquad i = 1,...,n$$

$$E(u_iu_j) = \begin{cases} 0 & \text{for } i \neq j; i, j = 1,...,n \\ \sigma_u^2 & \text{for } i = j; i, j = 1,...,n \end{cases}$$

where $E(\cdot)$ is the expected value operator, and b_0, b_1, and σ_u^2 are unknown parameters that must be estimated from the sample observations $X_1,...,X_n$ and $Y_1,...,Y_n$. It is usually assumed that $u_i \sim N(0, \sigma_u^2)$ (i.e., u_i is normally distributed with mean 0 and variance σ_u^2).

To begin, denote the regression line by

$$\hat{Y} = \hat{b}_0 + \hat{b}_1X$$

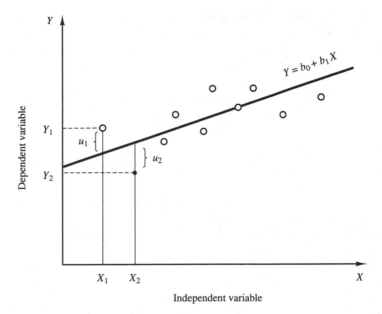

Figure 7-8   Typical scatter diagram.

where $\hat{b}_0$ and $\hat{b}_1$ are estimates of the unknown parameters b_0 and b_1, and $\hat{Y}$ is the value of the dependent variable for any given value of X. To fit such a line we must develop formulas for $\hat{b}_0$ and $\hat{b}_1$ in terms of the sample observations. This is done by the principle of least squares (Draper and Smith 1981), as discussed in Appendix 7A.

With some activities more than one independent variable is required to estimate the performance time. For example, consider the activity of populating a printed circuit board. The use of three independent variables might be appropriate, the first being the number of components to be inserted, the second being the number of setups or tool changes required, and the third being the type of equipment used (here a qualitative rather than a quantitative measure is called for).

In general, if we start with m independent variables, the regression line is

$$Y = b_0 + b_1X_1 + b_2X_2 + \cdots + b_mX_m + u$$

The coefficients $b_0, b_1, \ldots, b_m$ are also estimated by using the principle of least squares. Goodness of fit is measured by the R^2 value, which ranges from 0 (no correlation) to 1 (perfect correlation). The formula used in its calculation is given in Appendix 7A. However, some analysts prefer to use a normalized version of R^2 known as *adjusted R^2* given by

$$R_a^2 = 1 - (1 - R^2)\frac{n - 1}{n - m - 1}$$

where n is the total number of observations and $m + 1$ is the number of coefficients to be estimated. By working with the adjusted R^2 it is possible to compare regression models used to estimate the same dependent variable using different numbers of independent variables.

Guidelines for developing a regression equation include the following steps:

- Identify the independent variables that affect activity duration.
- Collect data on past performance time of the activity for different values of the independent variables.
- Check the correlation between the variables. If necessary, use appropriate transformations and only then generate the regression equation.

In the case that several potential independent variables are considered, a technique called *stepwise regression analysis* can be used. This technique is designed to select the independent variables to be included in the model. At each step, at most one independent variable is added to the model. In the first step a simple regression equation is developed with the independent variable that is the best predictor of the dependent variable (i.e., the one that yields the highest value of R^2). Next, a second variable is introduced. This process continues until no improvement in the regression equation is observed. The final form of the model includes only those independent variables that entered the regression equation during the stepwise iterations.

The quality of a regression model is assessed by analysis of residuals. These residuals ($e_i = Y_i - \hat{Y}_i$) are assumed to be normally distributed with a mean of zero. If this is not the case or a trend in the value of the residuals as a function of any independent variable exists, the dependent variable or some of the independent variables may require a transformation.

Example 7-1

An organization decides to use a regression equation to estimate the time required to develop a new software package. The candidate list of independent variables includes

$$X_1 = \text{number of subroutines in the program}$$
$$X_2 = \text{average number of lines of code in each subroutine}$$
$$X_3 = \text{number of modules or subprograms}$$

Table 7-1 summarizes the data collected on 10 software packages. The time required in person-months denoted by Y is the dependent variable (the duration is given by the number of person-months divided by the number of programmers assigned to the project). Running a stepwise regression on the data yields the following equation:

$$Y = -0.76 + 0.13X_1 + 0.045X_2$$

with $R^2 = 0.972$ and $R_a^2 = 0.964$. Figure 7-9 plots the data points and the fitted line. The value of R_a^2 is lower than R^2 since

$$R_a^2 = 1 - (1 - R^2)\frac{n - 1}{n - m - 1} = 1 - (1 - 0.972)\left(\frac{9}{7}\right) = 0.964$$

TABLE 7-1   DATA FOR REGRESSION ANALYSIS

Package number	Time required, Y	X_1	X_2	X_3
1	7.9	50	100	4
2	6.8	30	60	2
3	16.9	90	120	7
4	26.1	110	280	9
5	14.4	65	140	8
6	17.5	70	170	7
7	7.8	40	60	2
8	19.3	80	195	7
9	21.3	100	180	6
10	14.3	75	120	3

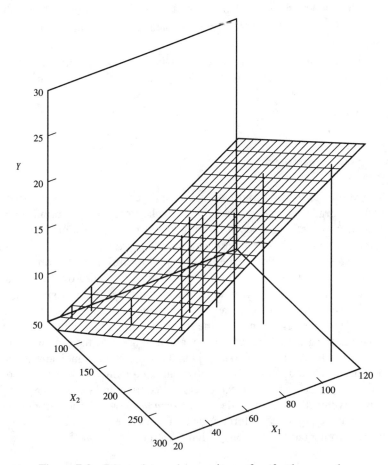

Figure 7-9   Data points and regression surface for the example.

By introducing the third candidate X_3 into the regression model the value of R_a^2 is reduced to 0.963; consequently, it is best to use only the independent variables X_1 and X_2 as predictors, although the difference is minimal.

If a new software package similar to the previous 10 is to be developed, and it contains $X_1 = 45$ subroutines with an average of $X_2 = 170$ lines of code in each, the estimated development time is

$$Y = -0.76 + (0.13)(45) + (0.045)(170) = 12.7 \text{ person-months}$$

∎

In general, the following points should be taken into account when using and evaluating the results of a regression analysis:

- For the activity under investigation, only data collected on similar activities performed by the same work methods should be used in the calculations.
- When the value of R^2 (or the adjusted R^2) is low (below 0.5), the independent variables may not be appropriate.
- If the distribution of the residuals is not close to normal, or there is a trend in the residuals as a function of any independent variable, the regression model may not be appropriate.

7.3 EFFECT OF LEARNING

The ability to learn is translated into improved performance as experience is gained, at both the organizational and individual levels. Improved performance can be measured by reductions in activity times or lower direct costs per repetition. Experience is usually measured by the number of repetitions for a given activity.

Most organizations have the potential to improve performance. This potential will only be realized, however, if sufficient motivation exists on the part of management and the workforce. Improvement at the individual level stems from the ability of a person to move faster and more accurately as experience is gained. Details of the work to be performed are memorized and the time spent on reading instructions, looking at drawings, and experimenting with different procedures decreases. At the organizational level, the potential for improvement is found largely in the areas of communications and logistics and may be achieved with the use of more efficient equipment and work methods.

The relationship between performance time and experience (number of repetitions) can conveniently be represented by a *learning curve*. The underlying model relates the direct labor required to perform an activity to the experience gained in its execution. The basic learning curve equation (Wright 1936) is

$$T(n) = T(1)n^{\beta} \tag{7-3}$$

where $T(n)$ = expected number of direct labor hours required to perform the activity in the nth repetition

n = repetition number

$T(1)$ = expected number of direct labor hours required to perform the activity the first time

β = learning coefficient

A common practice is to describe this learning curve by the percent decline of labor hours required for repetition $2n$ compared to the required labor hours for repetition n. A 90% learning curve means that the time required for repetition $2n$ is 90% of that required for n; thus

$$\frac{T(2n)}{T(n)} = \frac{T(1)(2n)^{\beta}}{T(1)(n)^{\beta}} = 2^{\beta} = 0.9$$

so

$$\beta \log_{10} 2 = \log_{10} 0.9$$

or

$$\beta = \frac{\log_{10} 0.9}{\log_{10} 2} = -0.15$$

If we assume a $100 \times L$ percent learning curve (where L is a fraction between 0 and 1), then

$$\beta = \frac{\log_{10} L}{\log_{10} 2} \tag{7-4}$$

Other learning curve models are discussed in Yelle (1979) and Smunt (1986).

The effect of learning is most important during startup when the cumulative number of repetitions is small. This is because the same relative improvement takes place whenever the number of repetitions is doubled; that is,

$$\frac{T(2n)}{T(n)} = 2^{\beta}$$

Thus the relative improvement between the first and second repetitions is the same as the improvement between the tenth and the twentieth repetitions.

This observation suggests that in projects where a small number of identical units is to be produced, the careful assignment of workers to activities is crucial. By assigning the same workers to perform an activity on all units, direct labor costs and time can be saved due to learning. The scheduling of projects under learning is discussed in detail by Shtub (1991) and LeBlanc et al. (1992).

The learning process in projects may be interrupted if there are long breaks between consecutive repetitions of activities. Such breaks may cause forgetting. The relationship between learning and forgetting was studied by Globerson et al. (1989), who developed an empirical learning–forgetting model based on the results of a field study.

To demonstrate the effect of learning in task planning, consider the following example: An activity is to be repeated four times in a project. Its duration is estimated to be 100 hours if performed by a single worker. The learning percentage defined is estimated as 80%. Solving eq. (7-4) for β, we get

$$\beta = \frac{\log_{10} 0.8}{\log_{10} 2} = -0.322$$

Based on the initial estimate, the time to perform this activity is as follows:

Repetition number	Performance time
1	100
2	$100 \times 2^{-0.322} = 80$
3	$100 \times 3^{-0.322} = 70$
4	$100 \times 4^{-0.322} = \underline{64}$
	314

Tables 7B-1 and 7B-2 in Appendix 7B can replace the calculations above. In Table 7B-1 the values of n^{β} are given for different values of n and $100 \times L\%$. Using Table 7B-1, the performance time for the activity when $n = 3$ (and assuming an 80% learning curve) is $100 \times 0.7021 = 70$. Using Table 7B-2, the total time for the four repetitions is $100 \times 3.142 = 314$.

Thus in this example the total time to perform the activity is 314 hours if the same worker is assigned to the activity and learning takes place. If, however, the four repetitions are assigned to four different workers, the total time required would be $100 \times 4 = 400$ hours.

The learning curve can also be used to update time and cost estimates. Suppose that the actual time for the first repetition was 105 hours, while the actual time for the second repetition was 90 hours. In this case $T(1) = 105$ and $T(2) = 90$, so from eqs. (7-3) and (7-4),

$$2^{\beta} = \frac{T(2)}{T(1)} = \frac{90}{105} = 0.857 \quad \text{or} \quad \beta = \frac{\log_{10} 0.857}{\log_2 2} = -0.22$$

By using the learning curve model for time and cost estimation, and by scheduling workers so that learning is maximized, the project manager can take advantage of the learning effect.

7.4 PRECEDENCE RELATIONS AMONG ACTIVITIES

The schedule of activities is constrained by the availability of resources required to perform each activity and by technological constraints known as *precedence relations*. Several types of precedence relations exist among activities. The most common, termed "start to end," requires that an activity can start only after its predecessor has been completed. For example, it is possible to lift a piece of equipment by a crane only after the equipment is secured to the hoist.

A "start to start" relationship exists when an activity can start only after a specified activity has already begun. For example, in projects where concurrent engineering is applied, logistic support analysis starts as soon as the detailed design phase begins. The "end to start" connection occurs when an activity cannot end until another activity has begun. This would be the case in a project of building a nuclear reactor and charging it with fuel, where one industrial robot transfers radioactive material to another. The first robot can release the material only after the second robot achieves a tight enough grip. The "end to end" connection is used when an activity cannot terminate unless another activity is completed. Quality control efforts, for example, cannot terminate before production ceases, although the two activities can be performed at the same time.

A lag or time delay can be added to any of these connections. In the case of the "end to end" arrangement, there might be a need to spend 2 days on testing and quality control after production shuts down. In the case of the "start to end" connection, a fixed setup may be required between the two activities. In some situations the relationship between activities is subject to uncertainty. For example, after testing a printed circuit board that is to be part of a prototype communications system, the succeeding activity might be either to install the board on its rack, to repair any defects found, or to scrap the board if it fails the functionality test.

In the following sections we concentrate on the analysis of "start to end" connections, the most commonly used. The analysis of other types of connections, as well as the effect of uncertainty on precedence relations (probabilistic networks), are discussed in Wiest and Levy (1977). The large number of precedence relations among activities makes it difficult to rely on verbal descriptions alone to convey the effect of technological constraints on scheduling, so graphical representations are frequently used. In subsequent sections, a number of such representations are illustrated with the help of an example project. Table 7-2 contains the relevant activity data.

In this project, only "start to end" precedence relations are considered. From the table we see that activities A, B, and E do not have any predecessors and thus can start at

TABLE 7-2   DATA FOR EXAMPLE PROJECT

Activity	Immediate predecessors	Duration (weeks)
A	—	5
B	—	3
C	A	8
D	A,B	7
E	—	7
F	C,E,D	4
G	F	5

any time. Activity C, however, can start only after A finishes, while D can start after the completion of A and B. Further examination reveals that F can start only after C, E, and D are finished, and that G must follow F. Because activity A precedes C, and C precedes F, A must also precede F by transitivity. Nevertheless, when using a network representation, it is only necessary to list immediate or direct precedence relations; implied relations are taken care of automatically.

The three models used to analyze precedence relations and their affect on the schedule are the Gantt chart, the critical path method, and the program evaluation and review technique. As mentioned, the latter two are based on network techniques in which the activities are placed either on the nodes or on the arrows, depending on which is more intuitive for the analyst.

7.5 GANTT CHART

The most widely used management tool for project scheduling and control is a version of the bar chart developed during World War I by Henry L. Gantt. The Gantt chart, as it is called, enumerates the activities to be performed on the vertical axis and their corresponding duration on the horizontal axis. It is possible to schedule activities by either early start or late start logic. In the early start approach, each activity is initiated as early as possible without violating the precedence relations. In the late start approach, each activity is delayed as much as possible as long as the earliest finish time of the project is not compromised.

A range of schedules is generated on the Gantt chart if a combination of early and late starts is applied. The early start schedule is performed first and yields the earliest finish time of the project. That time is then used as the required finish time for the late start schedule. Figure 7-10 depicts the early start Gantt chart schedule for the example above. The bars denote the activities; their location with respect to the time axis indicates the time over which the corresponding activity is performed. For example, activity D can start only after activities A and B finish, which happens at the end of week 5. A direct output of this schedule is the earliest finish time for the project (22 weeks for the example).

Based on the earliest finish time, the late start schedule can be generated. This is done by shifting each activity to the right as much as possible while still starting the project at time zero and completing it in 22 weeks. The resultant schedule is depicted in Fig. 7-11. The difference between the start (or the end) times of an activity on the two schedules is called the slack (or float) of the activity. Activities that do not have any slack are denoted by a black bar and are termed *critical*. The sequence of critical activities connecting the start and end points of the project is known as the *critical path*, which logically turns out to be the *longest path* in the network. A delay in any activity along the critical path delays the entire project. Put another way, the sum of durations for critical activities represents the *shortest* possible time to complete the project.

Gantt charts are simple to generate and interpret. In the construction there should be a one-to-one correspondence between the listed tasks and the work breakdown structure and its numbering scheme. As shown in Fig. 7-12, which depicts the Gantt chart for the microcomputer development project, a separate column can be added for this purpose. In

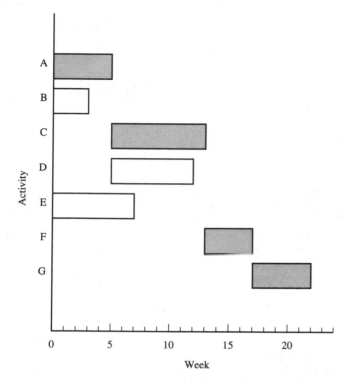

Figure 7-10 Gantt chart for an early start schedule.

fact, the schedule should not contain any tasks that do not appear in the WBS. Often, however, the Gantt chart includes milestones such as project kickoff and design review, which are listed along with the tasks.

   In addition to showing the critical path, Gantt charts can be modified to indicate project and activity status. In Fig. 7-12, a bold border is used to identify a critical activity, and a shaded area to indicate the approximate completion status at the August review. Accordingly, we see that tasks 2, 5, and 8 are critical, falling on the longest path. Task 2 is 100% complete, task 4 is 65% complete, task 7 is 50% complete, while tasks 5, 6, and 8 have not yet been started.

   Gantt charts can be modified further to show budget status by adding a column that lists planned and actual expenditures for each task. This is taken up in Chapter 8. Many variations of the original bar graph have been developed to provide more detailed information for the project manager. One commonly used variation is shown in Fig. 7-13, which replaces the bars with lines and triangles to indicate project status and revision points. To explain the features, let us examine No. 2, equipment design. According to the code given in the lower left-hand corner of the figure, this task has been rescheduled three times, finally starting in February, and was completed by the end of June.

   The problem with adding features to the bar graph is that they take away from the clarity and simplicity of the basic form. Nevertheless, the additional information conveyed

Activity
A
B
C
D
E
F
G
0
5
10
15
20
Week

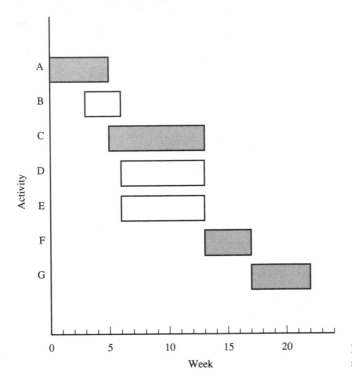

Figure 7-11 Gantt chart for a late start schedule.

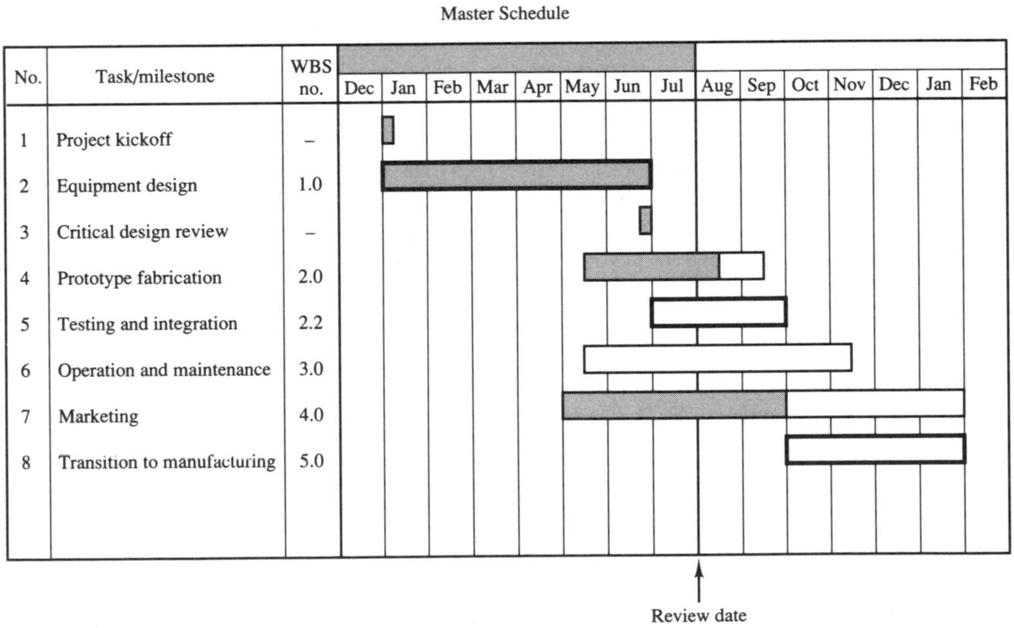

Master Schedule
No.
Task/milestone
WBS no.
Dec
Jan
Feb
Mar
Apr
May
Jun
Jul
Aug
Sep
Oct
Nov
Dec
Jan
Feb
1 Project kickoff –
2 Equipment design 1.0
3 Critical design review –
4 Prototype fabrication 2.0
5 Testing and integration 2.2
6 Operation and maintenance 3.0
7 Marketing 4.0
8 Transition to manufacturing 5.0
Review date

Figure 7-12 Gantt chart for the microcomputer development example. (From H. J. Thamhain, *Engineering Program Management*, copyright © 1984 by John Wiley & Sons, Inc.; reprinted by permission of the publisher.)

MASTER SCHEDULE

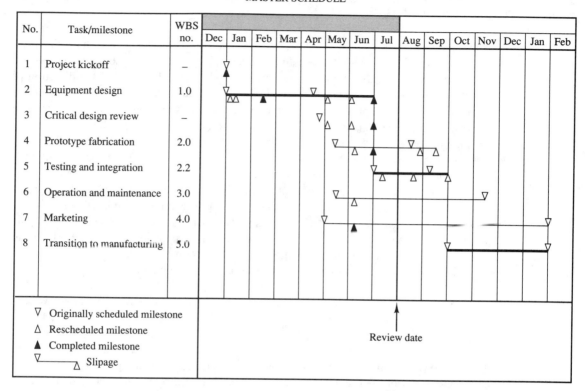

Figure 7-13   Extended Gantt chart with task details. (From H. J. Thamhain, *Engineering Program Management*, copyright © 1984 by John Wiley & Sons, Inc.; reprinted by permission of the publisher.)

to the user may offset the additional effort required in generating and interpreting the data. A common modification of the analysis is the case when a milestone has a contractual due date. Consider, for example, activity 8 (WBS No. 5.0) in Fig. 7-13. If management decides that the required due date for the termination of this activity is the end of February (instead of the end of January), a slack of 1 month will be added to each activity in the project. If, however, the due date of activity 8 is the end of December, the schedule in Fig. 7-13 is no longer feasible because the sequence of activities 2, 5, and 8 (the critical sequence) cannot be completed by the end of December. In Section 7.12, scheduling conflicts and their management are discussed in detail.

The major limitation of bar graph schedules is their inability to show task dependencies and time–resource trade-offs. Network techniques are often used in parallel with Gantt charts to compensate for these shortcomings.

7.6 ACTIVITY-ON-ARROW NETWORK APPROACH
FOR CPM ANALYSIS

Although the *activity-on-arrow* (AOA) model is most closely associated with PERT, it can be applied to CPM as well (it is sometimes called *activity on arc*). In constructing the network, an arrow is used to represent an activity, with its head indicating the direction of progress of the project. The precedence relations among activities are introduced by defining events. An event represents a point in time that signifies the completion of one or more activities and the beginning of new ones. The beginning and ending points of an activity are thus described by two events known as the head and the tail. Activities originating from a certain event cannot start until the activities terminating at the same event have been completed.

Figure 7-14a shows an example of a typical representation of an activity (i,j) with its tail event i and it head event j. Figure 7-14b depicts a second example where activities (1,3) and (2,3) must be completed before activity (3,4) can start. For computational purposes it is customary to number the events in ascending order so that, compared to the head event, a smaller number is always assigned to the tail event of an activity.

The rules for constructing a diagram are summarized below.

Rule 1.   Each activity is represented by one and only one arrow in the network.

No single activity can be represented twice in the network. This is to be differentiated from the case where one activity is broken down into segments wherein each segment may then be represented by separate arrows. For example, in designing a new computer architecture, the controller might first be developed followed by the arithmetic unit, the I/O processor, and so on.

Rule 2.   No two activities can be identified by the same head and tail events.

A situation like this may arise when two or more activities can be performed in parallel. As an example, consider Fig. 7-15a, which shows activities A and B running in parallel. The procedure used to circumvent this difficulty is to introduce a dummy activity either between A or B to erase the four equivalent ways of doing this are shown in Fig. 7-15b, where D_1 is the dummy activity. As a result of using D_1, activities A and B can now be identified by a unique set of events. It should be noted that dummy activities do not consume time or resources. Typically, they are represented by "broken" directed arrows in the network.

Dummy activities are also necessary in establishing logical relationships that cannot otherwise be represented correctly. Suppose that in a certain project, tasks A and B must precede C, while task E is preceded only by B. Figure 7-16a shows an incorrect, but quite common way that many beginners would draw this part of the network. The difficulty is that although the relationship among A, B, and C is correct, the diagram implies that E must be preceded by both A and B. The correct representation using dummy D_1 is depicted in Fig. 7-16b.

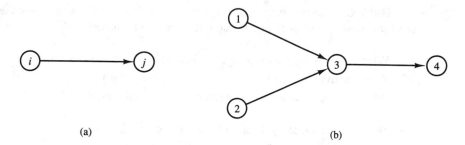

(a)   (b)

Figure 7-14   Network components.

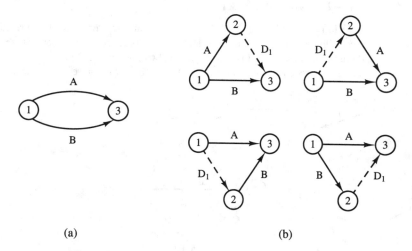

(a)   (b)

Figure 7-15   Use of a dummy arc between two nodes.

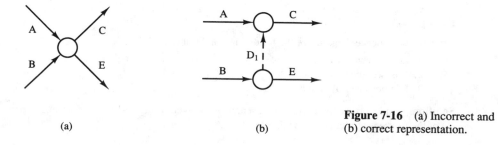

(a)   (b)

Figure 7-16   (a) Incorrect and (b) correct representation.

Rule 3. To ensure the correct representation in the AOA diagram, the following questions must be answered as each activity is added to the network:

1. Which activities must be completed immediately before this activity can start?
2. Which activities must immediately follow this activity?
3. Which activities must occur concurrently with this activity?

This rule is self-explanatory. It provides guidance for checking and rechecking the precedence relations as the network is constructed.

The following examples further illustrate the use of dummy activities.

Example 7-2

Draw the AOA diagram so that the following precedence relations are satisfied:

1. E is preceded by B and C.
2. F is preceded by A and B.

Solution Consider Fig. 7-17. Part (a) shows an incorrect precedence relation for activity E. According to the requirements, B and C are to precede E, and A and B are to precede F. The dummy D_1 is therefore inserted to allow B to precede E. Doing so, however, implies that A *also* must precede E, which is incorrect. Part (b) in the figure shows the correct relationships.

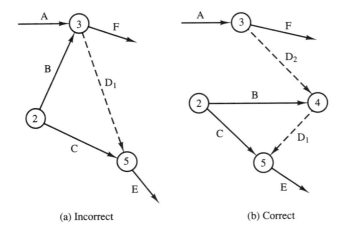

(a) Incorrect   (b) Correct

Figure 7-17 Subnetwork with two dummy arcs: (a) incorrect; (b) correct.

Example 7-3

Draw the precedence diagram for the following conditions:

1. G is preceded by A.
2. E is preceded by A and B.
3. F is preceded by B and C.

Solution An incorrect and correct representation is given in Fig. 7-18. The diagram in part (a) of the figure is wrong because it implies that A precedes F.

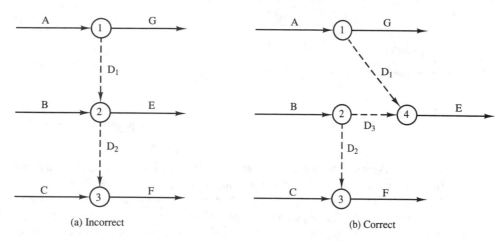

Figure 7-18   Subnetwork with complicated precedence relations: (a) incorrect; (b) correct.

It is good practice to have a single start event common to all activities that have no predecessors, and a single end event for all activities that have no successors. The actual mechanics of drawing the AOA network will be illustrated using the data in Table 7-2.

The process begins by identifying all activities that have no predecessors and joining them to a unique start node. This is shown in Fig. 7-19. Each activity terminates at a node. Only the first node in the network is assigned a number (1); all other nodes are labeled only when network construction is completed, as explained presently. Since activity C has only one predecessor (A), it can be added immediately to the diagram (see Fig. 7-19).

Activity D has both A and B as predecessors; thus there is a need for an event that represents the completion of A and B. We begin by adding two dummy activities D_1 and D_2. The common end event of D_1 and D_2 is now the start event of D, as depicted in Fig. 7-20. As we progress, it may happen that one or more dummy activities are added that really are not necessary. To correct this situation, a check will be made at completion and redundant dummies eliminated.

Prior to starting activity F, activities C, E, and D must be completed. Therefore, an event should be introduced that represents the terminal point of these activities. Notice that

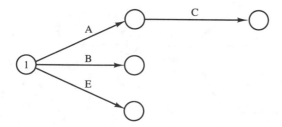

Figure 7-19   Partial plot of the example AOA network.

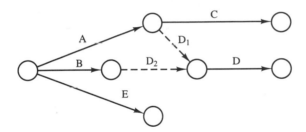

Figure 7-20   Using dummy activities to represent precedence relations.

C, E, and D are not predecessors of any other activity but F. This implies that we can have the three arrows representing these activities terminate at the same node (event)—the tail of F. Activity G which has only F as a predecessor can start from the head of F (see Fig. 7-21).

    Once all the activities and their precedence relations have been included in the network diagram, it is possible to eliminate redundant dummy activities. A dummy activity is redundant if it is the only activity starting or ending at a given event. Thus D_2 is redundant and is eliminated by connecting the head of activity B to the event that marked the end of D_2. The next step is to number the events in ascending order, making sure that the tail always has a lower number than the head. The resulting network is illustrated in Fig. 7-22. The duration of each activity is written next to the corresponding arrow. The dummy D_1 is shown like any other activity but with a duration of zero.

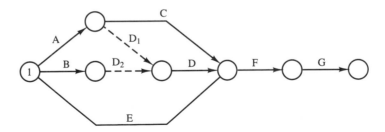

Figure 7-21   Network with activities F and G included.

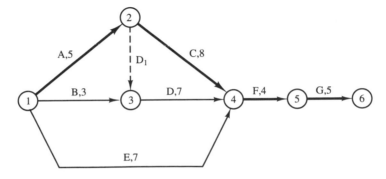

Figure 7-22   Complete AOA project network.

Example 7-4

Construct an AOA diagram comprising activities A,B,C,...,L such that the following relationships are satisfied.

1. A, B, and C, the first activities of the project, can start simultaneously.
2. A and B precede D.
3. B precedes E, F, and H.
4. F and C precede G.
5. E and H precede I and J.
6. C, D, F, and J precede K.
7. K precedes L.
8. I, G, and L are the terminal activities of the project.

Solution   The resulting diagram is shown in Fig. 7-23. The dummy activities D_1 and D_3 are needed to establish correct precedence relations. D_2 is introduced to assure that the parallel activities E and H have unique end events. Note that the events in the project are numbered in such a way that if there is a path connecting nodes i and j, then $i < j$. In fact, there is a basic result from graph theory which states that a directed graph is acyclic if and only if its nodes can be numbered so that for all arcs (i,j), $i < j$.

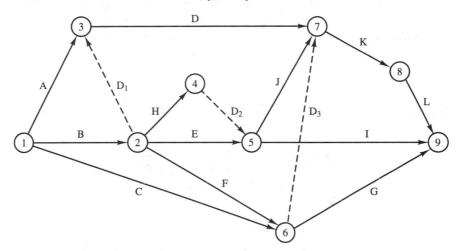

Figure 7-23   Network for Example 7-4.

In general, once the nodes in a graph are numbered, the network can be represented by a matrix whose respective rows and columns correspond to the start and end events of a particular activity. The matrix for the example in Fig. 7-22 is as follows:

		Ending Event					
		1	2	3	4	5	6
	1		×	×	×		
	2			×	×		
Starting	3				×		
event	4					×	
	5						×
	6						

where the entry "$\times$" means that there is an activity connecting the two events (instead of an $\times$, it may be more efficient to use the activity number or its duration). For example, the $\times$ in row 3, column 4 indicates that an activity starts at event 3 and ends at event 4, that is, activity D. The absence of an entry in the second row and fifth column means that no activity starts at event 2 and ends at event 5.

Because the numbering scheme used ensures that if activity (i,j) exists, then $i < j$, it is sufficient to store only the portion of the matrix that is above the diagonal. Alternatively, the lower portion of the matrix can be used to store other information about an activity, such as resource requirements or budget. This conveniently represents computer input.

For complex projects it may not be obvious how to label the nodes in the desired manner. Suppose that we have a graph which is described by its adjacency matrix A, where $a_{ij} = 1$ if node i immediately precedes j and 0 otherwise, and that the rows and columns of this matrix are ordered according to the given arbitrary numbering of the nodes. Let $v(j)$ denote the new number of node j, and define the in-degree of a node as the number of arcs entering it. Let $d_j^{(in)}$ be the in-degree of node j. Initially $d_j^{(in)}$ is computed for all nodes j, by forming the sum of the entries in column j of matrix A. A node k for which $d_k^{(in)} = 0$ is found, and $v(k)$ is set to 1. The in-degrees are revised by subtracting the entries in row k of A, and repeating the process. The accompanying algorithm is summarized below.

Step 0   (Start)

$$\text{Set } d_j^{(in)} = \sum_{i=1}^{n} a_{ij}, \quad j = 1,2,...,n.$$

Set $N = \{1,2,...,n\}$.
Set $m = 1$.

Step 1   (Detection of node with zero in-degree)
Find $k \in N$ such that $d_k^{(in)} = 0$. If there is no such k, stop; the network is not acyclic—it contains one or more cycles.
Set $v(k) = m$.
Set $m = m + 1$.
Set $N = N - \{k\}$.
If $N = \emptyset$, stop; all nodes have been correctly labeled.

Step 2   (Revision of in-degrees)
Set $d_j^{(in)} = d_j^{(in)} - a_{kj}$, for all $j \in N$.
Return to step 1.

If it is not possible to assign node numbers so that each activity starts at an event with a number lower than its end event, there is a logical error in the definition of precedence relations and a closed loop of activities exists in the network. This problem must be solved before the analysis can proceed.

From the network diagram it is easy to see the sequences of activities connecting the start of the project to its terminal node. As explained earlier, the longest sequence is called the *critical path*. The total time required to perform all the activities on the critical path is the minimum duration of the project since these activities cannot be performed in parallel due to precedence relations among them.

To simplify the analysis, it is recommended that in the case of multiple activities that have no predecessors, a common start event should be used for all of them. Similarly, in cases where multiple activities have no successors, a common end event should be defined.

In the example network of Fig. 7-22 there are four sequences of activities connecting the start and end nodes. Each is listed in Table 7-3. The last column of the table contains the duration of each sequence. As can be seen, the longest path (critical path) is sequence 1, which includes activities A, C, F, and G. A delay in completing any of these (critical) activities due to, say, a late start or a longer performance time than initially expected will cause a delay in project completion.

TABLE 7-3   SEQUENCES IN THE NETWORK

Sequence number	Events in the sequence	Activities in the sequence	Sum of activity times
1	1–2–4–5–6	A,C,F,G	22
2	1–2–3–4–5–6	A,D_1,D,F,G	21
3	1 3–4–5–6	B,D,F,G	19
4	1–4–5–6	E,F,G	16

Activities not on the critical path have slack and can be delayed temporarily on an individual basis. Two types of slack are possible: free slack (free float) and total slack (total float). *Free slack* denotes the time that an activity can be delayed without delaying both the start of any succeeding activity and the end of the project. *Total slack* is the time that the completion of an activity can be delayed without delaying the end of the project. A delay of an activity that has total slack but no free slack reduces the slack of other activities in the project.

A simple rule can be used to identify the type of slack. A noncritical activity whose end event is on the critical path has free slack. For example, activity E, which is noncritical, has free slack since event 4 (its head) is on the critical path. The head of noncritical activity B, however, is not on the critical path, so B has no free slack. The head of activity B is the start event of activity D, which is also noncritical. Thus the difference between the length of the critical sequence (A–C), and the noncritical sequence (B–D), which runs in parallel to (A–C), is the total slack of B and D and is equal to $(5 + 8) - (3 + 7) = 3$. Any delay in activity B will reduce the remaining slack for activity D. Therefore, the person responsible for performing D should be notified.

■

The roles of the total and free slacks in scheduling noncritical activities can be explained in terms of two general rules:

1. If the total slack *equals* the free slack, the noncritical activity can be scheduled anywhere between its early start and late finish times.

2. If the free slack is *less than* the total slack, the starting of the noncritical activity can be delayed relative to its early start time by no more than the amount of its free slack without affecting the schedule of those activities that immediately succeed it.

Further elaboration and an exact mathematical expression for calculating activity slacks is presented in the following subsections.

7.6.1 Calculating Event Times and Critical Path

Important scheduling information for the project manager is the earliest and latest times each event can take place without causing a schedule overrun. This information is needed to compute the critical path. The *early time* of an event i is determined by the length of the longest sequence from the start node (event 1) to event i. Denote t_i as the early time of event i, and let $t_1 = 0$, implying that activities without precedence constraints begin as early as possible. If a starting date is given, t_1 is adjusted accordingly.

To determine t_i for each event i, a *forward pass* is made through the network. Let L_{ij} be the duration or length of activity (i,j). The following formula is used for the calculations:

$$t_j = \max_i[t_i + L_{ij}] \qquad \text{for all } (i,j) \text{ activities defined} \qquad (7\text{-}5)$$

where $t_1 = 0$. Thus, to compute t_j for event j, t_i for the tail events of all incoming activities (i,j) must be computed first. In words, the early time of each event is the latest of the early times of its immediate predecessors plus the duration of the connecting activity.

The forward-pass calculations for the example network in Fig. 7-22 will now be given. The early time for event 2 is simply

$$t_2 = t_1 + L_{12} = 0 + 5 = 5$$

where $L_{12} = 5$ is the duration of the activity connecting event 1 to event 2 (activity A).

Early time calculations for event 3 are a bit more complicated since event 3 marks the completion of the two activities D_1 and B. By implication, there are two sequences connecting the start of the project to event 3. The first comprises activities A and D_1 and is of length 5; the second includes activity B only and has $L_{13} = 3$. Using eq. (7-5), we get

$$t_3 = \max\begin{Bmatrix} t_1 + L_{13} \\ t_2 + L_{23} \end{Bmatrix} = \max\begin{Bmatrix} 0 + 3 \\ 5 + 0 \end{Bmatrix} = 5$$

so the early time of event 3 is $t_3 = 5$.

The remaining calculations are performed as follows:

$$t_4 = \max\begin{Bmatrix} t_1 + L_{14} \\ t_2 + L_{24} \\ t_3 + L_{34} \end{Bmatrix} = \max\begin{Bmatrix} 0 + 7 \\ 5 + 8 \\ 5 + 7 \end{Bmatrix} = 13$$

$$t_5 = t_4 + L_{45} = 13 + 4 = 17$$

$$t_6 = t_5 + L_{56} = 17 + 5 = 22$$

This confirms that the earliest the project can finish is in 22 weeks.

The late time of each event is calculated next by making a *backward pass* through the network. Let T_i denote the late time of event i. If n is the end event, the calculations are generally initiated by setting $T_n = t_n$ and working backward toward the start event using the following formula:

$$T_i = \min_j[T_j - L_{ij}] \qquad \text{for all } (i,j) \text{ activities defined} \qquad (7\text{-}6)$$

If, however, a required project completion date is given which is later than the early time of event n, it is possible to assign that time as the late time for the end event. If the required date is earlier than the early time of the end event, no feasible schedule exists. This case is discussed later in the chapter.

In our example, $T_6 = t_6 = 22$. The late time for event 5 is calculated as follows:

$$T_5 = T_6 - L_{56} = 22 - 5 = 17$$

Similarly,

$$T_4 = T_5 - L_{45} = 17 - 4 = 13$$
$$T_3 = T_4 - L_{34} = 13 - 7 = 6$$

Event 2 is connected by sequences of activities to both events 3 and 4. Thus, applying eq. (7-6), the late time of event 2 is the minimum among the late times dictated by the two sequences; that is,

$$T_2 = \max \left\{ \begin{matrix} T_3 - L_{23} \\ T_4 - L_{24} \end{matrix} \right\} = \max \left\{ \begin{matrix} 6 - 0 - 6 \\ 13 - 8 = 5 \end{matrix} \right\} = 5$$

The late time of event 1 is calculated in a similar manner:

$$T_1 = \max \left\{ \begin{matrix} 6 - 3 = 3 \\ 5 - 5 = 0 \end{matrix} \right\}$$

The results are summarized in Table 7-4.

The critical activities can now be identified by using the results of the forward and backward passes. An activity (i,j) lies on the critical path if it satisfies the following three conditions:

$$t_i = T_i$$
$$t_j = T_j$$
$$t_j - T_i = T_j - T_i = L_{ij}$$

These conditions actually indicate that there is no float or slack time between the earliest start (completion) and the latest start (completion) of the critical activities. In Fig. 7-22, activities (1,2), (2,4), (4,5), and (5,6) define the critical path forming a chain that spans the network from node 1 (start) to node 6 (end).

TABLE 7-4   SUMMARY OF EVENT TIME CALCULATIONS

Event, i	Early time, t_i	Late time, T_i
1	0	0
2	5	5
3	5	6
4	13	13
5	17	17
6	22	22

7.6.2 Calculating Activity Start and Finish Times

In addition to scheduling the events of a project, detailed scheduling of activities is performed by calculating the following four times (or dates) for each activity (i,j):

ES_{ij} = *early start* time: the earliest time activity (i,j) can start without violating any precedence relations

EF_{ij} = *early finish* time: the earliest time activity (i,j) can finish without violating any precedence relations

LS_{ij} = *late start* time: the latest time activity (i,j) can start without delaying the completion of the project

LF_{ij} = *late finish* time: the latest time activity (i,j) can finish without delaying the completion of the project

The calculations proceed as follows:

$$ES_{ij} = t_i \qquad\qquad \text{for all } i$$

$$EF_{ij} = ES_{ij} + L_{ij} \qquad \text{for all } (i,j) \text{ defined}$$

$$LF_{ij} = T_j \qquad\qquad \text{for all } j$$

$$LS_{ij} = LF_{ij} - L_{ij} \qquad \text{for all } (i,j) \text{ defined}$$

Thus the earliest time an activity can begin is equal to the early time of its start event; the latest an activity can finish is equal to the late finish of its end event. For activity D in the example, which is denoted by arc (3,4) in the network, we have $ES_{34} = t_3 = 5$ and $LF_{34} = T_4 = 13$.

The earliest time an activity can finish is given by its ES plus its duration; the latest time an activity can start is equal to its LF minus its duration. For activity D, this implies that $EF_{34} = ES_{34} + L_{34} = 5 + 7 = 12$, and $LS_{34} = LF_{34} - L_{34} = 13 - 7 = 6$. The full set of calculations is presented in Table 7-5.

TABLE 7-5 SUMMARY OF SCHEDULE ANALYSIS

Activity	(i,j)	L_{ij}	$ES_{ij} = t_i$	$EF_{ij} =$ $ES_{ij} + L_{ij}$	$LF_{ij} = T_j$	$LS_{ij} =$ $LF_{ij} - L_{ij}$	$TS_{ij} =$ $LS_{ij} - ES_{ij}$	$FS_{ij} =$ $t_j - t_i - L_{ij}$
A	(1,2)	5	0	5	5	0	0	0
B	(1,3)	3	0	3	6	3	3	2
C	(2,4)	8	5	13	13	5	0	0
D	(3,4)	7	5	12	13	6	1	1
E	(1,4)	7	0	7	13	6	6	6
F	(4,5)	4	13	17	17	13	0	0
G	(5,6)	5	17	22	22	17	0	0
D_1	(2,3)	0	5	5	6	6	1	0

7.6.3 Calculating Slacks

As mentioned, there are two types of slack associated with an activity: total slack and free slack. Information about slack is important to the project manager, who may have to adjust budgets and resource allocations to stay on schedule. Knowing the amount of slack in an activity is essential if he or she is to do this without delaying the completion of the project. In a multiproject environment, slack in one project can be used temporarily to free up resources needed for other projects that are behind schedule or overly constrained.

Due to the importance of slack, project management is sometimes referred to as slack management. We will elaborate on slack management in the chapters dealing with resources and budgets. The total slack TS_{ij} (or total float TF_{ij}) of activity (i,j) is equal to the difference between its late start (LS_{ij}) and its early start (ES_{ij}) or the difference between its late finish (LF_{ij}) and its early finish (EF_{ij}); that is,

$$TS_{ij} = TF_{ij} = LS_{ij} - ES_{ij} = LF_{ij} - EF_{ij}$$

This is equivalent to the difference between the maximum time available to perform the activity ($T_j - t_i$) and its duration (L_{ij}). The total slack of activity D (3,4) in the example is $TS_{34} = LS_{34} - ES_{34} = 6 - 5 = 1$.

The free slack (or free float) is defined by assuming that all activities start as early as possible. In this case, the free slack, FS_{ij}, for activity (i,j) is the difference between the early time of its end event j and the sum of the early time of its start event i plus its length; that is,

$$FS_{ij} = t_j - (t_i + L_{ij})$$

For the example, the free slack for activity D (3,4) is $FS_{34} = t_4 - (t_3 + L_{34}) = 13 - (5 + 7) = 1$. Thus it is possible to delay activity D by 1 week without affecting the start of any other activity. The times and slacks for the events and activities of the example are summarized in Table 7-5.

Activities with a total slack equal to zero are critical, as any delay in these activities will cause a delay in the completion of the project. The total slack is either equal to or larger than the free slack since the total slack of an activity is composed of its free slack plus the slack shared with other activities. For example, activity B denoted by (1,3) has a free slack of 2 weeks. Thus it can be delayed up to 2 weeks without affecting its successor D. If, however, B is delayed by 3 weeks, the project can still be finished on time provided that D starts immediately after B finishes. This follows because activities B and D share 1 week of total slack. Finally, notice that activity D_1 has a total slack of 1 and a free slack of 0, implying that noncritical activities may have zero free slack. The converse is not true.

In the AOA network representation, the length of the arrows is not necessarily proportional to the duration of the activities. When developing a graphical representation of the problem, it is convenient to write the duration of each activity next to the corresponding arrow. As discussed in Chapter 13, most software packages that are based on the AOA model follow this convention. In addition, they typically provide the user with the option of placing a subset of activity parameters above or below the arrows. We have intentionally omitted placing this information on our diagrams because of the clutter that it occasions. Nevertheless, it is good practice when manually performing the forward and backward calculations to write the early and late start times above the corresponding nodes.

7.7 ACTIVITY-ON-NODE NETWORK APPROACH FOR CPM ANALYSIS

The activity-on-node (AON) model is an alternative approach to representing project activities and their interrelationships. It is most closely associated with CPM analysis and is the basis for most computer implementations. In the AON model, the arrows are used to denote the precedence relations among activities. Its basic advantage is that there is no need for dummy arrows and it is very easy to construct. In developing the network it is convenient to add a single start node and a single end node which uniquely identify these milestones. This is illustrated in Fig. 7-24 for the example.

Some additional network construction rules include:

1. All nodes, with the exception of the terminal node, must have at least one successor.
2. All nodes, except the first, must have at least one predecessor.
3. There should be only one initial and one terminal node.
4. No arrows should be left dangling. Notwithstanding rules 1 and 2, every arrow must have a head and a tail.
5. An arrow specifies only precedence relations; its length has no significance with respect to the time duration accompanying either of the activities it connects.
6. Cycles or closed-loop paths through the network are not permitted. They imply that an activity is a successor of another activity that depends on it.

As with the AOA model, the computational procedure involves forward and backward passes through the network. This is discussed next.

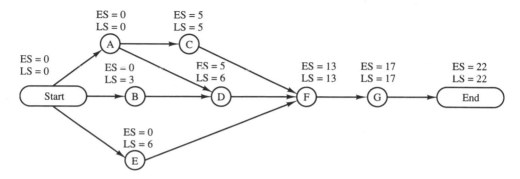

Figure 7-24   AON network for the example project.

7.7.1 Calculating Early Start and Early Finish Times

A forward pass is used to determine the earliest start time and the earliest finish time for each activity. During the forward pass, it is assumed that each activity begins as soon as possible, that is, as soon as the last of its predecessors is completed. Thus the early start (ES) time of an activity is equal to the maximum early finish (EF) time of all the activities immediately preceding it. The ES time of the initial activity is assumed to be zero as is its EF. For all other activities, the EF time is equal to its early start time plus its duration.

Using slightly different notation to distinguish the AON calculations from those prescribed for the AOA model, we have

$$ES(K) = \max[EF(J) : J \text{ an immediate predecessor of } K] \qquad (7\text{-}7)$$

$$EF(K) = ES(K) + L(K) \qquad (7\text{-}8)$$

where $L(K)$ denotes the duration of activity K.

Returning once again to the example, activities A, B, and E do not have predecessors (except the start node), and thus their early start times are zero; that is, $ES(A) = ES(B) = ES(E) = 0$. The early finish time of these activities is equal to their early start time plus their duration, so $EF(A) = 0 + 5 = 5$, $EF(B) = 0 + 3 = 3$, and $EF(E) = 0 + 7 = 7$.

From eq. (7-7) the early start of any other activity is determined by the latest (the maximum) early finish time of its predecessors. For activity D the calculations are

$$ES(D) = \max \left\{ \begin{array}{c} EF(A) \\ EF(B) \end{array} \right\} = \max \left\{ \begin{array}{c} 5 \\ 3 \end{array} \right\} = 5$$

The early start and early finish times of the remaining activities are computed in a similar manner. Table 7-6 summarizes the results.

TABLE 7-6   EARLY START AND
EARLY FINISH OF PROJECT
ACTIVITIES

Activity	Early start	Early finish
A	0	5
B	0	3
C	5	13
D	5	12
E	0	7
F	13	17
G	17	22

7.7.2 Calculating Late Times of Activities

The calculation of late times on the AON network is performed in the reverse order of the calculation of early times. As with the AOA model, a backward pass is made beginning at the expected completion time and concluding at the earliest start time. To complete the project as soon as possible, the late finish (LF) of the last activity is set equal to its early finish (EF) time calculated in the forward pass. Alternatively, the latest allowable completion time may be fixed by a contractual deadline, if one exists, or some other rationale.

In general, the late finish time of an activity with more than one successor is the earliest of the succeeding late start times. The late start (LS) time of an activity is its LF time minus its duration. Computational expressions for LF and LS are:

$$LF(K) = \min[LS(J) : J \text{ is a successor of } K] \tag{7-9}$$

$$LS(K) = LF(K) - L(K) \tag{7-10}$$

To begin the calculations for the example network in Fig. 7-24, we set $LF(G) = EF(G) = 22$, and apply eq. (7-10) to get $LS(G) = LF(G) - L(G) = 22 - 5 = 17$. The late finish of any other activity is equal to the earliest (or the minimum) among the late start time of its succeeding activities. Because activity F has only one successor (G), we get

$$LF(F) = LS(G) = 17 \quad \text{and} \quad LS(F) = 17 - 4 = 13$$

Continuing with activities C and D yields

$$LF(C) = LS(F) = 13 \quad \text{and} \quad LS(C) = 13 - 8 = 5$$

$$LF(D) = LS(F) = 13 \quad \text{and} \quad LS(D) = 13 - 7 = 6$$

Since A has two successors we get

$$LF(A) = \min \left\{ \begin{matrix} LS(C) \\ LS(D) \end{matrix} \right\} = \min \left\{ \begin{matrix} 5 \\ 6 \end{matrix} \right\} = 5 \quad \text{and} \quad LS(A) = LF(A)$$

$$- L(A) = 5 - 5 = 0$$

The late start and late finish times of activities in the example project are summarized in Table 7-7. As expected, these results are identical to those of the AOA model.

TABLE 7-7 LATE FINISH AND LATE START OF PROJECT ACTIVITIES

Activity	Late finish	Late start
A	5	0
B	6	3
C	13	5
D	13	6
E	13	6
F	17	13
G	22	17

The total slack of an activity is calculated as the difference between its late start (or finish) and its early start (or finish). The free slack of an activity is the difference between the earliest among the early start times of its successors and its early finish time. That is, for each activity K,

$$TS(K) = LS(K) - ES(K)$$

$$FS(K) = \min[ES(J) : J \text{ is successor of } K] - EF(K)$$

Activities with zero total slack fall on the critical path. When performing the calculations manually, it is convenient to write the corresponding ES and LS times above each node to help identify the critical path.

7.8 LINEAR PROGRAMMING APPROACH FOR CPM ANALYSIS

Many classical network problems can be formulated as linear programs and solved using related techniques. Finding the shortest and longest paths through a network are two such examples. Of course, the latter is exactly the problem that is solved in CPM analysis. To see its linear programming representation, we make use of the following notation, and assume an AOA model:

i,j = indices for nodes in the network; each node corresponds to an event; $i = 1$ is the unique project start node

N = set of nodes or events

n = number of events in the network; n is the unique node marking the end of the project

A = set of arcs in the network; each arc (i,j) corresponds to a project activity, where i denotes its start event and j its end event

L_{ij} = length of the activity that starts at node i and terminates at node j

t_i = decision variable associated with the start time of event $i \in N$

The following linear program (LP) schedules all events and all activities in a feasible manner such that the project finishes as early as possible, assuming that work begins at time $t_1 = 0$:

$$\min t_n \tag{7-11a}$$

subject to

$$t_j - t_i \geq L_{ij} \qquad \text{for all activities } (i,j) \in A \tag{7-11b}$$

$$t_1 = 0, \quad t_i \geq 0 \quad \text{for all } i \in N \tag{7-11c}$$

Note that the nonnegativity condition $t_i \geq 0$ is redundant, and that the last event t_n denotes the completion time of the project.

The slack associated with a nonbinding constraint in (7-11b) represents the slack of the corresponding activity given the start times, t_i, found by the LP. These values may not coincide with the CPM calculations. To find the total slack of an activity it is necessary to perform sensitivity (ranging) analysis on the LP solution. The amount that each right-hand side (L_{ij}) can be increased without changing the optimal solution is equivalent to the total slack of activity (i,j).

The linear programming formulation for the example project is:

$$\min t_6$$

subject to

$t_2 - t_1 \geq 5$	activity A
$t_3 - t_1 \geq 3$	activity B
$t_4 - t_2 \geq 8$	activity C
$t_4 - t_3 \geq 7$	activity D
$t_4 - t_1 \geq 7$	activity E
$t_5 - t_4 \geq 4$	activity F
$t_6 - t_5 \geq 5$	activity G
$t_3 - t_2 \geq 0$	dummy D_1
$t_1 = 0$	

which, as determined by the well-known software package LINDO (Schrage 1986), has solution $t = (0,5,6,13,17,22)$. The slack vector for the first eight rows is $(0,3,0,0,6,0,0,1)$. Notice that these results differ slightly from those in Tables 7-4 and 7-5. To guarantee that the LP (7-11) finds the earliest time each event can start, as was done in Section 7.6.1, the following penalty term must be added to the objective function (7-11a):

$$\varepsilon \left(\sum_{i=2}^{n-1} \right) t_i$$

where $\varepsilon > 0$ is an arbitrarily small constant. Conceptually, in the augmented formulation, the computations are done in two stages. First, t_n is found. Then, given this value, a search is conducted over the set of alternative optima to find the minimum values of $t_i, i = 2,...,$ $n - 1$. In reality, the computations are all done in one stage, not two.

7.9 AGGREGATING ACTIVITIES IN THE NETWORK

The detailed network model of a project is very useful in scheduling and monitoring progress at the operational (short-term) level. Management concerns at the tactical or strategic level, however, create a need for a focused presentation that eliminates unnecessary clutter. For projects that span a number of years and include hundreds of activities, it is likely that only a portion of those activities will be active or require close control at any point in time. To facilitate the management function, there is a need to condense information and aggregate tasks. The two common tools used for this purpose are hammock activities and milestones.

7.9.1 Hammock Activities

When a group of activities has a common start and a common end point, it is possible to replace the entire group with a single activity called a hammock activity. For example, in the network depicted in Fig. 7-25, it is possible to use a hammock activity between events 4 and 6. In so doing, activities F and G are collapsed into FG whose duration is the sum of L_{45} and L_{56}. In general, the duration of a hammock activity is equal to the duration of the longest sequence of activities that it replaces. If another hammock activity is used to represent A, B, C, D, and E, its length would be

$$\max\begin{Bmatrix} L_{12} + L_{24} \\ L_{13} + L_{34} \\ L_{12} + L_{23} + L_{34} \\ L_{14} \end{Bmatrix} = \max\begin{Bmatrix} 5 + 8 \\ 3 + 7 \\ 5 + 0 + 7 \\ 7 \end{Bmatrix} = 13$$

Hammock activities reduce the size of a network while preserving, in general, information on precedence relations and activity durations. By using hammock activities an upper-level network can be created that presents a synoptic view of the project. Such networks are useful for medium (tactical) and long-range (strategic) planning. The common practice is to develop a hierarchy of networks where the various levels correspond to the levels of either the work breakdown structure or the organizational breakdown structure. Higher-level networks contain many hammock activities and provide upper management with a general picture of flows, milestones, and overall status. Lower-level networks consist of single activities and provide detailed schedule information for team leaders. Proper use of hammock activities can help in providing the right level of detail to each participant in the project.

7.9.2 Milestones

A higher level of aggregation is also possible by introducing milestones to mark the completion of significant activities. As explained in Section 7.1.1, milestones are commonly

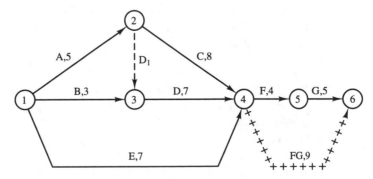

Figure 7-25   Example of a hammock activity.

used to mark the delivery of goods and services, to denote points in time where payments are due, and to flag important events such as the successful completion of a critical design review. In the simplest case, a milestone can mark the completion of a single activity, as event 2 in our example marks the completion of activity A. It can also mark the completion of several activities as exemplified by event 4, which denotes the completion of C, D, and E.

By using several levels of aggregation; that is, networks with various layers of hammock activities and milestones, it is possible to design the most appropriate decision support tool for each level of management. Such an exercise should take into account the WBS and the OBS. At the lowest levels of these structures a detailed network is essential; at higher levels, aggregation by hammock activities and milestones is the norm.

7.10 DEALING WITH UNCERTAINTY

The critical path method assumes that the duration of an activity is either known and deterministic, or that a point estimate such as the mean or mode can be used in its place. It makes no allowance for activity variance. When fluctuations in performance time are low, this assumption is logically justified and has empirically been shown to produce quite accurate results. When high levels of uncertainty exist, however, the critical path method may not provide a very good estimate of the project completion time. In these situations, there is a need to account explicitly for the effects of uncertainty. Monte Carlo simulation and PERT are the two most common approaches that have been developed for this purpose.

7.10.1 Simulation Approach

This approach is based on simulating the project by randomly generating performance times for each activity from their perceived distributions. In most cases it is assumed that activity times follow a beta distribution as discussed in Section 7.2.1. In each simulation run a sample of the performance time of each activity is taken and a CPM analysis is conducted to determine the critical path and the project finish time for that realization. By repeating the process a large number of times, it is possible to construct a frequency distribution or histogram of the project completion time. This distribution may then be used to calculate the probability that the project finishes by a given date, as well as the expected error of each such estimate.

A single simulation run would consist of the following steps:

1. Generate a random value for the duration of each activity from the appropriate distribution.
2. Determine the critical path and its duration using CPM.
3. Record the results.

The number of times this procedure must be repeated depends on the error tolerances deemed acceptable. Standard statistical tests can be used to verify the accuracy of the estimates.

To understand the calculations, let us focus on the AOA network in Fig. 7-22 for the example project and assume that each activity follows a beta distribution with parameter values given in Table 7-8. After performing 10 simulation runs the results listed in Table 7-9 for activity durations, critical path, and project completion time were obtained. Additional data collected but not presented include the earliest and latest start and completion times of each event, and activity slacks.

TABLE 7-8   STATISTICS FOR EXAMPLE ACTIVITIES

Activity	Optimistic time, a	Most likely time, m	Pessimistic time, b	Expected value, $\bar{d}$	Standard deviation, s
A	2	5	8	5	1
B	1	3	5	3	0.66
C	7	8	9	8	0.33
D	4	7	10	7	1
E	6	7	8	7	0.33
F	2	4	6	4	0.66
G	4	5	6	5	0.33

TABLE 7-9   SUMMARY OF SIMULATION RUNS FOR EXAMPLE PROJECT

Run number	Activity duration							Critical path	Completion time
	A	B	C	D	E	F	G		
1	6.3	2.2	8.8	6.6	7.6	5.7	4.6	A–C–F–G	25.4
2	2.1	1.8	7.4	8.0	6.6	2.7	4.6	A–D–F–G	17.4
3	7.8	4.9	8.8	7.0	6.7	5.0	4.9	A–C–F–G	26.5
4	5.3	2.3	8.9	9.5	6.2	4.8	5.4	A–D–F–G	25.0
5	4.5	2.6	7.6	7.2	7.2	5.3	5.6	A–C–F–G	23.0
6	7.1	0.4	7.2	5.8	6.1	2.8	5.2	A–C–F–G	22.3
7	5.2	4.7	8.9	6.6	7.3	4.6	5.5	A–C–F–G	24.2
8	6.2	4.4	8.9	4.0	6.7	3.0	4.0	A–C–F–G	22.1
9	2.7	1.1	7.4	5.9	7.9	2.9	5.9	A–C–F–G	18.9
10	4.0	3.6	8.3	4.3	7.1	3.1	4.3	A–C–F–G	19.7

As can be seen from Table 7-9 for the first run, the duration of activity A is 6.3, while the duration of activity B is 2.2. In the second run, the duration of A is 2.1, and so on. Note that the critical path differs from one replication to the next depending on the randomly generated durations of the activities. In the 10 runs reported, the sequence A–D–F–G is the longest (critical) in two replications, while the sequence A–C–F–G is critical in the other eight. Activities A, F, and G are critical in 100% of the replications, while activity C is critical in 80% and activity D is critical in 20%.

A principal output of the simulation runs is a frequency distribution of the project length (the length of the critical path). Figure 7-26 plots the results of some 50 replications

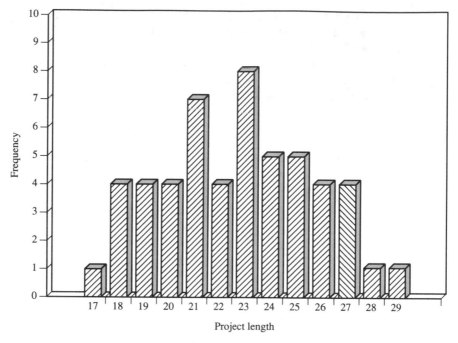

Figure 7-26   Distribution of project length for simulation runs.

for the example. As can be seen, the project length varied from 17 to 29 weeks, with a mean of 22.5 weeks and a standard deviation of 2.9 weeks. Now let X be a random variable associated with project completion time. The probability of finishing the project within, say, τ weeks can be estimated from the following ratio:

$$P(X \leq \tau) = \frac{\text{number of times project finished in less than or equal to } \tau \text{ weeks}}{\text{total number of replications}}$$

For the example, if $\tau = 20$ weeks, the number of runs in which the length of the critical path was less than or equal to 20 weeks is seen to be 13, so $P(X \leq 20) = \frac{13}{50} = 26\%$.

In addition, it is possible to estimate the criticality of each activity. The *criticality index* (CI) of an activity is defined as the proportion of runs in which the activity was on the critical path (i.e., it had a zero slack). Dodin and Elmaghraby (1985) provide some theoretical background on this problem as well as extensive test results for large PERT networks.

The simulation approach is easy to implement and has the advantage that it produces arbitrarily accurate results as the number of runs increases. However, for problems of realistic size, the computational burden may be significant for each run, so a balance must be reached between accuracy and effort.

7.10.2 PERT and Extensions

Two common analytical approaches are used to assess uncertainty in projects. Both are based on the *central limit theorem*, which states that the distribution of the sum of independent random variables is approximately normal when the number of terms in the sum is sufficiently large.

The first approach yields a rough estimate and assumes that the duration of each project activity is an independent random variable. Given probabilistic durations of activities along specific paths, it follows that elapsed times for achieving events along those paths are also probabilistic. Now, suppose that there are n activities in the project, k of which are critical. Denote the durations of the critical activities by the random variables d_i with mean $\overline{d}_i$ and variance s_i^2, $i = 1,...,k$. Then the total project length is the random variable

$$X = d_1 + d_2 + \cdots + d_k$$

It follows that the mean project length, $F(X)$, and the variance of the project length, $V(X)$, are given by

$$E(X) = \overline{d}_1 + \overline{d}_2 + \cdots + \overline{d}_k$$

$$V(X) = s_1^2 + s_2^2 + \cdots + s_k^2$$

These formulas are based on elementary probability theory, which tells us that the expected value of the sum of any set of random variables is the sum of their expected values, and the variance of the sum of independent random variables is the sum of the variances.

Now, invoking the central limit theorem, we can use normal distribution theory to find the probability of completing the project in less than or equal to some given time τ as follows:

$$P(X \leq \tau) = P\left(\frac{X - E(X)}{V(X)^{1/2}} \leq \frac{\tau - E(X)}{V(X)^{1/2}} \right) = P\left(Z \leq \frac{\tau - E(X)}{V(X)^{1/2}} \right) \quad (7\text{-}12)$$

where Z is the standard normal deviate with mean 0 and variance 1. The desired probability in eq. (7-12) can be looked up in Table 7C-1 in Appendix 7C.

Continuing with the example project, if (based on the simulation) the mean time of the critical path is 22.5 weeks and the variance is $(2.9)^2$, the probability of completing the project within 25 weeks is found by first calculating

$$Z = \frac{25 - 22.5}{2.9} = 0.86$$

and then looking up 0.86 in Table 7C-1. Doing so, we find that $P(Z \leq 0.86) = 0.805$, so the probability of finishing the project in 25 weeks or less is 80.5%. This solution is depicted in Fig. 7-27.

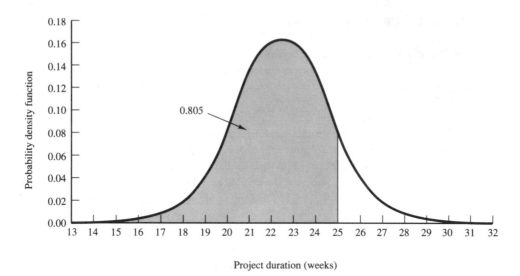

Figure 7-27   Example of probabilistic analysis with PERT.

If, however, the mean project length, $E(X)$, and the project length variance, $V(X)$, are calculated using the assumption that the critical activities are only those that have a zero slack in the deterministic CPM analysis (A–C–F–G), we get

$$E(X) = 5 + 8 + 4 + 5 = 22$$

$$V(X)^{1/2} = \sqrt{1^2 + 0.33^2 + 0.66^2 + 0.33^2} = 1.285$$

Based on this assumption, the probability of completing the project within 25 weeks is

$$P\left(Z \leq \frac{25 - 22}{1.285}\right) = P(Z \leq 2.33) = 0.99$$

This probability is higher than 0.805, as in the simulation both sequences A–C–F–G and A–D–F–G were critical.

The procedure above is, in essence, PERT. Summarizing for an AON network:

1. For each activity i, assess its probability distribution or assume a beta distribution and obtain estimates of a_i, b_i, and m_i. These values should be supplied by the project manager or experts working in the field.
2. If a beta distribution is assumed for activity i, use the estimates a, b, and m to compute the variance, s_i^2 and mean, $\bar{d}_i$ from eqs. (7-1) and (7-2) in Section 7.2.1.
3. Use CPM to determine the critical path given $\bar{d}_i$, $i = 1,...,n$.

4. Once the critical activities are identified, sum their means and variances to find the mean and variance of the project length.

5. Use eq. (7-12) with the statistics computed in step 4 to evaluate the probability that the project finishes within some desired time.

Using PERT, it is possible to estimate completion time for a desired completion probability. For example, for a 95% probability the corresponding Z value is $Z_{0.95} = 1.64$. Solving for the time τ for which the probability to complete the project is 95%, we get

$$Z_{0.95} = \frac{\tau - 22.5}{2.9} = 1.64 \quad \text{or} \quad \tau = (1.64)(2.9) + 22.5 = 27.256 \text{ weeks}$$

A shortcoming of the standard PERT calculations is that they ignore all activities not on the critical path. A more accurate analytical approach is to identify each sequence of activities leading from the start node of the project to the end event, and then to calculate separately the probability that the activities comprising each sequence will be completed by a given date. This step can be done as above by assuming that the central limit theorem holds for each sequence and then applying normal distribution theory to calculate the individual path probabilities. It is necessary, though, to make the additional assumption that the sequences themselves are statistically independent in order to proceed. This means that the time to traverse each path in the network is independent of what happens on the other paths. Although it is easy to see that this is rarely true because some activities are sure to be on more than one path, empirical evidence suggests that good results can be obtained if there is not too much overlap.

Once these calculations are performed, assuming that the various sequences are independent of each other, the probability of completing the project by a given date is set equal to the product of the individual probabilities that each sequence is finished by that date. That is, given n sequences with completion times $X_1, X_2, ..., X_n$, the probability that X is less than or equal to τ is found from

$$P(X \leq \tau) = P(X_1 \leq \tau)P(X_2 \leq \tau) \cdots (X_n \leq \tau) \tag{7-13}$$

where now the random variable $X = \max[X_1, X_2, ..., X_n]$.

Example 7-5

Consider the simple project in Fig. 7-28. If no uncertainty exists in activity durations, the critical path is A–B and exactly 17 weeks are required to finish the project. Now if we assume that the durations of all four activities are normally distributed (the corresponding means and standard deviations are listed under the arrows in Fig. 7-28), the durations of the two sequences are also normally distributed [i.e., $N(\mu, \sigma)$] with the following parameters:

$$\text{length(A–B)} = X_1 \sim N(17, 3.61)$$
$$\text{length(C–D)} = X_2 \sim N(16, 3.35)$$

[mean, standard deviation]

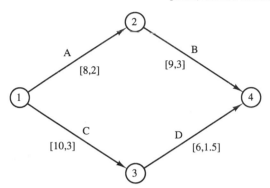

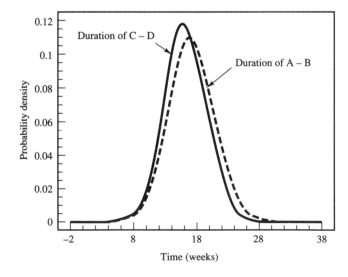

Figure 7-28   Stochastic network.

The accompanying probability density functions are plotted in Fig. 7-29. It should be clear that the project can end in 17 weeks only if both A–B and C–D are completed within that time. The probability that A–B finishes within 17 weeks is

$$P(X_1 \le 17) = P\left(Z \le \frac{17 - 17}{3.61}\right) = P(Z \le 0) = 0.5$$

and similarly for C–D,

$$P(X_2 \le 17) = P\left(Z \le \frac{17 - 16}{3.35}\right) = P(Z \le 0.299) = 0.62$$

Figure 7-29   Performance time distribution for the two sequences.

Using eq. (7-13), we can now find the probability that both sequences finish within 17 weeks:

$$P(X \leq 17) = P(X_1 \leq 17)P(X_2 \leq 17) = (0.5)(0.62) = 0.31$$

Thus the probability that the project will finish by week 17 is about 31%. A similar analysis for 20 weeks yields $P(X \leq 20) = 0.7 = 70\%$.

■

The approach based on calculating the probability of each sequence to complete by a given due date is accurate only if the sequences are independent. This is not the case when one or more activities are members of two or more sequences. Consider, for example, the project in Fig. 7-30. Here activity E is a member of the two sequences connecting the start of the project (event 1) to its termination node (event 5). The expected lengths and standard deviations of these sequences are:

Sequence	Expected length	Standard deviation
A–B–E	$8 + 9 + 3 = 20$	$\sqrt{2^2 + 3^2 + 4^2} = 5.39$
C–D–E	$10 + 6 + 3 = 19$	$\sqrt{3^2 + 1.5^2 + 4^2} = 5.22$

The probability that the sequence A–B–E will be completed in 17 days is calculated as follows:

$$Z = \frac{17 - 20}{5.39} = -0.5565 \quad \text{implying that } P = 0.29$$

which is obtained from Table 7C-1 by noting that

$$P(Z \leq -z) = 1 - P(Z \leq z)$$

Similarly, the probability that the sequence C–D–E will be completed in 17 days is calculated by determining $Z = (17 - 19)/5.22 = -0.383$ and then using Table 7C-1 to find $P = 0.35$.

[mean, standard deviation]

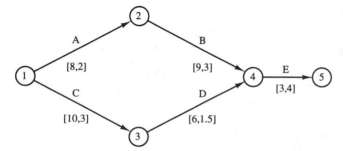

Figure 7-30   Stochastic network with dependent sequences.

Thus the simple PERT estimate (based on the critical sequence A–B–E) indicates that the probability of completing the project in 17 days is 29%. If both sequences A–B–E and C–D–E are taken into account, the probability of completing the project in 17 days is estimated as

$$P(X_{ABE} \leq 17)P(X_{CDE} \leq 17) = (0.29)(0.35) = 0.1 \quad \text{or} \quad 10\%$$

assuming that the two sequences are independent. However, since activity E is common to both sequences, the true probability of completing the project in 17 days is somewhere between 10 and 29%.

The next question that naturally arises is what to do if only the parameters of the distribution are known but not its form (e.g., beta, normal), and the number of activities is too small to rely on the central limit theorem to give accurate results. In this case, Chebyshev's inequality can be used to calculate project duration probabilities (see Bain and Engelhardt 1987). The underlying theorem states that if X is a random variable with mean μ and variance σ^2, then for any $k > 0$,

$$P(|X - \mu| \geq k\sigma) \leq \frac{1}{k^2}$$

An alternative form is

$$P(|X - \mu| < k\sigma) \geq 1 - \frac{1}{k^2}$$

Based on the second inequality, the probability of a random variable being within ± 3 standard deviations of its mean is at least $\frac{8}{9}$ or 89%. Although this might not be a tight bound in all cases, it is surprising that such a bound can be found to hold for all possible discrete and continuous distributions.

To illustrate the effect of uncertainty, consider the example project. Four sequences connect the start node to the end node. The mean length and the standard deviation of each sequence are summarized in Table 7-10.

The probability of completing each sequence in 22 weeks is computed next and summarized in Table 7-11.

Based on the simple PERT analysis, the probability to complete the project in 22 weeks is 0.5. If both sequences A–C–F–G and A–D–F–G are considered and assumed to be independent, the probability is reduced to $(0.5)(0.73) = 0.365$.

TABLE 7-10   MEAN LENGTH AND
STANDARD DEVIATION FOR SEQUENCES
IN EXAMPLE PROJECT

Sequence	Mean length	Standard deviation
A–C–F–G	22	1.285
A–D–F–G	21	1.595
B–D–F–G	19	1.407
E–F–G	16	0.808

TABLE 7-11 PROBABILITY OF COMPLETING
EACH SEQUENCE IN 22 WEEKS

Sequence	Z value	Probability
A–C–F–G	$\dfrac{22 - 22}{1.285} = 0$	0.5
A–D–F–G	$\dfrac{22 - 21}{1.595} = 0.626$	0.73
B–D–F–G	$\dfrac{22 - 19}{1.407} = 2.13$	0.98
E–F–G	$\dfrac{22 - 16}{0.808} = 7.42$	1.0

Since three activities (A,F,G) are common to both sequences, the actual probability to complete in 22 weeks is closer to 0.5 than to 0.365. Based on the data in Fig. 7-26, we see that in 24 out of 50 simulation runs the project duration was 22 weeks or less. This implies that the probability of completing the project in 22 weeks is 24/50 = 0.48 or 48%.

Continuing with this example, if the Chebyshev's inequality is used for the critical path (μ = 22, σ = 1.285), the probability of completing the project in (say) 22 + (2)(1.285) = 24.57 weeks is about

$$1 - \left(\frac{1}{2}\right)^2 = \frac{3}{4} = 0.75$$

By way of comparison, using the normal distribution assumption, the corresponding probability is

$$P\left(Z \le \frac{24.57 - 22}{1.285}\right) = P(Z \le 2) = 0.97$$

Of the two, the Chebyshev estimate is likely to be more reliable given the fact that there are only a few activities on the critical path.

Because uncertainty is bound to be present in most activities, it is quite possible that after determining the critical path with CPM, a noncritical activity may become critical as certain tasks are completed. From a practical point of view, this suggests the basic advantage of early start schedules. Starting each activity as soon as possible reduces the chances of a noncritical activity becoming critical and delaying the project.

7.11 ANALYSIS OF PERT AND CPM ASSUMPTIONS

PERT and CPM are models of projects and hence are open to a wide range of technical criticisms: difficulty in accurately estimating durations, variances, and costs; validity of using the beta distribution in representing durations; validity of applying the central limit theorem; and the heavy focus on the critical path for project control. Table 7-12 highlights some of the more significant assumptions and their criticisms. In addition, PERT and CPM analysis is based on the precedence graph, which contains only two types of information: activity times and precedence constraints. The results may be highly sensitive to the data estimates and defining relationships.

In addition to the points made in Table 7-12, Schonberger (1981) has shown that a PERT estimate based on the assumption that the variance of a sequence of activities is equal to the sum of the activity variances (i.e., that activities and sequences are independent) can lead to a consistent error in estimating the completion time of a project.

A related problem, investigated by Britney (1976), relates to the cost of over- and underestimating activity duration times. He found that underestimates precipitate the reallocation of resources and, in many cases, engender costly project delays. Overestimates, on the other hand, result in inactivity and tend to misdirect management's attention to relatively unfruitful areas, causing planning losses. (Britney recommends a modification of PERT called BPERT which employs concepts from Bayesian decision theory to consider these two categories of cost explicitly in deriving a project network plan.)

Another problem that sometimes arises, especially when PERT is used by subcontractors working with the government, is the attempt to "beat" the network in order to get on or off the critical path. Many government contracts provide cost incentives for finishing a project early or are negotiated on a "cost-plus-fixed-fee" basis. The contractor who is on the critical path generally has more leverage in obtaining additional funds from these contracts since he or she has a major influence in determining the duration of the project. In contrast, some contractors deem it desirable to be less "visible" and therefore adjust their time estimates and activity descriptions in such a way as to ensure that they will not be on the critical path. This criticism, of course, reflects more on the use of the method than on the method itself, but PERT and CPM, by virtue of their focus on the critical path, enable such ploys to be used.

Finally, the cost of applying critical path methods to a project is sometimes used as a basis for criticism. However, the cost of applying PERT or CPM rarely exceeds 2% of total project cost. Thus this added cost is generally outweighed by the savings from improved scheduling and reduced project time.

As with any analytic technique, it is important when using CPM and PERT to understand fully the underlying assumptions and limitations they impose. Management must be sure that the people charged with monitoring and controlling activity performance have a working knowledge of the statistical features of PERT as well as the general nature of critical path scheduling. Correct application of these techniques can provide a significant benefit in each phase of the project's life cycle as long as the above-mentioned pitfalls are avoided.

It must be remembered that models are simplifications of reality designed to support analysis and decision making by focusing on the most important aspects of the problem. They should be judged not so much by their fidelity with the actual system, but by the insight they provide, by the certainty with which they show the correct consequences of the working assumptions, and by the ease with which the problem structure can be communicated.

TABLE 7-12   PRINCIPAL ASSUMPTIONS AND CRITICISMS OF PERT/CPM

1. *Assumption:* Project activities can be identified as entities; that is, there is a clear beginning and ending point for each activity.

   Criticism: Projects, especially complex ones, change in content over time, and therefore a network constructed in the planning phase may be highly inaccurate later. Also, the very fact that activities are specified and a network formalized tends to limit the flexibility that is required to handle changing situations as the project progresses.

2. *Assumption:* Project activity-sequence relationships can be specified and arranged in a directed network.

   Criticism: Sequence relationships cannot always be specified beforehand. In some projects, in fact, the ordering of certain activities is conditional on previous activities. (PERT and CPM, in their basic form, have no provision for treating this problem, although some other techniques have been proposed that present the project manager with several contingency paths, given different outcomes from each activity.)

3. *Assumption:* Project control should focus on the critical path.

   Criticism: It is not necessarily true that the longest path obtained from summing activity expected duration values will ultimately determine project completion time. What often happens as the project progresses is that some activity not on the critical path becomes delayed to such a degree that it extends the entire project. For this reason it has been suggested that a critical activity concept replace the critical path concept as focus of managerial control. Under this approach, attention would center on those activities that have a high potential variation and lie on a *near-critical path*. A near-critical path is one that does not share any activities with the critical path and, though it has slack, could become critical if one or a few activities along it become delayed. Obviously, the more parallelism in a network, the more likely that one or more near-critical paths will exist. Conversely, the more a network approximates a single series of activities, the less likely it is to have near-critical paths.

4. *Assumption:* The activity times in PERT follow the beta distribution, with the variance of the project assumed to be equal to the sum of the variances along the critical path.

   Criticism: As mentioned in the discussion in Section 7.2.1, the beta distribution was selected for a variety of good reasons. Nevertheless, each component of the statistical treatment in PERT has been brought into question. First, the formulas are in reality a modification of the beta distribution mean and variance, which, when compared to the basic formulas, could be expected to lead to absolute errors on the order of 10% for the mean and 5% for the individual variances. Second, given that the activity-time distributions have the properties of unimodality, continuity, and finite positive endpoints, other distributions with the same properties would yield different means and variances. Third, obtaining three "valid" time estimates to put into the PERT formulas presents operational problems—it is often difficult to arrive at one activity-time estimate, let alone three, and the somewhat subjective definitions of a and b do not help the matter. (How optimistic and pessimistic should one be?)

Source: Adapted from Chase and Aquilano (1981).

7.12 SCHEDULING CONFLICTS

The discussion so far assumed that the only constraints on the schedule are precedence relations among activities. Based on these constraints the early and late time of each event, and the early and late start and finish of each activity, are calculated.

In most projects there are additional constraints that must be addressed such as those associated with resource availability and the budget. In some cases, ready time and due-date constraints also exist. These constraints specify a time window in which an activity must be performed. In addition, there may be a target completion date for the project or a due date for a milestone. If these due dates are earlier than the corresponding dates derived from the CPM analysis, the accompanying schedule will not be feasible.

There are several ways to handle these type of infeasibilities, including:

- Reducing some activity durations by allocating more resources to them. This approach is discussed in Chapter 8.
- Eliminating some activities or reducing their lengths by using a more effective technology. For example, conventional painting, which requires the application of several layers of paint and a long drying time, may be replaced by anodizing—a faster but more expensive process.
- Some precedence relations of the "start to end" type may be replaced by other precedence relations, such as "start to start," without affecting quality, cost, or performance. When this is the case, a significant amount of time may be saved.

It is common to start the scheduling analysis with each activity being performed in the most economical way and assuming "start to end" precedence relations. If infeasibility is detected, one or more of the foregoing courses of action can be used to ameliorate the problem.

TEAM PROJECT

Thermal Transfer Plant

A detailed schedule is now required for the project. Major milestones suggested by TMS's contract department follow:

Milestone	Time from project start (weeks)
Initial drawing	2
Order parts and materials	3
Initial drawing approval or revisions	4
Drawings revised and approved	5
Schedule production	5
Begin production	6
Document final testing procedures	6
Finish assembly/begin testing	9
Documentation, maintenance, and user manuals	9
Ship tested unit to site	11
Install on site	13
Final testing and operator personnel training	14
Customer satisfaction check	16

Your assignment is to prepare a list of activities and a detailed schedule (on a daily basis) for the project team and an upper-level schedule for TMS management. The detailed schedule should consist of up to 50 activities; the upper-level schedule should contain about 20 activities.

In your report, explain each task and activity, its corresponding WBS and OBS units, the type of precedence relations among tasks, the way task duration was estimated, and your confidence in these estimates. Use a network model to develop the schedule and a linear responsibility chart to identify its relationship to OBS units. Present the schedule as a Gantt chart and as a table of activities and events with their corresponding times and slacks.

Discuss the range of schedules that can be adopted for this project and explain the methodology by which your team has selected the most appropriate schedule. Present a "what if" analysis for your final choice, testing its sensitivity to important sources of uncertainty.

DISCUSSION QUESTIONS

1. What objectives, variables, and constraints should be considered in developing a project schedule?

2. If a project, by definition, is something that is not performed on a regular basis, how can one estimate activity times?

3. What are the advantages and disadvantages of the five project-activity-duration estimation techniques presented in Section 7.2?

4. What are the major characteristics that must be present in a project to use network techniques?

5. The "start to end" precedence relation is the most common found in projects. Give some examples where "start to start," "end to end," and "end to start" precedence relations arise.

6. Identify some projects where PERT and CPM are inappropriate. Explain.

7. How can the linear programming model in Section 7.8 be expanded to include resource constraints that might arise due to, say, the limited availability of equipment or technical personnel?

8. Discuss a project where scheduling is not important. Explain why this project is not sensitive to scheduling decisions.

9. Compare and list the relative advantages of (a) the Gantt chart, (b) CPM analysis, and (c) the basic PERT approach to scheduling.

10. Is it possible for a project team to achieve high efficiency without scheduling tasks and activities? Discuss.

11. "To excel in time-based competition, the early start schedule should always be implemented." Discuss.

12. "To maximize the net present value of a project, all cash-generating activities should begin on their early start, while all cost-generating activities should begin on their late start." Discuss.

EXERCISES

7-1. A project is defined by the list of activities in Table 7-13.
   (a) Draw the AOA network.
   (b) Draw the AON network.
   (c) Find the critical path.
   (d) Find the total slack and free slack of each activity.
   (e) Suppose that activities A, C, I are subject to uncertainty and that only the following time estimates are available:

Activity	a	m	b
A	2	4	5
C	1	3	4
I	8	11	15

Calculate the probability that the project will be completed in D days, for D = 10, 12, 14, 16, 18, 20. Plot the probability as a function of D.

7-2. Estimate the time it will take you to learn a new computer software package that combines a spreadsheet with statistical analysis. Explain how the estimate was made and what accuracy you think it has.

7-3. Use the modular technique to estimate the time required to prepare a proposal or business plan for manufacturing a new medical device that analyzes blood enzymes.

7-4. Use the benchmark job technique to estimate the time required to type a 50-page paper and prepare figures using a computer graphics package.

7-5. Develop a linear regression model to estimate the dependent variable "time to type a paper" as a function of two or more independent variables.

7-6. Develop a list of activities for the project "designing a new house." Estimate the duration of each activity and define the precedence relations among them. How much uncertainty exists in each activity?

TABLE 7-13

Activity	Immediate predecessors	Duration (days)
A	—	3
B	—	4
C	—	3
D	C	2
E	B	1
F	A	5
G	B	2
H	B	3
I	C	11
J	D,E	3
K	F,G	1
L	K	4
M	J,H	4

7-7. Develop an early start and a late start schedule for the project in Exercise 7-6 using a Gantt chart. Identify the critical path and calculate the slack of noncritical activities.

7-8. Develop the AOA network for the project in Exercise 7-6. Calculate the early time and the late time of each event, and the early start, early finish, late start, and late finish of each activity.

7-9. Develop an AON network model for the project in Exercise 7-6.

7-10. Develop a linear program that generates the schedule for the project in Exercise 7-6.

7-11. Develop a high-level AOA model for the project "designing and building a new house." The project of Exercise 7-6 should be one hammock activity in this new project.

7-12. Suppose that the project mentioned in Exercise 7-11 must be finished 2 months before the early finish time. How would you solve this scheduling conflict?

7-13. Caryn Johnson is in charge of relocating ("reconductoring") 1,700 ft of 13.8-kilovolt overhead primary line due to the widening of the road section in which the line is presently installed. Table 7-14 summarizes the activities for the project. Draw the network model for her and carry out the critical path computations.

7-14. Thomas Cruise wants to buy a new motor boat and has summarized the associated activities in Table 7-15. Draw the network model and carry out the critical path computations for him.

7-15. For Exercise 7-14, compute the total slacks and free slacks, and summarize the critical path calculations using the format in Table 7-5.

7-16. Determine the critical path(s) for projects (a) and (b) in Fig. 7-31.

7-17. For Exercise 7-16, compute the total slacks and free slacks, and summarize the critical path calculations in a tabular format.

TABLE 7-14

Activity	Description	Immediate predecessors	Duration (days)
A	Job review	—	1
B	Advise customers of temporary outage	A	0.5
C	Requisition stores	A	1
D	Scout job	A	0.5
E	Secure poles and materials	C,D	3
F	Distribute poles	E	3.5
G	Pole location coordination	D	0.5
H	Re-stake	G	0.5
I	Dig holes	H	3
J	Frame and set poles	F,I	4
K	Cover old conductors	F,I	1
L	Pull new conductors	J,K	2
M	Install remaining material	L	2
N	Sag conductor	L	2
O	Trim trees	D	2
P	Deenergize and switch lines	B,M,N,O	0.1
Q	Energize and phase new line	P	0.5
R	Clean up	Q	1
S	Remove old conductor	Q	1
T	Remove old poles	S	2
U	Return material to stores	R,T	2

TABLE 7-15

Activity	Description	Immediate predecessors	Duration (days)
A	Conduct feasibility study	—	3
B	Find potential customer for present boat	A	14
C	List possible models	A	1
D	Research all possible models	C	3
E	Conduct interviews with mechanics	C	1
F	Collect dealer propaganda	C	2
G	Compile and organize all pertinent information	D,E,F	1
H	Choose top three models	G	1
I	Test-drive all three choices	H	3
J	Gather warranty and financing information	H	2
K	Choose one boat	I,J	2
L	Compare dealers and choose dealer	K	2
M	Search for desired color and options	L	4
N	Test-drive chosen model once again	L	1
O	Purchase new boat	B,M,N	3

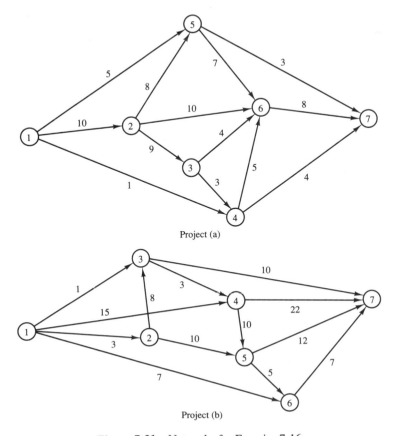

Project (a)

Project (b)

Figure 7-31 Networks for Exercise 7-16.

7-18. In Exercise 7-16, suppose that the estimates (a, b, m) are given as shown in Table 7-16. Find the probabilities that the events will occur without delay.

7-19. *Product Development.* Consider the simplified set of activities in Table 7-17 for the development of a consumer product from initiation through the market test phase.
(a) Draw the AOA network for this project.
(b) Calculate total slacks and free slacks, and interpret their meaning.
(c) Determine the critical path and interpret its meaning.
(d) Construct a Gantt chart and identify scheduling flexibilities.

7-20. For the product development project in Exercise 7-19, consider the detailed time estimates given in Table 7-18. Note that the time estimates in Exercise 7-19 are equivalent to modal time estimates in this exercise.
(a) Relabel your network in Exercise 7-19 to include $\overline{d}_{ij}$ (in place of d_{ij}) and s_{ij}. Use eqs. (7-1) and (7-2).

TABLE 7-16

Project (a)

Activity	(a, b, m)	Activity	(a, b, m)
1, 2	(5, 8, 6)	3, 6	(3, 5, 4)
1, 4	(1, 4, 3)	4, 6	(4, 10, 8)
1, 5	(2, 5, 4)	4, 7	(5, 8, 6)
2, 3	(4, 6, 5)	5, 6	(9, 15, 10)
2, 5	(7, 10, 8)	5, 7	(4, 8, 6)
2, 6	(8, 13, 9)	6, 7	(3, 5, 4)
3, 4	(5, 10, 9)		

Project (b)

Activity	(a, b, m)	Activity	(a, b, m)
1, 2	(1, 4, 3)	3, 7	(12, 14, 13)
1, 3	(5, 8, 7)	4, 5	(10, 15, 12)
1, 4	(6, 9, 7)	4, 7	(8, 12, 10)
1, 6	(1, 3, 2)	5, 6	(7, 11, 8)
2, 3	(3, 5, 4)	5, 7	(2, 8, 4)
2, 5	(7, 9, 8)	6, 7	(5, 7, 6)
3, 4	(10, 20, 15)		

TABLE 7-17

Activity	Symbol	Preceding activities	Time estimate (days)
Investigate demand	A	—	3
Develop pricing strategy	B	—	1
Design product	C	—	5
Conduct promotional cost analysis	D	A	1
Manufacture prototype models	E	C	6
Perform product cost analysis	F	E	1
Perform final pricing analysis	G	B,D,F	2
Conduct market test	H	G	8

TABLE 7-18

Activity	Time Estimate (weeks)		
	Optimistic	Most likely	Pessimistic
A	1	3	4
B	1	1	2
C	4	5	9
D	1	1	1
E	4	6	12
F	1	1	2
G	1	2	3
H	6	8	10

(b) Compare total slacks and free slacks to Exercise 7-19.

(c) Has the critical path changed?

(d) Determine the following probabilities:

  (1) That the project will be completed in 22 weeks or less.

  (2) That the project will be completed by its earliest expected completion date.

  (3) That the project takes more than 30 weeks to complete.

7-21. Criticism of the traditional PERT equations in Section 7.2.1 for estimating the means and standard deviations of activities has led to the development of alternative formulas by Perry and Greig (1975):

$$\bar{d}_{ij} = \frac{a_{ij} + 0.95m_{ij} + b_{ij}}{2.95} \tag{7-14}$$

$$s_{ij} = \frac{b_{ij} - a_{ij}}{3.25} \tag{7-15}$$

where a_{ij} and b_{ij} are estimates for the 5 and 95 percentiles of the probability distribution of activity (i,j), and m_{ij} is the mode. Use these equations to recalculate $\bar{d}_{ij}$ and s_{ij} and answer the same questions as in Exercise 7-20. Compare the results.

7-22. *Space Module Assembly.* An aerospace company has received a contract from NASA for the final assembly of a space module for an upcoming mission. A team of engineers has determined the activities, precedence constraints, and time estimates as given in Table 7-19.

(a) Draw the AOA network for this project. (*Hint:* You should have 10 events and two dummy activities.)

(b) Calculate total slacks and free slacks, and interpret their meaning.

(c) Determine the critical path and interpret its meaning.

(d) Construct a Gantt chart and identify scheduling flexibilities.

7-23. A more careful analysis of time estimates for the space module assembly of the preceding exercise is given in Table 7-20. Note that the "most likely estimates" are identical to the "time estimates" in Exercise 7-22.

(a) Relabel your network in Exercise 7-22 to include $\bar{d}_{ij}$ (in place of d_{ij}) and s_{ij}. Use eqs. (7-1) and (7-2).

(b) Compare total slacks and free slacks to Exercise 7-22.

(c) Has the critical path changed?

TABLE 7-19

Activity	Symbol	Preceding activities	Time estimate (days)
Construct shell of module	A	—	30
Order life support system and scientific experimentation package from same supplier	B	—	15
Order components of control and navigational system	C	—	25
Wire module	D	A	3
Assemble control and navigational system	E	C	7
Preliminary test of life support system	F	B	1
Install life support in module	G	D,F	5
Install scientific experimentation package in module	H	D,F	2
Preliminary test of control and navigational system	I	E,F	4
Install control and navigational system in module	J	H,I	10
Final testing and debugging	K	G,J	8

TABLE 7-20

Activity	Time estimate (weeks)		
	Optimistic	Most likely	Pessimistic
A	25	30	45
B	10	15	20
C	20	25	35
D	3	3	5
E	5	7	12
F	1	1	1
G	4	5	7
H	2	2	3
I	4	4	6
J	8	10	14
K	6	8	15

(d) Determine the following probabilities:
   (1) That the project will be completed in 54 days or less.
   (2) That the project will be completed by its earliest expected completion date.
   (3) That the project takes more than 70 days to complete.

7-24. Use eqs. (7-10) and (7-11) to recalculate $\bar{d}_{ij}$ and s_{ij} and answer the same questions as in Exercise 7-23. Compare the results.

7-25. As part of a R&D project, it is required to produce 60 circuit boards using a specific piece of equipment. According to the equipment specification, its design capacity is 0.4 board per hour. However, past experience indicates that significantly more time will be required. In particular, the following frequency data were collected over a 1-week period when the machine was working on other jobs.

Activity	Frequency
Machine is working on a job	67
Parts are being fed to the machine	6
Maintenance is being performed	9
Machine is waiting for parts	22

(a) Estimate the actual machine capacity.
(b) How long will it take to complete the 60 boards?
(c) If you want the capacity estimate to be within ±5% of the true value with a 95% level of confidence, what should the sample size be?

7-26. The project manager did not accept the approach you proposed in Exercise 7-25 and suggested the use of a parametric equation to estimate the machine's capacity.
(a) Specify the data that should be collected to develop such an equation.
(b) Furnish an example of such an equation and demonstrate how to use it.
(c) State the assumptions used in employing this approach.

7-27. Consider the precedence relations given in Table 7-21.
(a) Draw an early start Gantt chart.
(b) Draw the AON network for this project.
(c) Draw the AOA network.
(d) Generate all possible paths for the AOA network, calculate their duration, and analyze the findings.
(e) Calculate ES, EF, LF, and LS for each activity.
(f) Calculate the slacks for the activities.

TABLE 7-21

Activity	Immediate predecessors	Weeks
A	—	1
B	A	4
C	A	3
D	A	7
E	B	6
F	C,D	2
G	E,F	7
H	D	9
I	G,H	4

7-28. There is uncertainty regarding the duration of activities D and E in the project described in Exercise 7-27 expressed by the following data:

Activity	Time (weeks)		
	Optimistic	Likely	Pessimistic
D	6	7	8
E	5	6	9

(a) Using an early start approach, calculate the probability of completing the project within 22 weeks or less.

(b) Repeat part (a) using a late start approach. State your assumptions in both cases.

REFERENCES

Estimating the Duration of Project Activities

ABERNATHY, W. J., "Subjective Estimates and Scheduling Decisions," *Management Science*, Vol. 18, No. 2 (1971), pp. 80–88.

BAIN, L. J., and M. ENGELHARDT, *Introduction to Probability and Mathematical Statistics*, Duxbury Press, Boston, MA, 1987.

BRITNEY, R. R., "Bayesian Point Estimation and the PERT Scheduling of Stochastic Activities," *Management Science*, Vol. 22, No. 9 (1976), pp. 938–948.

DODIN, B., "Bounding the Project Completion Time Distribution in PERT Networks," *Operations Research*, Vol. 33, No. 4 (1985), pp. 862–881.

GRUBBS, F., "Attempts to Validate Certain PERT Statistics or 'Picking on PERT,'" *Operations Research*, Vol. 10 (1962), pp. 912–915.

HERSHAUER, J. C., and G. NABIELSKY, "Estimating Activity Times," *Journal of Systems Management*, Vol. 23, No. 9 (1972), pp. 17–21.

PERRY, C., and I. D. GREIG, "Estimating the Mean and Variance of Subjective Distributions in PERT and Decision Analysis," *Management Science*, Vol. 21, No. 12 (1975), pp. 1477–1480.

Effect of Learning

BADIRU, A. B., "Computational Survey of Univariate and Multivariate Learning Curve Models," *IEEE Transactions on Engineering Management*, Vol. 39, No. 2 (1992), pp. 176–188.

HANCOCK, W. M., and F. H. BAYHA, "The Learning Curve," in *Handbook of Industrial Engineering*, Wiley, New York, 1982.

SMUNT, T. L., "A Comparison of Learning Curve Analysis and Moving Average Ratio Analysis for Detailed Operations Planning," *Decision Science*, Vol. 17, No. 4 (1986), pp. 475–495.

WRIGHT, T. P., "Factors Affecting the Cost of Airplanes," *Journal of Aeronautical Sciences*, Vol. 3, No. 4 (1936), pp. 122–128.

YELLE, L. E., "The Learning Curve: Historical Review and Comprehensive Survey," *Decision Sciences*, Vol. 10 (1979), pp. 302–328.

Forgetting

GLOBERSON, S., N. LEVIN, and A. SHTUB, "The Impact of Breaks on Forgetting When Performing a Repetitive Task," *IIE Transactions*, Vol. 21, No. 4 (1989), pp. 376–381.

LEBLANC, L. J., A. SHTUB, and Z. CAI, *Project Planning with Learning: Models and Computational Testing*, Working Paper 91-2, Graduate School of Management, Vanderbilt University, Nashville, TN, 1992.

SHTUB, A., "Scheduling of Programs with Repetitive Projects," *Project Management Journal*, Vol. XXII, No. 6 (1991), pp. 49–53.

Project Scheduling

CLARK, K. B., and T. FUJIMOTO, "Overlapping Problem Solving in Product Development," in K. FERDOWS (editor), *Managing International Manufacturing*, North-Holland, New York, 1989.

HARTLEY, K. O., "The Project Schedule," in R. L. KIMMON and J. H. LOWREE (editors), *Project Management: A Reference for Professionals*, Marcel Dekker, New York, 1989.

HILLIER, F. S., and G. J. LIEBERMAN, *Introduction to Operations Research*, Fourth Edition, Holden-Day, Oakland, CA, 1986.

MEREDITH, J. R., and S. J. MANTEL, JR., *Project Management: A Managerial Approach*, Second Edition, Wiley, New York, 1989.

VAZSONYI, A., "The History of the Rise and Fall of the PERT Method," *Management Science*, Vol. 16, No. 8 (1970), pp. B449–B455.

WEBSTER, F. M., *Survey of CPM Scheduling Packages and Related Project Control Programs*, Project Management Institute, Drexel Hill, PA, 1991.

CPM Approach

CORNELL, D. G., C. C. GOTLIEB, and Y. M. LEE, "Minimal Event-Node Network of Project Precedence Relations," *Communications of the ACM*, Vol. 16, No. 5 (May 1973), pp. 296–298.

JEWELL, W. S., "Divisible Activities in Critical Path Analysis," *Operations Research*, Vol. 13, No. 5 (1965), pp. 747–760.

KELLEY, J. E., JR., and M. R. WALKER, "Critical Path Planning and Scheduling," *Proceedings of the Eastern Joint Computer Conference*, Boston, 1979, pp. 160–173.

PERT Approach

BURGHER, P. H., "PERT and the Auditor," *Accounting Review*, Vol. 39 (1964), pp. 103–120.

DODIN, M. B., "Determining the K Most Critical Paths in PERT Networks," *Operations Research*, Vol. 32, No. 4 (1984), pp. 859–877.

DODIN, M. B., and S. E. ELMAGHRABY, "Approximating the Criticality Indices of the Activities in PERT Networks," *Management Science*, Vol. 31, No. 2 (1985), pp. 207–223.

FAZAR, W., "Program Evaluation and Review Technique," *American Statistician*, Vol. 13, No. 2 (1959), p. 10.

FISHER, D. L., D. SAISI, and W. M. GOLDSTEIN, "Stochastic PERT Networks: OP Diagrams, Critical Paths and the Project Completion Time," *Computers & Operations Research*, Vol. 12, No. 5 (1985), pp. 471–482.

PERT, *Program Evaluation Research Task, Phase I Summary Report*, Vol. 7, Special Projects Office, Bureau of Ordinance, U.S. Department of the Navy, Washington, DC 1958, pp. 646–669.

VAN SLYKE, R. M., "Monte Carlo Methods and the PERT Problem," *Operations Research*, Vol. 11, No. 5 (1963), pp. 839–860.

PERT and CPM Assumptions

CHASE, R. B., and N. J. AQUILANO, *Production and Operations Management*, Third Edition, Richard D. Irwin, Homewood, IL, 1981.

GOLENKO-GINZBURG, D., "On the Distribution of Activity Time in PERT," *Journal of the Operational Research Society*, Vol. 39, No. 8 (1988), pp. 767–771.

LITTLEFIELD, T. K., and P. H. RANDOLPH, "PERT Duration Times: Mathematics or MBO," *Interfaces*, Vol. 21, No. 6 (1991), pp. 92–95.

SASIENI, M. W., "A Note on PERT Times," *Management Science*, Vol. 16, No. 8 (1986), pp. 1652–1653.

SCHONBERGER, R. J., "Why Projects Are Always Late: A Rationale Based on Manual Simulation of a PERT/CPM Network," *Interfaces*, Vol. 11, No. 5 (1981), pp. 66–70.

WIEST, J. D., and F. K. LEVY, *A Management Guide to PERT/CPM*, Second Edition, Prentice Hall, Englewood Cliffs, NJ, 1977.

Computational Issues

DRAPER, N., and H. SMITH, *Applied Regression Analysis*, Second Edition, Wiley, New York, 1981.

HINDELANG, T. J., and J. F. MUTH, "A Dynamic Programming Algorithm for Decision CPM Networks," *Operations Research*, Vol. 27, No. 2 (1979), pp. 225–241.

KULKARNI, V. G., and J. S. PROVAN, "An Improved Implementation of Conditional Monte Carlo Estimation of Path Lengths in Stochastic Networks," *Operations Research*, Vol. 33, No. 6 (1985), pp. 1389–1393.

SCHRAGE, L., *Linear, Integer and Quadratic Programming with LINDO*, Third Edition, Scientific Press, Palo Alto, CA, 1986.

APPENDIX 7A

Least-Squares Regression Analysis

In the least-squares method, we define the residual, e_i, or deviation from the estimated line, $\hat{Y} = \hat{b}_0 + \hat{b}_1 X$ for each point i as follows:

$$e_i = Y_i - \hat{Y}_i$$

These residuals will be positive or negative depending on whether the actual point lies above or below the line. If they are squared and summed, the resultant quantity must be nonnegative and will vary directly with the spread of the points from the line. Different pairs of values for $\hat{b}_0$ and $\hat{b}_1$ will give different lines and hence different values for the sum of the squared residuals about the line. Thus we have

$$\sum_{i=1}^{n} e_i^2 = f(\hat{b}_0, \hat{b}_1)$$

The principle of least squares is that the parameter estimates $\hat{b}_0$ and $\hat{b}_1$ should be chosen to make $\Sigma_i\, e_i^2$ as small as possible; that is,

$$\min \sum_{i=1}^{n} e_i^2 = \min \sum_{i=1}^{n} (Y_i - \hat{Y}_i)^2 = \min \sum_{i=1}^{n} (Y_i - \hat{b}_0 - \hat{b}_1 X_i)^2$$

From calculus we know that the first-order necessary (and sufficient, in this case) condition for optimality is that the partial derivatives with respect to $\hat{b}_0$ and $\hat{b}_1$ must be zero. Taking partial derivatives, setting the results to zero, and solving yields

$$\hat{b}_1 = \frac{\sum_{i=1}^{n} (X_i - \overline{X})(Y_i - \overline{Y})}{\sum_{i=1}^{n} (X_i - \overline{X})^2} \quad \text{and} \quad \hat{b}_0 = \overline{Y} - \hat{b}_1 \overline{X}$$

where

$$\overline{X} = \frac{1}{n} \sum_{i=1}^{n} X_i \quad \text{and} \quad \overline{Y} = \frac{1}{n} \sum_{i=1}^{n} Y_i$$

Given these estimates, an important question is: How good are they? Elementary treatment of the relationship between two variables usually emphasizes their *correlation coefficient*, R, which is computed as follows:

$$R = \frac{\sum_{i=1}^{n} (X_i - \overline{X})(Y_i - \overline{Y})}{\sqrt{\sum_{i=1}^{n} (X_i - \overline{X})^2 \sum_{i=1}^{n} (Y_i - \overline{Y})^2}}$$

This value can vary between -1 and $+1$. The closer it is to either extreme, the better the fit. A related value is R^2, sometimes known as the *coefficient of determination*, which can be calculated variously as

$$R^2 = \frac{\sum_{i=1}^{n} (\hat{Y}_i - \overline{Y}^2)}{\sum_{i=1}^{n} (Y_i - \overline{Y})^2} = 1 - \frac{\sum_{i=1}^{n} e_i^2}{\sum_{i=1}^{n} (Y_i - \overline{Y})^2}$$

From the right-hand-side expression it should be clear that the maximum value of R^2 is unity. This can occur only when $\sum_i e_i^2 = 0$, that is, when every e_i is zero, so that all the points on the scatter diagram line on a straight line. The minimum value of R^2 is zero, which occurs when $\sum_i e_i^2 = \sum_i (Y_i - \overline{Y})^2$, that is, when the regression line $\hat{Y} = \overline{Y}$ and the explained variation is zero.

The coefficient of determination is equivalent to the proportion of the Y variance explained by the linear influence of X. An R value of 0.9 therefore indicates that the least-squares regression of Y on X accounts for 81% of the variance in Y.

APPENDIX 7B

Learning Curve Tables

TABLE 7B-1 LEARNING CURVE VALUES FOR n^β

Repetitions	Percent learning curve							
	60%	65%	70%	75%	80%	85%	90%	95%
1	1.0000	1.0000	1.0000	1.0000	1.0000	1.0000	1.0000	1.0000
2	0.6000	0.6500	0.7000	0.7500	0.8000	0.8500	0.9000	0.9500
3	0.4450	0.5052	0.5682	0.6338	0.7021	0.7729	0.8462	0.9219
4	0.3600	0.4225	0.4900	0.5625	0.6400	0.7225	0.8100	0.9025
5	0.3054	0.3678	0.4368	0.5127	0.5956	0.6857	0.7830	0.8877
6	0.2670	0.3284	0.3977	0.4754	0.5617	0.6570	0.7616	0.8758
7	0.2383	0.2984	0.3674	0.4459	0.5345	0.6337	0.7439	0.8659
8	0.2160	0.2746	0.3430	0.4219	0.5120	0.6141	0.7290	0.8574
9	0.1980	0.2552	0.3228	0.4017	0.4930	0.5974	0.7161	0.8499
10	0.1832	0.2391	0.3058	0.3846	0.4765	0.5828	0.7047	0.8433
12	0.1602	0.2135	0.2784	0.3565	0.4493	0.5584	0.6854	0.8320
14	0.1430	0.1940	0.2572	0.3344	0.4276	0.5386	0.6696	0.8226
16	0.1296	0.1785	0.2401	0.3164	0.4096	0.5220	0.6561	0.8145
18	0.1188	0.1659	0.2260	0.3013	0.3944	0.5078	0.6445	0.8074
20	0.1099	0.1554	0.2141	0.2884	0.3812	0.4954	0.6342	0.8012
22	0.1025	0.1465	0.2038	0.2772	0.3697	0.4844	0.6251	0.7955
24	0.0961	0.1387	0.1949	0.2674	0.3595	0.4747	0.6169	0.7904
25	0.0933	0.1353	0.1908	0.2629	0.3548	0.4701	0.6131	0.7880
30	0.0815	0.1208	0.1737	0.2437	0.3346	0.4505	0.5963	0.7775
35	0.0728	0.1097	0.1605	0.2286	0.3184	0.4345	0.5825	0.7687
40	0.0660	0.1010	0.1498	0.2163	0.3050	0.4211	0.5708	0.7611
45	0.0605	0.0939	0.1410	0.2060	0.2936	0.4096	0.5607	0.7545
50	0.0560	0.0879	0.1336	0.1972	0.2838	0.3996	0.5518	0.7486
60	0.0489	0.0785	0.1216	0.1828	0.2676	0.3829	0.5367	0.7386
70	0.0437	0.0713	0.1123	0.1715	0.2547	0.3693	0.5243	0.7302
80	0.0396	0.0657	0.1049	0.1622	0.2440	0.3579	0.5137	0.7231
90	0.0363	0.0610	0.0987	0.1545	0.2349	0.3482	0.5046	0.7168
100	0.0336	0.0572	0.0935	0.1479	0.2271	0.3397	0.4966	0.7112
120	0.0294	0.0510	0.0851	0.1371	0.2141	0.3255	0.4830	0.7017
140	0.0262	0.0464	0.0786	0.1287	0.2038	0.3139	0.4718	0.6937
160	0.0237	0.0427	0.0734	0.1217	0.1952	0.3042	0.4623	0.6869
180	0.0218	0.0397	0.0691	0.1159	0.1879	0.2959	0.4541	0.6809

(Continues)

TABLE 7B-1   (CONTINUED)

Repetitions	Percent learning curve							
	60%	65%	70%	75%	80%	85%	90%	95%
200	0.0201	0.0371	0.0655	0.1109	0.1816	0.2887	0.4469	0.6757
250	0.0171	0.0323	0.0584	0.1011	0.1691	0.2740	0.4320	0.6646
300	0.0149	0.0289	0.0531	0.0937	0.1594	0.2625	0.4202	0.6557
350	0.0133	0.0262	0.0491	0.0879	0.1517	0.2532	0.4105	0.6482
400	0.0121	0.0241	0.0458	0.0832	0.1453	0.2454	0.4022	0.6419
450	0.0111	0.0224	0.0431	0.0792	0.1399	0.2387	0.3951	0.6363
500	0.0103	0.0210	0.0408	0.0758	0.1352	0.2329	0.3888	0.6314
600	0.0090	0.0188	0.0372	0.0703	0.1275	0.2232	0.3782	0.6229
700	0.0080	0.0171	0.0344	0.0659	0.1214	0.2152	0.3694	0.6158
800	0.0073	0.0157	0.0321	0.0624	0.1163	0.2086	0.3620	0.6098
900	0.0067	0.0146	0.0302	0.0594	0.1119	0.2029	0.3556	0.6045
1,000	0.0062	0.0137	0.0286	0.0569	0.1082	0.1980	0.3499	0.5998
1,200	0.0054	0.0122	0.0260	0.0527	0.1020	0.1897	0.3404	0.5918
1,400	0.0048	0.0111	0.0240	0.0495	0.0971	0.1830	0.3325	0.5850
1,600	0.0044	0.0102	0.0225	0.0468	0.0930	0.1773	0.3258	0.5793
1,800	0.0040	0.0095	0.0211	0.0446	0.0895	0.1725	0.3200	0.5743
2,000	0.0037	0.0089	0.0200	0.0427	0.0866	0.1683	0.3149	0.5698
2,500	0.0031	0.0077	0.0178	0.0389	0.0806	0.1597	0.3044	0.5605
3,000	0.0027	0.0069	0.0162	0.0360	0.0760	0.1530	0.2961	0.5530

TABLE 7B-2   CUMULATIVE LEARNING CURVE VALUES FOR n^{β}

Repetitions	Percent learning curve							
	60%	65%	70%	75%	80%	85%	90%	95%
1	1.000	1.000	1.000	1.000	1.000	1.000	1.000	1.000
2	1.600	1.650	1.700	1.750	1.800	1.850	1.900	1.950
3	2.045	2.155	2.268	2.384	2.502	2.623	2.746	2.872
4	2.405	2.578	2.758	2.946	3.142	3.345	3.556	3.774
5	2.710	2.946	3.195	3.459	3.738	4.031	4.339	4.662
6	2.977	3.274	3.593	3.934	4.299	4.688	5.101	5.538
7	3.216	3.572	3.960	4.380	4.834	5.322	5.845	6.404
8	3.432	3.847	4.303	4.802	5.346	5.936	6.574	7.261
9	3.630	4.102	4.626	5.204	5.839	6.533	7.290	8.111
10	3.813	4.341	4.931	5.589	6.315	7.116	7.994	8.955
12	4.144	4.780	5.501	6.315	7.227	8.244	9.374	10.62
14	4.438	5.177	6.026	6.994	8.092	9.331	10.72	12.27
16	4.704	5.541	6.514	7.635	8.920	10.38	12.04	13.91
18	4.946	5.879	6.972	8.245	9.716	11.41	13.33	15.52
20	5.171	6.195	7.407	8.828	10.48	12.40	14.61	17.13
22	5.379	6.492	7.819	9.388	11.23	13.38	15.86	18.72
24	5.574	6.773	8.213	9.928	11.95	14.33	17.10	20.31
25	5.668	6.909	8.404	10.19	12.31	14.80	17.71	21.10
30	6.097	7.540	9.305	11.45	14.02	17.09	20.73	25.00
35	6.478	8.109	10.13	12.72	15.64	19.29	23.67	28.86

(*Continues*)

TABLE 7B-2 (CONTINUED)

Repetitions	Percent learning curve							
	60%	65%	70%	75%	80%	85%	90%	95%
40	6.821	8.631	10.90	13.72	17.19	21.43	26.54	32.68
45	7.134	9.114	11.62	14.77	18.68	23.50	29.37	36.47
50	7.422	9.565	12.31	15.78	20.12	25.51	32.14	40.22
60	7.941	10.39	13.57	17.67	22.87	29.41	37.57	47.65
70	8.401	11.13	14.74	19.43	25.47	33.17	42.87	54.99
80	8.814	11.82	15.82	21.09	27.96	36.80	48.05	62.25
90	9.191	12.45	16.83	22.67	30.35	40.32	53.14	69.45
100	9.539	13.03	17.79	24.18	32.65	43.75	58.14	76.59
120	10.16	14.11	19.57	27.02	37.05	50.39	67.93	90.71
140	10.72	15.08	21.20	29.67	41.22	56.78	77.46	104.7
160	11.21	15.97	22.72	32.17	45.20	62.95	86.80	118.5
180	11.67	16.79	24.14	34.54	49.03	68.95	95.96	132.1
200	12.09	17.55	25.48	36.80	52.72	74.79	105.0	145.7
250	13.01	19.28	28.56	42.08	61.47	88.83	126.9	179.2
300	13.81	20.81	31.34	46.94	69.66	102.2	148.2	212.2
350	14.51	22.18	33.89	51.48	77.43	115.1	169.0	244.8
400	15.14	23.44	36.26	55.75	84.85	127.6	189.3	277.0
450	15.72	24.60	38.48	59.80	91.97	139.7	209.2	309.0
500	16.26	25.68	40.58	63.68	98.85	151.5	228.8	340.6
600	17.21	27.67	44.47	70.97	112.0	174.2	267.1	403.3
700	18.06	29.45	48.04	77.77	124.4	196.1	304.5	465.3
800	18.82	31.09	51.36	84.18	136.3	217.3	341.0	526.5
900	19.51	32.60	54.46	90.26	147.7	237.9	376.9	587.2
1,000	20.15	34.01	57.40	96.07	158.7	257.9	412.2	647.4
1,200	21.30	36.59	62.85	107.0	179.7	296.6	481.2	766.6
1,400	22.32	38.92	67.85	117.2	199.6	333.9	548.4	884.2
1,600	23.23	41.04	72.49	126.8	218.6	369.9	614.2	1,001.0
1,800	24.06	43.00	76.85	135.9	236.8	404.9	678.8	1,116.0
2,000	24.83	44.84	80.96	144.7	254.4	438.9	742.3	1,230.0
2,500	26.53	48.97	90.39	165.0	296.1	520.8	897.0	1,513.0
3,000	27.99	52.62	98.90	183.7	335.2	598.9	1,047.0	1,791.0

APPENDIX 7C

Normal Distribution Function Table

TABLE 7C-1 CUMULATIVE PROBABILITIES OF THE NORMAL DISTRIBUTION (AREAS UNDER THE STANDARDIZED NORMALIZED CURVE FROM $-\infty$ TO z)

z	0.00	0.01	0.02	0.03	0.04	0.05	0.06	0.07	0.08	0.09
0.0	0.5000	0.5040	0.5080	0.5120	0.5160	0.5199	0.5239	0.5279	0.5319	0.5359
0.1	0.5389	0.5438	0.5478	0.5517	0.5557	0.5596	0.5636	0.5675	0.5714	0.5753
0.2	0.5793	0.5832	0.5871	0.5910	0.5948	0.5987	0.6026	0.6064	0.6103	0.6141
0.3	0.6179	0.6217	0.6255	0.6293	0.6331	0.6368	0.6406	0.6443	0.6480	0.6517
0.4	0.6554	0.6591	0.6628	0.6664	0.6700	0.6736	0.6772	0.6808	0.6844	0.6879
0.5	0.6915	0.6950	0.6985	0.7019	0.7054	0.7088	0.7123	0.7157	0.7190	0.7224
0.6	0.7257	0.7291	0.7324	0.7357	0.7389	0.7422	0.7454	0.7486	0.7517	0.7549
0.7	0.7580	0.7611	0.7642	0.7673	0.7704	0.7734	0.7764	0.7794	0.7823	0.7852
0.8	0.7881	0.7910	0.7939	0.7967	0.7995	0.8023	0.8051	0.8078	0.8106	0.8133
0.9	0.8159	0.8186	0.8212	0.8238	0.8264	0.8289	0.8315	0.8340	0.8365	0.8389
1.0	0.8413	0.8438	0.8461	0.8485	0.8508	0.8531	0.8554	0.8577	0.8599	0.8621
1.1	0.8643	0.8665	0.8686	0.8708	0.8729	0.8749	0.8770	0.8790	0.8810	0.8830
1.2	0.8849	0.8869	0.8888	0.8907	0.8925	0.8944	0.8962	0.8980	0.8997	0.9015
1.3	0.9032	0.9049	0.9066	0.9082	0.9099	0.9115	0.9131	0.9147	0.9162	0.9177
1.4	0.9192	0.9207	0.9222	0.9236	0.9251	0.9265	0.9279	0.9292	0.9306	0.9319
1.5	0.9332	0.9345	0.9357	0.9370	0.9382	0.9394	0.9406	0.9418	0.9429	0.9441
1.6	0.9452	0.9463	0.9474	0.9484	0.9495	0.9505	0.9515	0.9525	0.9535	0.9545
1.7	0.9554	0.9564	0.9573	0.9582	0.9591	0.9599	0.9608	0.9616	0.9625	0.9633
1.8	0.9641	0.9649	0.9656	0.9664	0.9671	0.9678	0.9686	0.9693	0.9699	0.9706
1.9	0.9713	0.9719	0.9726	0.9732	0.9738	0.9744	0.9750	0.9756	0.9761	0.9767
2.0	0.9772	0.9778	0.9783	0.9788	0.9793	0.9798	0.9803	0.9808	0.9812	0.9817
2.1	0.9821	0.9826	0.9830	0.9834	0.9838	0.9842	0.9846	0.9850	0.9854	0.9857
2.2	0.9861	0.9864	0.9868	0.9871	0.9875	0.9878	0.9881	0.9884	0.9887	0.9890
2.3	0.9893	0.9896	0.9898	0.9901	0.9904	0.9906	0.9909	0.9911	0.9913	0.9916
2.4	0.9918	0.9920	0.9922	0.9925	0.9927	0.9929	0.9931	0.9932	0.9934	0.9936
2.5	0.9938	0.9940	0.9941	0.9943	0.9945	0.9946	0.9948	0.9949	0.9951	0.9952
2.6	0.9953	0.9955	0.9956	0.9957	0.9959	0.9960	0.9961	0.9962	0.9963	0.9964
2.7	0.9965	0.9966	0.9967	0.9968	0.9969	0.9970	0.9971	0.9972	0.9973	0.9974
2.8	0.9974	0.9975	0.9976	0.9977	0.9977	0.9978	0.9979	0.9979	0.9980	0.9981
2.9	0.9981	0.9982	0.9982	0.9983	0.9984	0.9984	0.9985	0.9985	0.9986	0.9986
3.0	0.9987	0.9987	0.9987	0.9988	0.9988	0.9989	0.9989	0.9989	0.9990	0.9990
3.1	0.9990	0.9991	0.9991	0.9991	0.9992	0.9992	0.9992	0.9992	0.9993	0.9993
3.2	0.9993	0.9993	0.9994	0.9994	0.9994	0.9994	0.9994	0.9995	0.9995	0.9995
3.3	0.9995	0.9995	0.9995	0.9996	0.9996	0.9996	0.9996	0.9996	0.9996	0.9997
3.4	0.9997	0.9997	0.9997	0.9997	0.9997	0.9997	0.9997	0.9997	0.9997	0.9998

8

Project Budget

8.1 INTRODUCTION

An organization's budget (usually expressed in dollars) represents management's long-range, midrange, and short-range plans. The budget should contain a statement of prospective investments, management goals, resources necessary to achieve those goals, and a timetable. Its structure should match that of the organization. In particular, a functional structure shows an organization's investments and expenditures grouped three ways: (1) development of new products (engineering), (2) production of existing products (manufacturing), and (3) campaigns for new or existing products (advertising, marketing). A project-oriented structure, on the other hand, reveals the organization's planned costs and expected revenues for each project, while a matrix structure partially supports both the functional and project-based component of an organization's budget. This is explained presently.

The budget of any specific project is tied to the organizational budget. In some organizations, a project budget includes only expenditures (e.g., government agencies such as the Department of Defense are engaged in projects strictly as clients). In other organizations, the project budget includes both income and expenditures (e.g., contractors whose expenditures for labor, materials, and subcontracting are covered by their clients). When an organization is involved in several projects, the budgets of these projects are coordinated centrally. It is important to combine the budget of each project to avoid the risk of steering the organization into financial difficulties. This issue should be considered when selecting new projects because it provides a hard constraint in the decision-making process.

In a matrix organization, the budget links the functional units to the projects. On a specific project the cost of resources invested by the functional unit is charged against the project's budget . This link is one of the interfaces between the functional structure and the project aspect of the matrix organization. In this chapter we discuss the principles used in developing, presenting, and using the budget in a project environment. The major focus is on the relationship between the organizational and the individual budgets of each project undertaken.

A well-designed budget is an efficient communication channel for management. Through the budget, managers (at all levels) are advised of their organizational goals and the resources allocated to their units. A detailed budget defines expected costs and expenditures, thus setting the framework of constraints within which each manager is expected to operate. These constraints represent organizational policy and goals. The well-structured budget is a yardstick that can be used to measure the performance of organizational units and their managers. Managers who participate in the budget development process commit themselves, their subordinates, and their unit's resources to the goals specified in the budget as well as the constraints implied by the negotiated funding levels. A successful manager is one who can achieve the budget goals with the resources allocated to his or her project, that is, one who can successfully execute the organization's policy. The well-structured budget is also a useful tool for identifying deviations from plans, the magnitude of these deviations, and their source. Therefore, it is part of the baseline for cost and schedule control systems. In addition, the budget's structure depends on the organizational structure, while its level of detail depends on the planning horizon for which it was prepared.

The *long-range*, or *strategic*, budget defines an aggregate level of activity for the organization over a period of several months to several years. For example, in a functional organization this budget might define a goal of selling 100,000 units in the coming year with a 15% increase in sales in each of the following 4 years. The expected marketing cost in the budget is $50,000 for the first year with 8% increases in each subsequent year. In an organization with a project structure, the strategic budget will define the total budget for each project. For example, assume that for project X the design stage has a 1-year completion due date and a $500,000 budget. A critical design review is scheduled accordingly. In 2 years a prototype will be tested in the lab. The associated budget is $600,000. The final product will be tested in the third year for a cost of $550,000. The long-range budget is typically updated annually.

By using the budgeting process, management establishes long-range goals, schedules to achieve these goals, and the available resources. When the actual expenditures, income and results are compared to the original budget, management can monitor the organization's performance. Also, when necessary, management can change the budget to control both goal setting and resource allocation.

A *midrange, tactical* budget is a detailed presentation of the long-range budget and covers 12 to 24 months. It is updated quarterly. The tasks to be performed provide the basis of the entries. A rolling planning horizon is used so that every time (e.g., quarterly) the midrange budget is updated, a budget for the ensuing quarter is added while the budget for the recently completed quarter is deleted. The tactical budget details the monthly expected costs of labor, materials, and overhead for each task. In a functional organization, the tactical budget projects the expected costs and revenues of each product family and the expected costs of each functional department.

A *short-range* or *operational* budget lists specific activities and their costs. This budget spans a period up to 1 year and covers the costs of resources (such as labor and material) required to perform each activity. For example, the short-range budget of a project might specify that the design of a prototype be done on a $10,000 CAD system which runs on a $5,000 piece of equipment. Lead times are 3 and 2 weeks, respectively, for the hardware and software. Installation starts as soon as both items are delivered. The expected cost of installation and training is $2,000. This short-term (operational) budget relates project costs to project activities through the project's lower-level network model.

A project's budget contains several dimensions. The first relates to the tasks and activities to be performed. The primary effort is to establish the relationship between cost and time for scheduled tasks and activities. The second dimension is based on the organizational breakdown structure. Each task is assigned to an organizational unit in the OBS. The third dimension is the work breakdown structure. Each task is assigned to a WBS element in the lowest level of the hierarchy. Over time, however, they are distributed among the WBS elements at their corresponding levels.

As each organization develops its own budgeting procedures, several points can help make the budget an efficient vehicle for planning, as well as a standard channel of communication:

- The budget should present management's objectives stated in terms of measurable outputs: for example, the successful completion of a test or the development of a new software module. These outputs should be presented with their budgetary constraints. Thus the budget presents available resources and the goals to be achieved using these resources. The presentation can be based on a functional structure, a project's organizational structure, or a combination of the two if a matrix structure is assumed.

- The budget should be presented quantitatively (e.g., in monetary units or sometimes in person-hours) as a function of time. The presentation should facilitate a periodic and cumulative comparison between actual and planned performance levels.

- The budget should be divided into long-range (strategic), midrange (tactical), and short-range (operational) levels. Each level should contain a detailed breakdown of the budget at the preceding level for the planning horizon. A rolling horizon approach should be used in developing the budgets of new periods and in updating the budgets of previous periods.

Management reserve may be included at strategic and tactical levels. This reserve acts as a buffer against uncertainty and should be consumed by transforming it into specific line items in the mid- and short-range budgets.

8.2 PROJECT BUDGET AND ORGANIZATIONAL GOALS

The budget of an organization reflects management's goals. These goals and organizational constraints determine decisions on project selection, resource allocation, and the desired rate of progress for each project. The budget depends on the perceived organizational mission and the sector to which the organization belongs (private, government, or nonprofit). It also depends on internal and external environmental factors. The following are seven common factors affecting project selection and budget structure:

1. *Competition.* Most organizations in the private sector need a competitive edge to survive. External challenges force continued improvement within the organization and occur in various ways, such as the following:

- *Time-based competition.* Spurs the implementation of concurrent engineering with the goal of shortening new product development cycles and improving customer service. It is also instrumental in reducing customer lead times. A major emphasis is on achieving project milestones and goals in a timely manner.

- *Cost-based competition.* In a cost-based environment, the project budget includes smaller, tightly controlled reserves.

- *Quality-based competition.* Total quality management is emphasized.

2. *Profit.* The ability to generate profits in the short and long run is essential to most organizations in the private sector. Selection decisions are frequently based on a project's expected profits. A project can be tentatively evaluated by any of the techniques discussed in Chapter 2, including net present value, internal rate of return, and payback period.

3. Cash flow. The organizational cash flow is an aggregate of all routine activities combined with other ongoing projects. When unexpected cash flow problems arise, projects that generate quick cash become high-priority items in the budget allocation process. In some cases an organization may prefer projects that begin to produce revenues immediately, albeit small, over projects that generate a slow cash flow and higher profits in the distant future. In the short run, to improve the cash position of the firm, activities that generate income (like payment milestones) may be budgeted earlier than other activities that have the same or an even shorter slack.

4. *Risk.* Uncertainty and risk may influence budgetary decisions. An organization that tries to avoid the risk of delays may budget its projects according to an early start schedule. This, in turn, may lead to early expenditures and cash flow problems. Organizations that try to minimize the risk of cost overruns sometime budget each activity at its lowest level. If a longer activity duration occurs, the lowered risk of a cost overrun can translate into an increased risk of delays.

The selection of new projects may also be influenced by risk assessment. In this case, the project's portfolio, to which the organization is committed, is affected by the organization's perceived risk level.

5. *Technological ability.* Some organizations in the public sector are willing to budget high-tech projects in order to acquire new, more advanced technologies. In the private sector (including such industries as computers, microelectronics, and aerospace), an organization's technological ability is an important aspect of its competitive edge. To outdistance competitors, technologically advanced projects are selected and budgeted to assure progress.

6. *Resources.* Each project's budget is a monetary representation of the value of resources allocated to perform that project. If adequate resources are not available, little can be accomplished, so whatever effort is expended will have negligible effect. Therefore, it is important to classify and track resources according to their availability. Chapter 9 presents a detailed classification scheme.

In the long- and midrange budgets, organizational plans for acquiring new resources are put forth. The short-range budget addresses plans to use these resources. In preparatory stages of a project, it is important to remember that some resources may not be available even if budgeted adequately. Therefore, in preparing the budget, resource availability (both inside and external to the organization) needs to be coordinated with the planned costs of these resources.

7. *Perceived Needs.* Project selection and budgeting depend largely on organizational goals. In the government sector, especially in defense, perceived needs (or new threats) are a driving force. Cost and risk considerations might be secondary when national security or public health are considered.

These seven factors link organizational goals and the internal and external aspects of the operational environment with each project's budget. Clearly, developing an organizational budget and a budget for each project requires a coordinated effort among management, accounting, marketing, and the other functional areas. This issue is the subject of the following section.

8.3 PREPARING THE BUDGET

Budget preparation is the process by which organizational goals are translated into a plan that specifies the allocated resources, the selected processes, and the desired schedule for achieving these goals. The budget must integrate information and objectives from all functional levels of the organization with information and objectives from the various project leaders. Although upper management sets the long-range (strategic) objectives, lower-level management is responsible for establishing the detailed (operational) plans and must clearly articulate and understand the short-range objectives before executing the budget.

In a project or a matrix organization, lower-level managers, who are concerned primarily with the daily operations, should be most knowledgeable in the technical details regarding the most appropriate way to perform each project. They should also be intimately familiar with expected activity durations and costs. Thus it is important to integrate upper-level management input with the knowledge of the functional and project managers.

The organizational budget consists of both ongoing activities, such as the production and marketing of existing products, and one-time efforts or projects. It is easier to budget ongoing activities, since past budgets for these activities can serve as a reference point for planning. By adjusting for anticipated demand, the expected inflation rate, and the effect of learning, the financial planners can develop the new budget based on past information. Project budgeting is more difficult, though, since previous budgets are often unavailable. Cost estimation (Chapter 10), the project schedule (Chapter 7), and the effect of resource availability (Chapter 9) should be considered in developing the project budget.

The building blocks of the project's budget are the work packages in which tasks performed on the lowest-level WBS elements are assigned to organizational units at the lowest level of the OBS. A budget is developed for each work package. Budgets are then developed for each WBS element at each level in the hierarchy and for each organizational unit at each OBS level.

Thus the process of integrating single project budgets and the budgets of ongoing activities into an acceptable organizational budget requires planning and coordination. The final budget should embody sound, workable programs for each functional area, and coordinate the efforts of functional units and project managers to achieve their goals. Three procedures are commonly used in budgeting: the top-down approach, the bottom-up approach, and the iterative-mixed approach.

8.3.1 Top-Down Budgeting

The trigger for the budgeting process is the strategic long-range plan that is developed by top management based on its experience and perception of the organization's goals and constraints. The long-range plan is then passed to the functional unit managers and the project managers who develop the tactical (midrange) and detailed operational (short-range) budgets, respectively.

One problem with top-down budgeting is the translation of long-range budgets into short-range budgets. The former can be spread in any number of ways over the budgets of projects and functional units. The best combination yielding the most efficient schedule for each of the projects involved is not easy to construct given the constraints imposed by the long-range budget. Therefore, the question is how to schedule projects in a "suboptimal way" to meet the strategic goals. This suboptimality is a result of top management's limited knowledge of the specifics of each project, task, and activity, knowledge that is unavailable when preparing the long-range budget using the top-down approach.

A second problem with this approach is the competition for funds among lower-level managers who try to secure adequate funding for their operations. But since top management fixes the total budget, the only way for lower-level managers to gain an advantage is to undercut their counterparts. Such a situation does not promote cooperation and understanding and does not guarantee the optimal allocation of funds. Table 8-1 illustrates the top-down budgeting process.

TABLE 8-1   TOP-DOWN APPROACH TO BUDGET PREPARATION

Step	Organizational level	Budget prepared at each step
1	Top management	Strategic budget based on organizational goals, constraints, and policies
2	Functional management	Tactical budget for each functional unit
3	Project managers	Detailed budgets for each project, including the cost of labor, material, subcontracting, overhead, etc.

8.3.2 Bottom-Up Budgeting

To overcome the disaggregation problem of top-down budgeting, many organizations adopt a budgeting approach starting at the project manager level. Each project manager is asked to prepare a budget proposal that supports efficient and on-schedule project execution. Based on these proposals, functional managers prepare the budgets for their units, considering the resources required in each period. Finally, top management streamlines and integrates the individual project and functional unit budgets into a strategic long-range organizational budget.

The advantages of this approach are the clear flow of information and the use of detailed data available at the project management level as the basic source of cost, schedule, and resource requirement information. The disadvantage of the approach is that top management has limited influence over the budgeting process, since the functional and project managers prepare most of the short- and midrange budgets. However, top management can influence the outcome by issuing a statement to the lower-level managers, as they prepare the short- and mid-range budgets, outlining organizational policies and goals. Also, top management can steer the budgeting process by selecting projects based on its perception of organizational needs and goals. Table 8-2 illustrates the bottom-up budgeting process.

As stated, the major problem with the bottom-up approach is the reduced level of control it offers top management. Since the aggregate budget is developed based on input obtained from the project and functional unit managers, the gap between strategic and operational objectives may be wide. This creates a need to fine tune the organizational budget. The process is carried out iteratively through adjustment and review until a satisfactory compromise is achieved.

TABLE 8-2   BOTTOM-UP APPROACH TO BUDGET PREPARATION

Step	Organizational level	Budget prepared at each step
1	Top management	Setting goals and selection of projects (a framework for budget)
2	Project management	Detailed budget proposals for projects including costs of material, labor, subcontracting, etc.
3	Functional management	Midrange budget for each functional unit
4	Top management	Adjustments and approval of the aggregate long plan budget resulting from the process

8.3.3 Iterative Budgeting

The two budgeting approaches presented above are "pure" in that the process flows in one direction, either bottom-up or top-down. Some of the shortcomings of these approaches can be eliminated by combining the information flows in an iterative fashion. A typical iterative approach starts with top management setting a budget framework for each year of a strategic plan. This framework then directs the selection of new projects and serves as a guideline for project managers as they prepare their budgets. Detailed project budgets are aggregated into functional unit budgets and finally, into an organizational budget that top management reviews and, if necessary, modifies. Based on the approved budget, functional unit and project managers modify their respective budgets. The process may undergo several iterations until convergence takes place at the strategic, tactical, and operational levels.

This process is based on input from all levels of management and usually produces better coordination between the different budgets (functional versus project and long range versus short- and midrange). Major disadvantages center on the relatively long duration needed for agreement and the excessive use of management time.

The process of adjusting a project budget to the framework of the organizational budget is based on the internal relationship between schedule, resources, and cost. This relationship can be exploited in several ways, as explained next.

8.4 TECHNIQUES FOR MANAGING THE BUDGET

The budget of a project represents scheduled expenditures and scheduled revenue as a function of time. The simplest approach to budgeting is to estimate the expected costs and income associated with each activity, task, and milestone. Based on the project schedule, these costs are assigned specific dates and a budget is generated; however, it may be only a partial budget because some of the indirect costs are usually not included at the preliminary stage. Typical indirect costs are those for management, facilities, and quality control and are not always related to specific activities. Adding these costs results in a more complete project budget. The product of this effort can serve as the basis for the decision-making process needed to develop a detailed, comprehensive budget. The development of detailed project budgets based on schedule and resource considerations is the first step in an iterative approach. The next step is to integrate them into an acceptable organizational budget.

8.4.1 Slack Management

One approach to integrating these projects is to change activity timing and the associated expenditure or income, an approach known as slack management. Noncritical activities that have free slack are usually the first candidates for this type of rescheduling. Activities with total slack are the next choices, and the final choices are critical activities that can be delayed only at the cost of delays in project completion time. Rescheduling activities makes the integration of single project budgets into an acceptable organizational budget easier.

To illustrate the relationship between a project's cash flow and its schedule, let us return to the example project. The length of the critical path in the project is 22 weeks. Critical activities are A, C, F, and G, while activities B, E, and D have either free or total slack that can be used for budget planning. Table 8-3 depicts the costs and durations of the project's activities.

An early start schedule results in relatively high expenditures in the project's earlier stages, while a late start schedule results in relatively high expenditures in the later stages. Table 8-4 presents this project's cash flow for the early start schedule assuming, for budgeting purposes, that the cost of each activity is evenly distributed throughout its duration. Table 8-5 enumerates the cash flow of the project for the late start case.

Figure 8-1 depicts the cash flows for the early and late start schedules; Fig. 8-2 depicts their cumulative cash flows. From Fig. 8-2 we see that if the strategic long-range organizational budget allocates only $4,913 to the project for weeks 1 through 5, then during this period only a late start schedule is feasible. Also, increasing the project's budget over $10,398 for the first 5 weeks makes an early start schedule feasible. Any budget in-between will force a delay of noncritical activities.

The choice between an early and a late start schedule affects the risk level associated with the project's on-time completion. Using a late start schedule means that all the activities are started as late as possible without any slack to buffer against uncertainty, increasing the probability of delays. Therefore, the budgeting process should resolve the conflict between a project budget that supports the organizational budgeting requirements versus the higher risk of a schedule overrun.

Projects with large numbers of activities tend to have a large choice of schedules with associated budgets. For example, in Fig. 8-1, any schedule that falls between the early and late start budget lines would be feasible from the point of view of meeting the critical milestones on time.

8.4.2 Crashing

In addition to using slack management as part of the budgeting process, another option may be available: change activity duration by selecting different technologies to perform the

TABLE 8-3 PROJECT ACTIVITY DURATIONS AND COSTS

Activity	Duration (weeks)	Cost ($1,000)
A	5	1.5
B	3	3.0
C	8	3.3
D	7	4.2
E	7	5.7
F	4	6.1
G	5	_7.2_
		31.0

TABLE 8-4   CASH FLOW OF AN EARLY START SCHEDULE

Week	Activity							Weekly cost	Cumulative cost
	A	B	C	D	E	F	G		
1	300	1,000			814.3			2,114	2,114
2	300	1,000			814.3			2,114	4,229
3	300	1,000			814.3			2,114	6,343
4	300				814.3			1,114	7,457
5	300				814.3			1,114	8,571
6			412.5	600	814.3			1,827	10,398
7			412.5	600	814.3			827	12,225
8			412.5	600				1,013	13,238
9			412.5	600				1,013	14,250
10			412.5	600				1,013	15,263
11			412.5	600				1,013	16,275
12			412.5	600				1,013	17,288
13			412.5					412	17,700
14						1,525		1,525	19,225
15						1,525		1,525	20,750
16						1,525		1,525	22,275
17						1,525		1,525	23,800
18							1,440	1,440	25,240
19							1,440	1,440	26,680
20							1,440	1,440	28,120
21							1,440	1,440	29,560
22							1,440	1,440	31,000
	1,500	3,000	3,300	4,200	5,700	6,100	7,200	31,000	

activity and by adding or deleting the necessary resources. So far we have assumed that each activity is performed in the most economical way. Thus the combination of resources assigned to each activity is assumed to be selected to minimize the total cost of performing that activity. However, in many cases it is possible to reduce an activity's duration by spending more money. Thus trade-offs exist between the minimum cost–longest duration option at one extreme and any other option that reduces an activity's duration at a higher cost.

This is the essence of the original version of CPM, which places equal emphasis on time and cost. The emphasis is achieved by constructing a time–cost curve for each activity, such as the one shown in Fig. 8-3. This curve plots the relationship between the direct cost for the activity and its resulting duration. In its simplest form, the plot is typically based on two points: the *normal* point and the *crash* point. The former gives the cost and time involved when the activity is performed in the normal way without extra resources such as overtime, special materials, or improved equipment that could speed things up. By contrast, the crash point gives the time and cost when the activity is fully expedited; that is, no cost is spared to reduce its duration as much as possible. As an approximation, it is then assumed that all intermediate time–cost trade-offs are possible and that they lie on the line segment between these two points (see the line segment in Fig. 8-3). Thus the only estimates needed are the cost and time for normal and crash points.

TABLE 8-5 CASH FLOW OF THE LATE START SCHEDULE

Week	A	B	C	D	E	F	G	Weekly cost	Cumulative cost
				Activity					
1	300							300	300
2	300							300	600
3	300	1,000						1,300	1,900
4	300	1,000						1,300	3,200
5	300	1,000						1,300	4,500
6			412.5					412	4,913
7			412.5	600	814.3			1,827	6,739
8			412.5	600	814.3			1,827	8,566
9			412.5	600	814.3			1,827	10,393
10			412.5	600	814.3			1,827	12,220
11			412.5	600	814.3			1,827	14,046
12			412.5	600	814.3			1,827	15,873
13			412.5	600	814.3			1,827	17,700
14						1,525		1,525	19,225
15						1,525		1,525	20,750
16						1,525		1,525	22,275
17						1,525		1,525	23,800
18							1,440	1,440	25,240
19							1,440	1,440	26,680
20							1,440	1,440	28,120
21							1,440	1,440	29,560
22							1,440	1,440	31,000
	1,500	3,000	3,300	4,200	5,700	6,100	7,200	31,000	

Consider, for example, a manual painting operation requiring 4 days at $400 per day. With a special compressed airflow system, however, two workers can complete the job in 2 days for $1,000 per day. Thus the activity can be performed in 4 days for $400 × 4 = $1,600 or in 2 days for $1,000 × 2 = $2,000. The normal duration is associated with the lowest-cost option for the activity. This value is used in a CPM analysis and in the preparation of the initial budget.

More formally, the normal duration of an activity is the duration that minimizes the direct cost. In some instances a schedule based on normal durations may produce high indirect costs, for example, when a project due date is given and a penalty is charged for completion after the due date. Even when the due date can initially be met by a normal schedule, uncertainty during the project execution may cause schedule overruns. The resultant penalties must be traded off with the cost of shortening the duration of some activities to minimize (or avoid completely) these late charges.

A similar situation occurs when a fixed overhead is charged for a project's duration. Rent for facilities would be such an example. In this case, management might consider shortening some activities to reduce the project's duration and hence save on indirect costs.

Crashing is the procedure whereby an activity's duration is shortened by adding resources and paying extra direct costs. A crashed program includes activities performed more

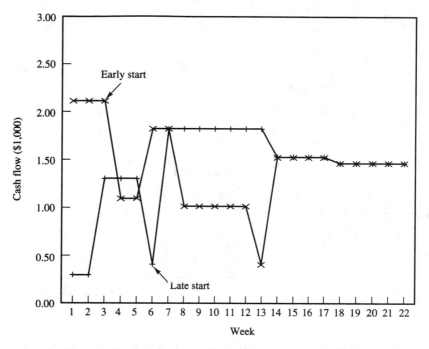

Figure 8-1 Cash flow for early start and late start schedules.

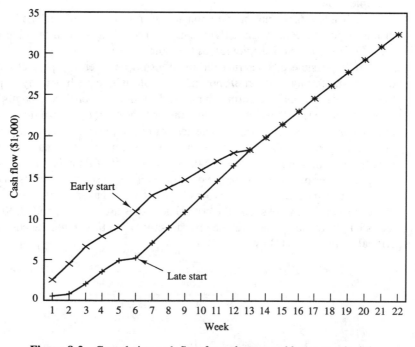

Figure 8-2 Cumulative cash flow for early start and late start schedules.

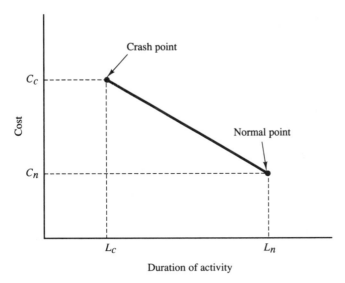

Figure 8-3 Typical time–cost trade-off curve.

quickly than they normally would be due to the allocation of additional resources. To plan a crashed program, management must decide which activities to crash and by how much. To illustrate this point, consider the crashing costs and durations list in Table 8-6 for the example project.

In Table 8-6, the normal duration and the normal cost of each activity are those used in the basic schedule. Each activity can be crashed at least once. Five of the activities (A,C,D,F,G) can be crashed twice, as the table shows.

It is possible to construct the relationship between the project's duration and its direct cost, starting with an all normal schedule where each activity is performed at its lowest direct cost and at a normal duration. To reduce the project's length, the critical path must be shortened. Thus, at each step, the critical path(s) are examined, and the activity that is least expensive to crash is selected for crashing on each critical path. These activities are crashed, and the process continues with the new critical paths being examined. In case the activities are not crashed in the same amount of time, it is possible to use the cost of crashing per period as a measure of attractiveness to select activities to crash.

To illustrate this heuristic process, consider the data in Table 8-6. The project's normal duration is 22 weeks, and the critical activities are A, C, F, and G. Reducing the project's length requires crashing one critical activity. At this stage, the cost of crashing each critical activity is as follows:

Activity	Cost to crash
A	$2,000
C	2,000
F	1,000
G	1,000

TABLE 8-6   DURATION AND COST FOR NORMAL AND CRASHED ACTIVITIES

Activity	Normal		Crashing activity the first time		Crashing activity a second time	
	Cost	Duration (weeks)	Additional cost	Duration (weeks)	Additional cost	Duration (weeks)
A	$1,500	5	$2,000	4	$1,000	3
B	3,000	3	2,000	2	—	—
C	3,300	8	2,000	7	1,000	6
D	4,200	7	2,000	6	2,000	5
E	5,700	7	1,000	6	—	—
F	6,100	4	1,000	3	2,000	2
G	7,200	5	1,000	4	1,000	3

Activities F and G are the least expensive to crash. In particular, the cost of activity F crashed from 4 to 3 weeks is $7,100, as illustrated in Table 8-7. The first column in the table represents the project with a normal duration (22 weeks). The second column represents the project after crashing F from 4 to 3 weeks; the project's duration is now 21 weeks and the crashed activity (F) is marked by an asterisk (*). The crashing procedure can continue until a 14-week span is obtained. At this point, two critical paths emerge: A–C–F–G, which lasts 14 weeks and contains no activities that can be crashed further; and B–D–F–G, which also lasts 14 weeks but contains two activities, B and D, which can be crashed for $2,000 and $3,000, respectively. Since the length of sequence A–C–F–G cannot be reduced, the project's minimum duration is 14 weeks. Table 8-7 summarizes the results.

Budgeting decisions are easier when the time–cost relationship for a project is known. The following example analyzes the trade-off between direct and indirect costs. Suppose that a fixed overhead of $500 per week is charged for a project's duration. Further, assume that the project is due in 18 weeks and that a penalty of $1,000 per week is imposed starting the nineteenth week. The budget problem in this case translates into a trade-off between the cost of crashing and the overhead plus penalty. Table 8-8 summarizes these cost components, accompanied by the project's total costs as a function of its length.

Based on Table 8-8, the minimum cost occurs at a project length of 19 weeks; that is, it is more economical to pay a penalty of $1,000 and an overhead of $500 for the nineteenth week than to crash activity F for $2,000. The total project cost may not be the only criterion for budget planning. If, for example, customer satisfaction depends on project completion within 18 weeks, the $500 savings should be evaluated against the customer goodwill that might be lost. Budgeting decisions should be based on this evaluation.

Figure 8-4 graphically depicts the different cost components and the total cost of the project as a function of its duration. The crashing problem can be modeled as either a linear or mixed-integer linear program, depending on whether a continuous trade-off exists between an activity's duration and cost (as assumed in Fig. 8-3), or whether only certain combinations are possible. The formulation presented below reflects the latter case, where

TABLE 8-7 CRASHING THE PROJECT (COST IN $1,000, DURATION IN WEEKS)

Activity	22 Weeks		21 Weeks		20 Weeks		19 Weeks		18 Weeks		17 Weeks		16 Weeks		15 Weeks		14 Weeks	
	Cost	Dur.	Cost	Dur.	Cost	Dur.	Cost	Dur.	Cost	Dur.	Cost	Dur.	Cost	Dur.	Cost	Dur.	Cost	Dur.
A	1.5	5	1.5	5	1.5	5	1.5	5	1.5	5	3.5[a]	4	4.5[a]	3	4.5	3	4.5	3
B	3.0	3	3.0	3	3.0	3	3.0	3	3.0	3	3.0	3	3.0	3	3.0	3	3.0	3
C	3.3	8	3.3	8	3.3	8	3.3	8	3.3	8	3.3	8	3.3	8	5.3[a]	7	6.3	6
D	4.2	7	4.2	7	4.2	7	4.2	7	4.2	7	4.2	7	4.2	7	4.2	7	6.2[a]	6
E	5.7	7	5.7	7	5.7	7	5.7	7	5.7	7	5.7	7	5.7	7	5.7	7	5.7	7
F	6.1	4	7.1[a]	3	7.1	3	7.1	3	9.1[a]	2	9.1	2	9.1	2	9.1	2	9.1	2
G	7.2	5	7.2	5	8.2[a]	4	9.2[a]	3	9.2	3	9.2	3	9.2	3	9.2	3	9.2	3
Total cost of activities	31		32		33		34		36		38		39		41		44	

[a]Crashed activity.

TABLE 8-8   PROJECT COSTS AS A FUNCTION OF ITS DURATION

Project length (weeks)	Direct cost of activities	Late completion penalty	Overhead cost	Total project cost
22	$31,000	$4,000	$11,000	$46,000
21	32,000	3,000	10,000	45,500
20	33,000	2,000	10,500	45,000
19	34,000	1,000	9,500	44,500
18	36,000	0	9,000	45,000
17	38,000	0	8,500	46,500
16	39,000	0	8,000	47,000
15	41,000	0	7,500	48,500
14	44,000	0	7,000	51,000

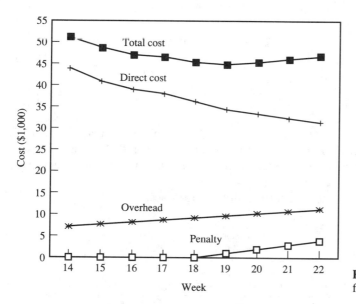

Figure 8-4   Example project cost as a function of its duration.

only a finite number of time–cost combinations are available for each activity. Assuming an activities-on-arrow network, the following notation is used:

i = index for events on the AOA network model; $i \in N = \{1,...,n\}$, where $i = 1$ is the unique "start" event that has no predecessors and $i = n$ is the unique "end" event that has no successors

(i,j) = project activity that starts at event i and ends at event j; $(i,j) \in A$

k = index for a particular time–cost combination

$K(ij)$ = index set of possible time–cost combinations for activity (i,j)

L_{ijk} = duration of activity (i,j) when it is performed at time–cost combination k

C_{ijk} = direct cost of activity (i,j) if performed at time–cost combination k

C_o = overhead cost per period of time

Decision variables

t_i — time event i takes place

y_{ijk} = (binary) equal to 1 if time–cost combination $k \in K(ij)$ is selected for activity (i,j); 0 otherwise

The problem of minimizing total cost is

$$\min \left[C_o t_n + \sum_{(i,j) \in A} \sum_{k \in K(ij)} C_{ijk} y_{ijk} \right] \tag{8-1a}$$

subject to

$$t_j - t_i \geq \sum_{k \in K(ij)} L_{ijk} y_{ijk} \qquad \text{for all } (i,j) \in A \tag{8-1b}$$

$$\sum_{k \in K(ij)} Y_{ijk} = 1 \qquad \text{for all } (i,j) \in A \tag{8-1c}$$

$$t_1 = 0, \quad t_i \geq 0, \qquad y_{ijk} = 0 \text{ or } 1 \tag{8-1d}$$

The objective function (8-1a) represents the project's total cost, which is composed of a direct and an indirect component. The first set of constraints (8-1b) maintains the precedence relations in the network; the second set (8-1c) ensures that each activity is performed at one of its time–cost combinations.

As an example of the model, consider the following project:

Activity (start event, end event)	Time–cost combination 1 Time (weeks)	Cost	Time–cost combination 2 Time (weeks)	Cost
(1,2)	5	$100	3	$150
(1,3)	4	70	3	100
(2,4)	4	200	3	300
(3,4)	6	500	3	900

The effect of the overhead cost per period (C_o) on the optimal schedule can be analyzed by solving (8-1) for the example and varying the value of C_o. The specific model for this example follows:

$$\min C_o t_4 + 100 y_{121} + 150 y_{122} + 70 y_{131} + 100 y_{132}$$
$$+ 200 y_{241} + 300 y_{242} + 500 y_{341} + 900 y_{342}$$

subject to

$$t_2 - t_1 \geq 5y_{121} + 3y_{122}$$
$$t_3 - t_1 \geq 4y_{131} + 3y_{132}$$
$$t_4 - t_2 \geq 4y_{241} + 3y_{242}$$
$$t_4 - t_3 \geq 6y_{341} + 3y_{342}$$

$$y_{121} + y_{122} = 1$$
$$y_{131} + y_{132} = 1$$
$$y_{241} + y_{242} = 1$$
$$y_{341} + y_{342} = 1$$

$$t_1 = 0, \quad t_i \geq 0, \quad i = 2,...,4;$$
$$y_{ijk} = 0 \text{ or } 1 \text{ for } (i,j) \in A = \{(1,2),(1,3),(2,4),(3,4)\}, \quad k = 1,2$$

Solving the model for different values of the overhead cost, C_o, gives the solutions presented in Table 8-9. The trade-off between the overhead cost and the cost of crashing activities is clear from these results. It is not justified to crash any activity whose overhead cost is below $20 per period. The first activity to be crashed is (1,3) and it is the only one for a wide range of C_o values. Only when the overhead per period is between $180 and $190 is it justified to crash all four activities.

TABLE 8-9 PARAMETRIC SOLUTION TO TIME–COST TRADE-OFF EXAMPLE (COST IN DOLLARS, DURATION IN WEEKS)

Overhead Cost C_o	Activity (1,2) Cost	Duration	Activity (1,3) Cost	Duration	Activity (2,4) Cost	Duration	Activity (3,4) Cost	Duration	Project Cost	Duration
10	100	5	70	4	200	4	500	6	970	10
20	100	5	70	4	200	4	500	6	1,070	10
30	100	5	100	3	200	4	500	6	1,170	9
40	100	5	100	3	200	4	500	6	1,260	9
50	100	5	100	3	200	4	500	6	1,350	9
60	100	5	100	3	200	4	500	6	1,440	9
70	100	5	100	3	200	4	500	6	1,530	9
80	100	5	100	3	200	4	500	6	1,620	9
90	100	5	100	3	200	4	500	6	1,710	9
100	100	5	100	3	200	4	500	6	1,800	9
110	100	5	100	3	200	4	500	6	1,890	9
120	100	5	100	3	200	4	500	6	1,980	9
130	100	5	100	3	200	4	500	6	2,070	9
140	100	5	100	3	200	4	500	6	2,160	9
150	100	5	100	3	200	4	500	6	2,250	9
160	100	5	100	3	200	4	500	6	2,340	9
170	100	5	100	3	200	4	500	6	2,430	9
180	100	5	100	3	200	4	500	6	2,520	9
190	150	3	100	3	300	3	900	3	2,590	6
200	150	3	100	3	300	3	900	3	2,650	6

Budgeting decisions are influenced by external factors such as the cost of money. If this cost is high, the net present value of the project may become an important criterion in budgeting. The reference list at the end of the chapter contains several papers dealing with this subject. The intuitive approach to project budgeting under the NPV criterion is to delay activities that require a capital outlay and to start, as early as possible, activities that generate cash. Since most activities lead to customer payment (i.e., cash generation) but require a capital outlay, a trade-off analysis is required to schedule these activities in the best possible way from the perspective of the budget.

8.4.3 PERT/Cost

PERT/Cost is a cost accounting technique for achieving realistic estimates of costs associated with activities and for providing an information system that allows good control of interim project costs. The federal government, which originally published the technique, uses it regularly for controlling cost overruns in governmental contracts. The need for an accounting system that is conceptually consistent with project management becomes evident when one considers that traditional systems group costs not by activities but by organizational areas, flows of materials, and time periods. PERT/Cost provides a means for structuring costs that is consistent with project management models.

With respect to planning and scheduling, PERT/Cost generates cumulative and average expenditures on a period-by-period basis for alternative schedules. This feature is useful in deciding when activities should be started between their earliest and latest starting dates. Additionally, the information system for PERT/Cost provides reports that allow project managers to control costs and evaluate performance with respect to the schedule. For example, suppose that at some point in time actual cost for an activity is 80% of budgeted cost at some point in time. One might think that cost is under control; however, the activity may be only 50% complete. To overcome this problem, a set of criteria called *cost/schedule control systems criteria* were developed, which we discuss in Chapter 11.

8.5 PRESENTING THE BUDGET

The project budget is a communications channel that must serve both internal and organizational planning and control needs. Two dimensions are used to measure the quality of a project's budget: the budget's ability (1) to advance organizational goals within the imposed constraints, and (2) to communicate the proposed plan to the project team and organization, and sometimes to subcontractors and the client.

The budget is easier to understand and use if it is presented clearly and concisely. Consider the following recommendations when preparing and presenting a project's budget:

1. Incorporate a schedule indicating the time that expenditures and revenues are expected to be realized.

2. Present the budget in quantitative, measurable units such as dollars or person-hours. If you use different units in the same budget, clearly define the conversion between units.

3. Make an effort to define milestones corresponding to the achievement of measurable goals. Typical milestones for R&D projects are system design review, preliminary design review, critical design review, and the passing of prototype performance tests. In contractor–client projects, the achievement of such milestones can serve as the basis for client payments. It is important to budget milestones according to the costs of activities leading up to them. In the example project (see Fig. 7-22), if event 3 is defined as a milestone representing the completion of activities A and B, its budget is based on the costs of activities A and B ($4,500). Assuming an early start schedule, these activities are scheduled to terminate 5 weeks after the project initiation. Assuming a $2,500 overhead (or $500 per week), the total payment of this milestone is likely to be above $4,500 and close to $7,000.

4. Use the budget as a baseline for progress monitoring and control. If a weekly progress report is required, plan the budget at the weekly level. On the other hand, if weekly progress reports are issued but the budget is prepared on a monthly basis, a meaningful comparison between planned progress and actual progress is possible only once every four progress reports. Similarly, break down the budget to enable a direct comparison with the progress reports. The cost breakdown used in preparing the budget should be the same as the breakdown used to collect and analyze data for both the project and organizational control systems.

5. The budget should translate short-range objectives into work orders, purchasing orders, and so on. This links the design and development phases to the production phase through the budgeting and work authorization processes.

6. Break down the budget by the organizational units responsible for its execution and the work content assigned to such units. For example, Table 8-10 itemizes the activities by assigned departments of the example project.

7. Whenever you use a specific standard in budgeting, reference it. For example, suppose that activity C is welding the pressure tank of a submarine and is budgeted at $3,300. This figure might have been derived from the company standard, which says that it costs $300 per inch to perform a weld. The estimated welding length is 11 inches. Such information should be referenced in the budget. By referencing the standard used, you can later trace any deviations in actual cost to the deviation's source (i.e., the cost per inch or the length of the welding) and, if necessary, update the standard.

TABLE 8-10 BREAKDOWN OF THE BUDGET BY ORGANIZATIONAL UNITS

Activity	Department 1	Department 2
A		$1,500
B		3,000
C	$3,300	
D	4,200	
E		5,700
F	6,100	
G	7,200	
Department total	$20,800	$10,200

8. Include five components in the short-term (operational) budget:

    (a) *Work packages of discrete effort.* Each work package defines the organizational element responsible for a task and the task's WBS element. Identifying the work package this way allows you to present the budget along WBS and OBS lines. Such an identification also serves as a baseline for a control system capable of tracing the sources of deviations between planned and actual progress, as explained in Chapter 11.

    (b) *Level of effort.* This category, which includes the cost of efforts related to more than one work package, occurs as the activities progress over time.

    (c) *Apportioned effort.* This category includes the cost of efforts based on a factor of a discrete effort (work package) as exemplified by such activities as inspection and quality control.

    (d) *Cost of material.* These costs include the WBS element for which it will be used and the OBS element that will use it.

    (e) *Other costs.* Costs such as those associated with subcontracting must be included.

9. Budget planners should try to define most of the project's effort in discrete terms as part of the work packages. These packages present units of work at levels where the work is performed and where the effort is assignable to a single organizational element.

10. Budget overhead costs for each organizational element with a clear definition of the procedures used for allocating these costs. One option is to include a management reserve in the long- and midrange budgets as a buffer against uncertainty. The level of management reserve depends on the amount of uncertainty involved in estimating the actual cost, timing, and technological maturity of the effort required. This reserve should be factored into the budget, once again in discrete terms, as work progresses and information becomes available.

11. Define a target budget at completion as the total budget costs plus management reserve and undistributed monies.

Following this list of recommendations will make it easier to prepare and use the budget. Much can be gained by presenting a financial plan quantitatively in terms that relate the required effort to cost, timing, responsible organizational elements, and project components. Nevertheless, each organization has its own guidelines for budget preparation and presentation, so the recommendations above should be used advisedly to supplement such guidelines in areas where they are unclear or incomplete.

8.6 PROJECT EXECUTION: CONSUMING THE BUDGET

During the project's production phase, three processes occur simultaneously:

1. The short-range budget is translated into work and purchasing authorizations. This process generates work orders, purchase orders, and contracts with suppliers and subcontractors. It requires a feedback system that facilitates a comparison between actual progress

and the original plans, and compares the actual cost of the effort performed with the budgeted cost. The exact structure of a feedback system used for project control depends on the project's structure and the organization's needs. This is explained in Chapter 11.

2. The tactical (midrange) budget is translated into a short-range budget through a rolling horizon mechanism. Cost estimates and schedules are accumulated into cost accounts as well as into apportioned effort and level of effort. This is a multistage process since the tactical budget for each period contains several short-range (operational) budget periods. Developing a new, realistic short-range budget requires detailed planning involving the integration of original project plans with reports on actual progress. The short-range budget should detail the midrange budget and in case of cost or schedule deviations, present a detailed plan for corrective action. Thus development of the operational budget is based on knowledge regarding the planned execution of activities and the project's actual status.

3. The long-range budget is gradually converted into the midrange budget. This process involves the distribution of accumulated funds, the allocation of management reserves to specific work packages, and the handling of engineering changes. Such changes are frequent in long projects. During the project execution phase, new market requirements (client needs) or new technological developments may call for modifications in the project's technological aspects. The configuration management system handles all these change requests. This system keeps track of change requests and the steps followed that lead to approval or rejection. An approved technological change may have both cost and schedule consequences. Thus the process of translating long-range budgets into midrange budgets should address all approved technological changes and their impact on the project.

Management reserve, designed to buffer uncertainty, should be consumed as soon as the results of tests and studies are available. Such results provide the basis for developing a detailed project plan translating management reserve budgets into work packages, thus reducing the level of uncertainty.

The budgeting process is ongoing. Long-term plans are translated into detailed short-term budgets, and short-term budgets are translated into work orders, purchase orders, and contracts with subcontractors and suppliers.

8.7 IMPORTANT POINTS IN THE BUDGETING PROCESS

The budgeting process provides an interface between organizational goals as perceived by top management, and the project managers' actions to achieve those goals. The techniques for budget preparation link the project's schedule, required resources, and net present value. The outcome provides an action framework for each organizational element. This framework integrates the budgets of the individual functional units and projects, as well as those of routine, ongoing activities into the total organizational budget.

Each project's budget is important in transforming goals into both plans and actions while providing guidelines for integration across the organizational and work breakdown structures. Management uses the budget as a communications channel to inform organizational elements of resource allocation decisions and the level of performance that is expected of them over time. This channel should be designed with rapid response in mind so that approved changes and deviations from the plan can be communicated quickly. The clearer the budget presentation, the easier it is for management at all levels to win over resistant elements in the organization. Thus not only is the quality of the budget important but also the planning that goes into its presentation.

TEAM PROJECT

Thermal Transfer Plant

Your proposed schedule has been reviewed by the contract department at TMS and has been given tentative approval. However, the vice-president of finance has requested a detailed budget for the project of designing and manufacturing the rotary combustor. The budget should tie the OBS and the WBS to the project's activities. Use the following format:

Week	Direct cost			Indirect cost		
	Labor	Material	Other	Labor	Material	Other
1						
2						
.						
.						
.						

For each line item in the budget identify its OBS and WBS relationship, and specify the *expected* cost and corresponding variance. Along with the budget, discuss the effect of an early start schedule and a late start schedule on cash flow, and explain why the selected schedule is the best from the cash flow point of view. (Is it?)

DISCUSSION QUESTIONS

1. Develop a budgeting procedure for a university. Explain the role of each management level together with its input and output.
2. Develop a budgeting procedure for a contractor who works on small housing projects.
3. Develop a budget for the project "getting an undergraduate degree." Explain your assumptions and your analysis.

4. Assume that you are in charge of developing your state's department of transportation budget. Write specific instructions to project managers in your department to facilitate a bottom-up budgeting process.

5. What kind of logic is used in the budgeting process of the federal government?

6. Give an example of a project in which a late start schedule is used because of budgeting and cash flow considerations.

7. Give a detailed example of an activity that can be performed in several modes. Describe each mode, the technology required, and the associated cost.

8. Develop a flowchart for a computerized project budgeting program. Explain the input and output of each element and the data processing required.

9. Identify two projects where the top-down budgeting approach would be most appropriate. What advantages does it provide?

10. Assume that you have crashed a project as much as possible but that the length of the critical path is still not acceptable. What other options are available?

11. Most computer codes developed to solve the crashing problem assume a linear relationship between the time and cost for an activity. This leads to a linear program. What does this assumption say in terms of resource allocation, and when might it be acceptable?

12. When a project leader tries to perform slack management, what difficulties might he or she encounter?

EXERCISES

8-1. Develop a budget for the project described in Table 8-11 assuming that the cost of each activity is linearly distributed over its duration.
   (a) Assuming an early start schedule.
   (b) Assuming a late start schedule.
   (c) Assuming that a "leveled budget" is desired (i.e., the same daily cost is desired for each day of the project).

TABLE 8-11

Activity	Duration (days)	Immediate predecessors	Cost
A	3	—	$ 3,000
B	4	—	2,000
C	3	—	6,000
D	2	C	2,000
E	1	B	1,000
F	5	A	10,000
G	2	B	4,000
H	3	B	9,000
I	11	C	11,000
J	3	D,E	3,000
K	1	F,G	1,000
L	4	K	2,000
M	4	J,H	8,000

8-2. In Exercise 8-1, assume that the activities can be crashed as shown in Table 8-12. Develop the functional relationship between the direct cost of this project and its duration.

8-3. In Exercise 8-2, assume that the overhead for the project is given by

$$\text{overhead} = 2,000 + \alpha \times 1,000 \text{ per day}$$

What is the project duration that minimizes its total cost if:
(a) $\alpha = 1$?
(b) $\alpha = 3$?

8-4. Assume that a continuous time–cost trade-off exists for each activity, as shown in Fig. 8-3. Write out the corresponding linear program for minimizing the total project cost, defining all notation used. What constraint would you add to ensure that the project is completed within T time periods?

8-5. For the project data given in Table 8-13, assuming an overhead of $350 per period, find the minimum cost schedule. What are the critical activities? How much total slack and free slack exist for the noncritical activities? Find the cost of the early start and late start schedules. Re-solve the problem with a deadline of 9 weeks using the linear program developed in Exercise 8-4.

8-6. In Exercise 7-16, given the data shown in Table 8-14 for the direct costs of the normal and crash durations, find the different minimum cost schedules between the normal and crash points.

8-7. Consider the cost–time estimates for the product development project of Exercise 7-19 as given in Table 8-15. Indirect cost is made up of two components: a fixed cost of $5,000 and a variable cost of $1,000 per week of elapsed time. Also, for each week that the project exceeds 17 weeks, an opportunity cost of $2,000 per week is assessed.
(a) Construct a table that enumerates the critical path and corresponding *direct cost* and *duration* for each possible funding strategy. The first two entries should be the "normal" and "all crash" strategies. Then either crash or compress (one week at a time) all activities on the critical path and calculate the corresponding direct cost and duration for the resulting strategies. Use the data in the table to construct a bar graph of completion time versus total cost (direct + indirect + opportunity).

TABLE 8-12

Activity	Normal time	Crash time	Cost of crashing per day
A	3	2	$1,000
B	4	2	500
C	3	2	500
D	2	1	1,000
E	1	1	—
F	5	4	500
G	2	1	1,500
H	3	2	1,000
I	11	8	1,500
J	3	1	1,000
K	1	1	—
L	4	3	1,000
M	4	4	—

TABLE 8-13

Activity	Immediate predecessors	Normal Duration	Normal Cost	Crash Duration	Crash Cost
A	—	4	$100	2	$300
B	—	3	200	1	200
C	A,B	2	50	1	100
D	A,B	3	100	2	300
E	A	4	150	1	400
F	C,D	4	250	1	100
G	D,E	2	300	1	200
H	F,G	3	200	2	100

TABLE 8-14

Project (a):

Activity (i,j)	Normal Duration	Normal Cost	Crash Duration	Crash Cost
(1,2)	5	$100	2	$200
(1,4)	2	50	1	80
(1,5)	2	150	1	180
(2,3)	7	200	5	250
(2,5)	5	20	2	40
(2,6)	4	20	2	40
(3,4)	3	60	1	80
(3,6)	10	30	6	60
(4,6)	5	10	2	20
(4,7)	9	70	5	90
(5,6)	4	100	1	130
(5,7)	3	140	1	160
(6,7)	3	200	1	240

Project (b):

Activity (i,j)	Normal Duration	Normal Cost	Crash Duration	Crash Cost
(1,2)	4	$100	1	$400
(1,3)	8	400	5	640
(1,4)	9	120	6	180
(1,6)	3	20	1	60
(2,3)	5	60	3	100
(2,5)	9	210	7	270
(3,4)	12	400	8	800
(3,7)	14	120	12	140
(4,5)	15	500	10	750
(4,7)	10	200	6	220
(5,6)	11	160	8	240
(5,7)	8	70	5	110
(6,7)	10	100	2	180

TABLE 8-15

	Time estimates (weeks)		Direct cost estimates ($1,000)	
Activity	Normal	Crash	Normal	Crash
A	3	1.0	3.5	10.0
B	1	0.5	1.2	2.0
C	5	3.0	9.0	18.0
D	1	0.7	1.0	2.0
E	6	3.0	20.0	50.0
F	1	0.5	2.2	3.0
G	2	1.0	4.0	9.0
H	8	6.0	100.0	150.0

(b) Construct the Gantt chart for the minimum total cost schedule.

(c) Construct a two-part schedule of direct costs (of the type illustrated in Figs. 7-10 and 7-11) based on the time schedule in part (b). Of the two, which schedule yields the lowest peak cost? Also, which of the two levels cost the most based on variance?

8-8. Develop a mathematical programming formulation for the problem of minimizing the total cost of completing the project discussed in Exercise 8-7. Use a commercial optimization package to find the solution.

8-9. Panmatics is undertaking a modernization program. The set of activities listed in Table 8-16 has been defined for refurbishing one of its wave-soldering machines.

(a) Find the critical path, total slacks, and free slacks.

(b) Find the probability of completion within 45 days.

(c) Find the minimum cost increase to reduce the expected project duration by 1 day.

(d) Find the minimum cost increase to reduce the expected project duration by 2 days.

(e) Find the minimum project duration and the expected cost increase.

8-10. Consider the project information given in Table 8-17.

(a) Calculate the project cost based just on the costs of the activities.

TABLE 8-16

			Crash date	
Task	$\overline{d}_{ij}$ (days)	s_{ij} (days)	Maximum possible compression (days)	Expediting cost per day ($)
A	6	2	0	—
B	2	0	1	50
C	12	3	2	80
D	8	1	2	175
E	7	2	1	100
F	16	4	0	—
G	23	2	1	100
H	25	5	3	300
I	4	1	1	1,000

TABLE 8-17

Activity	Immediate predecessors	Expected time (weeks)	Normal cost ($)	Expediting cost per week ($/week)	Minimum time (weeks)
A	—	3	3,000	1,500	2
B	—	6	7,200	1,000	4
C	A	2	2,000	2,000	1
D	A	7	7,000	2,000	3
E	C,B	1	4,000	—	1
F	B	3	3,000	1,500	2

   (b) Generate the weekly and cumulative cash flow charts, once for an early start schedule and once for a late start schedule.

   (c) Discuss the implications of the charts generated in part (b).

8-11. For the project described in Exercise 8-10:

   (a) Generate the time–cost chart.

   (b) What is the shortest completion time for the project, and what are the bottleneck activities prevent further time reduction?

8-12. A managerial fee of $1,400 per week is to be paid as long as the project in Exercise 8-10 has not been completed.

   (a) Calculate the optimum project duration.

   (b) You have been offered a bonus of $5,000 if you complete the project within 8 weeks. Will you make it? Explain.

8-13. Each activity in the project described in Exercise 8-10 has a duration variance of 1 week. For example, the expected time for activity A is 3 weeks, with a variance of 1 week. Assuming that the normal cost of each activity is to be used, discuss the possible impact of the activity variance on the project cash flow.

8-14. You have signed a contract to complete the project described in Exercise 8-10 within 10 weeks. The weekly managerial fee is $2,000.

   (a) Generate the schedule that will delay expenses to the last possible moment and indicate its associated cash flow.

   (b) Generate the cash flow requirement resulting from the objective to increase the probability that the project will be completed on schedule.

REFERENCES

Budgeting Process

BARD, J. F., "Coordination of a Multidivisional Firm through Two Levels of Management," *Omega*, Vol. 11, No. 5 (1983), pp. 457–465.

FIELDS, M. A., "Effect of the Learning Curve on the Capital Budgeting Process," *Managerial Finance*, Vol. 17, No. 2–3 (1991), pp. 29–41.

GRINOLD, R. C., "The Payment Scheduling Problem," *Naval Research Logistics Quarterly*, Vol. 19 (1972), pp. 123–136.

REGINLAND, J. L., and G. H. TRENIN, *Budgeting: Key to Planning and Control, Practical Guidelines for Managers*, American Management Association, New York, 1970.

TAVARES, L. V., "Stochastic Planning and Control of Program Budgeting: The Model Macao," in A. COELHO and L. V. TAVARES (editors), *OR Models on Microcomputers*, Elsevier Science Publishers, North-Holland, Amsterdam, 1986.

WILDAUSKY, A. B., *Budgeting: A Comparative Theory of Budgetary Processes*, Transaction Books, New Brunswick, NJ, 1986.

Time–Cost Trade-off Models

ELMAGHRABY, S. E., "The Determination of Optimum Activity Duration in Project Scheduling," *Journal of Industrial Engineering*, Vol. 19, No. 1 (1968), pp. 48–51.

ELMAGHRABY, S. E., and S. ARISAWA, "Optimal Time–Cost Trade-offs in GERT Networks," *Management Science*, Vol. 18, No. 11 (1972), pp. 589–599.

FALK, J. E., and J. L. HOROWITZ, "Critical Path Problems with Concave Cost–Time Curves," *Management Science*, Vol. 19, No. 4 (1974), pp. 446–455.

FULKERSON, D. R., "A Network Flow Computation for Project Cost Curve," *Management Science*, Vol. 7, No. 2 (1961), pp. 167–178.

GOYAL, S. K., "A Note on a Simple CPM Time–Cost Tradeoff Algorithm," *Management Science*, Vol. 21, No. 6 (1975), pp. 718–722.

LAMBERSON, L. R., and R. R. HOCKING, "Optimum Time Compression in Project Scheduling," *Management Science*, Vol. 16, No. 10 (1970), pp. B597–B606.

Cash Flow and Net Present Value Models

BEY, R. B., R. H. DOERSCH, and J. H. PATTERSON, "The Net Present Value Criterion: Its Impact on Project Scheduling," *Project Management Quarterly*, Vol. 12, No. 2 (1981), pp. 35–45.

DOERSCH, R. H., and J. H. PATTERSON, "Scheduling a Project to Maximize Its Present Value: A Zero One Programming Approach," *Management Science*, Vol. 23, No. 8 (1977), pp. 882–889.

ELMAGHRABY, S. E., and S. HERROELEN, "The Scheduling of Activities to Minimize the Net Present Value of Projects," *European Journal of Operational Research*, Vol. 49, No. 11 (1990), pp. 35–49.

RUSSEL, R. A., "Cash Flow in Networks," *Management Science*, Vol. 16, No. 5 (1970), pp. 357–373.

RUSSEL, R. A., "A Comparison of Heuristics for Scheduling Projects with Cash Flows and Resource Restrictions," *Management Science*, Vol. 32, No. 10 (1986), pp. 1291–1300.

SHTUB, A., "The Trade-off between the Net Present Cost of a Project and the Probability to Complete It on Schedule," *Journal of Operations Management*, Vol. 6, No. 4 (1987), pp. 461–470.

SMITH-DANIELS, D. E., and N. J. AQUILANO, "Using a Late-Start Resource-Constrained Project Schedule to Improve Project Net Present Value," *Decision Sciences*, Vol. 18 (1987), pp. 617–630.

SMITH-DANIELS, D.E., and V. L. SMITH-DANIELS, "Maximizing the Net Present Value of a Project Subject to Materials and Capital Constraints," *Journal of Operations Management*, Vol. 7, No. 1–2 (1987), pp. 33–45.

TAVAKOLI, A., "Construction Project Cash Flow Analysis," *Cost Engineering*, Vol. 30, No. 3 (1988), pp. 18–20.

9

Resource Management

9.1 EFFECT OF RESOURCES ON PROJECT PLANNING

So far we have assumed that in scheduling project activities the precedence relations among activities and the project budget are the sole constraints. Based on this assumption, each activity could start as soon as all its predecessors are completed (assuming end-to-start precedence relations and an adequate budget). When cash flow constraints are present, the slack of noncritical activities could be used to generate a feasible schedule, as discussed in Chapter 8. This type of analysis is based on the hidden assumption that there are enough resources available to permit any number of activities to be scheduled simultaneously. As we will see, this is rarely the case.

Resource planning is the process by which the project manager decides which resources to obtain, from what source, when to obtain them, and how to use them. Therefore, project resource planning is mainly concerned with the trade-off analysis between (1) the cost of alternative schedules designed to accommodate resources shortages, and (2) the cost of using alternative resources: for example, overtime to meet a schedule or subcontracting to accommodate a schedule change. This analysis may be subject to constraints on resource availability, budget allocations, and task deadlines.

An important function of the project manager is to monitor and control resource use and performance during project execution. If technical personnel are scarce, or if materials and equipment are in short supply, rescheduling becomes a top management priority. Shortages and uncertainty can wreak havoc on the best of plans. However, the efficient use of resources goes a long way in reducing both the cost and duration of a project at each stage of its life cycle.

Project resources are aggregated through the budget and expended over time. The relationship between the project budget and schedule was discussed in Chapter 8. Money is used to acquire the resources needed for the project, but in some financial organizations such as banks and insurance companies, money itself is a resource used for operations. This can be confusing, so to plan for and track resource use, some type of classification system is needed. This is taken up next.

9.2 CLASSIFICATION OF RESOURCES USED IN PROJECTS

Project resources can be classified in several ways. One approach is based on accounting principles, which distinguish between labor costs (human resources), material costs, and other "production" costs, such as subcontracting and borrowing. This classification scheme is very useful for budgeting and accounting. Its major drawbacks are that it does not specifically include the cost of the less tangible resources such as information (blueprints, databases), and it does not capture the main aspect of project resource management (i.e., the availability of resources).

A second approach is based on resource availability. Some resources are available at the same level every time period (e.g., a fixed workforce). These are *renewable* resources. A second class consists of resources that come in a lump sum at the beginning of the project and are used up over time. These are *depletable* resources such as material or computer time. A third class of resources is available in limited quantities each period. However, their total availability throughout the project is also circumscribed. These are called *doubly constrained* resources. The cash available for a project is a typical example of a doubly constrained resource.

A third classification scheme is similarly based on resource availability. The first class includes all "nonconstrained" resources—those available in unlimited quantities for a cost. A typical example is untrained labor or general-purpose equipment. The second class includes resources that are very expensive or impossible to obtain within the time span of the project. Special facilities such as a test range that is open only 4 hours a day, or technical experts who work on many projects, are two such examples. This class also includes resources of which a given quantity is available for the whole project, such as a rare type of material that has a long lead time. The quantity ordered at the beginning of the project must last throughout, due to its limited supply.

This scheme is characteristic of an ABC inventory management system. Resources of the first class (C category) are available in unlimited quantities and so do not require continuous monitoring. Nevertheless, they still might be expensive, so their efficient use will contribute to the cost-effectiveness of the project. Resources in the second class (A category) have high priority and should be monitored closely because shortages might significantly affect the project schedule.

In general, depletable resources, and those limited by periodic availability, should be considered individually during the planning process. This means that project schedules should be designed to ensure efficient use of nonconstrained resources, and that tight controls should be placed on the consumption of constrained resources.

In addition to availability considerations, the cost of resources should be weighed when developing project schedules. This is very important whenever activities can be performed by different sources. The combination of resources (often called the "mode") assigned to activities affects both the schedule and the cost of the project.

Quite often, it is not possible to allocate resources to activities accurately at the early stages of a project. This is because of the underlying uncertainty that initially shrouds resource requirements. Therefore, resource planning is a continuous process that takes place throughout the life cycle of the project.

In a multiproject environment the specific resource alternative selected also affects other ongoing projects. It is common wisdom to start the planning process by assuming that each activity is performed by the minimum cost resource alternative. To identify this alternative, the following points should be considered:

- The selection of resources should be designed for maximum flexibility so that resources not essential for one project can be used simultaneously on other projects. This flexibility can be achieved by buying general-purpose equipment and by broadly training employees.

- Up to a certain point, the more of a particular resource used, the less expensive it is per period of time (due to savings in setup cost, greater learning, and economies of scale).

- The marginal contribution of a resource decreases with usage. Frequently, when increasing the quantity of a resource type assigned to an activity, a point is reached where additional resources do not shorten the activity's duration. That is, inefficiencies and diminishing returns set in.

- Some resources are discrete. When this is the case, decreasing resource levels, necessarily in integer quantities, could result in a sharp decline in productivity and efficiency.

- Resources are organizational assets. Resource planning should take into consideration not only what is best for an individual project, but what is best for the organization as a whole.

- The organization has better control over its own resources. When the choice of acquiring or subcontracting for a resource exists, the degree of availability and control should be weighed against cost considerations.

The output of each resource is measured by its capacity, which is commonly defined in either one of two ways:

1. *Nominal capacity:* maximum output achieved under ideal conditions. The nominal capacity of equipment is usually contained in its technical manual. Nominal capacity of labor can be estimated with standard work measurement techniques commonly used by industrial engineers.

2. *Effective capacity:* maximum output taking into account the mixture of activities assigned, scheduling and sequencing constraints, maintenance aspects, the operating environment, and other resources used in combination.

Resource planning is relatively easy when a single resource is used in a single project. When the coordinate use of multiple resources is called for, planning and scheduling become more complicated, especially when dependencies exist among several projects. In some cases it is justified to use excessive levels of inexpensive resources to maximize the utilization of resources that are expensive or in limited supply.

The life cycle of a project affects its resource requirements. In the early stages, the focus is on design. Thus highly trained personnel such as system analysts, design engineers, and financial planners are needed. In subsequent stages, execution becomes dominant, and machines and material requirements increase. A graph of resource requirements as a function of time is called a *profile*. An example of labor and material profiles as a function of a project's life-cycle stages is presented in Fig. 9-1. Curve (a) depicts the

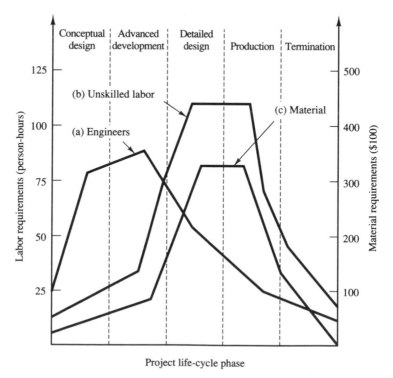

Figure 9-1   Typical resource requirement profiles.

requirements for engineers as a function of time. As can be seen, demand peaks during the advanced development phase of the project. Curve (b) displays the requirements for technicians. In this case, the maximum is reached during the detailed design and production phases. This is also true for material requirements, as shown in curve (c).

The general shape of the profiles depicted in Fig. 9-1 can be modified somewhat by careful planning and control. Slack management is one way to reshape resource requirements. Because it is always possible to start an activity within the range defined by its early and late start schedules, it may be possible to achieve higher resource utilization and lower costs by exploring different assignment patterns. In some projects, limited resource availability forces the delay of activities beyond their late start. When this happens, project delays are inevitable unless corrective action can be taken immediately.

9.3 RESOURCE LEVELING SUBJECT TO PROJECT DUE-DATE CONSTRAINTS

To discuss the relationship between resource requirements and the scheduling of activities, consider the example project that was introduced in Table 7-2. Assuming that only a single resource is used (unskilled labor) in the project, Table 9-1 lists the resource requirements for each of the seven activities.

The data in Table 9-1 are based on the assumption that performing an activity requires that the resource be used at a constant rate. Thus activity A requires 8 unskilled labor-days in each of its 5 weeks. When the usage rate is not constant, resource requirements should be specified for each time period (a week in our example).

The Gantt chart for the early start schedule is shown in Fig. 9-2a; the corresponding resource requirement profile is depicted in Fig. 9-2b. As can be seen, the early start schedule produces a high level of resource use at the early stages of the project. During the first 3 weeks there is a need for 17 labor-days each week. Assuming 5 working days per week, the requirement during the first 3 weeks is $17/5 = 3.4$ unskilled workers per day. The fractional component of demand can be met with overtime, second-shift, or part-time workers. The lowest resource requirements occur in week 13, where only 3 labor-days are needed. Thus the early start schedule generates a widely varying profile, with a high of 17 labor-days per week and low of 3 labor-days per week; the range is $17 - 3 = 14$.

TABLE 9-1   RESOURCE REQUIREMENTS FOR THE EXAMPLE PROJECT

Activity	Duration (weeks)	Required labor days per week	Total labor days required
A	5	8	40
B	3	4	12
C	8	3	24
D	7	2	14
E	7	5	35
F	4	9	36
G	5	7	35

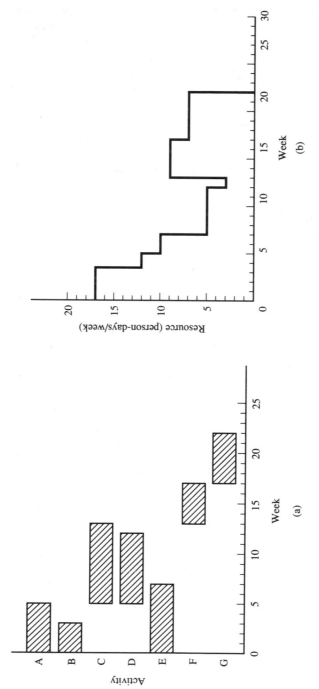

Figure 9-2 (a) Gantt chart and (b) resource profile for the early start schedule.

The Gantt chart and resource requirement profile associated with the late start schedule are illustrated in Fig. 9-3. Due to the effects that scheduling decisions have on resource requirements, there is a difference between the profiles associated with the late start and early start schedules. In the example, the late start schedule moves the maximum resource usage from weeks 1 through 3 to weeks 3 through 5. Furthermore, maximum usage is reduced from 17 labor-days per week to 12 labor-days per week, giving a range of $12 - 3 = 9$. It is important to note that the reduction in range while moving from the early start to the late start schedule is not necessarily uniform over the intermediate cases.

Resource leveling can be defined as the reallocation of total or free slack in activities to minimize fluctuations in the resource requirement profile. It is assumed that a more steady usage rate leads to lower resource costs. For labor, this assumption is based on the proposition that costs increase with the need to hire, fire, and train personnel. For materials, it is assumed that fluctuating consumption rates mean an increase in storage requirement (perhaps to accommodate the maximum expected inventory), and more effort invested in material planning and control.

Resource leveling can be performed in a variety of ways, some of which are described in the references listed at the end of the chapter. A generic resource-leveling procedure is illustrated next and used to solve the example project.

1. Calculate the *average* number of resource-days per period (e.g., week). In the example, a total of 196 resource-days or labor-days are required. Since the project duration is 22 weeks, $196/22 = 8.9$ or about 9 labor-days per week are required on the average.

2. With reference to the early start schedule and noncritical activities, gradually delay activities one at a time, starting with those activities that have the largest free slack. Check the emerging resource requirement profile after each delay. Select the schedule that minimizes resource fluctuations by generating daily resource requirements close to the calculated average.

Continuing with the example, we see from Table 7-5 that activity E has the largest free slack (6 weeks). The first step is to delay the start of E by 3 weeks until the end of activity B. This reduces resource requirements in weeks 1 through 3 by 5 units. The emerging resource profile is:

Week	1	2	3	4	5	6	7	8	9	10	11
Load	12	12	12	13	13	10	10	10	10	10	5

Week	12	13	14	15	16	17	18	19	20	21	22
Load	5	3	9	9	9	9	7	7	7	7	7

This profile has a maximum of 13 and a minimum of 3 labor-days per week. Since the maximum occurs in weeks 4 and 5 and activity E can be delayed further, consider a

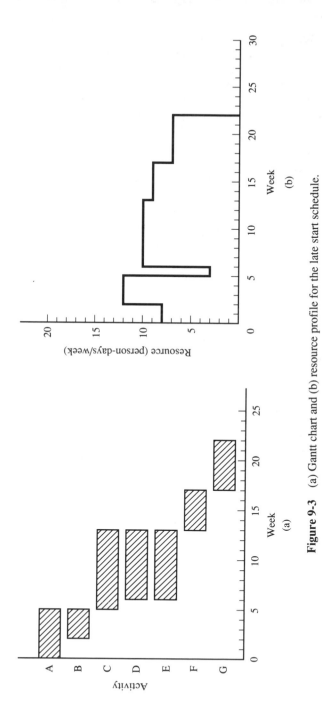

Figure 9-3 (a) Gantt chart and (b) resource profile for the late start schedule.

schedule where E starts after A is finished (after week 5). The resource requirements profile in this case is:

Week	1	2	3	4	5	6	7	8	9	10	11
Load	12	12	12	8	8	10	10	10	10	10	10

Week	12	13	14	15	16	17	18	19	20	21	22
Load	10	3	9	9	9	9	7	7	7	7	7

The maximum resource requirement is now 12 and occurs in weeks 1 through 3. The minimum is still 3, giving a range of $12 - 3 = 9$. The next candidate for adjustment is activity B, with a free slack of 2 weeks. However, delaying B by 1 or 2 weeks will only increase the load in weeks 4 and 5 from 8 to 12, yielding a net gain of zero. Therefore, we turn to the last activity with a positive free slack—activity D, which is scheduled to start at week 5. Delaying D by 1 week results in the following resource requirement profile:

Week	1	2	3	4	5	6	7	8	9	10	11
Load	12	12	12	8	8	8	10	10	10	10	10

Week	12	13	14	15	16	17	18	19	20	21	22
Load	10	5	9	9	9	9	7	7	7	7	7

The corresponding graph and Gantt chart are depicted in Fig. 9-4. Note that this profile has a range of $12 - 5 = 7$, which is smaller than that associated with any of the other candidates, including the early start and late start schedules. This is as far as we can go in minimizing fluctuations without causing a delay in the entire project.

For small projects, the foregoing procedure works well but cannot always be relied upon to find the optimal profile. To improve on the results, a similar procedure can be activated by starting with the late start schedule and checking the effect of moving activities with slack toward the start of the project. In some projects the objective may be to keep the maximum resource utilization below a certain ceiling rather than merely leveling the resources. If this objective cannot be met by rescheduling the critical activities, one or more of them would have to be expanded to reduce the daily resource requirements.

The analysis is more complicated when several types of resources are used, the number of activities is large, and several projects share the same resources. Sophisticated heuristic procedures have been developed for these cases. Most project management software packages employ such procedures for resource leveling.

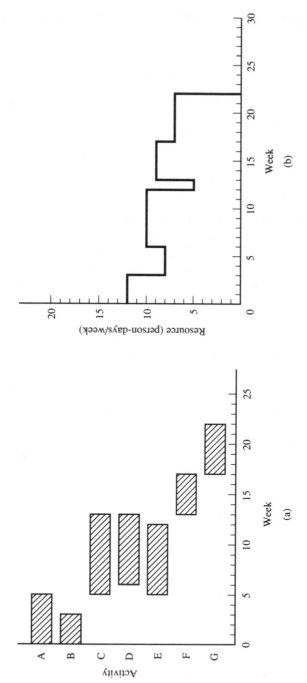

Figure 9-4 (a) Gantt chart and (b) leveled resource profile for the example project.

9.4 RESOURCE ALLOCATION SUBJECT TO RESOURCE AVAILABILITY CONSTRAINTS

Most projects are subject to resource availability constraints. This is common when resources are limited and good substitutes cannot be found. As a consequence, any delay or disruption in an activity may render the original project schedule infeasible. Cash flow difficulties may cause limited availability of all resource types: renewable, depletable, and nonconstrained. Some resources may be available in unlimited quantities, but due to cash flow problems, their use may have to be cut back in a specific project or over a specific period of time.

Under resource availability constraints, the project completion date calculated in the critical path analysis may not be achieved. This is the case when the resources required exceed the available resources in one or more time periods, and the slack of noncritical activities is not sufficient to solve the problem.

Of course, resource availability constraints are not always binding on the schedule. This can be illustrated with the example project. If 17 or more labor-days are available every week, then either an early start or a late start schedule can be employed to complete the project within 22 weeks. The leveled resource profile derived above requires at most 12 labor-days per week. Therefore, as long as this number is available, no delays will be experienced. If fewer resources are available in some weeks, however, the project may have to be extended beyond its earliest completion date. Activities A and B require a total of 12 labor-days per week when performed in parallel. To avoid an extension, despite a low resource availability, the project manager can try using one or more of the following techniques:

1. *Performing activities at a lower rate using available resource levels.* This technique is effective only if the duration of an activity can be extended by performing it with fewer resources. Consider activity B in the example. Assuming that only 11 labor-days are available each week and activity A (which is critical) is scheduled to be performed using 8 of those days, only 3 days a week are left for activity B. Since B requires a total of (3 weeks) $\times$ (4 labor-days per week) $=$ 12 labor-days of the resource, it may be possible to schedule B 3 days per week for 4 weeks. If this is not satisfactory, extending B to 5 weeks at 3 days per week may provide the solution.

This technique may not be applicable if a minimum level of resources is required each period (week) in which the activity is performed. Such a requirement might result from technological or safety considerations.

2. *Activity splitting.* It might be possible to split some activities into subactivities without significantly altering the original precedence relations. For example, consider splitting activity A into two subactivities: A_1, which is performed during weeks 1 and 2, and A_2, which is performed after a break of 4 weeks. It is possible then to complete the project within 22 weeks, using only 11 labor-days each week. This technique is attractive whenever an activity can be split, the setup time after the break is relatively short, and the activities succeeding the first subactivity can be performed in accordance with the original plan, that is, the second subactivity has no effect on the original precedence relations.

3. *Modifying the network.* Whenever the network is based solely on end-to-start precedence relations, the introduction of other types of precedence relations might help manage the constrained resources. For example, if an end-to-start connection on the critical path is replaced by a start-to-start connection, the delay caused by lack of resources may be eliminated. By considering the real precedence constraints among activities and modeling these constraints using all types of precedence relations defined in Chapter 7, some conflicts can be resolved.

4. *Use of alternative resources.* This option is available for some resources. Subcontractors or personnel agencies, for example, are possible sources of additional labor. However, the corresponding costs may be relatively high, so a cost overrun versus a schedule overrun trade-off analysis may be appropriate.

If these techniques cannot solve the problem, one or more activities will have to be delayed beyond their total slack, causing a delay in the completion of the project. To illustrate, consider the example project under a resource constraint of 11 labor-days per week. Because activity A requires 8 of these 11 days, activity B can start only when A finishes. The precedence relations force a delay of activity D—the successor of B, as well as F and G. The new schedule and resource profile are depicted in Fig. 9-5.

It is interesting to note that the maximum level of resources used in the new schedule is 10 labor-days. Thus in the example project, a reduction of the available resource level from 11 to 10 labor-days per week does not result in a change in the schedule. A further reduction to 9 labor-days each week will cause a further delay of the project since the concurrent scheduling of activities C, D, and E requires a total of 10 resource-days. A feasible schedule in this case and the accompanying resource profile are shown in Fig. 9-6.

It is impossible to reduce the resource level below 9 labor-days per week since activity F must be performed at that level. Table 9-2 summarizes the relationship between the resource level available and the project duration.

Resource utilization is defined as the proportion of time that a resource is used. For example, if 12 labor-days are available each week and the project duration is 22 weeks, a total of $12 \times 22 = 254$ resource days are available. Because only 196 days are used to perform all the project's activities, the utilization of this resource is $196/254 = 0.74$. Resource utilization is an important performance measure, particularly in a multiproject environment. Resource leveling and resource allocation techniques can be used to achieve high levels of utilization over all projects and resources. Matrix organizational structures help organizations achieve high utilization by taking advantage of pooled resources.

The analysis of multiple projects where several types of resources are used in each is a complicated scheduling problem. In most real-life applications the problem is solved with heuristics, using priority rules to make the allocations among activities. Some of these rules are discussed in the following section.

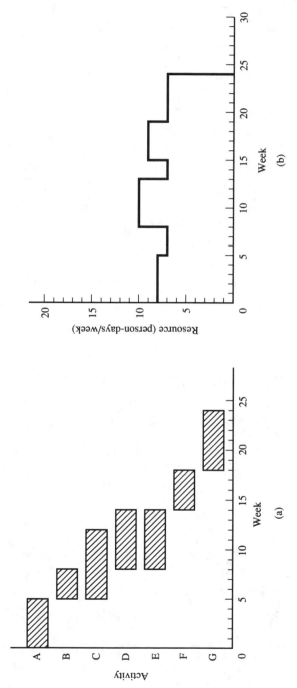

Figure 9-5 Scheduling under the 11 resource days/week constraint: (a) Gantt chart; (b) resource profile.

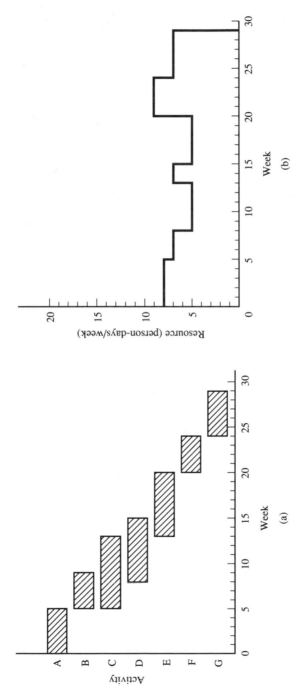

Figure 9-6 Scheduling under the 9 resource days/week constraint: (a) Gantt chart; (b) resource profile.

TABLE 9-2   IMPLICATIONS OF RESOURCE
AVAILABILITY

Resource availability (work days/week)	Project duration (weeks)	Resource utilization
12	22	0.74
11	24	0.74
10	24	0.82
9	29	0.75

9.5 PRIORITY RULES FOR RESOURCE ALLOCATION

A common approach to resource allocation is to begin with a simple critical path analysis assuming unlimited resource availability. Next, a check is made to see if the resultant schedule is infeasible. This would be the case whenever a resource requirement exceeds its availability. Infeasibilities are addressed one at a time, starting with the first activity in the precedence graph, and making a forward pass toward the last. A priority measure is calculated for each activity competing for a scarce resource. The activity with the lowest priority is delayed until sufficient resources are available. This procedure is used to resolve each infeasibility.

Examples of common priority rules are:

- Activity with the smallest slack
- Activity with minimum late finish time (as determined by critical path analysis)
- Activity that requires the greatest number of resource units (or the smallest number of resource units)
- Shorter activities (or longer activities)

A priority rule based on the late start of the activity and the project duration calculated by a critical path analysis is also possible. For example, define:

$$\text{CPT} = \text{earliest completion time of the project (based on critical path analysis)}$$
$$\text{LS}(i) = \text{late start of activity } i \text{ (based on critical path analysis)}$$
$$\text{PT}(i) = \text{priority of activity } i, \text{ where } \text{PT}(i) \equiv \text{CPT} - \text{LS}(i)$$

This rule gives high priority to activities that should start early in the project life cycle. In the case of multiple-project scheduling, the value of CPT is calculated for each project.

Next, we look at a priority rule based on each activity's resource requirements. Let

$$\text{AT}(i) = \text{duration of activity } i$$
$$\text{R}(i,k) = \text{level of resource } k \text{ required per unit of time for activity } i$$
$$\text{PR}(i,k) = \text{priority of activity } i \text{ with respect to resource type } k, \text{ where } \text{PR}(i,k) \equiv \text{AT}(i) \times \text{R}(i,k)$$

In this rule, high priority is given to the activity that requires the maximum use of resource k.

A rule based on aggregated resources is used when some activities require more than a single resource. Define

$$PSUMR(i) = \text{priority of activity } i \text{ based on all its required resources}$$
$$\equiv AT(i) \times \sum_k R(i,k)$$

To operationalize this rule, it is necessary to define a common resource unit such as a resource-day.

A weighted time-resource requirement priority rule can be fashioned from two of the previous rules; for example, let

$$\omega = \text{weight between 0 and 1}$$
$$PTR(i) = \text{weighted priority of activity } i, \text{ where } PTR(i) \equiv \omega PT(i) +$$
$$(1 - \omega) PSUMR(i)$$

By controlling the value of ω, emphasis can be shifted from the time dimension, $PT(i)$, to the resource dimension, $PSUMR(i)$.

Many of the priority rules above can be modified to take into account a variety of additional factors, including:

- Slack of the activity (total slack, free slack)
- Early start, late start, early finish, and late finish of the activity
- Duration of the activity
- Number of succeeding/preceding activities
- Length of the longest sequence of activities containing the activity
- Maximum resource requirement sequence of activities containing the activity

9.6 PROJECT MANAGEMENT BY CONSTRAINTS

The notion of bottlenecks used in job-shop and flow-shop scheduling can be extended to project resource management. Bottleneck resources delay activities due to their limited availability. In a multiresource project, bottlenecks whose capacity is relatively inexpensive to increase may cause low utilization of expensive or scarce resources. For example, a leased crane is an expensive resource that might be idle if an operator is not available. This is because to perform an activity, both resources are required simultaneously. From an economic point of view, it is preferable to maximize the utilization of the expensive resource at the risk of underutilizing the inexpensive one. Therefore, if the leased crane is available and needed 14 hours each day but an operator can only work between 8 and 10 hours a day, it would be advisable to hire two operators for a total of 16 hours a day, allowing for 2 hours of operator idle time.

Of course, idle resources signal inefficiencies that should be brought to the attention of management to see if they can be put to alternative use. Resource utilization is a key factor, sharing center stage with cost and on-time performance during project evaluation. Each of these factors figures prominently in the planning and review process.

9.7 MATHEMATICAL MODELS FOR RESOURCE ALLOCATION

Project scheduling under resource availability constraints has been the subject of much research (e.g., see Davis 1973, Patterson 1984, Tavares 1990). Most of the related studies assume that the scheduling objective is to complete the project as early as possible (the scheduling approach) or to maximize the net present value (minimize the net present cost) of the project (the budgeting approach). A model proposed by Patterson et al. (1989, 1990) can handle both objectives.

The following notation is used to describe the model; an AON network is assumed:

Indices and sets

d = index for number of time periods that an activity is in progress

j = index for project activity ($j = 1,2,..., J$)

k = index identifying resources that are available in a fixed quantity each period [i.e., renewable resources ($k = 1,2,...,K$)]

m = index for mode of an activity (i.e., the combination of resources assigned to perform a particular project activity)

t = index for time periods ($t = 1,2,...,T$)

P = set of all pairs of immediate predecessor relations; $(a,b) \in P$ denotes that activity a is an immediate predecessor of activity b

Parameters

C_{jmd} = cash flow of activity j if performed in mode m during its dth period in progress ($d = 1,2,...,D_{jm}$); if $C_{jmd} < 0$, there is a cash withdrawal; if $C_{jmd} > 0$, there is a cash inflow

C_{jmv}^* = nonnegative cash inflow v periods after the completion of activity j ($v \geq 1$) (completion of a payment milestone)

C_t = net cash position in period t; C_0 is the cash available at the start of the project

D_{jm} = duration of activity j if performed in mode m

$E_j(L_j)$ = earliest (latest) completion time for activity j determined from critical path analysis based on shortest (longest) completion time mode for activities in the network

J = unique terminal activity (may be a dummy) which has only one mode ($m = 1$; J also represents the number of activities in the project)

M_j = number of modes associated with activity j ($m = 1,2,..., M_j$)

R_{kt} = amount of resource k available in period t

r_{jmk} = per period amount of renewable resource k required to perform activity j at mode m

T = due date for project

α_t = single-payment present value discount factor for t periods at interest rate i;

$$\alpha_t = \left(\frac{1}{i+1}\right)^{t-1}$$

Decision variables

$$x_{jmt} = \begin{cases} 1 & \text{if activity } j \text{ at mode } m \text{ is completed in period } t \\ 0 & \text{otherwise} \end{cases}$$

The problem formulation for the case where project duration is minimized as follows:

$$\min \sum_{t=E_J}^{L_J} t x_{Jlt} \tag{9-1a}$$

subject to

$$\sum_{m=1}^{M_j} \sum_{t=E_j}^{L_j} x_{jmt} = 1, \qquad j = 1,...,J \tag{9-1b}$$

$$-\sum_{m=1}^{M_a} \sum_{t=E_a}^{L_a} t x_{amt} + \sum_{m=1}^{M_b} \sum_{t=E_b}^{L_b} (t - D_{bm}) x_{bmt} \geq 0 \qquad \text{for all } (a,b) \in P \tag{9-1c}$$

$$\sum_{j=1}^{J} \sum_{m=1}^{M_j} \sum_{q=t}^{t+D_{jm}-1} r_{jmk} x_{jmq} \leq R_{kt}, \qquad k = 1,...,K; \quad t = 1,...,T \tag{9-1d}$$

$$C_{t-1} + \sum_{j=1}^{J} \sum_{m=1}^{M_j} \sum_{q=t}^{t+D_{jm}-1} \left(C_{jm(D_{jm}+t-q)} x_{jmq} + \sum_{v=1}^{t-1} C^*_{jmv} x_{jm(t-v)} \right) = C_t, \qquad t = 1,...,T \tag{9-1e}$$

$$x_{jmt} = 0 \text{ or } 1 \qquad \text{for all } j, m, t \tag{9-1f}$$

In the model, the objective [eq. (9-1a)] of minimizing the duration of the project is achieved by scheduling the unique terminal activity J as early as possible subject to the following constraints:

- Ensuring that each activity will be completed in exactly one time period using only one activity mode [eq. (9-1b)]
- Maintaining the precedence relations among activities [eq. (9-1c)]
- Imposing resource restrictions [eq. (9-1d)]
- Ensuring that an activity mode is selected only if sufficient cash is available during each period of its duration [eq. (9-1e)]

To maximize the net present value of the project, the objective function (9-1a) is replaced by

$$\max \sum_{t=1}^{T} \alpha_t (C_t - C_{t-1}) + C_0 \qquad (9\text{-}2)$$

Problem (9-1) and its variant (9-2) are formally known as *zero–one integer programs*. In practice, it is not realistic to try to solve this type of problem to optimality when projects with several hundred activities are considered or when several projects that share the same resources are scheduled in parallel. Nevertheless, good solutions can be obtained with a variety of heuristics. For example, Patterson et al. (1990) have developed a backtracking algorithm that makes initial allocations and then tries to improve on the solution by shifting around resources, starting with the last node and working backward.

9.8 PROJECTS PERFORMED IN PARALLEL

The resource allocation and resource leveling techniques discussed so far are based on the assumption that each project undertaken by an organization is managed separately. This assumption is problematic if one or more of the following conditions exist:

- Technological dependency between projects
- Resource dependency between projects
- Budget dependency between projects

1. *Technological dependency.* Technological dependencies arise when precedence relations among projects are present. Consider, for example, an electronics firm involved in two projects: (1) the development of a new microprocessor, and (2) the development of a notebook computer. If a decision is made to use the new microprocessor in the notebook, the success of the computer project is dependent on the completion of the microprocessor. If this seems too risky, the new computer might be designed alternatively with an existing microprocessor as well as with the new one. This reduces the degree of dependency between the two original parallel projects.

2. *Resource dependency.* Resource dependencies occur when two or more projects compete for the same resources. In the previous example, an electrical engineer might be involved in both projects, so management must decide how best to allocate her time. One way to make this decision is to examine the the priority rules discussed earlier. Other factors that should be considered are technological dependencies, the due date of each project, and the economic consequences attending late completion.

3. *Budget dependency.* Budget dependencies exist when several projects compete for the same dollars or when the income from one group of projects is expected to cover the costs of some other group. In this case, coordination among the various projects is required.

The techniques developed for single-project scheduling can usually be used when dealing with parallel projects. A single network constructed by connecting all projects according to the precedence relations among them, or by assuming that all projects have the same start node and the same end node, may be used as a single project model for the multiproject situation. Once all projects are combined into a single network, the techniques developed for resource management in a single project are applicable.

TEAM PROJECT

Thermal Transfer Plant

With the approved schedule and budget, it is time to get the required resources to execute the rotary combustor project. Your team should submit a detailed plan indicating all the resources required for the project and the schedule for each resource. First, be sure to define the various resources (e.g., electrical engineers, mechanical engineers, material, a crane and operator) and their unit costs. Then for each resource, show how the respective costs fit the budget.

Assume that resources are available but that management's policy is to level resource use throughout the life cycle of each project. Develop a leveled resource plan. Explain the differences between your initial resource plan and the new one. In particular, discuss the benefits and costs associated with the leveled plan.

DISCUSSION QUESTIONS

1. Consider a project that you are familiar with and describe it briefly. Classify each resource used in that project by each classification scheme discussed in Section 9.2.

2. Discuss an example of a project that is not subject to resource constraints. Is this project subject to other constraints?

3. Discuss the importance of information as a resource in a technological project. Give an example where availability of information is a major constraint.

4. Select a classification scheme and classify the resource "information" required by a technologically advanced country trying to develop a manned space program.

5. Develop a flow diagram for a resource leveling procedure that can be translated into a computer program. What are your objectives, and what are the input, output, and data processing requirements?

6. Modify the flow diagram developed in Question 5 so that it can handle resource allocation problems.

7. Give an example of a bottleneck resource in a project. Under what conditions should this constraint be removed?

8. In the fall of 1990, a coalition force under the auspices of the United Nations moved massive amounts of equipment, matériel, and troops into the Persian Gulf area in response to Iraq's invasion of Kuwait. The operation was called Desert Shield. The military action that followed was called Desert Storm. Discuss the dependencies between these two projects.

9. What are the difficulties involved in leveling a schedule, particularly when the activities consume multiple resources?

10. How much does a project manager need to know about a scheduling or resource leveling computer program to use the output intelligently?

11. Why is the impact of scheduling and resource allocation generally more significant in multi-project organizations? How do large fluctuations in demand affect the situation?

12. What difficulties do you foresee in assigning technical personnel such as software engineers to multiple projects?

EXERCISES

9-1. The following project is performed with a single type of resource (labor), which is assumed to be available in unlimited quantities. The resource usage rate is constant throughout the duration of each activity. Thus if the duration of an activity is 5 days and it requires 60 hours of the resource, $60 \div 5 = 12$ hours of the resource are required each day the activity is performed. The project data are shown in Table 9-3. Develop a schedule that minimizes resource fluctuations.

9-2. Assume that daily resource availability is 2 hours less than the daily resource requirement indicated by the schedule derived in Exercise 9-1.

(a) Use two different priority rules to allocate the available resources to activities.

(b) Comment on the performance of the rules selected.

TABLE 9-3

Activity	Duration (days)	Immediate predecessors	Resource requirements (hours)
A	3	—	12
B	4	—	16
C	3	—	9
D	2	C	10
E	1	B	6
F	5	A	15
G	2	B	16
H	3	B	12
I	11	C	44
J	3	D,E	30
K	1	F,G	10
L	4	K	16
M	4	J,H	8

9-3. Each activity in a project can be performed by two different resource combinations (Table 9-4). Assume that the usage rate of each resource is constant throughout the duration of each activity. Now find a schedule that minimizes the duration of the project. Resources I and II are both available at a level of 12 hours each day.

9-4. Find a schedule for Exercise 9-3 that minimizes the cost of the project assuming that resource I costs $10/hour, resource II costs $15/hour, and there is an overhead of $150/day for the project.

9-5. Develop a resource plan and a schedule for the project "cleaning and resupplying a passenger plane between flight legs." Which resource is the bottleneck?

9-6. The precedence relations and crew size required to complete a project are given in Table 9-5. For example, activity E, which comes after activity C, requires 10 weeks for its completion by a crew of six people.
   (a) Construct an early start Gantt chart and identify the critical path.
   (b) Calculate and chart the labor profile required to complete the project for both an early start and a late start schedule.
   (c) Level the required labor as much as possible with the goal of completing the project within the time period specified in part (a).

TABLE 9-4

	Activity					
	A	B	C	D	E	F
Immediate predecessors	—	A	A	—	B,C	D,E
Mode 1						
Duration (days)	2	3	5	3	2	1
Resource I required (hours/activity)	0	9	10	6	8	4
Resource II required (hours/activity)	5	6	5	9	6	3
Mode 2						
Duration (days)	1	2	4	2	1	1
Resource I required (hours/activity)	12	12	8	6	9	4
Resource II required (hours/activity)	7	8	16	12	5	3

TABLE 9-5

Activity	Immediate predecessors	Time (weeks)	Crew size
A	—	4	4
B	A	2	5
C	A	6	3
D	B	3	7
E	C	10	6
F	—	2	5
G	D	5	6
H	F	7	2
I	D,E,G	1	8
J	H	10	2

9-7. (a) Referring to Exercise 9-6, assume that 10 persons are assigned to work on the project until it is finished. In light of the following assumptions, schedule the project and calculate labor utilization.

   1. No activities are allowed to be interrupted.
   2. The crew size that performs an activity cannot be reduced, but it is possible to increase the project's completion time.
   3. It is impossible to change the network.

   (b) Repeat part (a) now assuming that you can reduce the crew size and increase its duration (the number of person-weeks required for each activity is constant).

   (c) Repeat part (b) now assuming that you may interrupt each activity before it is completed and reschedule the remaining tasks at a later time.

9-8. (a) Assuming that the weekly labor cost per employee is $1,200 and that the fringe benefit rate is 25%, determine the cumulative cash flow requirement for the project described in Exercise 9-6 (1) for an early start schedule, and (2) for a latest start schedule.

   (b) What if the allocated budget is below the late start cash flow line?

   (c) What if the allocated budget is above the early start cash flow line?

9-9. The required labor profile for Exercise 9-6 is not of constant rate but resembles a symmetric trapezoid, with the peak lasting 1 week. As an example, consider activity G. Since a crew of six must work for a period of 5 weeks to complete this activity, a total of 30 person-weeks is required. To calculate the labor requirement during the peak period, assuming a trapezoid profile, one should substitute the proper values into the following equation:

$$\text{lbrq} = \text{peak} \times \left(\frac{\text{dur}}{2} + 0.5 \right)$$

where lbrq = total labor required to perform the activity
       peak = peak labor required during the 1-week peak time
       dur = activity duration

Solving the equation for activity G, we obtain

$$\text{peak} = \frac{\text{lbrq}}{(\text{dur}/2) + 0.5} = \frac{30}{5/2 + 0.5} = 10$$

That is, during the 1-week peak period, there is need for a crew of 10 employees. Moreover, the required labor profile for the first 2 weeks is linear starting from zero and ending at 10. The labor profile for the last 2 weeks is in the opposite direction; it starts at 10 and ends at zero at the end of the fifth week. Assuming a symmetric trapezoid profile for each activity, generate an early start resource profile for the project.

9-10. A trapezoid profile is a common shape used to describe labor requirements over time. Assuming that the permanent crew size is equal to the peak requirement, develop a model to calculate the crew utilization for an activity as a function of the peak duration. In so doing, assume that each activity i should be completed within a prespecified duration, say D_i days, and requires L_i labor-days.

9-11. A second project, identical to the one described in Exercise 9-6, is planned to start 1 week after the first. That is, the company intends to work on the two projects at the same time.

   (a) Generate the early start resource profile for the two projects.

   (b) Schedule the two projects so that the required labor profile will be as level as possible.

   (c) Discuss the significance of the differences observed in the schedules found in parts (a) and (b).

9-12. The following data concern an activity that has to be performed as part of a project:

> Expected duration (days) 10
> Standard deviation of the duration 2
> Expected labor-days 30
> Standard deviation of labor-days 3

(a) What is the probability that completing the activity on time will require at least a 10% addition to the expected labor-days?

(b) A crew of three workers is assigned to this activity. What is the probability that it will be completed in less than 11 days? State your assumptions.

9-13. Suppose that in Exercise 7-16 personnel requirements are specified for the various activities in projects (a) and (b) as shown in Table 9-6.

(a) Draw the early start Gantt chart for projects (a) and (b) and plot the required number of workers as a function of time.

(b) Level the resources for projects (a) and (b) as much as possible without extending their durations. Plot the corresponding manpower requirements over time.

9-14. Table 9-7 gives the results of a critical path analysis; Table 9-8 lists worker requirements for each of the project's activities.

(a) Draw the precedence graph for the project.

(b) Draw the Gantt charts for the early and late start schedules. What is the maximum number of workers required?

(c) Draw the resource requirement profiles for the early and late start schedules.

TABLE 9-6

Project (a):

Activity (i,j)	Number of workers	Activity (i,j)	Number of workers
(1,2)	5	(3,6)	9
(1,4)	4	(4,6)	1
(1,5)	3	(4,7)	10
(2,3)	1	(5,6)	4
(2,5)	2	(5,7)	5
(2,6)	3	(6,7)	2
(3,4)	7		

Project (b):

Activity (i,j)	Number of workers	Activity (i,j)	Number of workers
(1,2)	1	(3,7)	9
(1,3)	2	(4,5)	8
(1,4)	5	(4,7)	7
(1,6)	3	(5,6)	2
(2,3)	1	(5,7)	5
(2,5)	4	(6,7)	3
(3,4)	10		

TABLE 9-7

Activity, (i,j)	Duration, L_{ij}	Earliest Start, ES_{ij}	Earliest Finish, EF_{ij}	Latest Start, LS_{ij}	Latest Finish, LF_{ij}	Total slack, TS_{ij}	Free slack, FS_{ij}
(0,1)	2	0	2	2	4	2	0
(0,2)	3	0	3	0	3	0	0
(1,3)	2	2	4	4	6	2	2
(2,3)	3	3	6	3	6	0	0
(2,4)	2	3	5	4	6	1	1
(3,4)	0	6	6	6	6	0	0
(3,5)	3	6	9	10	13	4	4
(3,6)	2	6	8	17	19	11	11
(4,5)	7	6	13	6	13	0	0
(4,6)	5	6	11	14	19	8	8
(5,6)	6	13	19	13	19	0	0

TABLE 9-8

Activity	Number of workers	Activity	Number of workers
(0,1)	0	(3,5)	2
(0,2)	5	(3,6)	1
(1,3)	0	(4,5)	2
(2,3)	7	(4,6)	5
(2,4)	3	(5,6)	6

(d) Try to level the resource requirements (workers needed) as much as possible by applying the leveling procedure discussed in the text. [Note that activities (0,1) and (1,3) require no manual labor, which is indicated by assigning zero workers to each activity. As a result, the scheduling of (0,1) and (1,3) can be made independent of the resource leveling procedure.]

(e) Suppose that activities (0,1) and (1,3) require eight and two workers, respectively. Perform resource leveling and redraw the Gantt chart and profile graph.

9-15. A project has 11 activities that can be accomplished either by one person working alone or by several persons working together. The activities, precedence constraints, and time estimates are given in Table 9-9. Suppose that you have up to five people that can be assigned on any given day. A person must work full days on each activity, but the number of persons working on an activity can vary from day to day.

(a) Prepare an AOA network diagram, calculate the critical path, total slacks, and free slacks assuming that one person (independently) is working on each task.

(b) Prepare an early start Gantt chart.

(c) Prepare a daily assignment sheet for personnel with the goal of finishing the project in the minimum amount of time.

(d) Prepare a daily assignment sheet to "best" balance the workforce assigned to the project.

(e) How many days could the project be compressed if unlimited personnel resources were available?

TABLE 9-9

Activity	Immediate predecessors	Person-days required
A	—	10
B	A	8
C	A	5
D	B	6
E	D	8
F	C	7
G	E,F	4
H	F	2
I	F	3
J	H,I	3
K	J,G	2

REFERENCES

Resource Allocation and Leveling

BOCTOR, F. F., "Some Efficient Multi-Heuristic Procedures for Resource-Constrained Project Scheduling," *European Journal of Operational Research*, Vol. 49 (1990), pp. 3–13.

CHRISTOFIDES, N., R. ALVAREX-VALDES, and J. M. TAMARIT, "Project Scheduling with Resource Constraints: A Branch and Bound Approach," *European Journal of Operational Research*, Vol. 29 (1987), pp. 262–273.

DAVIS, E. W., "Project Scheduling under Resource Constraints: Historical Review and Categorization of Procedures," *AIIE Transactions*, Vol. 5, No. 4 (1973), pp. 297–313.

DAVIS, E. W., "Networks: Resource Allocation," *Industrial Engineering*, Vol. 6, No. 4 (1974), pp. 22–32.

DAVIS, E. W., "Project Network Summary Measures Constrained-Resource Scheduling," *AIIE Transactions*, Vol. 7, No. 2 (1975), pp. 132–142.

DREXEL, A., "Scheduling of Project Networks by Job Assignment," *Management Science*, Vol. 37, No. 12 (1991), pp. 1590–1602.

HERROELEN, W. S., "Resource Constrained Project Scheduling: The State of the Art," *Operations Research Quarterly*, Vol. 23, No. 3 (1972), pp. 261–275.

KHATTAB, M., and F. CHOOBINEH, "A New Heuristic for Project Scheduling with a Single Resource Constraint," *Computers and Industrial Engineering*, Vol. 20, No. 3 (1991), pp. 381–387.

PATTERSON, J. H., "A Comparison of Exact Approaches for Solving the Multiple Constrained Resource, Project Scheduling Problem," *Management Science*, Vol. 30, No. 7 (1984), pp. 854–867.

PATTERSON, J. H., and W. D. HUBER, "A Horizon-Varying, Zero–One Approach to Project Scheduling," *Management Science*, Vol. 20, No. 6 (1974), pp. 990–998.

PATTERSON, J. H., and G. W. ROTH, "Scheduling a Project under Multiple Resource Constraints: A Zero–One Programming Approach," *AIIE Transactions*, Vol. 8, No. 4 (1976), pp. 449–455.

PATTERSON, J. H., R. SLOWINSKI, F. B. TALBOT, and J. WEGLARZ, "An Algorithm for a General Class of Precedence and Resource Constrained Scheduling Problems," in R. SLOWINSKI and J. WEGLARZ (editors), *Advances in Project Scheduling*, Elsevier, Amsterdam, 1989, pp. 3–28.

PATTERSON, J. H., F. B. TALBOT, R. SLOWINSKI, and J. WEGLARZ, "Computational Experience with a Backtracking Algorithm for Solving a General Class of Precedence and Resource-Constrained Scheduling Problems," *European Journal of Operational Research*, Vol. 49 (1990), pp. 68–79.

PRITSKER, A. A. B., L. J. WATTERS, and P. M. WOLFE, "Multiproject Scheduling with Limited Resources: A Zero–One Programming Approach," *Management Science*, Vol. 16, No. 1 (1969), pp. 93–108.

RUSSELL, R. A., "A Comparison of Heuristics for Scheduling Projects with Cash Flows and Resources Restrictions," *Management Science*, Vol. 32, No. 10 (1986), pp. 1291–1300.

SHTUB, A., "The Integration of CPM and Material Management in Project Management," *Construction Management and Economics*, Vol. 6 (1988) pp. 261–272.

SLOWINSKI, R., "Two Approaches to the Problem of Resource Allocation among Project Activities: A Comparative Study," *Journal of the Operational Research Society*, Vol. 31, No. 8 (1980), pp. 711–723.

SLOWINSKI, R., "Multiobjective Network Scheduling with Efficient Use of Renewable and Non-renewable Resources," *European Journal of Operational Research*, Vol. 7, No. 3 (1981), pp. 265–273.

SLOWINSKI, R., and J. WEGLARZ (editors), *Advances in Project Scheduling*, Elsevier, Amsterdam, 1989.

STINSON, J. B., E. W. DAVIS, and B. M. KHUMAWALA, "Multiple Resource-Constrained Scheduling Using Branch and Bound," *AIIE Transactions*, Vol. 10, No. 3 (1978), pp. 252–259.

TALBOT, F. B., and J. H. PATTERSON, "An Efficient Integer Programming Algorithm with Network Cuts for Solving Resource-Constrained Scheduling Problems," *Management Science*, Vol. 24, No. 11 (1978), pp. 1163–1174.

TAVARES, V. L., "A Multi-stage Non-deterministic Model for Project Scheduling under Resources Constraints," *European Journal of Operational Research*, Vol. 49 (1990), pp. 92–101.

ULUSOY, G., and L. OZDAMAR, "Heuristic Performance and Network/Resource Characteristics in Resource-Constrained Project Scheduling," *Journal of the Operational Research Society*, Vol. 40, No. 12 (1989), pp. 1145–1152.

WIEST, J. D., "A Heuristic Model for Scheduling Large Projects with Limited Resources," *Management Science*, Vol. 13, No. 6 (1967), pp. B359–B377.

Multiple Projects

DEAN, B. V. (editor), *Project Management: Methods and Studies*, Elsevier Science Publishers, North-Holland, Amsterdam, 1985.

DEAN, B. V., D. R. DENZLER, and J. J. WATKINS, "Multiproject Staff Scheduling with Variable Resource Constraints," *IEEE Transactions on Engineering Management*, Vol. EM-39, No. 1 (1992), pp. 59–72.

DECKRO, R. F., E. P. WINKOFSKY, J. E. HEBERT, and R. GAGNON, "A Decomposition Approach to Multiproject Scheduling," *European Journal of Operational Research*, Vol. 51, No. 1 (1991), pp. 110–118.

FENDLEY, L. G., "Toward the Development of a Complete Multiproject Scheduling System," *Journal of Industrial Engineering*, Vol. 19, No. 10 (1968), pp. 505–515.

KIM, S. O., and M. J. SCHNIEDERJANS, "Heuristics Framework for the Resource Constrained Multiproject Scheduling Problem," *Computers & Operations Research*, Vol. 16, No. 6 (1989), pp. 541–556.

KURTULUS, I. S., and E. W. DAVIS, "Multi-Project Scheduling: Categorization of Heuristic Rules Performance," *Management Science*, Vol. 28, No. 2 (1982), pp. 161–172.

KURTULUS, I. S. and S. C. NARULA, "Multi-Project Scheduling: Analysis of Project Performance," *IIE Transactions*, Vol. 17, No. 1 (1985), pp. 58–66.

LEVY, F. K., G. L. THOMPSON, and J. D. WIEST, "Multi-ship, Multi-shop, Workload Smoothing Program," *Naval Research Logistics Quarterly*, Vol. 9, No. 1 (1962), pp. 37–44.

PRITSKER, A. A. B., L. J. WATTERS, and P. M. WOLFE, "Multi-Project Scheduling with Limited Resources: A Zero–One Programming Approach," *Management Science*, Vol. 16, No. 1 (1969), pp. 93–108.

WEGLARZ, J., "Project Scheduling with Continuously-Divisible Double Constrained Resources," *Management Science*, Vol. 27, No. 9 (1981), pp. 1040–1053.

APPENDIX 9A

Estimating Peak Resource Requirements

Overall resource requirements are estimated in the initial stage of a project's life cycle, even before there is a commitment to undertake the project. When responding to an RFP or pitching a new idea to management, rough estimates must be included in the presentation. After the project has been approved, there is a need for more accurate planning. At this point a usage profile is developed for each resource. For example, an aggregate estimate may specify that a certain design activity or work package requires 50 person-months over a 10-month period. With this information, the usage profile over the activity's duration can be developed. If one assumes a constant rate of depletion, there is a need for a permanent crew of 50/10 = 5 workers. However, constant usage is rarely the case. Labor requirements typically increase gradually until they hit a peak, remain at the peak for some time, and then decrease back to zero. This pattern is represented by the trapezoid in Fig. 9A-1.

Continuing with the example, assume that the manager responsible for the foregoing activity estimates that it will take 3 months to reach the peak usage rate, and that the

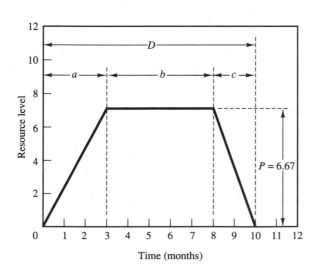

Figure 9A-1 Required labor profile with a trapezoidal pattern.

decline back to zero will take 2 months. The total labor required is equal to the area under the trapezoid; that is,

$$\text{TOT} = \frac{P \times a}{2} + P \times b + \frac{P \times c}{2}$$

where TOT = total labor (person-months) required
          P = peak person-months per month
          a = time taken to reach peak usage
          b = length of time resource level is at its peak
          c = time taken to reduce the resource level from its peak back to zero
Rearranging terms gives

$$P = \frac{\text{TOT}}{(a/2) + b + (c/2)}$$

Thus, knowing the total labor required for an activity, TOT, the duration of the activity, D, and the shape of the labor profile, it is possible to estimate peak labor requirements. To demonstrate the calculations for the trapezoidal case, let TOT = 50 person-months, D = 10 months, a = 3 months, and c = 2 months. We can now determine b, the length of the peak period, as follows:

$$b = D - (a + c) = 10 - (3 + 2) = 5 \text{ months}$$

so

$$P = \frac{50}{(3/2) + 5 + (2/2)} = 6.67 \text{ workers/month}$$

That is, during the five peak months, there will be a need for 6.67 workers every month.

10

Life-Cycle Costing

10.1 NEED FOR LCC ANALYSIS

Life-cycle cost (LCC) is the total cost of ownership of a product, structure, or system over its useful life. For products purchased off the shelf, the major factors are the cost of acquisition, operation, service, and disposal. For products or systems that are not available for immediate purchase, it may be necessary to include the costs associated with conceptual analysis, feasibility studies, development and design, logistics support analysis, manufacturing, and testing.

In discussing the life-cycle cost of a system or product versus a project, a distinction is often made between the various phases of the two. The main difference is that the project usually terminates when the system or product enters its operational life. The life cycle of the system or product, however, may continue far beyond that point. In Chapter 1 we introduced the five life-cycle phases of a project. Here we introduce the five life-cycle phases of a system or product.

1. Conceptual design phase
2. Advanced development and detailed design phase
3. Production phase
4. Project termination and system operation and maintenance phase
5. System divestment phase

The need for life-cycle costing arises because decisions made during the early stages of a project inevitably have an impact on future outlays. This need was recognized in the mid-1960s by the Logistics Management Institute, which issued a report stating that "the use of predicted logistics costs, despite their uncertainty, is preferable to the traditional practice of ignoring logistic's costs because the absolute accuracy of their quantitative values cannot be assured in advance."

An LCC analysis is designed to help managers identify and evaluate the economic consequences of their decisions. In 1978, the Massachusetts Institute of Technology Center for Policy Alternatives published one of the first studies on life-cycle cost estimates. The focus was on appliances; some of the estimates are summarized in Table 10-1. As can be seen, the cost of acquisition was between 40.9 and 60.2% of the whole. Nevertheless, the decisions made at the acquisition stage affect 100% of the life-cycle cost.

The MIT research demonstrated the importance of costs that are incurred during the operational stage of a system or product. This led the principal investigators to propose the establishment of consumer LCC data banks. Today, information on the operational costs of appliances such as energy consumption of refrigerators is posted on the units in the retail outlets. Similarly, the Environmental Protection Agency makes data on gasoline mileage of passenger cars readily available to the public.

A parallel situation exists for purchased commodities, as well as for research, development, and construction projects where decisions made in the early stages have a significant impact on the entire life-cycle cost. Engineering projects where a new system or product is being designed, developed, manufactured, and tested may span years, as in the case of a new automobile, or over a decade in the case of a nuclear power plant. New product development takes anywhere from several months to several years. In lengthy processes of this type, decisions made at the outset may have substantial, long-term effects that are frequently difficult to analyze. The trade-off between current objectives and long-term consequences of each decision is therefore a strategic aspect of project management that should be integrated into the project management system.

A typical example of a decision that has a long-term effect deals with the selection of components and parts for a new system at the advanced development and detailed design phase. Often, the short-term cost of manufacturing can be reduced by selecting less expensive components and parts at the expense of a higher probability of failures during the operational life of the system. Another example is the decision regarding inspection and

TABLE 10-1 LIFE-CYCLE COST ESTIMATES FOR APPLIANCES

Useful life:	Air conditioners 10 years		Refrigerators 15 years		Televisions 12 years		Gas ranges 15 years	
Cost element								
Acquisition	$204	(58.7%)	$295	(40.9%)	$400	(60.2%)	$211	(50.8%)
Operations	131	(37.8%)	392	(54.3%)	178	(26.8%)	159	(38.3%)
Service	4	(1.2%)	19	(2.6%)	79	(11.9%)	35	(8.5%)
Disposal	8	(2.3%)	16	(2.2%)	7	(1.1%)	10	(2.4%)
	$347	(100%)	$722	(100%)	$664	(100%)	$415	(100%)

testing of components and subassemblies. Time and money can be saved at the early stages of a project by minimizing these efforts. A possible consequence is undetected design errors and faulty components that might surface later during operation.

A third example relates to the need for logistics support. In this regard, consider the maintenance costs during the operational phase of a system. These costs can be reduced by including in the design built-in test equipment that identifies problems, locates their source, and recommends a corrective course of action. Systems of this type that combine sensors with automated checklists and expert systems logic are expensive to develop, but in the long run decrease maintenance costs and increase availability.

The National Bureau of Standards (*Handbook 135*) defines life-cycle costing as "a general method of economic evaluation which takes into account all relevant costs of a building design, system, component, material, or practice over a given period of time adjusting for differences in the timing of those costs." LCC models track the costs of development, design, manufacturing, operations, maintenance, and disposal of a system over its useful life. They relate estimates of these cost components to independent (or explanatory) decision variables. By developing a functional representation [known as a cost estimating relationship (CER)] of the cost components in terms of the decision variables, the expected effect of changing any of the decision variables on one or more of the cost components can be analyzed.

A typical example of a CER is the effect of work design on the cost of labor. One aspect of this effect is the learning phenomenon discussed in Chapter 7. Since the slope of the learning curve depends on the type of manufacturing technology employed, a CER can help the design engineers select the most appropriate technology. This situation is depicted in Fig. 10-1, where two manufacturing technologies are considered. Technology I requires lower labor cost for the first unit produced but has a slower learning rate than that of

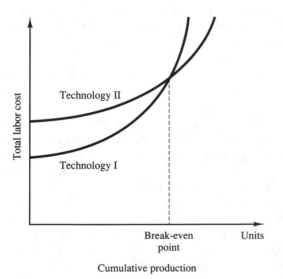

Figure 10-1 Learning curves for two technologies.

technology II. The decision to adopt either technology depends on the number of units required and the cost of capital (assuming that everything else is equal). For a small number of units, technology I is better, as labor costs are lower in the early stages of the corresponding learning curve. Also, if the cost of money is high, technology I might be preferred since it displaces a substantial portion of the labor cost into the future. Finally, for a large number of units, technology II is preferred. In Fig. 10-1, the point where the two technologies yield the same total cost is called the break-even point. In this example (as in many others) the importance of the LCC model increases when the proportion of manufacturing, operations, and maintenance costs is greater than the proportion of design and development costs over the lifetime of the product or system.

The development and widespread use of LCC models is particularly justified when a number of alternatives exist in the early stages of a project's life cycle and the selection of an alternative has a noticeable influence on the total life-cycle cost. At the outset of a project, they provide a means of evaluating alternative designs; as work progresses, they may be called on to evaluate proposed engineering changes. These models are also used in logistics planning, where it is necessary, for example, to compare different maintenance concepts, training approaches, and replenishment policies. At a higher level, model results support decisions regarding logistic and configuration issues, the selection of manufacturing processes, and the formulation of maintenance procedures. By proper use, engineers and managers can choose alternatives so that the life-cycle cost is minimized while the required system effectiveness is maintained. The development and application of life-cycle cost models is therefore an essential part of most engineering projects.

10.2 UNCERTAINTIES IN LCC MODELS

In the conceptual design phase where LCC models are usually developed, little is known about the system, the activities required to design and manufacture it, its modes of operation, and the maintenance philosophy to be employed. Consequently, LCC models are subject to the highest degrees of uncertainty at the beginning of a project. This uncertainty declines as progress is made and additional information becomes available.

Because decisions taken in the early stages of a project's life cycle have the potential to affect the overall costs more than decisions taken later, the project team faces a situation where the most critical decisions are made when uncertainty is highest. This is illustrated in Figs. 10-2 and 10-3, where the potential effect of decisions on cost and the corresponding level of uncertainty are plotted as a function of time. From these graphs, the importance of a good LCC model in the early phases of a system's life cycle is evident.

There are two principal types of uncertainty that LCC model builders are advised to consider: (1) uncertainty regarding the cost-generating activities during the system's life cycle, and (2) uncertainty regarding the expected cost of each of these activities. The first type of uncertainty is typically present when a new system is being developed and few historical data exist. The equipment used on board several of the early earth-orbiting satellites and the first space shuttle, *Columbia*, fall into this category. There was a high level of uncertainty with respect to maintenance requirements for this equipment as well as the

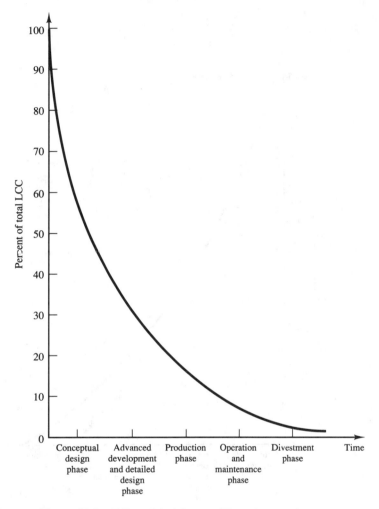

Figure 10-2   Effect of decisions on life-cycle cost of a system.

procedures for operating and maintaining the launch vehicles and supporting facilities. Maintenance practices were finalized only after sufficient operational experience was accumulated. The reliability and dependability of these systems were studied carefully to determine the required frequency of scheduled maintenance.

Nevertheless, the accuracy of LCC models where this type of uncertainty is present is relatively low, implying that their benefits may be somewhat limited to providing a framework for enumerating all possible cost drivers and promoting consistent data collection efforts throughout the life of the system. But even if this were the only use of the model, benefits would accrue from the available data when the time came to upgrade or build a second-generation system.

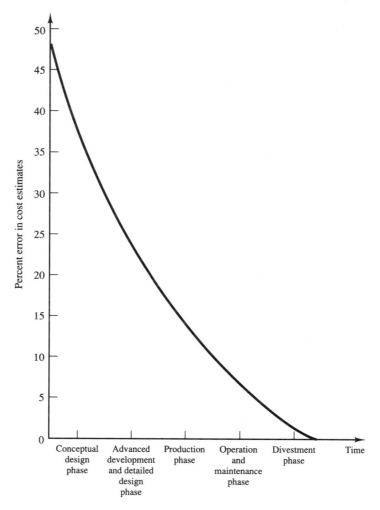

Figure 10-3 Cost estimate errors over time.

The second type of uncertainty, estimating the magnitude of a specific cost-generating activity, is common to all LCC models. There are multiple sources of this type of uncertainty, such as future inflation rates, the expected efficiency and utilization of resources, and the failure rate of system components. Each affects the accuracy of the cost estimates. To obtain better results, sophisticated forecasting techniques are often exploited, fed by an array of diverse data sources. Analysts building LCC models should always trade off the desired level of accuracy with the cost of achieving that level. Most engineering projects deal with the improvement in existing systems or the development of new generations of existing systems. For such projects it is frequently possible to increase the accuracy of cost estimates by investing more effort in collecting and analyzing the underlying data. Therefore, it is important to determine when the point of diminishing returns has been reached. More sophisticated models may pose an increasingly problematic challenge to their intended users and may become more expensive or complicated than the quality of the input data can justify.

The accuracy of cost estimates changes over the life cycle of the system. During the conceptual design phase a tolerance of −30% to +50% may be acceptable for some factors. By the end of the advanced development and detailed design phase, more reliable estimates are expected. Further improvement is realized during the production and system operation phases when field data become available.

10.3 CLASSIFICATION OF COST COMPONENTS

The selection of a specific design alternative, the adoption of a maintenance or training policy, or the analysis of the impact of a proposed engineering change are based on the trade-off between the expected costs and the expected benefits of each candidate. To ensure that the economic analysis is complete, the LCC model should include all significant costs that are likely to arise over the system's life cycle. In this effort it is essential for the model builder to consider the type of problem being investigated. Based on the logical design of the project, familiar management concerns, and supporting data requirements the cost classifications and structures can be developed.

Many ways of classifying costs are possible in LCC analysis. Some are generic while others are tailored to meet the individual circumstance. In the following discussion we present several commonly used schemes. Each can be modified to fit a specific situation, but a particular application may require a unique approach.

One way to classify costs is by the five life-cycle phases:

1. *Cost of the conceptual design phase.* This category highlights the costs associated with early efforts in the life cycle. These efforts include feasibility studies, configuration analysis and selection, systems engineering, initial logistic analysis, and initial design.

The cost of the conceptual design phase usually increases with the degree of innovation involved. In projects aimed at developing new technologies, this phase tends to be long and expensive. For example, consider the development of a new drug for AIDS or the development of a permanently manned space station. In such projects, high levels of uncertainty motivate in-depth feasibility studies, including the development of models, laboratory tests, and detailed analyses of alternatives. When a modification or improvement of an existing system is being weighed, the level of uncertainty is lower, and consequently the cost associated with the conceptual design phase is lower. This is the case, for example, with many construction projects where the use of new techniques or technologies is not the main issue.

The life-cycle cost model can be used in this phase to support benefit-cost analyses. One must proceed with caution, however, because initial LCC estimates may be subject to large errors. A comparison of alternatives is appropriate only when the cost difference between them is measurably larger then the estimation errors, and hence can be detected by the LCC models.

2. *Cost of the advanced development and detailed design phases.* Here the cost of planning and detailed design is presented. This includes product and process design, preparation of final performance requirements; preparation of the work breakdown structure, schedule, budget, and resource management plans; and the definition of procedures and management tools to be used throughout the life cycle of the project.

These phases are labor intensive. Engineers and managers design the product and plan the project for smooth execution. Attempts to save time and money by starting implementation prior to a satisfactory completion of these phases can lead to future failures. The development of a good product design and an efficient project plan are preconditions for successful implementation.

In the advanced development and detailed design phase of the LCC analysis, accurate estimates of cost components are made. These estimates can be used to support decisions regarding the selection of alternative technologies or the preferred logistic support of the product.

3. *Cost of the production phase.* This category consists of the costs associated with constructing new facilities or remodeling existing facilities for assembly, testing, production, and repair. Also included are the actual costs of equipment, labor, and material required for operations, as well as blueprint reproduction costs for engineering drawings and the costs associated with documenting production, assembly, and testing procedures.

In many projects and systems this is the highest-cost phase. The quality of decisions made earlier in the project determine the actual cost of production. By accumulating and storing the actual costs in appropriate databases, life-cycle cost analysis can be improved for similar future projects. The LCC model in this phase becomes increasingly accurate, making detailed cost analysis of alternative maintenance and operation policies possible.

4. *Cost of operating and maintaining the system.* This category identifies the costs surrounding the activities performed during the operational life of a system. These include the cost of personnel required for operations and maintenance, together with the cost of energy, spare parts, facilities, transportation, and inventory management. Design changes and system upgrade costs also fall into this category.

5. *Cost of divestment phase.* When the useful life of a system has expired, it must be phased out. Parts and subassemblies must be inventoried, sold for scrap, or discarded. In some cases it is necessary to take the system apart and dispose of its components safely.

The phasing out of a system might have a negative cost (i.e., produce revenue) when it is sold at the end of its useful life, or it might have a positive cost (often high), as in the case of a nuclear reactor that has to be carefully dismantled and its radioactive components safely discarded.

The relative importance of each phase in the total life-cycle cost model is system specific. Figure 10-4 presents a comparison for two generic systems by life-cycle phase. In general, when alternative projects are being considered, the relative magnitude and timing of the different cost components figures prominently in the analysis. In Fig. 10-4, system A requires substantial research and development efforts. The conceptual design phase and the advanced development phase account for 50% of the LCC. In system B, these two

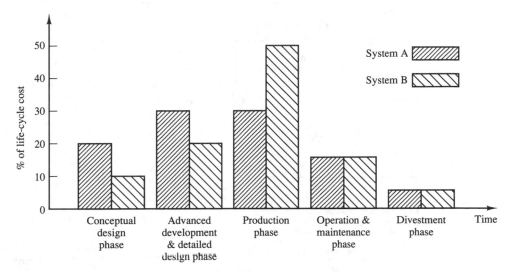

Figure 10-4   Cost classification by life-cycle phase.

phases account for only 30% of the total cost. Thus system B can be thought of more as a production/implementation project, while system A represents more of a design/development project.

A second classification scheme has its origins in manufacturing and is based on cost type, that is, direct labor versus indirect labor, subcontracting, overhead allocations, and material (direct and indirect), as illustrated in Fig. 10-5. These categories parallel those traditionally found in cost accounting, so data should be readily available for many applications.

A third means of classification is based on the time period in which each cost component is realized. To make this scheme operational, it is necessary to define a minimum time period, such as 1 month or 1 quarter, in the system's life cycle. All costs that occur in this predetermined time period are grouped together. This is illustrated in Fig. 10-6, where the graphs provide a 12-month history of costs. The classification in Fig. 10-6 is important when cash flow constraints are considered. Two projects with the same total cost may have a different cost distribution over time. In this case, due to cash flow considerations (the time value of money), the project for which cost outlays are delayed may be preferred.

A fourth classification scheme is by work breakdown structure. In this approach, the cost of each element is estimated at the lowest level of the WBS. If more detail is desired, each element can be further disaggregated by life-cycle phase (first classification), cost type (second classification), or time period (third classification).

As the situation dictates, other schemes, perhaps based on the bill of materials, the product structure, or the organization breakdown structure, might be employed. In particular, OBS classification has proven quite useful as a bridge between the life-cycle cost model and the project budget, which traditionally is prepared along organizational lines.

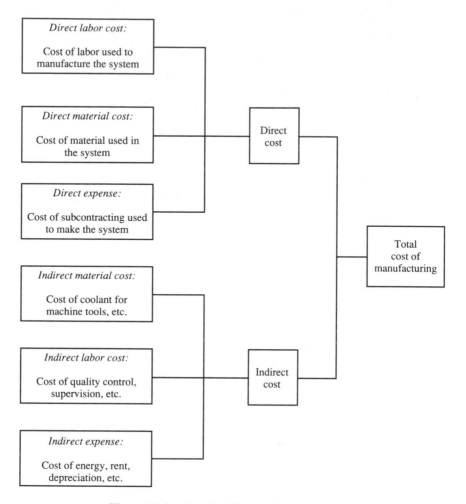

Figure 10-5 Cost classification for manufacturing.

It goes without saying that the scheme chosen should directly support the kinds of analyses to be undertaken. Thus, if future cash flow analyses are required, the timing of each cost component is important. If, however, a system is developed by one organization (a contractor) for use by another (the client), and the customer is scheduled to deliver some of the subsystems, as in the case of *government-furnished equipment* in government contracts, classification of cost based on the organization responsible for each cost component might be appropriate.

Sophisticated LCC models apply several classification schemes in the cost breakdown structure so that each cost component may be categorized by the life-cycle phase and time period in which it arises, the work breakdown structure element in which it appears, and the class type from an accounting point of view. The cost of developing and maintaining such models depends on the desired resolution (number of subcategories in each classification

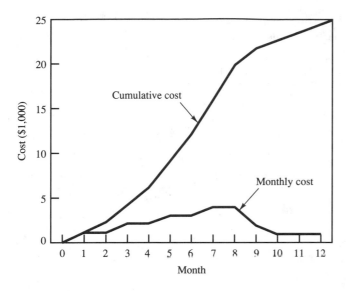

Figure 10-6 LCC as a function of time.

scheme) and accuracy of the cost estimates, the updating frequency, and the number of classification schemes employed. LCC model builders should strive to balance development costs with maintenance and data collection requirements.

An example of an LCC model for a hypothetical system in which a simple three-dimensional cost structure is used is given next. In this classification scheme, costs are broken down by (1) the life-cycle phase, (2) the quarter in which they occur, and (3) labor and material. The data are presented in Table 10-2.

In the example we assume that three different models of the same system are being developed during the first two years (eight quarters). Production starts on the first model before detailed design of the other two is finalized. Thus during quarters 6 through 8, advanced development and detailed design as well as production costs are present. Similarly, the first model becomes operational prior to the completion of the production phase of the other models, implying overlapping costs in these categories in quarters 9 and 10. The three models are phased out in quarters 14, 15, and 17, as noted by divestment costs and reduced operation and maintenance costs in these periods.

The life-cycle cost data in Table 10-2 can be used to produce several views, each giving a different perspective and highlighting different aspects of the project. For example, in Fig. 10-7 we plot the cumulative life-cycle cost of the system over time, as well as the cost that is incurred in each quarter. The life-cycle cost can also be presented by life-cycle phase. This is illustrated in Fig. 10-8. A third possibility is labor cost versus material cost, as shown in Fig. 10-9. Although the periodic and total life-cycle costs are the same in Figs. 10-8 and 10-9, the breakdown of these costs is different and can serve different purposes, as discussed in the next section.

In the example, a fourth classification (or dimension) might correspond to the work breakdown structure and a fifth to the organizational structure. By using a five-dimensional grid, questions such as "What is the expected cost of software development by the main contractor for the real-time control system during the third quarter of the project?"

can be answered. The type of questions and scenarios for which the LCC model is to be exercised is the principal consideration in its design.

TABLE 10-2   EXAMPLE OF AN LCC MODEL ($1,000)

| | System life-cycle phase | | | | | | | | | | |
| | Conceptual design | | Advanced development and detailed design | | Production | | Project termination/ operation and maintenance | | Divestment | | |
Quarter	Labor	Mat'l	Labor	Mat'l	Labor	Mat'l	Labor	Mat'l	Labor	Mat'l	Total
1	2										2
2	3										3
3	3										3
4	1		3								4
5			4	1							5
6			5	1	10	3					19
7			5	1	12	4					22
8			3	1	15	6					25
9					10	5	3	1			19
10					7	3	4	2			16
11							5	3			8
12							5	3			8
13							5	3			8
14							5	3	1		9
15							4	2	1		7
16							4	2			6
17							3	1	1		5
18											
	9	—	20	4	54	21	38	20	3	—	169

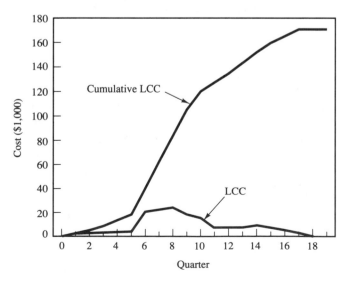

Figure 10-7   Total life-cycle cost of the system.

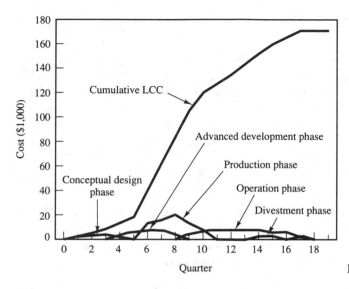

Figure 10-8   Life-cycle cost by phase.

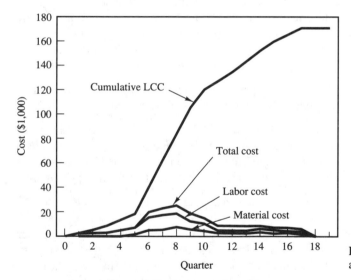

Figure 10-9   Cost breakdown by labor and material.

10.4 DEVELOPING THE LCC MODEL

The first step in the design of an LCC model is to identify the type of analysis it is intended to support. The following is a list of several common applications.

■ *Strategic or long-range budgeting.* Because the LCC model covers the entire life cycle of a system, it can be used to coordinate investment expenditures over the system's useful life or to adjust the requirement for capital for one system or project with capital needed or generated by other systems or projects. Such long-range budget planning is important for strategic investment decisions.

■ *Strategic or long-range technical decisions.* Strategic decision making as it relates to such issues as the redesign of a system or the early termination of an R&D project is difficult to support. The LCC model can be used to monitor changes in cost estimates as the project evolves. Revised estimates of production or operations, or maintenance costs that are substantially higher than the baseline figures, may serve as a trigger for unscheduled design reviews, major changes in system engineering, or even a complete shutdown of the project. Because LCC estimates improve over time, rough projections made in the early phases of a project's life cycle may be updated later and provide managers with more accurate data to support the technical decision-making process.

■ *Data analysis and processing.* LCC models routinely serve as a framework for the collection, storage, and retrieval of cost data. By using an appropriate data structure (e.g., life-cycle cost breakdown structure), the cost components of current or retired systems can be analyzed simultaneously to yield better estimates for future systems.

■ *Logistic support analysis.* Logistics is generally concerned with transportation, inventory and spare parts management, database systems, maintenance, and training. Questions such as what maintenance operations should be performed and at what frequency, how much to invest in spare parts, how to package and ship systems and parts, what training facilities are required, and what type of courses should be offered to the operators and maintenance personnel are examples of decisions supported by LCC analyses.

Once agreement is reached on the types of analyses that will be conducted, LCC model development can proceed. The following steps should be carried out:

1. *Classification.* In this step the classification schemes are developed. Major activities that generate cost are listed and major cost categories (labor, material, etc.) are identified. For example, the LCC data presented in Table 10-2 can be classified by the organizational unit responsible for each cost component and the activities performed by that unit.

2. *Cost breakdown structure (CBS).* Next, a coding system is selected to keep track of each cost component. To gain further insights, the latter may be organized in a multidimensional hierarchical structure based on the system chosen in step 1. Each component at each level of the hierarchy is assigned an identification number. The CBS enables the cost components to be aggregated based on the classification scheme. Thus with the proper scheme, the labor cost of a specific activity in a given period, or the cost of a specific subsystem during its operational phase, can be determined. The cost breakdown structure links cost components to organizational units, to work breakdown structure elements, and to the system's bill of material.

For example, consider the cost breakdown structure of a project aimed at developing a new radar system. The system is composed of a transmitter, receiver, antenna, and computer. The computer and its software are subcontracted as well as part of the antenna servo, while the rest of the components are developed in-house. The CBS is coded as shown in Table 10-3.

TABLE 10-3   CODING AND CLASSIFICATION SCHEME FOR LCC

Digit	Classification	Code assignment	
1	Who performs the work	Performed in-house	1
		Subcontracted	2
2	System part	Transmitter	1
		Receiver	2
		Antenna	3
		Computer	4
3	Life-cycle phase	Conceptual	1
		Detailed design	2
		Production	3
		Operations and maintenance	4
		Divestment	5
4	Type of cost	Direct labor	1
		Direct material	2
		Overhead	3

Using this simple four-digit code, a question such as "What is the direct cost of material to be used during the production phase of the receiver?" can be answered by retrieving all cost components with the following LCC codes:

first digit	1 or 2
second digit	2
third digit	3
fourth digit	2

Thus we would search for the LCC codes 1232 and 2232. The corresponding cost components might represent the cost at different months of the project, assuming that cost is estimated on a monthly basis. Other situations are possible.

3. *Cost estimates.* After the various cost components are identified and organized within the chosen classification scheme, the final step is to estimate each cost component. The American Association of Cost Engineers (AACE 1986) has proposed three classifications for this purpose:

- *Order of magnitude:* accuracy of -30% to $+50\%$. An estimate that is made without any detailed engineering data.
- *Budget:* accuracy of -15% to $+30\%$. This estimate is based on preliminary layout design and equipment details, and is performed by the client to establish a budget for a new project (at the RFP stage).
- *Definitive:* accuracy of -5% to $+15\%$. This cost estimate is based on well-defined engineering data and a complete set of specifications.

The work involved in preparing cost estimates is a function of the required accuracy and the size and cost of the project. In the process industries, the typical costs for preparing estimates were estimated by Pikulik and Diaz (1977):

■ *Order-of-magnitude estimates*

Project cost ($ million)	Cost of estimate ($ thousand)
Up to 1	7.5 to 20
1 to 5	17.5 to 45
5 to 50	30 to 60

■ *Budget estimates*

Project cost ($ million)	Cost of estimate ($ thousand)
Up to 1	20 to 50
1 to 5	45 to 85
5 to 50	70 to 130

■ *Definitive estimates*

Project cost ($ million)	Cost of estimate ($ thousand)
Up to 1	35 to 85
1 to 5	85 to 175
5 to 50	150 to 330

A variety of estimation procedures are used in industry, all of which are based on the assumption that past experience is a valid predictor of future performance. Estimation procedures fall into one of two categories: (1) causal, where the aim is to derive cost estimating relationships; and (2) non-causal or direct. Causal estimates follow from an assumed functional relationship between the cost component and one or more explanatory variables. For example, the cost of fuel required during the operational life of a car might be estimated as a function of the distance driven, the weight of the car, its engine size, and the expected road conditions. An equation relating the cost of fuel to the explanatory variables can be developed by using regression analysis or any other curve fitting technique (see Section 7.2.5). With CERs the expected effect of changing any explanatory variable on the LCC can be analyzed. To develop CERs, past data on the values of the cost component under investigation and the explanatory variables are required.

As an example, consider the equipment cost CER proposed by Fabrycky and Blanchard (1991),

$$C = C_r \times \left(\frac{Q_c}{Q_r}\right)^m \tag{10-1}$$

where

$$C = \text{Cost for a new design size } Q_c$$
$$C_r = \text{Cost for existing reference design } Q_r$$
$$Q_c = \text{Design size—new design}$$
$$Q_r = \text{Design size—existing reference design}$$
$$m = \text{correlation parameter; } 0 < m < 1$$

In linearized form, eq. 10-1 becomes:

$$\log C - \log C_r = m(\log Q_c - \log Q_r) \tag{10-2}$$

Suppose that a cost estimate for a new 750 gallon nuclear reactor is required and that information on the actual cost of five reactors is available. This data is presented below.

Reactor	Cost	Size (gallons)
1	$14,000	200
2	18,000	300
3	21,500	400
4	25,000	500
5	28,000	600

A pairwise comparison between the five reactors yields the following data in the form needed for eq. (10-2).

C_r	C	Q_r	Q_c	$\log C - \log C_r$	$\log Q_c - \log Q_r$
14,000	18,000	200	300	0.109	0.176
18,000	21,500	300	400	0.077	0.125
21,500	25,000	400	500	0.066	0.096
25,000	28,000	500	600	0.049	0.079

The regression equation is

$$\log C - \log C_r = 0.628(\log Q_c - \log Q_r)$$

with $R^2 = 0.983$. Now, using the fourth reactor as a reference (Q_r), the estimated cost for a new 750-gallon (Q_c) reactor (same type) is

$$C = 25,000 \times \left(\frac{750}{500}\right)^{0.628} = \$32,249$$

This type of CER is useful for a company that has to estimate the cost of new reactors that differ from existing reactors mainly by size.

Cost estimates can alternatively be derived using noncausal methods, such as:

- Judgment and experience; rules of thumb or the use of organizational standards for similar activities. These techniques are informal, inexpensive, and therefore appropriate when formal LCC models and cost estimates with high levels of accuracy are not essential.
- Analogy to a similar system or component, and appropriate adjustment of cost components according to the difference between the systems.
- Technical estimation based on drawings, specifications, time standards, and values of parameters such as MTBF and MTTR.
- Value of contracts for similar systems, such as maintenance contracts. It is also possible to estimate costs based on bids from contractors responding to RFPs.

Each technique requires a combination of resources, such as time, data, equipment, and software, and may call on the expertise and experience of many people within or perhaps external to the organization. From the data and resources available, the required accuracy, and the cost of using each cost estimating technique, the most suitable approach for each application can be selected. For each cost component, one or more cost estimating techniques might be appropriate. In the early stages of the life cycle, technical estimation is usually not feasible, as drawings and other information are not available. For new systems, analogy might not be feasible if similar systems have not been developed or previously deployed.

The selection of a cost estimating procedure depends on feasibility, required accuracy, and cost. The LCC analyst should consider all three aspects in the process of model design and application. To demonstrate the process of developing an LCC model, consider the problem of estimating energy costs in residential buildings. It is possible to reduce the cost of energy by proper design, the use of insulation and improved ventilation, and the selection of efficient heating and cooling devices. The following is an example of a basic LCC model for such a project. The model has only two classifications: the first centers on the activities that generate cost and the second is based on time. Table 10-4 depicts levels 1 and 2 of the cost breakdown structure for the cost-generating activities.

TABLE 10-4   PARTIAL COST BREAKDOWN STRUCTURE FOR RESIDENTIAL BUILDING EXAMPLE

1. Cost of engineering	2. Cost of construction	3. Cost of operations
1.1 Structural design	2.1 Equipment	3.1 Energy
1.2 Interior design	2.2 Contractors	3.2 Maintenance
1.3 Drawing preparation	2.3 Material	3.3 Consumables
1.4 Supervision	2.4 Labor	3.4 Subcontractors
1.5 Management	2.5 Energy	
	2.6 Inspection	
	2.7 Management	

A time dimension is added to the model by introducing the timing of each cost component. For example, the structural design (1.1) may take 3 months. Assuming that the cost of the first month is $500, the cost of the second month is $1,100, and the cost of the last month is $400, the total cost of structural design is $500 + 1,100 + 400 = \$2,000$ over a 3-month period. By assigning the cost of each cost component in Table 10-3 to a specific month, the time aspect of this LCC model is introduced.

If more detail is needed, the model can be expanded to three or four levels. For example, consider item 2.1. Equipment can be further broken down by air conditioning system, heater, and so on. Once the lowest level is constructed and the data elements defined, the model can be used to estimate the cost of each component for each design alternative on a periodic basis, if necessary. Alternatives might differ in their total life-cycle cost, in the allocation of costs over the life-cycle, and in the allocation of costs among different system components. As discussed in Chapters 3 and 4, selection of the best alternative depends on the criteria and performance measures specified. For instance, if the minimum net present cost is the criterion chosen, between two design alternatives with the same total life-cycle cost, the alternative that delays monetary outlays the longest is preferred. In the example, this might lead to an energy-inefficient house—one that is less expensive to build but more expensive to maintain.

A possible CER for the example might be a linear equation relating the cost of heating to the insulation used and the difference between the required temperature inside the house and the ambient temperature outside. Additional explanatory variables that might be included in the equation are the area of windows and the type of glass used.

The cost breakdown structure can be as detailed as required to capture the impacts of decisions on overall cost and performance. Continuing with item 2.1, equipment can be broken down further to the level of components used in the air-conditioning system if it were thought that the selection of these components would measurably affect the life-cycle cost.

10.5 USING THE LCC MODEL

The integration of the cost breakdown structure with estimates for each component produces the aggregate life-cycle cost model for the system. This model (distributed over time) is the basis for several types of analyses and decision making.

1. *Design evaluations.* In the planning stages of a project, alternative designs for the entire system or its components have to be evaluated. The LCC model combined with a measure for system effectiveness produce a basis for cost-effectiveness analysis during various stages of the development cycle. Methodological details are provided in Chapters 3 and 4, where issues related to risk, benefit estimation, and criteria selection are discussed.

2. *Evaluation of engineering change requests (ECRs).* As explained in Chapter 6, the process of ECR approval or rejection is based on estimates of cost and effectiveness with and without the proposed change. The LCC model provides the foundation for conducting the analysis.

3. *Sensitivity analysis and risk assessment.* In the development of CERs, parameters that affect the life-cycle cost of the system are used as the explanatory variables. A sensitivity analysis should always be conducted to see how the life-cycle cost changes as each parameter is varied over its feasible range. Depending on the nature of the project and the time horizon, some typical explanatory variables might be the rate of inflation, the cost of energy, and the minimum attractive rate of return.

4. *Logistic support analysis.* The evaluation of policies for maintenance, training, stocking of spare parts, inventory management, shipping, and packaging are all supported by appropriate LCC models. By estimating the cost of different alternatives for logistic support, decision makers can trade off the cost and benefits of each scenario under consideration.

5. *Pareto or ABC analysis.* This analysis is used to identify the most important cost components of a project. The first step is to sort each component by cost and then to place them into one of the following three groups:

Group A: small percentage of the top cost components (10 to 15%), which together account for roughly 60% or more of the total cost

Group B: all cost components not members of group A or C

Group C: large percentage of the bottom cost components (about 50%), which account for 10% or less of the total cost

In the sorted list, the first 10 to 15% of the cost components are members of group A and the last 50% are members of group C. The remaining components in the middle range of the list are assigned to group B. This clustering scheme is the basis for management control. The strategy is to monitor closely those items that account for the largest percentage of the total life-cycle cost (i.e., group A components). Conversely, group C components, which represent a relatively large number of items but account for a relatively small portion of the total cost, require the least amount of attention.

6. *Budget and cash flow analysis.* Here the concern is staying within budget and cash flow constraints, and estimating future capital investment needs. By combining the LCC models of all projects in an organization, the net cash flow for each future period can be forecast. The results may then be used to support feasibility analyses, decisions regarding the acceptance of new projects, and recommendations for rescheduling or abandoning ongoing projects.

The LCC model is an important project management tool for strategic financial planning, logistic analysis, and technological decision making. Properly designed and maintained LCC models help the project manager in both planning and control by linking together the cost and technological aspects of a project. By using CERs, the effect of technological decisions of the system's life-cycle cost can be analyzed and used as a basis for alternative evaluation and selection, resource acquisition, and configuration management.

TEAM PROJECT

Thermal Transfer Plant

Your plans for the prototype rotary combustor project have been approved. TMS management is now weighing the possibility of investing in a plant for manufacturing the combustors. There is a feeling, however, that the degree of subcontracting associated with producing the prototype may not be appropriate for the repetitive manufacturing environment of the new plant.

Your team has been requested to perform a life-cycle cost analysis to help determine which parts and components of the rotary combustor to manufacture in-house and which to buy or subcontract. Design your models to answer these "make or buy" questions, keeping in mind that the expected life of a rotary combustor is about 25 years and TMS would like to support these units throughout their life cycle.

Make any assumptions necessary to estimate costs and risks. Present your assumptions explicitly and discuss the sensitivity of your results.

DISCUSSION QUESTIONS

1. Estimate the life-cycle cost for a passenger car. In so doing, select an appropriate cost breakdown structure and explain your cost estimates.

2. Explain how the design of a car affects its life-cycle cost.

3. Compare the cost of ownership of a new car with that of a used car of similar type.

4. Explain the design factors that affect the life-cycle cost of an elevator in a New York City office building.

5. What are the sources of uncertainty in Question 4?

6. What do you think are the principal cost drivers in designing a permanently manned lunar base? What noncost factors would you want to consider?

7. Identify a potential consumer product that is not yet on the market, such as video telephones, and list the major costs in each phase of its life cycle. How might these costs be estimated?

8. Pick an R&D project of national scope, such as mapping all the genes on a human chromosome (the human genome project). First, sketch a potential organizational breakdown structure for the project and identify the tasks that might fall within each organizational unit. Then develop a cost breakdown structure and relate it to the OBS.

9. Develop a life-cycle cost model to assist you in selecting the best heating system for your house. Discuss the alternatives and explain the cost structure you have selected.

10. Discuss the effect of taxes on the life-cycle cost of passenger cars. Compare domestic and imported cars.

11. Discuss the effect of life-cycle cost on the decision to locate a new warehouse.

12. Discuss a project in which the first phase of the life cycle accounts for more than 50% of the life-cycle cost.

13. Discuss a project in which the detailed design phase accounts for more than 50% of the life-cycle cost.

EXERCISES

10-1. The cost of a used car is highly correlated with the following variables:

$$t = \text{age of the car} \qquad 1 \le t \le 5 \text{ (years)}$$
$$V = \text{volume of engine} \qquad 1000 \le V \le 2500 \text{ (cubic centimeters)}$$
$$D = \text{number of doors} \qquad D = 2, 3, 4, 5$$
$$A = \text{accessories and style} \quad A = 1, 2, 3, 4, 5, 6 \text{ (qualitative)}$$

Using regression analysis, the following relationship between the cost of a car and the four independent variables was found:

$$\text{purchase cost} = \left(1 + \frac{1}{t}\right) \times V \times \left(\frac{D}{2} + A\right)$$

(a) Plot the purchase cost as a function of the four variables.

(b) Which variable has the greatest effect on cost?

(c) You have a total of $5,000. List the different types of cars (combinations of the parameters) you can afford.

(d) Develop a model by which you select the best car for your needs.

(e) Maintenance and operation cost of the car are estimated as follows:

$$\text{annual maintenance cost} = \frac{t}{2} \times V \times \frac{s}{1000}$$

$$\text{annual operating cost} = \left(D \times t + \frac{V}{1000}\right) \times \frac{s}{250}$$

where s is the number of miles driven annually. What is the best car (combination of parameters) for a person who drives 12,000 miles every year?

10-2. A construction project consists of 10 identical units. The cost of the first unit is $25,000 and a learning curve of 90% applies to the cost and duration of consecutive units. Assume that the first unit takes 6 months to finish and that the project is financed by a loan taken at the beginning of the project at an annual interest rate of 10%.

(a) Should the units be constructed in sequence (to maximize learning) or in parallel (to minimize the cost of the loan)?

(b) Find the schedule for the 10 units that minimizes the total cost of the project.

10-3. Develop three cost classifications for the life-cycle cost of an office building.

10-4. Develop a cost breakdown structure for the cost of an office building. Estimate the cost of each component.

10-5. Show a cash flow analysis for the life-cycle cost of an office building.

10-6. Perform a Pareto (ABC) analysis on the data of the life-cycle cost of an office building.

10-7. Develop an estimate for the cost of a 3-week vacation in Europe.

10-8. Develop a life-cycle cost model to support the decision to buy or rent a car.

10-9. Natasha Gurdin is debating which of two possible models of a car to buy (A or B), being indifferent with regard to their technical performance. She has been told that the average monthly cost of owning model A, based on LCC analysis, is $500.

   (a) Using the following data for model B, calculate its LCC and determine which model is the better choice for Natasha.

Purchase price	$23,000
Life expectancy	4 years
Resale value	$13,000
Maintenance-free insurance	$1,100 per year
Operational cost (gas, etc.)	$90 per month
Car insurance	$1,400 per year
Mean time between failures	14 months
Personal damage per failure	$650

   (b) Develop a general model that can be used to calculate LCC for a car.

10-10. Your company has just taken over an old apartment building and is renovating it. You have been appointed manager and must decide which brand of refrigerator to install in each apartment unit. Your analysis should consider expenses such as purchase price, delivery charges, operational costs, insurance for service, and selling price after 6 years of service. Identify two brands of 18-cubic foot refrigerators and compare them.

10-11. You have been told that even in warehouse location decisions LCC analysis should be used. Discuss this issue.

10-12. Maurice Micklewhite has decided to replant his garden. Show him what the cost is of making an erroneous decision at various stages of the project, starting with conceptual design and ending with the ongoing maintenance of the garden.

10-13. The relative cost of each stage in the project life cycle is a function of the nature of the project, or product. Generate a list of possible projects and group them by the similarity of their relative cost profile.

10-14. Different organizations and customers look at different aspects of the LCC. Select five projects and identify the relevant LCC aspects for each organization and customer involved.

10-15. Enumerate a list of cost components for two projects and estimate their values. Identify the components that constitute about 80% of the projects' costs and analyze the results.

REFERENCES

Life-Cycle Cost

BLANCHARD, B. S., *Design and Manage to Life Cycle Cost*, Matrix Press, Chesterland, OH, 1978.

Center for Policy Alternatives, Massachusetts Institute of Technology, "Consumer Durables: Warranties, Service Contracts and Alternatives," *Analysis of Consumer Product and Warranty Relationships*, Vol. IV, Cambridge, MA, 1978.

COLLIER, C. A., and W. B. LEDBETTER, *Engineering Economic and Cost Analysis*, Second Edition, Harper & Row, New York, 1988.

DHILLON, B. S., *Life Cycle Costing: Techniques, Models and Applications*, Gordon and Breach, New York, 1989.

EARLS, U. E., *Factors, Formulas and Structures for Life Cycle Costing*, Second Edition, Eddins-Earles, Concord, MA, 1981.

FABRYCKY, J. W., and B. S. BLANCHARD, *Life Cycle Cost and Economic Analysis*, Prentice Hall, Englewood Cliffs, NJ, 1991.

FISHER, G. H., *Cost Considerations in System Analysis*, Elsevier, New York, 1971.

Logistics Management Institute, *Supplemental Report on Life Cycle Costing in Equipment Procurement*, LMI, Washington, DC, 1967.

RIGGS, J. L., and D. JONES, "Flowgraph Representation of Life Cycle Cost Methodology: A New Perspective for Project Managers," *IEEE Transactions on Engineering Management*, Vol. 37, No. 2 (1990), pp. 147–152.

SPENCE, G., "Designing for Total Life Cycle Costs," *Printed Circuit Design*, Vol. 6, No. 8 (1989), pp. 14–17.

U.S. Department of Commerce, *Life Cycle Cost Manual for the Federal Energy Management Program*, National Bureau of Standards Handbook 135, Washington, DC, 1980.

U.S. Department of Defense, *Life Cycle Costing (LCC-3) Department of Defense Life Cycle Costing Guide for System Acquisition*, U.S. Government Printing Office, Washington, DC, 1973.

U.S. Department of Defense, *Joint-Design-to-Cost Guide: Life Cycle Cost as a Design Parameter*, DARCOM P700-6 (Army), NAVMAT P5242 (Navy), AFLCP/AFSCP 800-19 (Air Force), Washington, DC, 1977.

U.S. Department of Defense, *Uniform Budget Cost Terms and Definitions*, Instruction 5000.33, Washington, DC, 1977.

U.S. Department of Defense, *Transition from Development to Production*, DOD Directive 4245.7-M, Washington, DC, 1985.

Cost Estimation

AACE, *Standard Cost Engineering Terminology*, American Association of Cost Engineers, Morgantown, WV, 1986.

ABRAHAM, C. T., R. PRASAD, and M. GHOSH, *A Probabilistic Approach to Cost Estimation*, Report 68-10-001-IBM, IBM Research Center, Yorktown Heights, NY, 1968.

HUMPHREYS, K. K. (editor), *Project and Cost Engineers' Handbook*, Second Edition, Marcel Dekker, New York, 1984.

HUSIC, F. J., *Cost Uncertainty Analysis*, RAC-P-29, Research Analysis Corporation, Arlington, VA, 1968.

NEIL, J. M., *Construction Estimating for Project Control*, Prentice Hall, Englewood Cliffs, NJ, 1982.

NEIL, J. M. (editor), *Skills and Knowledge of Cost Engineering*, Second Edition, American Association of Cost Engineers, Morgantown, WV, 1988.

OSTWALD, P. F., *Cost Estimating*, Second Edition, Prentice Hall, Englewood Cliffs, NJ, 1984.

PAGE, J. S., *Conceptual Cost Estimating Manual*, Gulf Publishing Company, Houston, TX, 1984

PIKULIK, A., and H. E. DIAZ, "Cost Estimating for Major Process Equipment," *Chemical Engineering*, Vol. 84 (1977), p. 106.

POPESCU, C., and A. HAMIANI, *Directory of Microcomputer Software for Cost Engineering*, Marcel Dekker, New York, 1985.

STEWART, R. D., and R. M. WYSKIDA, *Cost Estimator's Reference Manual*, Wiley, New York, 1987.

11

Project Control

11.1 INTRODUCTION

Planning is a fundamental component of project management. A detailed plan covering the technological, budgetary, scheduling, organizational, and risk-related aspects is essential to facilitate coordination among the participants. Those with major roles and influence most often include the various support departments, such as marketing and sales, outside contractors, and the technical disciplines. Unfortunately, even the best of plans cannot guarantee success. Uncertainty and changing environmental conditions are bound to intervene in unforeseen ways, sometimes positively, sometimes negatively. Plans are based on assessments of needs and the estimation of such factors as activity durations, resource availability, labor efficiency, and cost, each of which may be subject to a high degree of variability. Furthermore, needs and goals are dynamic, changing over time. New technologies developed during the life cycle of the project, a rethinking of corporate strategy, the replacement of key personnel, and new market or legal circumstances may all conspire to make the original plans obsolete. Thus a fundamental need exists to monitor actual progress and to update the original plans continually. This need is more evident in complex, technically advanced projects where the likelihood of technological, environmental, and economic changes occurring during the projects' lifetimes is greatest.

The design and implementation of a project control system is therefore an important part of the project management effort. The basis of any control system is a statement of the project goals and their relative importance. For each such goal one or more performance measures are needed. For example, a common goal is to keep the project on schedule. Appropriate performance measures can be based on the actual start or finish times of critical activities, the completion of milestones, or the timing of acceptance tests. The selection of a performance measure depends on the corresponding goal and the level of management to which actual values of the performance measure will be reported. Thus a low-level manager who is responsible for a specific set of activities may be interested in detailed information on those activities. The project manager may be interested in monitoring the actual completion time of critical activities, while upper management might be interested in information on the completion time of major milestones.

Once performance measures are selected, the information required to report the actual value of each performance measure must be defined. For example, the completion of a milestone may be reported at the successful completion of an acceptance test and the issuance of an appropriate report by quality control. The same milestone may be reported as completed only after the customer payment based on the completion of that milestone has been made. The selection of appropriate performance measures and data collection is not trivial, and an effort to use data available from existing reporting systems is always justified. This reduces the cost of data collection and minimizes the problem of conflicting data in the project control system and other management information systems.

The data collected are used as a basis for estimating the current value of the performance measures and to forecast their future values based on past performance. Estimates of current values are the basis of "real-time" control. This type of control is achieved by comparing the actual value of a performance measure to its planned value. Control limits are set to assess the severity of deviations. Deviations that are larger than a predetermined value are used to trigger corrective action. This type of control is based on the philosophy of management by exception, where actual deviations from plans alter management to a particular problem that needs attention. A second mode of control is trend control, which is based on forecasts of future performance measures. Actual values of the performance measures are extrapolated into the future in an effort to detect deviations before they occur. Forecasts of future deviations trigger preventive actions designed to minimize future problems. This mode of trend control is important, as information on present values of performance measures may not reveal irregularities, but data trends over the last few control periods may indicate a high likelihood of future problems.

The designer of a project control system should therefore address the following questions:

1. What performance measures should be selected?
2. What data should be used to estimate the current value of each performance measure?

3. How should raw data be collected, from which sources, and in what frequency?

4. How should the data be analyzed to detect current and future deviations?

5. How should the results of the analysis be reported, in what format, to whom, and how often?

The answers to these questions underlie the design of the control system's data collection, data processing, and information distribution functions. Project control throughout the project life cycle is exercised by management, who use the information from the control system as a basis for the ongoing decision-making process aimed at keeping the project on track.

Several measurements can be taken in support of project control. These can be classified into four categories: schedule, cost, resources, and performance. Table 11-1 elaborates on each with an eye toward understanding the difficulties that may arise.

Some of the measurements in Table 11-1 are commonly used by industrial and service organizations to manage their routine functions, such as inventory tracking, accounting/auditing, quality control, production scheduling, and data processing. There are issues, however, unique to the project environment that inform a need for special control systems to handle the one-time nonrepetitive effort typical of projects. In this chapter we concentrate on techniques specifically developed for cost and schedule control and integration with quality control and control of technological changes as discussed in Chapter 6, which deals with configuration management.

TABLE 11-1   MEASUREMENTS FOR PROJECT CONTROL

Measurement	Category affected
Critical tasks not started on time	Schedule
Critical tasks not finished on time	Schedule
Noncritical tasks becoming critical	Schedule
Milestones missed	Schedule
Due date changes	Schedule
Price changes	Cost
Cost overruns	Cost
Insufficient cash flow	Cost
High overhead rates	Cost
Long supply lead time for material required	Resources, schedule
Low utilization of resources	Resources, cost
Resources availability problems	Resources, schedule, cost
Changes in labor cost	Resources, cost
Changes in scope of project	Performance, cost, schedule, resources
Lack of technical information	Performance, cost, schedule
Failure in tests	Performance, cost, schedule
Delays in client approvals of configuration changes	Performance, schedule
Errors in records (inventories, configuration, etc.)	Performance, cost, schedule

11.2 COMMON FORMS OF PROJECT CONTROL

Project control can be exercised through formal or informal mechanisms. Small projects performed by small teams located in the same place under a single organizational unit may not need a formal control system. This is frequently the case when members of the team are highly motivated and communicate well with each other. Examples of such projects can be found in churches, schools, social clubs, and among members of small communities performing short-range technically simple projects.

The decision to introduce a formal control system and the selection of a specific system should be based largely on two aspects of the project: (1) the risk involved, and (2) the cost of the control system and its expected benefits. High-risk situations where the probability of undesired outcomes is significant due to the complexity of the project, the environmental conditions, or other factors, and where the cost associated with such undesired outcomes is high, justify the investment in a formal, well-designed control system.

The selection of a control system depends on many factors, such as the characteristics of the project structure (OBS and WBS), the technological aspects, the schedule, the budget, and the personality of the members of the project team. Project control systems can be very simple, taking the form of weekly team meetings to discuss current status, or can be very sophisticated, comprising a battery of hardware, software, and personnel. The cost of control should never exceed the expected benefits (i.e., savings) due to the control system.

Schedule control in its simplest form is based on a comparison between the planned schedule, as depicted by a Gantt chart or the results of a critical path analysis, and actual performance. Data on actual progress are collected periodically (every week, every month, etc.) or continuously (as soon as an activity is completed or a milestone is achieved) and are used as input to the control system. By comparing the initial schedule (the baseline) to the current updated schedule, deviations are detected. These deviations are used to trigger corrective action, such as reallocation of resources to expedite late activities.

Simple cost control is achieved by comparing the actual cost of project activities to the planned budget. Although most organizations have some form of cost monitoring system and cost control system, it might be difficult to get the data required for project cost control from these systems. For example, the direct labor and material cost of specific project activities or the cost of subcontracting for a specific WBS element may not be available. Actual cost may be accumulated by department or by work orders. To facilitate cost control by project activities or by WBS elements, a special project cost control system may be required. Once the information on actual costs of project activities is available, cost overruns can be detected, trends can be analyzed, and management's attention can be requested when current or future costs are considered out of control.

The idea of cost and schedule control based on a simple comparison between planned and actual performance is illustrated next using the example project. Suppose that a weekly report is issued detailing cost and schedule performance. As discussed in Section 7.5, three activities (A, B, and E) are scheduled to start the first week of the project, assuming an early start schedule. The duration and cost of these activities are summarized in Table 11-2. Actual performance for the first month of the project (weeks 1 through 4) is

TABLE 11-2 DURATION AND COST FOR
ACTIVITIES PERFORMED IN MONTH 1

Activity	Duration (weeks)	Cost	Cost per week
A	5	$1,500	$ 300
B	3	3,000	1,000
E	7	5,700	814

summarized in Table 11-3. Based on the information in Tables 11-2 and 11-3, the following observations can be made:

- *Week 1.* All three activities started on schedule. Activity A shows a cost overrun of $200 ($500 − $300); activities B and E are exactly on target.
- *Week 2.* All three activities are in process as scheduled. Activity A has a cumulative cost overrun of $400 ($1,000 − 2 × $300), while the overrun for week 2 is $200 ($500 − $300). The actual cost of activity B is as planned, and the actual cost of activity E for the period ($1,500 − 2 × $814 = $128) is below the budget.
- *Week 3.* Activity B is late since it was scheduled to be completed by the end of week 3. Activities A and E are in process as scheduled. The cumulative actual cost of activity A is $1,300, while its planned cost was only 3 × $300 = $900; the difference of $1,300 − $900 = $400 is the same as in week 2. The actual cost of activity B is only $2,500 compared to a budget of $3,000, and the actual cost of E is $2,500 compared to a budget of 3 × $814 = $2,442.
- *Week 4.* Activity A is completed 1 week earlier than planned, activity B is completed 1 week late, and the total cost of both activities is exactly as planned. Activity E is in process and its total cost of $2,900 is below the budget of $3,256 (4 × $814).

Analysis.   The information on the actual start and end of activities may cause late detection of schedule delays. For example, activity B was not completed on time (week 3), but only at the end of week 3 did this fact become known.

When not related to actual progress, information on actual cost may not be a sufficient measure for cost control. The cost overrun associated with activity A was due to the fact that it was ahead of schedule (completed in 4 weeks instead of 5). This situation could not be observed from the cost and schedule information above.

TABLE 11-3 ACTUAL PERFORMANCES IN MONTH 1

	Week 1		Week 2		Week 3		Week 4	
Activity	Activity status	Actual cost	Activity status	Actual cost	Activity status	Actual cost	Activity status	Actual cost
A	Started	$ 500	In process	$1,000	In process	$1,300	Completed	$1,500
B	Started	1,000	In process	2,000	In process	2,500	Completed	3,000
E	Started	814	In process	1,500	In process	2,500	In process	2,900

In addition to cost and schedule control, performance control is the third aspect common to control systems of most projects. Simple control of performance can be achieved by using the organization's quality control and quality assurance system. Reports issued by these systems provide information on the level on performance achieved. A major problem with performance control stems from the one-time nature of projects. Engineering changes throughout the project life cycle make quality control a difficult task, primarily because it is not possible to use past data as a basis for statistical process control. In addition, quality control is dependent on the availability of an updated project configuration, something that is difficult to achieve in a timely manner. To monitor and control engineering changes, a configuration management system is needed.

Although "stand-alone" independent control systems for cost, schedule, and performance are quite common, these three dimensions are not entirely independent in most projects. To integrate the three control systems, project review meetings are frequently held. In such a meeting, representatives from the various groups and organizations participating in the project can discuss progress and decide on necessary corrective action. Project review meetings can be scheduled periodically or when a predetermined milestone is reached. Typical examples are the *preliminary design review* and the *critical design review*, which are major milestones in the engineering project design phase.

Milestone-related review meetings are frequently scheduled to demonstrate prototypes and major subsystems. The integrative nature of a project review meeting where progress is assessed and problems are aired is the essential advantage of this form of control; however, the need to bring together experts from different functional areas (and sometimes from different organizations) for such meetings makes this form of control expensive. In some projects performed in several locations, frequent meetings may be impossible. There is a need for a project control system that integrates information on cost, schedule, and performance to help management monitor and control technically complex projects performed by several organizational units. The basic building blocks of such a system are taken up next.

11.3 INTEGRATING PROJECT STRUCTURE WITH COST AND SCHEDULE CONTROL

As discussed earlier, the project control system is designed to give management the assurance that the project is proceeding according to plan. Its chief function is to monitor progress and to detect deviations between the original plan and current conditions and trends that indicate a high probability that deviations may occur in the future. Control limits are established for the important parameters, and any deviations outside these limits are flagged. Corrective action is taken when the deviations are considered significant. A major problem in project control is the lack of standards deriving from past performance. The ad hoc nature of projects motivates the adoption of control limits that are based on intuition and risk analysis rather than on historical data, as is the case in statistical process control.

The idea of control limits is depicted in Fig. 11-1. In Fig. 11-1a the cumulative budget for activity A in the example project is plotted along with actual cost as a function of

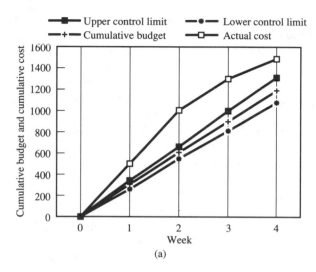

(a)

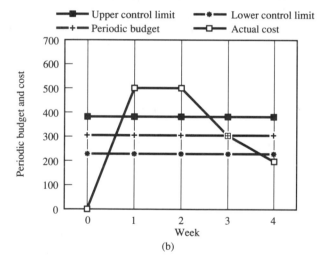

(b)

Figure 11-1 Control limits and actual cost for activity A, weeks 1 through 4.

time for weeks 1 through 4. The control limits for actual cost are set at ±10% of the cumulative budget. The need for an upper control limit is obvious, as the prevention of budget overruns is a common goal for projects. Actual expenditures below budget are also monitored since they might signal a delay in performing some activities (this lower control limit is important for the detection of delays in noncritical activities that are not monitored by CPM-based schedule control systems).

In Fig. 11-1b the weekly budget for activity A in the example project is plotted as a function of time for weeks 1 through 4. Again the control limits for actual cost are set at ±10% of the weekly budget. By monitoring the variance between actual cost and planned cost, corrective actions may be taken whenever this variance is considered too high.

A similar report for several activities or for the entire project can be constructed. Each report should be designed according to the needs of the management level for which it is produced. By introducing the cost variance and the control limits, automatic detection of problematic deviations is made possible. Based on predetermined control limits, management can be supplied with the cost variance of activities whose periodic or cumulative variance exceeds the acceptable range and therefore may require attention.

The effectiveness of a project control system can be measured by its average response time, that is, the average time between the occurrence of a deviation outside the control limits and its detection. Another performance measure is traceability, the ability of the control system to identify the source of the problem causing the deviations. It is important that the relationship between the source of the problem and the project components affected be established and the responsible organizational units notified. The time when the problem occurred should also be recorded as a third dimension of this measure.

An appropriate data structure is required to achieve traceability. This structure must relate plans and corresponding progress reports to the relevant time periods, to the appropriate segments of the project work content, and to the organizational units responsible for these segments. Two hierarchical structures are commonly employed in an integrated fashion to facilitate traceability: (1) the *organizational breakdown structure* (OBS), and (2) the *work breakdown structure* (WBS).

11.3.1 Hierarchical Structures

As discussed in Chapter 5, the OBS is a model of the project's organizational structure. In this model, each entity responsible for one or more project tasks is represented. At the lowest level, the operational units engaged in the execution of activities are present. Higher levels represent various management layers, such as foremen, department managers, up to the vice president of operations and the chief executive officer. Along with the OBS, authority and responsibility have to be clearly defined, as well as the policies and procedures promulgated for reporting and authorizing work. The OBS defines the communication lines used for reporting progress (from the bottom up), and for issuing work orders and technical instructions (from the top down). An OBS for the example project is illustrated in Fig. 11-2. Activities are assigned to organizational units as follows:

Organizational unit	Activities performed
Department 1	C,D,F,G
Department 2	A,B,E

The OBS is integrated with the WBS, which is a product-oriented hierarchy composed of hardware, software, services, and tasks required to complete the project. The WBS organizes, defines, and displays the product to be produced as well as the work to be accomplished in the project. At the lowest level of the WBS, specific tasks are listed. These tasks are integrated through the higher levels into subsystems, systems, and at the top level into the whole project. A simplified WBS that consists of three elements is

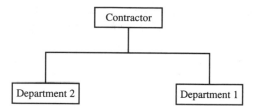

Figure 11-2 Organizational break-down structure.

illustrated in Fig. 11-3. The upper level in the figure represents the entire project, while the lowest level comprises the three major elements of the WBS. Assuming that this is the WBS of the example project, the following relationships exist between the project activities and the WBS elements:

WBS element	Activities related to WBS element
I	A,C,D
II	B,F
III	E,G

The same principles apply to larger projects. For example, the upper three levels of a work breakdown structure for an electronic system are presented in Appendix 11A (based on MIL-STD-881A).

By integrating the OBS and the WBS, each activity in the project is linked to both structures at their lowest levels, as illustrated in Fig. 11-4. From the figure it is clear that department 1 performs activities C and D required for element I in the WBS. In general, we define a *work package* as the work content of one element in the lowest level of the WBS performed by one organizational unit at the lowest level of the OBS. The cost associated with each work package is accumulated and controlled by the corresponding *cost account*. Work packages and cost accounts form the basic building blocks of a project control system that supports traceability in both the OBS and WBS dimensions.

Design of the control system is initiated during the conceptual design phase of the project as goals and performance measures are defined. Later, the OBS and WBS are

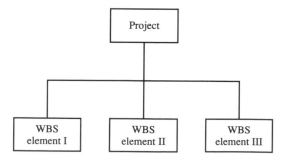

Figure 11-3 Simple work breakdown structure.

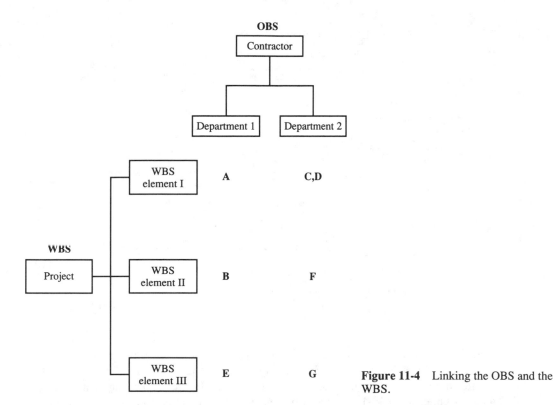

Figure 11-4 Linking the OBS and the WBS.

developed together with the activities to be performed, related costs, durations, and precedence relations. By the end of the planning phase, the detailed OBS, WBS, schedule, and budget serve as the control system baseline.

During project implementation, at the end of each control period, comparisons are made between the work content completed and the work content scheduled for that period. One comparison focuses on the ability of the organization to complete the project within schedule. An effort is made to detect overruns, and if present, to reduce them to a minimum by adjusting the original plan. Simultaneously, at the end of each control period, a comparison between the budgeted cost for that period and the actual costs is performed. Both exercises are done on a regular basis for the most recent period, and on a cumulative basis for the time elapsed since the start of the project.

In the example project, under an early start schedule, activities A, B, and E are scheduled for weeks 1 through 4. Assuming for simplicity that the budget of each activity is linear in time, the weekly budget of activity A is $300 (a total budget of $1,500 divided by a duration of 5 weeks), as shown in Table 11-2. Similarly, activity B is budgeted for $3,000/3 is $1,000 per week, and activity E for $5,700/7 is about $814/week. The budget for the first week is therefore $300 + $1,000 + $814 = $2,114.

Based on the original schedule, activity A should be completed 5 weeks after the start of the project, activity B after 3 weeks, and activity E after 7 weeks. Figure 11-5 presents

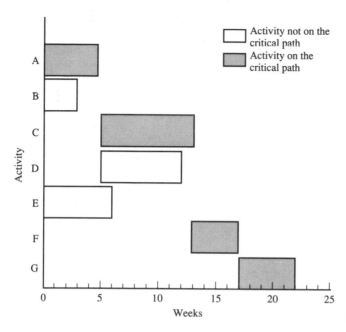

Activity not on the critical path

Activity on the critical path

Activity

Weeks

Figure 11-5   Gantt chart for an early start.

a Gantt chart for the original plan. A summary report of actual progress after 4 weeks together with the planned and actual costs is presented in Table 11-4. The Gantt chart in Fig. 11-5 illustrates the early start schedule for activities A, B, and E. The summary report in Table 11-4 indicates actual progress measured by work content performed, actual cost as reported by the accounting system, and the original budget for these activities.

Actual progress made can be estimated by several methods. In many instances it is a simple matter of measuring output. For example, assuming that activity E involves forging 70 bevel gears for a batch of transmissions and that by the end of the fourth week only 20 have been finished, then $\frac{2}{7}$ or 28.6% of the work content has been accomplished. Here the estimate of actual work completed is unbiased and exact. In other cases it may be more subjective, based on the opinion or observation of an expert such as a foreman, an engineer, the client representative, or the quality control group. A rough estimate can be used when the duration of activities is about the same as the length of the control period. In this

TABLE 11-4   SUMMARY REPORT FOR WEEKS 1–4

Activity	Actual cost	Budgeted cost	Work performed as % of work content
A	$1,500	$300 \times 4 = \$1,200$	100
B	3,000	3,000	100
E	$\dfrac{2,900}{\$7,400}$	$814 \times 4 = \dfrac{3,256}{\$7,456}$	$\frac{2}{7} = 28.6$

case an activity can be assumed 50% completed when it starts and 100% completed at its finish. This estimate is easy to perform and eliminates the need for a subjective measure.

Continuing with the example, a simple analysis of the costs for the first month does not identify any problems since actual costs ($7,400) are a bit less than budgeted costs ($7,456). Further, a critical path analysis based on actual progress reveals that the free slack of activity E (6 weeks) is shortened by 2 weeks due to delays but that activity E is still not on the critical path. Nevertheless, none of the analytic techniques discussed thus far is capable of detecting the deviations between the project plan and actual progress. More detail is needed to assess the situation accurately. In particular, an exhaustive cost/schedule control analysis that integrates cost data with information on actual progress reveals that the project is not only behind schedule but also over budget. This is due to the fact that in 4 weeks the actual progress on activity E is equal to the work content planned for the first 2 weeks. Thus activity E is subject to a 50% delay. Furthermore, the budgeted cost of $\frac{2}{7}$ of E is only $2 \times \$814 = \$1,628$, while its actual cost is $2,900 for the first 4 weeks.

This example illustrates the close relationship between cost, schedule and work content, and the need for an integrative measure that ties all three components together in the control system.

11.3.2 Earned Value Approach

The *earned value* (EV) concept integrates cost, schedule, and work performed by ascribing monetary values to each. In EV-based control systems only three variables are used as the basic building blocks. Each is discussed below.

1. *Budgeted cost of work scheduled* (BCWS) is defined as the value (in monetary units) of the work scheduled to be accomplished in a given period of time (a single control period, or an ordered sequence beginning with the first). The BCWS values of activities A, B, and E in the example project for the first month are as follows:

Activity	BCWS
A	$4 \times 300 = \$1,200$
B	$3,000
E	$4 \times 814 = \underline{\$3,256}$
	$7,456

Thus the work content scheduled to be accomplished during the first 4 weeks of the project is budgeted at $7,456.

2. *Actual cost of work performed* (ACWP) is defined as the cost actually incurred and recorded in accomplishing the work performed within the control period. In the example, these costs are:

Activity	ACWP
A	$1,500
B	3,000
E	2,900
	$7,400

As can be seen, a total of $7,400 was spent during the first 4 weeks to accomplish the work performed.

3. *Budgeted cost of work performed* (BCWP) is defined as the monetary value of the work actually accomplished within the control period. It is also known as the earned value. In the example, 100% of activity A is accomplished. Therefore, its BCWP is equal to the total budget of activity A, which is $1,500. Similarly, for activity B, BCWP = $3,000. However, for activity E, the work performed is only $\frac{2}{7}$ of the activity's estimated work content, so its BCWP = $5700 \times \frac{2}{7} = $1,628$. The BCWP values are summarized as follows:

Activity	BCWP
A	$1,500
B	3,000
E	1,628
	$6,128

The three measures BCWS, ACWP, and BCWP are the basis of the control analysis by which deviations in time and schedule are detected. In particular, we are concerned with the following.

1. *Schedule deviations*. The difference between the budgeted cost of work performed (BCWP) and the budgeted cost of work scheduled (BCWS) indicates (in monetary units) the deviation between the work content performed and the work content scheduled for the control period. If the absolute value of the difference is very small, then in terms of work content, the project is on schedule. A positive difference indicates that the project is ahead of schedule, and a negative difference implies that the project is late. Defining the schedule variance (SV) as the difference between BCWP and BCWS, we get:

Activity	BCWP − BCWS = SV
A	$1,500 − $1,200 = $300
B	$3,000 − $3,000 = $0
E	$1,628 − $3,256 = −$1,628
	Cumulative variance = −$1,328

Based on the SV values, we conclude that in activity A, the work performed is worth $300 more than what was planned for the control period; in activity B, the work performed is exactly equal to what was planned; and in activity E, the work performed is worth $1,628 less than what was planned for the period.

The cumulative variance is an indication in terms of work content performed, that the project is already late 4 weeks after its start. This measure, together with a simple CPM analysis, provides the means for tracking critical activities and for detecting overall trends in schedule performance. Although the delay in noncritical activities may not cause immediate project delays, the fact that these activities are not performed on schedule means that the resources required to perform them will be needed in a later period. This shift in resource requirements may cause a problem if the load on resources exceeds the available capacity.

The schedule delays detected by the earned value analysis should be monitored closely. When the delay extends beyond the control level, analysis of resource requirements should be initiated to test whether, due to resource limits, the entire project may be delayed. By combining CPM analysis to detect delays in critical activities with earned value analysis, the two major sources of schedule delays are monitored (i.e., delays in critical activities and delays caused by resource shortages).

2. *Cost deviations.* Deviations in cost are calculated based on the work content actually performed during the control period. Therefore, the *cost variance* (CV) is defined as the difference between the budgeted cost of work performed (BCWP) and the actual cost of work performed (ACWP). A positive CV indicates a lower actual cost than budgeted for the control period, while a negative CV indicates a cost overrun. The cost variance of activities A, B, and E for the first month of the example project is presented next.

Activity	BCWP − ACWP = CV
A	$1,500 − $1,500 = $0
B	$3,000 − $3,000 = $0
E	$1,628 − $2,900 = −$1,272
	Cumulative variance = −$1,272

Activities A and B are exactly on budget; the actual cost of performing these activities is equal to the budgeted cost for the accomplished work content. Activity E, however, shows a cost overrun of $1,272, since the work performed on this activity was budgeted at $1,628 whereas the actual cost turned out to be $2,900.

The schedule variance and the cost variance are absolute measures indicating deviations between planned performance and actual progress, in monetary units. Based on these measures, however, it is difficult to judge the relative schedule or cost deviation. A relative measure is important because a $1,000 cost overrun of an activity originally budgeted for $500 is clearly more troublesome than the same overrun on an activity originally budgeted for $50,000. A *schedule index* (SI) and a *cost index* (CI) are designed to be proportional measures of schedule and cost performance, respectively.

The schedule index is defined as the ratio BCWP/BCWS. Thus an SI value equal to 1 indicates that the associated activity is on schedule. Values larger than 1 suggest that the activity is ahead of schedule, and values smaller than 1 indicate a schedule overrun.

The cost index is defined as the ratio BCWP/ACWP, implying that when CI equals 1, the activity is on budget. CI values larger than 1 indicate better-than-planned cost performance, and values smaller than 1 indicate cost overruns.

Following are CI and SI values for the example project after 4 weeks:

Activity	$\dfrac{\text{BCWP}}{\text{BCWS}} = \text{SI}$	$\dfrac{\text{BCWP}}{\text{ACWP}} = \text{CI}$
A	$\dfrac{1,500}{1,200} = 1.25$	$\dfrac{1,500}{1,500} = 1$
B	$\dfrac{3,000}{3,000} = 1$	$\dfrac{3,000}{3,000} = 1$
E	$\dfrac{1,628}{3,256} = 0.5$	$\dfrac{1,628}{2,900} = 0.56$

These values indicate that during the control period, 25% more work was performed for activity A than planned (SI $=$ 1.25) but at the exact cost budgeted for that work content (CI $=$ 1). For activity B, the planned work content was performed at the planned cost, and for activity E only half of the planned work content was performed (SI $=$ 0.5). The planned cost of performing that work content was only 56% (CI $=$ 0.56) of the actual cost.

The schedule index and the cost index can be calculated for a single activity, for a group of activities, or for the whole project. This is done by accumulating the values of BCWS, BCWP, and ACWP for the appropriate activities and calculating the values of SI and CI based on these totals. For our example, the project schedule index after 4 weeks is

$$SI = \frac{1,500 + 3,000 + 1,628}{1,200 + 3,000 + 3,256} = 0.82$$

and the cost index is

$$CI = \frac{1,500 + 3,000 + 1,628}{1,500 + 3,000 + 2,900} = 0.83$$

The earned value analysis can be performed on a periodic or on a cumulative basis. Table 11-5 summarizes the three values (BCWS, BCWP, and ACWP) for activities A, B, and E for weeks 1 through 4. This information can also be presented graphically for each activity or for the entire project. Figure 11-6 depicts the cumulative values of BCWS, BCWP, and ACWP for each activity, and Fig. 11-7 presents these values for the entire project.

TABLE 11-5   VALUES (IN DOLLARS) OF BCWS, BCWP, AND ACWP FOR Weeks 1–4

Activity	Week 1			Week 2			Week 3			Week 4		
	BCWS	BCWP	ACWP	BCWS	BCWP	ACWP	BCWS	BCWP	ACWP	BCWS	BCWP	ACWP
A	300	500	500	300	500	500	300	300	300	300	200	200
B	1,000	1,000	1,000	1,000	1,000	1,000	1,000	500	500	0	500	500
E	814	300	814	814	400	686	814	500	1,000	814	428	400
	2,114	1,800	2,314	2,114	1,900	2,186	2,114	1,300	1,800	1,114	1,128	1,100

Depending on the activity, Fig. 11-6 illustrates three different situations:

a. *Activity A.* The earned value (BCWP) and the actual cost (ACWP) are the same and both are above BCWS. This implies that activity A is performed at cost and ahead of schedule.

b. *Activity B.* BCWP and ACWP are the same. During weeks 1 and 2 they are equal to BCWP (i.e., activity B is on budget and on schedule). In week 3 BCWP and ACWP are below BCWS, indicating a delay that causes activity B to finish in week 4 instead of week 3.

c. *Activity E.* The value of BCWP is consistently below BCWS and ACWP. Therefore, activity E is late and experiences a budget overrun.

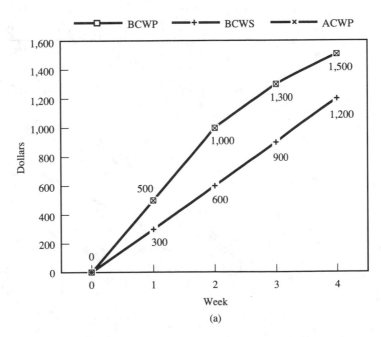

Figure 11-6   Earned value analysis: (a) activity A . . . (Continued on page 474).

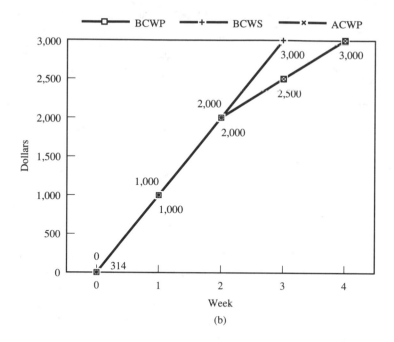

(b)

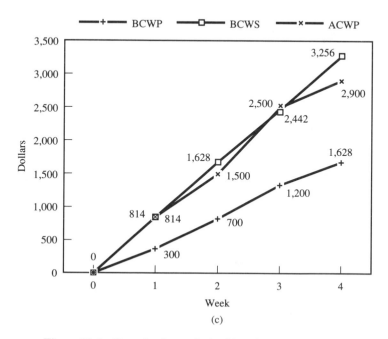

(c)

Figure 11-6 Earned value analysis: (b) activity B; (c) activity E.

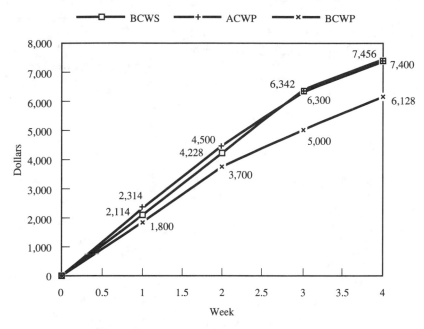

Figure 11-7   Earned value analysis for the project.

Figure 11-7 illustrates the project cost and schedule situation. BCWP is below BCWS and ACWP; thus the entire project is late and over budget. The schedule index and the cost index of the project for the first four weeks are summarized in Table 11-6.

An alternative view of the data in Fig. 11-7 is presented in Figs. 11-8 and 11-9, where the values of SI and CI are plotted as a function of time. Both SI and CI are below 1, which means that the project is late and suffers from budget overruns. Furthermore, there is no clear trend of improvement in SI and CI.

To integrate schedule and cost information, the values of SI and CI are plotted together in Fig. 11-10. Each point on the graph corresponds to a control period. By observing the time associated with each point, it is possible to see the trend in the cost and schedule indices.

TABLE 11-6   VALUES OF SI AND CI FOR WEEKS 1–4

Week	BCWS	BCWP	ACWP	$CI = \dfrac{BCWP}{ACWP}$	$SI = \dfrac{BCWP}{BCWS}$
1	$2,114	$1,800	$2,314	0.78	0.85
2	4,228	3,700	4,500	0.82	0.88
3	6,342	5,000	6,300	0.79	0.79
4	7,456	6,128	7,400	0.83	0.82

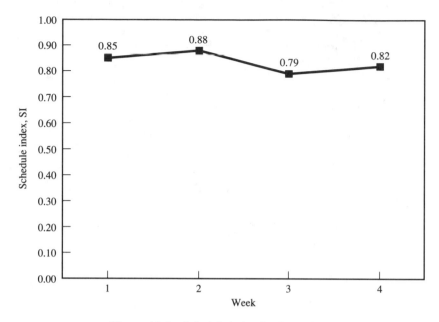

Figure 11-8   Schedule index for the project.

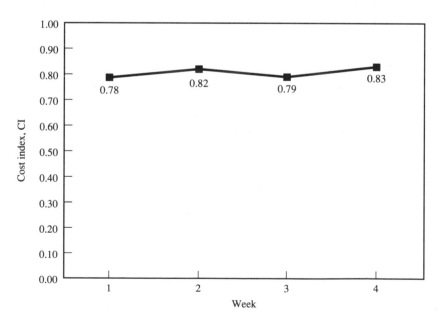

Figure 11-9   Cost index for the project.

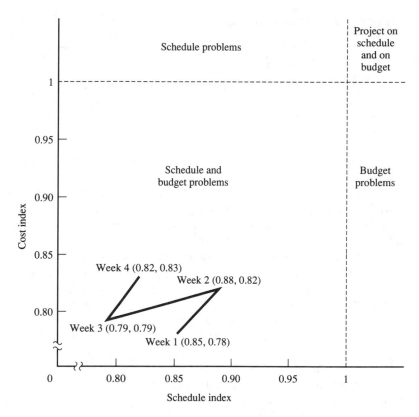

Figure 11-10   Integrating CI and SI.

The objective of project management is to have both CI and SI larger than or equal to 1, which would place them in the upper right quadrant of Fig. 11-10. Analysis of Fig. 11-10 indicates that in week 4 both CI and SI show improved performance. This is after a similar improvement in week 2, which was followed by poor performance in week 3.

11.4 REPORTING PROGRESS

The values of BCWP and ACWP for each activity are the building blocks in a progress report. The OBS–WBS matrix that relates each activity to a bottom-level OBS unit and to a bottom-level WBS element facilitates analysis at any OBS level, WBS level, or combination of the two. For example, from Fig. 11-4 we see that activities A, B, and E are performed by department 2. None of the activities assigned to department 1 is scheduled for the first month. Thus the OBS-based progress report given in Table 11-7 for the first month shows no activity for department 1 and a summary of activities A, B, and E for department 2.

TABLE 11-7   CUMULATIVE COST AND SCHEDULE CONTROL REPORT BY OBS ELEMENT
(WEEKS 1–4)

Organizational unit	BCWS	BCWP	ACWP	SV	CV	SI	CI
Department 1	0	0	0	0	0	—	—
Department 2	$7,456	$6,128	$7,400	−$1,328	−$1,272	0.82	0.83
Total project	$7,456	$6,128	$7,400	−$1,328	−$1,272	0.82	0.83

Based on the data in Table 11-7, it is clear that department 1 was not scheduled to work on the project during the first month and indeed did not perform any activities. Department 2 was scheduled to perform work content budgeted at $7,456 but completed only $6,128 worth of work, while the actual cost for the period was $7,400. One can therefore conclude that department 2 ran into trouble performing its work content, precipitating a budget overrun of $1,328 during the first month.

The WBS report for the example project is contained in Table 11-8. The similarity in structure between this report and the OBS report stems from the fact that both are based on the same data—the project plan and the same three measures: BCWS, BCWP, and ACWP. The WBS report reveals that element III should be monitored carefully since it is experiencing both a schedule delay and a budget overrun.

The reports in Tables 11-6 and 11-7 can be produced for each control period or on a cumulative basis from the start of the project. Many computer packages that support EV provide this information. The totals in Tables 11-6 and 11-7 are identical. This is because both represent the total project performance for the first 4 weeks. The accumulation of information from lower-level OBS or WBS elements to the project level (or any other higher level) is called *roll-up*. This roll-up mechanism is possible in both the OBS and WBS because of their hierarchical nature. At each level of the OBS or WBS the values of ACWP, BCWS, and BCWP associated with each organizational unit or WBS element are calculated as the sum of the corresponding values of the organizational units or WBS elements under it. Using the roll-up mechanism it is possible to generate reports at different OBS and WBS levels according to management needs. Based on the cumulative values of BCWS, BCWP, and ACWP, the cost and schedule variances can be calculated. Thus the integration of the two hierarchical structures (OBS and WBS) with the earned value concept provides the foundation for an information system that supports cost and schedule control at each managerial level.

TABLE 11-8   COST AND SCHEDULE CONTROL REPORT BY WBS ELEMENT

WBS Element	BCWS	BCWP	ACWP	SV	CV	SI	CI
I	$1,200	$1,500	$1,500	$300	0	1.25	1
II	3,000	3,000	3,000	0	0	1	1
III	3,256	1,628	2,900	−1,628	−1,272	0.5	0.56
	$7,456	$6,128	$7,400	−$1,328	−$1,272	0.82	0.83

11.5 UPDATING COST AND SCHEDULE ESTIMATES

When data reflecting the current status of tasks and actual costs are collected, it is only logical to update previous estimates of the project's completion time and budgetary requirements. Estimates tend to improve as actual progress is made. This is due to the completion of activities for which actual duration and cost become known, as well as to better information on workforce productivity and the availability and cost of resources. Bear in mind that the original estimates are usually based on historical records of similar projects and may be quite problematic. When new data become available, a new critical path analysis can be performed in which the actual duration of past activities and updated estimates for the duration of future activities are incorporated.

New estimates derived from current information are the basis of trend analysis. If, for example, a recent estimate indicates that the expected total cost of the project is (much) higher than the original budget, a management decision may be needed. The revised estimate may cause a change in specification so that the expected total cost will not exceed the budget, a change in schedule aimed at replanning future cash flow according to available budgets, or in the extreme, abandonment of the project. Other options may also exist. The task of the control system is to focus management's attention on potential problems as soon as the likelihood of such problems, based on trend analysis and updated estimates, is deemed high.

To reestimate the cost of the project, acceptable accounting procedures must be defined together with the necessary data elements. The following notation will be used for this purpose:

BAC   *Budget at completion:* total budget of the project activities based on the original project plan
     = sum of BCWS values over lower-level OBS elements, or
     = sum of BCWS values over lower-level WBS elements
WR   *Work remaining:* budgeted cost of the work not yet accomplished by the end of the reporting period; WR = BAC − BCWP
ETC   *Estimate to complete:* updated estimate of the cost of the work remaining (WR)
EAC   *Estimate at completion:* updated estimate of the total project cost; EAC = ACWP + ETC

Because the value of ACWP is known, only a revised estimate of ETC is required to update the EAC estimate.

Estimating EAC: original estimate approach.   This approach is based on the assumption that the original estimate of the cost of work remaining is valid and therefore only the original estimate of the work that was already performed should be replaced by the actual cost of that work content.

$$EAC = ACWP + ETC$$

and since

$$ETC = BAC - BCWP$$

we get

$$EAC = ACWP + (BAC - BCWP)$$
$$= BAC - (BCWP - ACWP) = BAC - CV$$

Thus in the revised budget, the estimate at completion (EAC) is equal to the original budget BAC adjusted by the cost variance (CV).

Estimating EAC: revised estimate approach. Assuming that data collected on past performance can be used to improve the estimates of future costs, this approach is based on (1) updates of the actual costs of work content performed, and (2) the estimated cost of the remaining work content. The updated estimate of work remaining is based on the assumption that the relative deviation in the cost of the work completed is a good estimate for the relative deviation of the cost of remaining work.

The relative deviation of cost of work completed is defined as follows:

$$\frac{ACWP}{BCWP} = \frac{1}{CI}$$

Assuming the same deviation factor for WR, we get

$$ETC = WR \times \frac{1}{CI} = (BAC - BCWP) \times \frac{1}{CI}$$

Therefore, we can write

$$EAC = ACWP + (BAC - BCWP) \times \frac{1}{CI} = ACWP + \frac{BAC}{CI} - \frac{BCWP}{CI}$$

Substituting

$$ACWP = \frac{BCWP}{CI}$$

we get

$$EAC = \frac{BAC}{CI} = BAC \times \frac{ACWP}{BCWP}$$

The two estimation procedures can be applied at each OBS level, at each WBS level, or at the total project level. In the example project, the report after 1 month shows the following results:

$$
\begin{array}{lll}
BCWS = \$7,456 & CV = -\$1,272 & CI = 0.83 \\
BCWP = \$6,128 & SV = -\$1,328 & SI = 0.82 \\
ACWP = \$7,400 & &
\end{array}
$$

Thus revised costs using (1) the original estimate approach, and (2) the revised estimate approach are:

$$(1) \ \text{EAC} = \text{ACWP} + \text{BAC} - \text{ACWP} = \text{BAC} - \text{CV}$$
$$= \$31,000 - (-\$1,272) = \$32,272$$

$$(2) \ \text{EAC} = \text{BAC} \times \frac{1}{\text{CI}} = \$31,000 \times \frac{1}{0.83} = \$37,349$$

The difference between the two values stems from the fact that in the first approach we assume that past cost performance is not a predictor for future performance, while in the second, we assume that the past deviations are a good predictor of the cost deviations in the remaining work.

The two estimating procedures are not the only alternatives available. Other techniques, such as those based on time-series or regression analysis, may be used as well. The important point is to be consistent and use the same estimation procedure for the entire project throughout its life cycle.

The selection of an estimation procedure is a management decision that should be made in the conceptual design phase of the project. Consistency in predicting total costs results in the ability to show at each control period the current cost status together with the trend of cost predictions from the start of the project. Such consistency enables comparisons of performance between OBS and WBS elements at different time periods, as well as the monitoring of cost trends that foreshadow future problems.

11.6 TECHNOLOGICAL CONTROL: QUALITY AND CONFIGURATION

Cost and schedule control are important management responsibilities. Technological control is required to detect any deviations from specifications and standards that may change during the life cycle of the project. To achieve a satisfactory level of performance, an integrated quality control and quality assurance program with well-established procedures must be designed and implemented.

The concept of total quality control is relevant for the success of a project. Based on this concept, quality becomes the focal point of any organizational unit (OBS element) performing work on any element of the project (WBS element) at any point in the project life cycle. In the early stages, systems engineers evaluate various design alternatives based on performance, quality, and reliability measures, as well as cost and schedule. It is well known among these engineers that the bitter taste of a low-quality, unreliable product lingers long after the sweet taste of low cost and fast delivery.

The alternative selected in the initial stages of a project is designated the "baseline" for purposes of configuration management and control. Recall that configuration management (CM) is a system designed to ensure that the product delivered at the end of the project is built according to the specifications laid out in the baseline and all subsequent engineering change requests (ECRs). The components, procedures, and logic of the configuration management system are discussed in Chapter 6.

An effective control system integrates all four aspects of control throughout the life cycle of the project. This is illustrated in Table 11-9. The interface between cost control and schedule control is the earned value concept measured by BCWP and compared to ACWP and BCWS, as discussed earlier. The interface between schedule/cost control and quality control is achieved by an appropriate procedure that guarantees that only the work approved by the quality control group will be recognized as earned value. Thus by equating BCWP with the earned value of work performed within accepted and approved standards, full integration of cost control, schedule control, and quality control is achieved.

TABLE 11-9   INTEGRATING THE FOUR CONTROL SYSTEMS

	Cost	Configuration	Quality
Schedule	Earned value system	Configuration control board	Earned value system
Cost		Configuration control board	Earned value system
Configuration			Configuration review and audit board

The last component—configuration control—is integrated with quality control by a mechanism called *configuration test and audit*. This component of the CM system is designed to guarantee that the quality control activity is based on the most recent configuration composed of the baseline design and all approved ECRs. The integration of configuration management with cost and schedule control is done at the *configuration control board* (CCB). The CCB is the focal point of configuration control, as explained in Chapter 6. Members of the CCB are representatives of the project and the functional areas that might be affected by proposed design changes. At the CCB, engineering change requests are evaluated based on their impact on cost, schedule, and performance. By linking all four control systems together, deviations in cost, schedule, quality, or design can be detected and addressed in a timely manner.

The four basic control systems associated with cost, schedule, quality, and configuration operate throughout the project life cycle within the framework of the OBS–WBS matrix. Together they are used to detect deviations, to identify their organizational source and their effect on various elements of the work breakdown structure, and to assist in developing solutions to problems caused by such deviations.

A threshold value may be used to trigger management-by-exception activities. For example, it is possible to have a 5% threshold on CI and a 10% threshold on SI, so that any deviation larger than the threshold value is reported to higher levels of management along with a correction plan. The exact values of each threshold as well as the procedures for reporting and planning are specific issues that each organization has to deal with individually. In the following section we discuss some general guidelines for control systems.

11.7 COST/SCHEDULE CONTROL SYSTEMS CRITERIA

Control systems are part of an organization's management information system (MIS). Each organization tends to develop or adopt an MIS that fits its needs, its structure, and the environment in which it operates.

Clients who fund R&D projects, and say, due to technological uncertainties, agree to pay the actual cost of the project plus a predetermined contractor fee (cost plus fixed fee contract) face the problem of controlling the activities of different contractors, each of whom may be using a different control system. Major clients such as the U.S. Department of Defense (DOD), the U.S. Department of Energy (DOE), and the National Aeronautic and Space Administration (NASA), have developed guidelines or requirements that their contractors must incorporate in their respective control systems. The common approach is to let contractors choose the MIS and control system that best suit their needs, subject to a set of criteria called *cost/schedule control systems criteria* (C/SCSC). Rules are given for the following five aspects of the project: organization, planning and budgeting, accounting, analysis, and revisions and access to data. Appendix 11B lists the criteria used by the Department of Energy. Similar criteria are used by the DOD and NASA.

11.8 LINE OF BALANCE

Project management techniques discussed so far are designed for the one-time effort where a specific, unique set of goals related to a single project has to be met. There are, however, projects that involve more than one product unit. Examples might include the construction of multiple power plants at the same site, or an order to build identical ships at a shipyard. In the latter case, it is possible to view each ship as a project (though no longer unique), or to define the total order as a single project with repetitive activities. Because such projects are not uncommon, a special technique called the *line of balance* (LOB) has been developed to support their management and control.

The LOB technique is based on control points or milestones in the production process of the product. These control points are related to critical activities and resources that are identified during the planning cycle. A typical control point is the successful completion (including test and inspection) of an activity on the critical path. The elapsed time between consecutive control points is estimated and a milestone schedule for a single product unit is developed.

The master production schedule (MPS) in such projects specifies the planned delivery time of each product unit, based on the contractual agreement with the client. As the project starts, control is exercised by comparing the number of units that pass each control point with the number that should have passed that point according to the MPS. Any deviations trigger a detailed analysis aimed at identifying the cause of the deviation and the appropriate corrective action.

To illustrate the LOB approach, consider a manufacturer of communication systems. Each system is tailormade for the customer, who may place one or several identical orders. Suppose that a customer orders a total of 110 systems in a specific configuration. It is estimated that 6 weeks is required to complete 1 unit. Four milestones are selected as control points (see Table 11-10):

  A. *End of rack installation:* 2 weeks after the start of work on a system
  B. *End of subsystems (modules) installation:* 3 weeks after the start of work on a system

TABLE 11-10   SCHEDULE OF MILESTONES OR CONTROL POINTS

Control point	Description	Week after start of unit production	Lead time to delivery (weeks)
A	Rack installation	2	$6 - 2 = 4$
B	Subsystems installation	3	$6 - 3 = 3$
C	Subsystems integration	5	$6 - 5 = 1$
D	Acceptance test	6	$6 - 6 = 0$

C. *End of subsystems integration:* 5 weeks after the start of work on a system

D. *End of acceptance tests and delivery:* 6 weeks after the start of work on a system

The MPS specifies delivery dates for the 110 systems in accordance with the data in Table 11-11. Based on the MPS and the list of control points (or milestones) it is possible to forecast the number of systems expected to pass through each milestone at the end of each week. For example, the number of systems expected to pass each milestone by the end of the fifth week is as follows:

- The 20 systems scheduled for delivery on week 10 should be 5 weeks from delivery. Because it takes 6 weeks to complete a system, these systems should be 1 week in process, not having passed any milestone yet.
- The 30 systems scheduled for delivery on week 9 should be 4 weeks from delivery or 2 weeks into the process, and should have completed milestone A only.
- The 10 systems scheduled for delivery on week 8 should be 3 weeks from delivery or 3 weeks into the process, and should have completed milestones A and B.
- The 20 systems scheduled for delivery on week 7 should be 2 weeks from delivery or 4 weeks into the process, and should have finished milestones A and B.
- The 30 systems scheduled for delivery on week 6 should be 1 week from delivery or 5 weeks into the process, and should have finished milestones A, B, and C.

These results are summarized in Table 11-12. Thus 90 systems should have completed milestone A, 60 milestone B, and 30 milestone C. Figure 11-11 displays this information graphically.

TABLE 11-11   DELIVERY SCHEDULE FOR THE 110 SYSTEMS

Delivery date as of week:	Systems scheduled for delivery	Cumulative number of systems
6	30	30
7	20	50
8	10	60
9	30	90
10	20	110

TABLE 11-12   SCHEDULED MILESTONES AT THE END OF WEEK 5

Deliveries scheduled on week:	Number of systems	Time to delivery (weeks)	Number of systems scheduled to finish at milestone:			
			A	B	C	D
5	—	—	—	—	—	—
6	30	1	30	30	30	—
7	20	2	20	20	—	—
8	10	3	10	10	—	—
9	30	4	30	—	—	—
10	20	5				
			90	60	30	0

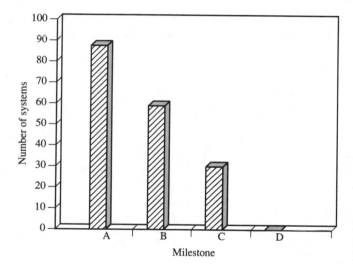

Figure 11-11   Planned number of systems to finish each milestone after 5 weeks.

It is possible to use a graphical procedure to control a repetitive project by combining the milestone information with the master production schedule. To construct the control chart, we use a graphic display of the MPS, where the cumulative number of systems is plotted versus time. On this display, start at the current control period (week 5), and for each milestone, add its corresponding lead time. For milestone A, the lead time is 4 weeks. Adding 4 weeks to the current control period (week 5), we get 9 weeks. The cumulative number of units corresponding to 9 weeks on the MPS is 90 units, as illustrated in Fig. 11-12. Thus the expected number of systems to complete milestone A is 90 systems. In a similar way, the expected number of systems can be constructed for each milestone.

The LOB displays the work that should be accomplished to ensure delivery according to the MPS. Suppose that after 5 weeks, 80 systems completed milestone A, 60 completed milestone B, 40 completed milestone C, and 20 systems completed milestone D. The deviations between the plan (LOB) and actual achievement are as follows:

Milestone	LOB	Actual	Deviation
A	90	80	−10
B	60	60	0
C	30	40	−10
D	0	20	20

Thus milestones A and C are late with respect to the MPS. At A, 10 systems late corresponds to a 10/90 × 100% = 11% delay. Milestone B is exactly on schedule, while at C and D actual performance is ahead of schedule.

A detailed analysis of the activities performed prior to milestone A should be initiated. In case an increase in the workforce is required to catch up with the MPS, the necessary resources may be obtained from some of the activities that precede milestones C and D, which are ahead of schedule.

A graphical display of the LOB and the actual performance gives a clear indication of the project's status. Figure 11-13 depicts the situation for week 5 in the example.

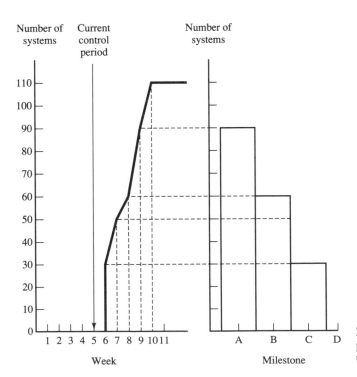

Figure 11-12 Constructing the planned status from the master production schedule.

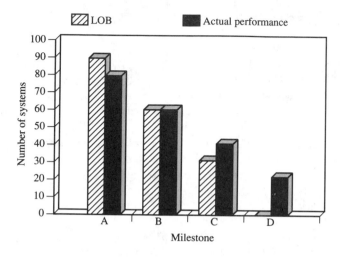

Figure 11-13   LOB and actual performance.

TEAM PROJECT

Thermal Transfer Plant

With the approval of the rotary combustor project, a detailed plan for project control is required. In developing the plan, your team should address the following issues:

1. Which aspects of the project should be monitored (e.g., cost, schedule)?
2. Where will the data come from?
3. What is the original source of data?
4. How often should data be collected?
5. How should the data be processed? (Distinguish between trend analysis and identification of exceptions.)
6. What kind of reports will be issued? Who should get the reports? How often?
7. What kind of ad hoc questions should the control system support?

Be specific as possible. Present a flow diagram for data processing and the format of each report suggested. Be careful not to produce too many reports or to collect data that will not be used later. Explain and justify your approach to the control of the project.

DISCUSSION QUESTIONS

1. Describe the control systems used in one organization with which you are familiar.
2. Referring to Question 1, explain how each control system deals with uncertainty.
3. Give an example of an organization that does not use any control systems. Is this justified?
4. Suppose that you have decided to build a new house; explain what kind of project control you will consider, and why.
5. Why is it important to integrate cost and schedule control? Give an example where separate cost and schedule control systems may not function properly.
6. Explain how you would measure the earned value of the following activities.
   (a) Writing a term paper
   (b) Building a nuclear power plant
   (c) Designing a new car
   (d) Developing a new training program
7. Is there a need for "technological control" in developing a new insurance policy? Explain.
8. Explain what the responsibilities of "quality control" are in a project associated with making a Hollywood-style movie. How would these responsibilities differ if the movie were a documentary on, say, the Iran-Contra scandal?
9. Is there a need for a control system in projects performed by nonprofit organizations? Explain and give examples.
10. Explain the advantages and disadvantages of the line-of-balance technique as opposed to using several PERT networks.
11. How would you build TQM principles into a project control system?

EXERCISES

11-1. The National Institutes of Health (NIH) support research and development of new treatments for AIDS (acquired immune deficiency syndrome). Develop a project control system by which the agency will be able to control projects that it supports.
   (a) What are the objectives of the control system?
   (b) What are the performance measures?
   (c) What data are required?
   (d) How should raw data be collected?
   (e) How should the data be analyzed?
   (f) How should the results be reported, and how often?

11-2. Consider the project plan defined by Table 11-13. A cost schedule control system produces weekly reports. The reports for weeks 1, 2, and 3 are shown in Table 11-14. (Assume 5 working days each week).
   (a) Write a weekly progress report for each activity based on the above information.
   (b) Comment on the level of control that can be achieved based on the given information.

11-3. In Exercise 11-2, an estimate of the "percent complete" for each activity each week is reported in Table 11-15. Redo parts (a) and (b).

TABLE 11-13

Activity	Scheduled start day	Scheduled finish day	Cost/day
A	1	3	$1,000
B	1	5	5,000
C	3	7	3,000
D	5	15	1,000
E	7	22	2,000
F	7	25	4,000

TABLE 11-14

Activity	Week 1 Status	Week 1 Cost	Week 2 Status	Week 2 Cost	Week 3 Status	Week 3 Cost
A	In process	$ 1,500	Finished	$ 3,000	Finished	$ 3,000
B	In process	25,000	Finished	30,000	Finished	30,000
C	In process	7,000	Finished	10,000	Finished	10,000
D	Not started	0	In process	5,000	In process	7,000
E	Not started	0	Not started	0	In process	10,000
F	Not started	0	In process	10,000	In process	20,000

TABLE 11-15

Activity	Percent complete Week 1	Week 2	Week 3
A	50	100	100
B	30	100	100
C	10	100	100
D	0	20	60
E	0	0	25
F	0	30	40

11-4. An activity on the critical path of a project was scheduled to be completed within 12 weeks with a budget of $8,000. During a performance review that took place 7 weeks after the activity was initiated, it was found that 50% of the work had already been completed and that the actual cost was $4,500.

(a) Calculate the earned value of the activity.
(b) Calculate the cost and schedule indices for the activity.
(c) Calculate the expected budget at completion using the original estimate approach.
(d) Calculate the expected budget at completion using two other approaches.
(e) Compare and discuss the results obtained in parts (c) and (d).

11-5. The performance of a project was evaluated 10 weeks after its start. The relevant information concerning the project is shown in Table 11-16.

TABLE 11-16

Activity	Immediate predecessors	Normal time	Budget	Organization unit	Percent complete	Money spent
A	—	4	$ 90	U1	100	$110
B	A	2	35	U2	100	20
C	A	6	75	U2	40	40
D	B	3	60	U1	80	90
E	C	10	80	U1	0	0
F	—	2	40	U2	100	40
G	F	5	55	U1	50	30
H	F	7	80	U2	100	60
I	D,E,G	1	40	U2	0	0
J	H	10	100	U1	0	0

   (a) On the same Gantt chart, show the project plan and the project progress, and discuss the two.

   (b) Calculate the schedule index for each organizational unit U1 and U2 and for the project as a whole. Discuss.

   (c) Repeat part (b) for the cost index.

   (d) Based on past performance, update the expected completion time and budget. State your assumptions.

11-6. For the project described in Exercise 11-5, calculate and chart the following values: BCWS, BCWP, and ACWP. Assume linearity of cost versus time. State any additional assumptions that you feel are needed.

11-7. Your university has decided to start a new program for executives called "Management of Technology." Your task is to design the control system for this project. Discuss the following issues.

   (a) The performance measure that should be used.

   (b) Ways to collect the relevant data for evaluating the present situation.

   (c) How should raw data be selected for evaluating the project?

   (d) How should the data be analyzed?

   (e) How should the results be reported?

11-8. In designing the new program outlined in Exercise 11-7, identify the work packages and the organizations to be responsible for their implementation.

11-9. Explain configuration management and control within the curriculum of your school. Give three examples that demonstrate a good configuration control process and three that identify poor configuration management.

REFERENCES

BADIRU, A. B., *Project Management in Manufacturing and High Technology Operations*, Wiley, New York, 1988.

HILL, L. S., "Some Cost Accounting Problems in PERT/Cost," *Journal of Industrial Engineering*, Vol. 17, No. 2 (1966), pp. 87–99.

HOWARD, D. C., "Cost/Schedule Control Systems," *Management Accounting*, Vol. 58, No. 4 (1976), pp. 21–25.

MUELLER, F. W., *Integrated Cost and Schedule Control for Construction Projects*, Van Nostrand Reinhold, New York, 1986.

PETERSON, P., "Project Control Systems," *Datamation* (1979), pp. 147–163.

PRYOR, S., "Project Control, Part 1: Planning and Budgeting," *Management Accounting*, Vol. 66, No. 5 (1988), pp. 16–17.

PRYOR, S., "Project Control, Part 2: Measuring, Analyzing and Reporting," *Management Accounting*, Vol. 66, No. 6 (1988), pp. 18–19.

SHTUB, A., "Evaluation of Two Schedule Control Techniques for the Development and Implementation of New Technologies: A Simulation Study," *R&D Management*, Vol. 22, No. 1 (1992) pp. 81–87

SINGER, A., "When the Stakes Are High (Cost Overruns in Capital Projects)," *Accountancy*, Vol. 98 (1986), pp. 92–93.

TIONG, R. L. K., "Effective Controls for Large Scale Construction Projects," *Project Management Journal*, Vol. 21, No. 1 (1990), pp. 32–42.

TURBAN, E., "The Line of Balance: A Management by Exception Tool," *Journal of Industrial Engineering*, Vol. 19 (1968), pp. 440–448.

TURNER, W. S., III, *Project Auditing Methodology*, North-Holland, Amsterdam, 1980.

U. S. Department of Defense, *Cost/Schedule Control Systems Criteria Joint Implementation Guide*, Washington, DC. Several publication numbers issued by various DOD agencies since October 1976.

U. S. Department of Defense, *Performance Measurement for Selected Acquisitions*, DOD 7000.2, Washington, DC, June 1977.

U. S. Department of Energy, *Cost/Schedule Control Systems Criteria for Contract Performance Measurement: Work Breakdown Structure Guide*, Office of Project and Facilities Management, Washington, DC, 1981.

U. S. Department of Energy, *Cost/Schedule Control Systems Criteria for Contract Performance Measurement*, DOE/2250.1A, Office of Project and Facilities Management, Washington, DC, September 1982.

WARD, S. A., and T. LITCHFIELD, *Cost Control in Design and Construction*, McGraw-Hill, New York, 1980.

ZELDMAN, M., *Keeping Technical Projects on Target*, American Management Association, New York, 1978.

APPENDIX 11A

Example of a
Work Breakdown Structure

The work breakdown structure of engineering projects is an important building block of the project management system. Organizations that are frequently engaged in engineering projects have developed guidelines for designing the WBS. The following is a summary WBS for an electronic system. This is one of several WBSs presented in MIL-STD-881-A, *Work Breakdown Structures for Defense Material Items*, April 25, 1975.

Level 1	Level 2	Level 3
Electronic system	Prime mission equipment	Integration and assembly
		Sensors
		Communications
		Automatic data processing equipment
		Computer programs
		Data displays
		Auxiliary equipment
	Training	Equipment
		Services
		Facilities
	Peculiar support equipment	Organizational/intermediate (including equipment common to depot)
		Depot (only)
	Systems test and evaluation	Development test and evaluation
		Operational test and evaluation
		Mockups
		Test and evaluation support
		Test facilities
	System/program management	Systems engineering
		Project management
	Data	Technical publications
		Engineering data
		Management data
		Support data
		Data depository

(continues)

Level 1	Level 2	Level 3
	Operational/site activation	Contractor technical
		Support
		Site/construction
		Site/ship/vehicle
		Conversion
		System assembly
		Installation and checkout on site
	Common support equipment	Organizational/intermediate (including equipment common to depot)
		Depot (only)
	Industrial facilities	Construction/conversion/expansion
		Equipment acquisition or modernization
		Maintenance
	Initial spares and initial repair parts	(Specify by allowance list, grouping, or hardware element)

Source: MIL-STD-881-A, April 25, 1975.

APPENDIX 11B

Department of Energy Cost/Schedule Control Systems Criteria

1. *General*
   a. The management control systems used by the contractor in planning and controlling the performance of the contract shall meet the criteria set forth in paragraph 2 below. Nothing in these criteria is intended to affect the basis on which costs are reimbursed and progress payments are made, and nothing herein will be construed as requiring the use of any single system, or specific method of management control or evaluation of performance. The contractor's systems need not be changed, provided they satisfy the criteria.
   b. An element in the evaluation of proposals will be proposer's systems for planning and controlling contract performance. The proposer will fully describe the system to be used. The prospective contractor's cost and schedule control system proposal will be evaluated to determine whether it meets the criteria. The prospective contractor will agree to operate compliant systems throughout the period of contract performance if awarded the contract. DOE will rely on the contractor's compliant systems and, therefore, will not impose separate management control systems.
2. *The Criteria.* The contractor's management control systems will include policies, procedures, and methods which are designed to ensure that they will accomplish the following:
   a. *Organization*
      (1) Define all authorized work and related resources to meet the requirements of the contract, using the framework of the contract work breakdown structure.
      (2) Identify the internal organizational elements and the major subcontractors responsible for accomplishing the authorized work.
      (3) Provide for integration of the contractor's planning, scheduling, budgeting, estimating, work authorization, and cost accumulation systems with each other, the contract work breakdown structure, and the organizational structure.
      (4) Identify the managerial positions responsible for controlling overhead (indirect costs).

(5) Provide for integration of the contract work breakdown structure with the contractor's functional organizational structure in a manner that permits cost and schedule performance measurement for contractor work breakdown structure and organizational elements.

b. *Planning and budgeting*

(1) Schedule the authorized work in a manner which describes the sequence of work and identifies the significant task interdependencies required to meet the development, production, construction, installation, and delivery requirements of the contract.

(2) Identify physical products, milestones, technical performance goals, or other indicators that will be used to measure output.

(3) Establish and maintain a time-phased budget baseline at the cost account level against which contract performance can be measured. Initial budgets established for this purpose will be based on the negotiated target cost. Any other account used for performance measurement purposes must be formally recognized by both the contractor and the Government.

(4) Establish budgets for all authorized work with separate identification of cost elements (labor, material, and so forth).

(5) To the extent the authorized work can be identified in discrete, short-span work packages, establish budgets for this work in terms of dollars, hours, or other measurable units. Where the entire cost account cannot be subdivided into detailed work packages, identify the long-term effort in larger planning packages for budget and scheduling purposes.

(6) Provide that the sum of all work packages budgets, plus planning package budgets within a cost account equals the cost account budget.

(7) Identify relationships of budgets or standards in underlying work authorization systems to budgets for work packages.

(8) Identify and control level of effort activity by time-phased budgets established for this purpose. Only that effort which cannot be identified as discrete, short-span work packages or as apportioned effort will be classed as level of effort.

(9) Establish overhead budgets for the total costs of each significant organizational component whose expenses will become indirect costs. Reflect in the contract budgets at the appropriate level the amounts in overhead pools that will be allocated to the contract as indirect costs.

(10) Identify management reserve and undistributed budget.

(11) Provide that the contract target cost plus estimated cost of authorized but unpriced work is reconciled with the sum of all internal contract budgets and management reserve.

c. *Accounting*

(1) Record direct costs on an applied or other acceptable basis in a formal system that is controlled by the general books of account.

(2) Summarize direct costs from cost accounts into the work breakdown structure without allocation of a single cost account to two or more work breakdown structure elements.

(3) Summarize direct costs from the cost accounts into the contractor's functional organizational elements without allocation of a single cost account to two or more organizational elements.

(4) Record all indirect costs which will be allocated to the contract.

(5) Identify the bases for allocating the cost of apportioned effort.

(6) Identify unit costs, equivalent unit costs, or lot costs as applicable.

(7) The contractor's material accounting system shall provide for:

    (a) Accurate cost accumulation and assignment of costs to cost accounts in a manner consistent with the budgets, using recognized, acceptable costing techniques.

    (b) Determination of price variances by comparing planned versus actual commitments.

    (c) Cost performance measurement at the point in time most suitable for the category of material involved, but no earlier than the time of actual receipt of material.

    (d) Determination of cost variances attributable to the excess usage of material.

    (e) Determination of unit or lot costs when applicable.

    (f) Full accountability for all material purchased for the contract, including the residual inventory.

d. *Analysis*

(1) Identify at the cost account level on a monthly basis using data from, or reconcilable with, the accounting and budgeting systems:

    (a) Budgeted cost for work scheduled and budgeted cost for work performed.

    (b) Budgeted cost for work performed and applied (actual where appropriate) direct costs for the same work.

    (c) Estimates at completion and budgets at completion.

    (d) Variances resulting from the above comparisons classified in terms of labor, material, or other appropriate elements together with the reasons for significant variances, including technical problems.

(2) Identify on a monthly basis, in the detail needed by management for effective control, budgeted indirect costs, actual indirect costs, and variances along with the reasons.

(3) Summarize the data elements and associated variances listed in paragraph 2d(1) and (2) above through the contractor organization and contract work breakdown structure to the reporting level specified in the contract.

(4) Identify significant differences on a monthly basis between planned and actual schedule accomplishment together with the reasons.

(5) Identify managerial actions taken as a result of paragraph 2d(1) through (4) above.

(6) Based on performance to date and on estimates of future conditions, develop revised estimates of cost at completion for work breakdown structure elements identified in the contract and compare these with the contract budget base and the latest statement of funds requirements reported to the Government.

e. *Revisions and access to data*

(1) Incorporate contractual changes in a timely manner recording the effects of such changes in budgets and schedules. In the directed effort before negotiation of a change, base such revisions on the amount estimated and budgeted to the functional organizations.

(2) Reconcile original budgets for those elements of the work breakdown structure identified as priced line items in the contract, and for those elements at the lowest level of the project summary work breakdown structure, with current performance measurement budgets in terms of changes to the authorized work and internal replanning in the detail needed by management for effective control.

(3) Prohibit retroactive changes to records pertaining to work performed that will change previously reported amounts for direct costs, indirect costs, or budgets, except for correction of errors and routine accounting adjustments.

(4) Prevent revisions to the contract budget base except for Government-directed changes to contractual effort.

(5) Document, internally, changes to the performance measurement baseline, and on a timely basis, notify the Government project management through prescribed procedures.

(6) Provide the contracting officer and his or her duly authorized representatives access to all of the foregoing information and supporting documents.

12

Research and Development Projects

12.1 INTRODUCTION

Over the last decade 40% of the Fortune 500 have dropped off this imperial list, victims of complacency, poor financial management, and a failure to keep pace with the competition. It should come as no surprise that today's organizations, especially the behemoths, are not designed for innovation. They are the by-products of a more orderly and regulated environment. Courting change, acting opportunistically, and shifting direction at a moment's notice were not, until recently, required for survival, never mind excellence. Not only were such traits not required, but to have emphasized them would have detracted from performance! Doing yesterday's job just a little better—at most—was the prescription for success. Indeed, this is the saga of the post–World War II U.S. automobile industry, steel industry, chemical industry, and even the first two decades of the computer industry.

In the field of high technology, the key to staying competitive is product innovation supported by a strong commitment to research and development (R&D). But how to do this? One school of thought says that we have to be much faster at developing new products. Proponents provide airtight schemes for reducing cycle times and filling niches as they appear. The complexity of these schemes is often stunning, but who could argue? As Tom Peters (1990) says, "It's a complex world." New approaches are required to slash product development cycles by at least an order of magnitude in many industries. How-

498

ever, the rigidity of several of the most popular approaches with their "one size fits all" character leaves something to be desired.

The alternative to airtight formulas, some say, is a "ten-man band of lunatics cast adrift." Although this idea, realized in what are sometimes called "skunkworks," has worked well for Lockheed in its development of high-altitude reconnaissance aircraft, and Data General when it developed the Eagle line of super minicomputers (Kidder 1981), there is a wealth of evidence suggesting that the significant breakthroughs of the 1980s, as exemplified by the products and services of Federal Express, Apple, Dell, BancOne, Wal-Mart, CNN, and MCI, will not come from orderly plans alone or the right company at the right time. Formulas are questionable and ten-man bands alone are not up to the innovation task. To begin, what is needed is a strategic plan with R&D prominently featured.

Companies vary in the degree of sophistication with which they accomplish planning. Gluck et al. (1980) present a four-stage evolutionary model for carrying out this task:

- Stage I companies have the basic financial planning system in which everything is reduced to a financial problem and the value standard is to meet the budget.
- Stage II companies extend basic financial planning by means of long-range forecasts.
- In stage III companies, planners try to understand the market phenomena that are forcing change, look for opportunities that may lead to a more attractive portfolio, and devise alternative strategies for top management consideration.
- In stage IV companies, management is involved in strategic planning that stimulates entrepreneurial thinking and promotes all-around commitment to the corporate plan.

This type of classification offers an easy way to segregate and evaluate where companies are in the planning process. Top management should give deliberate thought to the degree of sophistication that they have reached and how R&D fits into the overall plan. In general, planning is a two-pronged effort. The first prong centers on the development of the strategic plan that defines and communicates longer-term business directions; the second involves the development of an operating plan that specifically identifies tasks or projects to be undertaken in pursuit of corporate goals. At this point a distinction needs to be drawn between traditional capital budgeting and R&D planning. R&D, along with new product development, is a low-probability game, no matter how much you plan, survey, consult with customers, or align yourself with the competition. There are literally thousands of variables that must be juggled at once. There are variables that deal with *technology* (design, engineering, manufacturability, quality, serviceability), variables that deal with *distribution* (who, through which channels, level of interest in the product, when), and there are variables that deal with *customer use* (the lag time between development of a new product and its routine adoption, even when dramatic and unmistakable benefits are evident from the outset, often runs decades—and almost always occurs via a convoluted, totally unpredictable path), not to mention variables that involve *competitors* (big, small, domestic, foreign) and new entries into the marketplace.

In the remainder of this chapter we present some of the unique aspects of the R&D project. In so doing, we reflect and extend many of the ideas discussed previously. To be successful it is necessary for an organization to instill a project orientation everywhere. To be speedy, to practice innovation on every product and process, and to develop new and scintillating products quickly requires that all functional boundaries between design, engineering, manufacturing, operations, purchasing, sales, marketing, and distribution be destroyed—not broken down or softened, but destroyed. A second guideline is for virtually every person in the company to spend a fair amount of his or her day on project teams with people from other functions. The essence of perpetual quality improvement, service improvement, rapid product development, and increased operational efficiency is getting people from multiple, warring factions working together on output-oriented activities that generally go unmanaged in traditional "vertical" organizations.

12.2 RISK FACTORS

The crucial elements of risk for a venture based on advanced technology, which are probably the elements unique to this type of business, occur up front during the development and introduction phases of the technology. As Bower (1970) put it succinctly, "The most important and expensive decisions must be made well before the last word is heard from technology." There are some ideas, now well documented, that will help an enterprise reduce the risks of undertaking an R&D project or bringing a new product to market (Cooper 1987). Even so, these are still broad concepts that apply equally to all types of organizations and do not really get to the heart of the issues facing high-tech businesses. The following series of steps provides a useful guide for executives attempting to construct a risk profile:

1. Be aware of the problem; that is, recognize that risk is a factor that needs to be built into each life-cycle stage.
2. Formalize the process of identifying the potential sources of risk and judging the extent to which these apply in particular business situations.
3. Provide an assessment of probabilities so that sensible financial appraisals can take place.
4. Formulate a business plan that takes account of risk.

Most executives regard risk as a way of life. Having accepted this proposition in step 1, they then plow ahead using a "you can't get anywhere without taking risks" type of approach. This means that subsequent steps are not dealt with particularly well, so that when financial appraisals are required, interpretations and judgments are built into plans that are short on rigor. In the following subsections we discuss many of the issues that can have a substantial impact on R&D projects, particularly as they relate to new product development.

12.2.1 Technical Success versus Commercial Success

New technology may provide a product that does not prove acceptable as a substitute for existing technology. Although the product may deliver benefits beyond what is currently available, its benefit-to-cost ratio may be too low to justify adoption. This does not mean that all potential customers will stay home—there are always some innovators who are willing to try anything. It simply means that initial market penetration is likely to be slow and greater investment may be required to convert technical success into commercial success.

A good example of this exists in the welding field, where automated welding systems are becoming more widely available but are unlikely to gain significant penetration for a number of years (Meldrum and Millman 1991). Some customers will be forced into using them for safety or manufacturing reasons. Others will adopt them because their customers are placing quality demands that can be met only by automated systems. For the vast majority, though, welding is a "black art" or, at best, a low-profile activity within their companies. A similar example is found in large area displays for public information systems where electromechanical devices and LEDs (light-emitting diodes) have proven to be quite acceptable technologically for this application and have doggedly resisted replacement by liquid crystal developments.

The continued embrace of familiar techniques is a situation that has long been recognized by those doing research in diffusion of innovation theory (Rogers 1976), but for marketers of advanced technology, it is a special problem. Their task is to judge just how fast and how far their product will be received as an acceptable substitute by the various sectors of the market. This will depend on how well existing technology solves the customers' problems and how far the extra benefits supplied by the new technology are perceived to offer competitive advantages.

12.2.2 Changing Expectations

Many high-tech products are designed and developed against a customer specification. But specifications are prone to change during the development cycle, causing costs and schedules to deviate measurably from the original plan. Government-linked contracts have attracted media attention—none more so in Great Britain than GEC-Marconi's efforts to win the Nimrod contract with the U.K. Ministry of Defence. Although there were numerous problems surrounding this project, the performance requirements of the ministry shifted continually over its life, thwarting GEC's ability to come up with a suitable product on time.

Turning again to the example of automated welding systems, a similar story can be found. One prospective customer who manufactured components for the automobile industry returned to the developer four times to request a redesign of the system. The problem was that each time a new specification appeared, the customer realized a little better the potential of the product and other areas where the system might have an application. Unwilling to lose the development opportunity, unable to charge for quotation services, and rather desperate for customers, the supplier found himself involved in a significant amount of redesign work, for little reward in the long run. The customer eventually went back to a system close to the original specification, being unable to afford the more complex system that had taken his fancy.

12.2.3 Technology Leapfrogging

Substitute technologies or new generations of products based on existing technologies may appear just as a company is pushing its existing range of products into the marketplace. This is a particular problem in the high-tech field, where rapid innovation can turn the products that you have spent so much time developing, obsolete overnight. As a consequence, sustained investment is necessary to stay in the race.

Engines for wide-bodied passenger aircraft provide a useful illustration: The first generation of engines, such as the Pratt & Whitney JT9, Rolls-Royce RB211, and General Electric CF6, represented high-risk "discontinuous" product innovation of the make-or-break variety. Indeed, without U.K. government intervention following the placing of the Lockheed Tri-Star contract, Rolls-Royce would not have survived. Later generations of these large turbo-fan engines have been based on incremental innovation, typically offering higher thrust ratings, improved fuel consumption, lower noise levels, and so on. Similar patterns of discontinuous innovation and subsequent leapfrogging via incremental innovation are to be found in other industries. Witness the vying for leadership by Intel and Motorola in microprocessors, by Kodak and Fuji in photographic products, and by IBM and SUN in computer workstations.

An example may again be drawn from the electronic information display market, where the cathode ray tube is currently the most widely used medium, but is recognized as having a number of disadvantages. Many companies are therefore investing in the development of alternative technologies that can provide a flat-screen or panel replacement. The competing technologies include LEDs, liquid crystal, vacuum fluorescent, electroluminescent, plasma, gas discharge, and incandescent displays. Each technology has associated with it a number of well-known and not so well-known names. For several of them, the risk that the substitute technology will substantially reduce the potential market for their product will become a reality.

12.2.4 Standards

Both the existence and nonexistence of performance and quality standards for technology-based ventures can be a challenge in marketing innovative products. If formal standards do not exist, customers have nothing against which to evaluate their potential purchase. This has the effect of making the product difficult to sell because it will be a higher-risk purchase and the process of writing specifications will take a lot longer. On the other hand, in the absence of formal standards, informal or *de facto* standards may appear that can lead to a mismatch between the proposed technology and the requirements of the customer base.

Airship Industries, a U.K. company that has led efforts to reintroduce airships as a mode of transport and surveillance, provides a classic example of the risk associated with the nonexistence of standards. In their efforts to establish airships as a credible mode of transport, they felt that it would be essential to obtain U.K. Civil Aviation Authority certification, which would have worldwide acceptability. Their problem was that no standards existed, and certification, in any event, proved very hard to come by. As the commercial manager of the company noted, the first production model flew in 1981 but did not gain U.K. certification until 1984. Full U.S. Federal Aviation Administration certification was granted in 1989.

Another example of informal standards or industry-established norms preventing a technology from becoming a profitable commercial venture is the experience of JVC during their early attempts to establish a position in the video recorder market. The first commercial videotape recorder (VTR) was marketed in 1955 by the U.S. firm Ampex for use in film and television productions (Nayak and Ketteringham 1987). The first Japanese version of the Ampex system arrived in 1958. JVC, later to become the world leader in home video with the VHS system, produced their version of a similar VTR in 1959. This was a better and simpler product, but it failed commercially, as their machine was incompatible with the standards then established, which were derived from the Ampex and Sony technology. Although this provided an important lesson for JVC that proved useful in the now famous battle for home video standards, the failure nearly resulted in JVC's premature withdrawal from this market (Rosenbloom and Cusumano 1987).

A related example centers on the development of standards for high-definition television (HDTV). Since the early 1980s, a battle has been raging between the U.S. Federal Communications Commission and its Japanese and European counterparts. The contentious issues revolve around picture format and compatibility with existing systems. Resolution is not expected until the mid-1990s. As a result, the introduction of HDTV into the United States has been delayed by at least 5 years and has given the Japanese electronics industry a breathing space for perfecting the technology, virtually guaranteeing its dominance of the market.

On the other hand, formal standards can exhibit their ambiguous effects in other ways. Where standards do exist, they provide the supplier and the customer with a reference against which to manufacture and evaluate. However, they often vary among industries and countries, making it difficult for a competitive company to expand its sphere of operation. Instances in which formal standards have created problems for entry and expansion in export markets are not hard to find. For example, a company selling connectors for optical fiber cabling in the telecommunications market, having developed a good business in the United Kingdom based on British Telecom standards, found it hard to sell in Germany, where DIN standards operate. To gain approval, the company sought a collaborative arrangement with another connector company but ended up supplying them components only, thereby deriving reduced added value from this market.

12.2.5 Cost and Time Overruns

Some products cannot be produced to the specification originally envisaged without running into substantial cost or time overruns. Although the research, design, and development work may have progressed to a satisfactory stage, and the market may supply commercial potential, it is sometimes the case that the production technology or component availability necessary to commercialize the product is inadequate. This is a problem that occurs mainly in the earlier stages of new product development and prototype testing. The situation is exemplified by a manufacturer of lithium batteries which had to recall substantial numbers of its initial product because the casing technology used was not secure enough for this dangerous chemical. Another example, uncovered by Meldrum and Millman, relates to a large U.K. multinational that announced a product to the press and then found that the thin-film coating technology that they had assumed would meet the requirements of their product was not up to the task, thereby delaying entry into the market.

In some instances it has been argued that in the context of two primary risk factors, cost and time, cost overruns will have less impact than time. As noted by John Doyle, vice-president of Hewlett-Packard: "If we over-spend by 50% on our engineering budget, but deliver on time, it impacts 10% on revenue. But if we are late, it can impact up to 30% on revenues."

Postlaunch problems in manufacture still occur, although these are more likely to be associated with difficulties of supply and are not unique to high-tech enterprises. For example, a small electronics manufacturer had experienced a problem in obtaining the correct type of wound component and ended up frantically winding their own in a back room. Another, more public example is the recent purchase by Amstrad of a stake in Micron Technology to ensure long-term supplies of 256K memory chips, which they were having trouble sourcing.

12.2.6 Lack of Infrastructure

Another area of risk is concerned with support technology that is not adequate to make a proposed product a worthwhile investment. Often this means a product concept awaiting an "enabling" technology.

After almost two decades of promoting the "factory of the future" and the "paperless office," these concepts are only now approaching reality. Molins System 24, for example, is generally regarded as the forerunner of modern flexible manufacturing systems. Although several were installed in the late 1960s and 1970s, the concept was ahead of its time. Computing power and software development were inadequate and the step was too great for users to take. Islands of automation rather than fully integrated systems were the result; similarly with office automation. There will always be a market for stand-alone equipment, but extensive displacement of manual/paper-based tasks has little chance of acceptance without electronic integration. It is only with the growing provision of sophisticated communications technology and workstations such as multiplexing, networking products, and protocols that real market opportunities arise.

A similar problem was noted by Meldrum and Millman regarding a new optical storage medium, which is an inexpensive plastic film that stores vast amounts of optical information and can be formed into a sheet, disk, tape, or cylinder. One tape of 500 meters can store 1 terabyte (1 billion bytes) of information, but full commercialization and the opportunities that this technology can address are not yet realizable. The development of suitable hardware systems and some further advances in laser technology are still required before the potential market achieves any real size.

The enumeration of potential sources of risk described above covers most of those likely to be faced by high-tech organizations. As with any business problem, the first steps on the long trail to commercialization are recognizing the existence of the problem and identifying its parameters. Only if the range of potential threats is identified successfully can management hope to develop strategies to deal with them. Similarly, the identification of potential risk is the first step in risk analysis that will produce a clear picture of the risk profile for decisions on investment policy and R&D portfolio management.

12.3 MANAGING TECHNOLOGY

The meaning of technology is straightforward: knowing how to do something well. A more elaborate definition would be: the ability to create a reproducible way to generate improved products, processes, and services. In fact, a modern manufacturing business must have a substantial portfolio of individual technologies. The management of technology should ensure that the firm maintains command of the technologies relevant to its purposes and that these technologies support the firm's business strategy and shareholder value.

Technology management for strategic advantage is difficult and often frustrating. As Erikson et al. (1990) point out, the central issue is the need to reconcile risk and the unpredictability of discovery with the desire to fit technical programs into orderly management of the business. The traditional approach to managing technology has been largely intuitive. Research and development is treated as an overhead item, with budgets set in relation to some business measure (e.g., sales) and at a level deemed reasonable by industry practice. Budgets may be projected several years ahead but are usually set annually. Within this budget framework, decisions about areas of concentration and project continuations may be left largely to R&D management. There is no assurance that the R&D organization, left to its own devices, will pursue programs related to corporate strategy, either in focus or in degree of innovation and risk.

In response to this unsatisfactory situation, many firms have become somewhat more sophisticated. Managers outside the technology area participate in suggesting or reviewing projects, but the connection to company strategy is still casual or haphazard. Some firms subject R&D programs to a rigorous financial justification process based on net present value. Arguing that research and development projects are investments—as in a sense they are—corporate management seeks justification based on rate of return or payout. It is difficult, though, to predict financial returns for an R&D project, especially if it is focused on achieving a significant innovation. As a consequence, new activities may be limited to conservative, incremental projects; the results will be more predictable but will have marginal strategic impact.

Clearly, then, there is a need for a measured, sophisticated approach to R&D management. Interest in a better approach has been stimulated by various developments (Erikson et al. 1990, Jain and Triandis 1990, Roussel et al. 1990). First, many corporate leaders have moved beyond the financially driven planning characteristic of the 1970s and 1980s. Second, the success of entrepreneurial, high-technology companies has excited interest in the potential of technology to build company value. Third, firms have seen that industry leaders give high priority to technology management. Fourth, quality and manufacturing capability are now considered strategic business weapons. Together these developments have helped to create a desire to manage technology in a way that is congruent with business strategy.

The first step in the strategic management of technology is to answer the following question: For our firm, what mix of products and markets will best sustain and enhance our cash flow? The next step is to test how well the firm's technologies support the ideal product and market mix. The third step is to focus technology investments so that they better support the firm's strategy.

It is often useful to examine a firm's technologies in light of two questions:

1. What is the significance of the technologies in the firm's portfolio, as measured by their competitive impact and maturity?
2. In each product area or business, how strong is the firm's technological competitive position?

12.3.1 Classification of Technologies

In general, it is possible to identify three broad classes of technologies in a typical firm's technological portfolio.

1. *Base technologies.* These are technologies that a firm must master to be an effective competitor in its chosen product–market mix. They are necessary, but not sufficient, to achieve competitive advantage. These technologies are widely known and readily available. Electronic ignition systems for automobiles is an example.

The trick for R&D management here is to invest enough, but only enough, effort to maintain competence. The danger is that inertia will sustain programs in these base technologies longer and at greater scale than they deserve, perhaps because these are the traditional areas where the research and development organizations feel at home. The U.S. auto industry in the 1960s and 1970s, for example, invested too heavily in familiar areas of product technology rather than in new, less comfortable areas where opportunities to develop new process technology existed.

2. *Key technologies.* These technologies provide competitive advantage. They may permit the producer to embed differentiating features or functions in the product or to attain greater efficiencies in the production process. An example is food-packaging technology that enables the purchaser to use microwave cooking.

The primary focus of industrial R&D is on extending and applying the key technologies at the firm's disposal; they should be given the highest priority when contemplating investments opportunities. Unwilling to invest in key process technologies in the 1950s and 1960s, the U.S. steel industry paid the price in the 1970s; foreign competitors, whose entry into the U.S. market had been encouraged by consumer goods manufacturers, far outstripped their domestic counterparts in productivity.

3. *Pacing technologies.* These technologies could become tomorrow's key technologies. Not every participant in an industry can afford to invest in pacing technologies; this is typically what differentiates the leaders (who do) from the followers (who do not). The critical issue in technology management is balancing support of key technologies to sustain current competitive position and support of pacing technologies to create future vitality. Commitments to pacing technologies or potential breakthroughs are hard to justify in conventional, return-on-investment terms. Indeed, these commitments can be thought of more accurately as buying options on opportunity. Relatively modest commitments—and thus modest downside risk—can give the potential for large upside reward. Realizing that potential depends on still-unresolved technical and market contingencies. If the option is not pursued, the potential does not exist. Smith, Kline & French supported pursuit of receptor modeling in the 1960s, a pacing technology in the pharmaceutical industry at that time. This work led ultimately to the development of Tagamet and the establishment of the company as an industrial leader.

An effective R&D program must include some investment to build a core of competence in pacing technologies and some effort to gain intelligence from sources such as customers, universities, and scientific literature to help identify and evaluate these technologies. At the same time, disciplined judgments about commitments to pacing technologies are necessary; enthusiastic overspending on advanced technology can undercut essential support of key technologies.

12.3.2 Exploiting Mature Technologies

Technologies mature, just as industries and product lines do. The younger the technology, the greater the potential for further development, but the less certain the benefits. However, a mature technology can often be a key technology. Many Japanese firms use mature technologies as a major competitive weapon. The Sony Walkman, for example, was a wildly successful new product based on comparatively mature technologies. The Walkman fortuitously combined Sony's work on the miniaturization of its tape recorder line and its work on lightweight headphones. Company engineers were trying to make a miniature stereo tape player-recorder, but they could not fit the recording mechanism into the target package size. A senior officer realized that combining headphones with a non-recording tape "player" would eliminate the need for speakers, reduce battery requirements, and result in a small stereo tape player with outstanding sound (Nayak and Ketteringham 1987).

Sometimes a mature technology becomes a key technology when it is applied in a new context. Empire Pencil gained a major cost and quality advantage by using mature plastic extrusion technology as the basis of a new way to manufacture lead pencils. Conventional lead pencil manufacturing requires the use of fine-grained, high-quality wood, such as cedar, and a good deal of hand labor for assembly. Materials are becoming more expensive, and damage to the graphite core during the assembly process causes quality problems. A development team was confronted with this question: How can we improve quality and cut costs? The team realized that wood powder in a plastic binder could simulate the fine-grained wood. From there it was a straightforward step to produce pencil stock in a continuous extrusion process, with wood powder and a core of graphite powder in a plastic binder.

Other mature technologies may be protected (e.g., by patents or proprietary treatment) and thus give their owners a key competitive advantage. A Japanese grinding machine manufacturer successfully diversified into the manufacture of integrated-circuit wafer equipment. A critical factor in its success was its proprietary mature machine technology. Examples such as the latter may tempt a firm in a mature line of business to diversify into new products and markets where its proprietary but mature technology could have a key competitive impact, but this strategy is risky. The better alternative is to look, as Empire Pencil did, for new technology to invigorate a mature or aging product line.

A business or product line whose key technologies are mature faces a serious threat of being blindsided by a competitor employing new key technologies. This is what Xerox did to the established copier manufacturers and what word processing did to the typewriter industry.

As an industry or product sector matures, the key technologies often become *manufacturing process* technologies rather than product feature technologies. This is the case in many mature industries, including chemicals, machine tools, consumer appliances, and food products.

12.3.3 Relationship between Technology and Projects

Defining projects by type provides useful information on the role of existing technology in their development and how resources should be allocated. Wheelwright and Clark (1992) suggest a two-dimensional qualitative scale for classifying projects: (1) the degree of change in the product, and (2) the degree of change in the underlying manufacturing process. The greater the change along either dimension, the more resources are needed. They have also identified five project types. The first three—derivative, breakthrough, and platform—are associated with the marketplace; the remaining two—research and development—precede commercialization.

Each of these five project types requires a unique combination of development resources and management styles. Understanding how the categories differ helps managers predict the distribution of resources accurately and allows for better planning and sequencing of projects over time. A brief description of the first three categories follows.

Derivative projects range from less expensive versions of existing products to add-ons or enhancements to established production processes. For example, Kodak's wide-angle, single-use 35mm camera, the Stretch, was derived from the no-frills Fun Saver introduced in 1990. Designing the Stretch was primarily a matter of modifying the lens.

Development work on derivative projects typically falls into three categories: incremental product changes, say, new packaging or a new feature, with little or no manufacturing process change; incremental process changes, such as a lower-cost assembly technique, improved reliability, or a minor change in materials used, with little or no product change; and incremental changes on both dimensions. Because design changes are usually minor, incremental projects are more clearly bounded and require substantially fewer resources than do the other categories. Because derivative projects are completed in a few months, ongoing management involvement is minimal.

Breakthrough projects are at the other end of the development spectrum because they involve significant changes to existing products and processes. Successful breakthrough projects establish core products and processes that differ fundamentally from previous generations. Like compact disks and superconducting ceramics, they create an entirely new product area that can define a new market.

Breakthrough products often incorporate revolutionary technologies or materials and hence usually require revolutionary manufacturing processes. Management should give development teams considerable latitude in designing new processes, rather than force them to work with outdated or marginally efficient equipment, operating techniques, or supplier networks.

Platform projects are the middle of the development spectrum and are thus harder to define. They entail more product or process changes than do derivatives, but they do not introduce the untried technologies or materials that are found in breakthrough products. Honda's 1990 Accord line is an example of a new platform in the auto industry. Computer-integrated manufacturing techniques were successfully exploited to improve assembly operations, but no fundamentally new technologies were introduced. In the computer market, IBM's RISC/System 6000 is a workstation platform; in consumer products, Proctor & Gamble's Liquid Tide is the platform for a full line of Tide brand products.

Well-planned and well-executed platform products typically offer fundamental improvements in cost, quality, and performance over preceding generations. They introduce improvements across a range of dimensions: speed, functionality, size, and weight. (Derivatives, on the other hand, usually introduce changes along only one or two dimensions.) Platforms also represent a significantly better system solution for the customer. Because of the extent of changes involved, successful platforms require considerable up-front planning and the participation of marketing, manufacturing, and senior management, as well as engineering.

Companies target new platforms to meet the needs of a core group of customers but design them for easy modification into derivatives through the addition, subtraction, or removal of features. Well-designed platforms also provide a smooth migration path between generations so that neither the customer nor the distribution channel is disrupted. Consider Intel's 80486 microprocessor, the fourth in a series. This product was aimed at a core customer group—the high-end PC/workstation user—but variations addressed the needs of most other users. Moreover, software compatibility with the 386 permitted existing customers to make the transition to the 486 with minimal effort. Over the life of the 486 platform, Intel will introduce a host of derivative products, each offering some variation on speed, cost, and performance and each able to leverage the process and product innovations of the original platform.

Platforms offer considerable competitive leverage and the potential to increase market penetration, yet many companies underinvest in them systematically. The reasons vary, but the most common is that management lacks an awareness of the strategic value of platforms and fails to conceive projects that exploit their capabilities.

12.4 STRATEGIC R&D PLANNING

All corporate departments, operating divisions, and companies must develop plans. In each division, R&D, engineering, manufacturing, marketing, sales, and the various support groups should participate to produce the division's strategic plan. The purpose of this plan is to define how each unit will carry out relevant corporate goals. The relationship between corporate planning with R&D planning is shown in Fig. 12-1.

It may seem obvious that R&D portfolios should be aligned with corporate goals, but too often R&D groups are left without the top management guidelines and collaboration to know what is best for the company. Successful planning depends on a dialogue between top management and the R&D leader, mission, goals, strategies regarding and means of implementation. These are important aspects of participative R&D management.

12.4.1 Role of R&D Manager

An R&D manager fulfills corporate strategy by planning for change throughout the planning exercise. He or she must include uncertainties of innovation (probabilities of technical and market success) and uncertainties of the environment (effects of public policy, consumer mood, actions by the competition) in his or her deliberations. The manager must

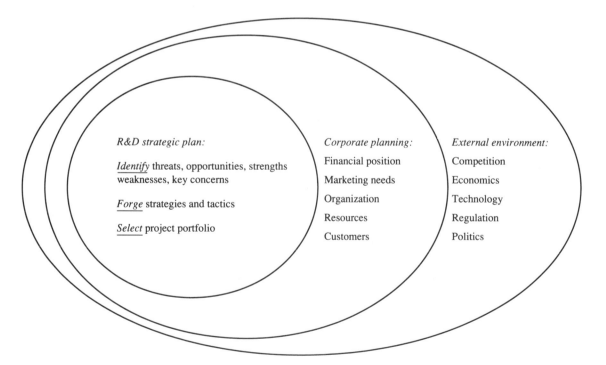

Figure 12-1   Relationship between corporate and R&D planning.

recognize technology push (the brilliant idea seeking a market) and market pull (a market need seeking a product) and what the general corporate climate or attitude is on projects based on either.

If you are a manager of an operational R&D group, you must recognize the needs of the parent business unit; if you are in a central R&D group, you must recognize the needs of the corporation as a whole. In either case, you must have the means and the ability to monitor technology and to forecast change. A key requirement is to keep your eyes on horizons well beyond current technology. Also, you must recognize where and who the entrepreneurs and project champions are in your company.

Finally, top management must understand the sources and effects of uncertainty, be receptive to innovation, and be the stimulus for strategic planning and the agitator for an innovative environment. If they are not, strategic technical planning will never evoke empathy and the group will flounder and fail.

12.4.2 Planning Team

The head of the R&D group in a business unit (a *unit* may be a section, department, division, combinations of these, or a company) and the managers of the various R&D areas in the group should be the planning team members. How deeply the team draws its members from the organization depends on company size, commonality of interests among business units, and questions such as what is a reasonable team size. Ideally, senior professionals, managers of various functions, and planners in the business unit will assist.

Research managers form the planning team.   To simplify the discussion, consider a corporate R&D planning group. In this example the vice president of R&D is the team leader, with other members being managers of the group's various R&D areas. Managers of relevant operations will assist or be asked for assistance. Corporate officers and staff from selected functions may be asked to review critical points in the developing plan.

If you are the head of R&D and thereby the team leader, you cannot delegate the thought processes required for the planning process and the derivation of results. You can, of course, use every fact-finding function available, but the team does the actual manipulation of inputs and produces the output results. This may seem like a lot of work, but once accomplished, you will know more about your operation than ever before and be equipped to manage your assigned functions.

Good managers do not delegate the planning process.   When the planning team is assembled, the leader should remind members of the unit's mission (also called *charter* or *definition* of business) and the mission of the R&D group. If the team is in an operations unit, the mission statement emphasizes upgrades and means to advance market share of present products; if the team is in central R&D, emphasis is on new products, new technologies, and new opportunities. These mission statements are important because they define the business and give its scope in clear, concise language. Two typical mission statements would be:

- Division X designs, manufactures, and sells sensors and monitoring equipment to meet the severe environments within the mining industry. The mission of the R&D group is to enhance the performance of current products and to discover and develop new products that will aid in maintaining and advancing market leadership of the division. In so doing, the R&D group will provide technical surveillance over current and emerging relevant technologies, monitor competitors' products and services, maintain and advance market share through upgrades and extensions to the product line, and develop selected new products within the scope of the business.
- Company Y designs, manufactures, and sells hardware and services to energy producers. The mission of the R&D group is to discover and develop products that will give the company commanding leadership in its selected business areas. In so doing, the corporate R&D group will provide surveillance over current and emerging relevant technologies, conceive and develop new products to meet future change, and provide problem-solving research and services to operations as needed.

Such mission statements focus the team on the issues important to the business of its parent unit. The team leader presents the needs of top management to the team and discusses specifically the goals that management wants R&D to meet. (These needs should reflect, in part, the inputs from R&D.) The goals of top management may be cast as general statements such as:

- Look at area X over the next few months and see if you can conceive an advanced method.
- Create a new generation of products in the near future from emerging technology A.

Alternatively, the goals may be more specific:

- Provide division Y with an upgraded product Z using your materials technology and let's see where you are in 6 months.
- Reduce materials costs of product B this year.

The planning team reviews any goals previously set by the R&D group to determine their compatibility with current goals. The team identifies what can be done within available resources, what employee and equipment resources are needed, and so on. Goals are reviewed, refined and revised during the planning phase (A *goal* is usually defined as something to be accomplished within a specified period.) Finally, the team discusses how to accomplish the six stages of planning enumerated in Table 12-1 and sets out tasks and schedules. The team leader also discusses the methods to be used in fulfilling the assigned tasks.

Planning is a multistage process. The strategic technical planning task required of the team can be facilitated by use of the six planning stages (Englert 1990). In the first stage, the team collects information. In the second stage, the team consolidates (categorizes,

TABLE 12-1   STAGES OF THE STRATEGIC TECHNICAL
PLANNING PROCESS

1. Information-gathering stage
   - Determine status of the business unit.
   - Ascertain needs of operations.
   - Determine status of competition.
   - Conduct technical planning studies.
   - Consider key concerns and issues.

2. Consolidation stage
   - Derive scenarios of possible futures.
   - List needs, opportunities, threats, impacts.
   - List key concerns and issues.
   - List strengths and weaknesses.

3. Strategy formulation stage
   - Analyze and evaluate lists of needs, etc., against lists of key concerns and issues.
   - Evaluate maturity of present technologies and possible use of new technologies.
   - Match lists with strengths and weaknesses.
   - Develop preliminary alternative strategies.
   - Develop candidate tactics.
   - Evaluate and suggest priority of strategies.

4. Selection stage
   - Select one set of strategies, or
   - Look again at some new technologies and then decide.

5. Implementation stage
   - Consider project candidates (tactics) in depth.
   - Test tactics against best and worst scenarios.
   - Consider funding limitations.
   - Suggest priorities of specific projects.
   - Describe the group's R&D areas.
   - Set goals.
   - Draft strategic and operational plans.

6. Review stage
   - Submit plans for review.
   - Adjust plans as necessary.

digests, and assimilates) this information into various lists. These lists are used in the succeeding stages of planning, so their comprehensiveness is critical to the overall effort. The next three stages are progressive refinements of current findings.

12.5 PARALLEL FUNDING: DEALING WITH UNCERTAINTY

A primary role of the R&D project manager is to narrow the range of technological choices facing the organization without sacrificing market or performance goals. Because of the inherent uncertainty at each stage of project development, it is not uncommon to identify and explore several alternatives to facilitate selection of the most promising candidates. During the development of the Airborne Warning and Control System by the U.S. Air Force in the mid-1970s, for example, both Hughes and Westinghouse were awarded multimillion-dollar contracts to design and build prototype radars for the Boeing aircraft. Considering the extent of the technological unknowns, the Air Force felt that the additional money spent in a runoff competition was justified given the rigorous technical requirements and tight timetables surrounding the program.

The use of parallel strategies is one means by which experienced managers cope with the uncertain nature of the R&D environment (Abernathy and Rosenbloom 1969). Such an approach has the threefold advantage of avoiding the difficulty of trying to predetermine which ideas or technologies will succeed, hedging against the risk of outright failure, and building a broader technological base. The decision to fund more than one alternative at each juncture, though, must be tempered by the potential trade-offs between increased probability of success and increased cost, as well as the behavioral issues associated with parallel choice (Balthasar et al. 1978). When a particular alternative evidences clear superiority, however, a sequential strategy may be called for wherein other candidates are pursued only if the preferred candidate fails to met expectations.

A stream of technical choices, made by project managers, group leaders, and their clients, determines the cost of an R&D project and the value of its outcome. Choices between competing approaches to the solution of technical problems must be made in the face of substantial uncertainty in situations where time and resources are limited.

By a "parallel strategy" we mean the simultaneous pursuit of two or more distinct approaches to a single objective, when successful completion of any one would satisfy the stated requirements. Nevertheless, the sequential strategy, that is, commitment to the best evident approach, is most common in practice. In a majority of situations the benefits of a parallel strategy may seem obscure, while its additional costs are quite real.

12.5.1 Categorizing Strategies

Despite the advantages of a parallel strategy, the manager's task is one of sequential choice under uncertainty. Any complex development project must be undertaken in steps or stages, the later stages depending on the results of the earlier ones. It is therefore a problem in sequential decision making, as discussed in Section 3.7. At each stage it is possible to make use of economic calculus in comparing the costs and gains from various alternative ways of proceeding. The costs in this case are those of the next stage of development, not the costs of procuring and operating the full system.

In practice, however, there is a significant difference between the calculus of sequential decision making in planning the definition of a program and in reducing a solution to practice. Thus one can generalize more advantageously about the structure of the decision problem by distinguishing two broad categories for the use of parallel strategies (Abernathy and Rosenbloom 1969). In the first category, called a *parallel synthesis strategy*, the uncertainty is broad, the cost of information is relatively low, and there may be only a limited commitment to further work. In the contrasting case, the *parallel engineering strategy*, the bounds of uncertainty are more definite, the information cost is relatively high, and there is a strong commitment to satisfy developmental objectives.

The parallel synthesis strategy is most often found in the first phase of a program. At that point substantial uncertainty exists concerning the types of needs the developmental product is to satisfy, the potential of each alternative to satisfy those needs, and the probable cost of each alternative. Information that can reduce those uncertainties can be obtained by means of analytical studies, special tests, and limited development of the several prototypes. This sort of activity frequently serves to synthesize an approach to the larger problem and defines many of the outcome characteristics. A parallel synthesis strategy typically is a means of gaining information and maintaining options so that the best path may be selected for subsequent development.

In the synthesis phase, definition of the program is incomplete. Attempting to map a "decision tree" specifying a sequence of possible acts and consequences would be misguided. At this stage the manager may still be ignorant of factors that will prove to be the most significant sources of later uncertainty. For example, Admiral Rickover said in reference to the nuclear submarine program: "In the beginning of Naval development neither the technical problems nor their solutions were well understood. Many of the problems were not even known." The various approaches to development are seldom independent, and a new approach may be synthesized from elements of those initially defined. In general, the history of R&D projects shows that initial judgments of cost, performance, and value are highly inaccurate.

The situation in which a parallel engineering strategy might be appropriate has a different economic structure and offers a different decision problem. It occurs in a later stage of the development process, when a great deal of information has already been acquired and there is little chance that the total program will be abandoned. In fact, one of the principal uncertainties of the earlier formative stages of the program—the gross worth and cost of the development—can now be estimated more accurately. With knowledge of the potential value of success, the consequences of failure or delay can also be made explicit.

With the parallel engineering strategy, in contrast to the synthesis strategy, the decision maker usually is committed to bringing the development project to successful completion. If he or she chooses only the preferred alternative and it does not prove acceptable, management must seek a new solution. This implies time delays and higher costs, however, since the development will continue at its high expenditure rate until a solution is found. Thus the basic cost structure of engineering development work influences the characteristics of a parallel engineering strategy. Additional costs stem from loss of reputation, penalty charges, and out-of-pocket and opportunity costs that result from not having the product available when it is needed or can be sold. Studies have shown that the cost of late completion is often the major component of the cost of following a single, unsuccessful approach. In the contrasting case of the synthesis strategy, the consequences are somewhat different. An incorrect choice may mean that the program is discontinued, since the benefits that would be offered by a different alternative may never be demonstrated.

12.5.2 Analytic Framework

For complex situations where many technological alternatives exist, an analytic methodology can be quite helpful in selecting among those project tasks whose outcomes can only be described in probabilistic terms. What makes the underlying problem exceptionally difficult is the fact that both systemic and statistical dependencies are likely to exist among these tasks. Typical dependencies include an overlap in resource use, technical interrelationships among task outcomes, and externalities where the value contributions or joint performance of several tasks may be nonadditive.

To address the combination of uncertain outcomes and task dependencies, analysts have relied on Monte Carlo–based simulation models such as SIMRAND developed by the Jet Propulsion Laboratory (Miles 1984), and Q-GERT developed by Pritsker (1979). In a similar vein, Bard (1985) formulated the decision problem as a probabilistic network and used a heuristic embodying simulation within a dynamic program as a solution methodology. In particular, he divided the R&D project into a number of different parts or stages, such that it was possible to complete each stage by undertaking one or more competing tasks. The corresponding problem can be represented diagrammatically as a directed network comprising sets of parallel arcs linked in series. An example of such a network is depicted in Fig. 12-2, where each arc represents a specific task whose outcome is characterized by an empirical probability distribution or random variable. Typical outcomes or performance measures might be eventual unit production costs, mean time to failure, or technical probability of success. In the model, Bard assumed that each task is defined by an algebraic expression consisting of one or more input (random) variables. As a consequence, outcome distributions are difficult to obtain in closed form, hence the need for simulation.

The full methodology was demonstrated by a three-stage project centering on the development of a flat-plate photovoltaic solar module. The three stages consisted of silicon purification, cell production, and module fabrication. In the first two stages, three tasks were considered, each one at two different funding levels; in the third stage, two tasks were considered, each at a single funding level. The results provided the necessary guidance for the project manager to initiate full-scale development.

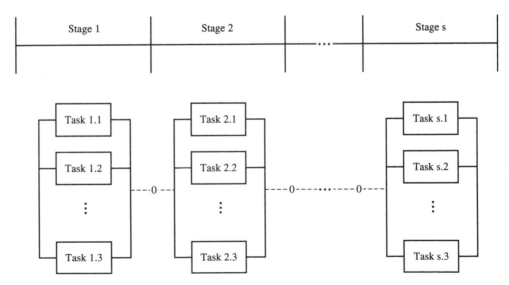

Figure 12-2   Network representation of parallel funding problem.

12.5.3 *Q*-GERT

The most heralded use of PERT and CPM network techniques since their inception has consistently been in R&D planning and control. These techniques, however, are somewhat limited in that they are unable to reflect many of the real complexities associated with R&D projects. Many situations, such as multiple branching (e.g., the success or failure of a task), probabilistic branching, and repeating activities via feedback loops, that are frequently part of the R&D process cannot be modeled in a PERT/CPM network. These limitations gave rise to GERT (graphical evaluation and review technique), a simulation methodology designed to accommodate the interdependencies and uncertain nature of project tasks (Moore and Clayton 1976).

An additional aspect of R&D management that occasions even greater complexity and difficulty is the scheduling and planning of several projects when more than one research team is involved. This problem has also been explored with GERT and promising results have been reported for applications of modest scope. Nevertheless, a limiting factor in GERT is that as the number of R&D teams and projects increases, the accompanying network becomes impossible to construct and decipher, thus defeating the value of the methodology. In response, *Q*-GERT was developed to provide even greater potential for planning and scheduling in a multiteam, multiproject environment.

Q-GERT is an extension of the GERT modeling procedure and, as such, contains most of the capabilities and features of the latter, including probabilistic branching, network looping, multiple sink nodes, multiple node realizations, and multiple probability distributions. *Q*-GERT derives its name from the special queue nodes it has available for modeling situations in which queues build up prior to service activities. However, *Q*-GERT contains other unique and innovative features for handling specific and complex networks that are particularly applicable in R&D planning. The most outstanding of these features is the ability to assign unique network attributes such as activity time and node branching probabilities to each individual project, and then process each project through a single generalized network.

In addition to the relative advantages *Q*-GERT offers with respect to other simulation and network techniques, Taylor and Moore (1980) attest to its ease of use. The methodology requires only that the R&D projects under consideration be diagrammed in network form and then converted into a standard input format for the *Q*-GERT simulation package. To demonstrate the power of the approach, Taylor and Moore present two case studies centering on an R&D subsidiary of a large textile manufacturer in the southeastern United States.

12.6 MANAGING THE R&D PORTFOLIO

Research and development is an investment that must compete for corporate support with other investment opportunities, such as plant modernization, advertising, and market expansion. Program and laboratory directors must continually defend the value of their research to top management as well as decide what mix of projects is best for the firm.

Project managers must determine whether their projects are on schedule and whether expected payoffs outweigh costs.

As part of their normal functions, upper management periodically reviews research programs, projects, and staff to assess progress and determine the contribution that each is making to the corporation's goals. The R&D management and review process, shown in Fig. 12-3, is nearly identical to that discussed in Chapter 3 for more conventional projects. The information gathered from the four basic reviews identified can be used to justify research expenditures, assist in budget and program planning, and provide a means of evaluating individual performance. The consequences of continuing to fund an R&D project when failure is imminent go beyond the actual dollars lost. The additional waste in human and material resources may have far-reaching effects: marginal projects may fail to receive the extra boost needed to move them beyond a critical stage, apparently healthy projects

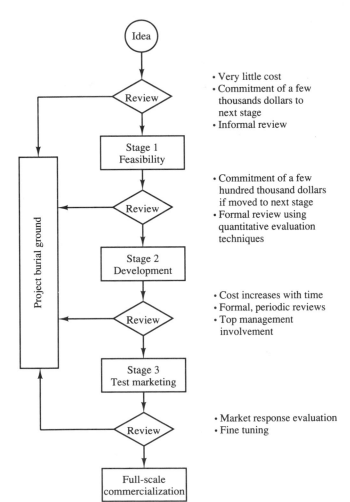

Figure 12-3   Stages of the industrial R&D process.

may begin to deteriorate when additional resources are not forthcoming, and promising new projects may have to be deferred as the competition moves ahead.

These points are underscored by Liberatore (1987), who attributes the importance of the R&D project management decision to two factors. First, R&D spending represents a sizable investment for many firms and may have a significant impact on their current and future financial position as well as on their ability to compete technologically. Second, projects often entail companywide commitments that translate into large opportunity costs if managed improperly.

Most projects do not begin until an in-depth assessment of their probability of success is made and the outcome appears favorable. As the project evolves, uncertainties may develop that jeopardize completion. In some instances, the market for the end product may change, falling below acceptable levels and calling into question overall profitability. Alternatively, technological problems may arise that become either too expensive or too difficult to solve. This is most critical during the early stages of development, where quality and cost decisions are made and research directions are forged.

Although much work has been done in project selection and resource allocation over the last 30 years (see Liberatore and Titus 1983; Schmidt and Freeland 1992 for surveys), the examination of decisions involved in project termination is a more recent phenomenon (Balachandra and Raelin 1980, Balachandra 1984). To help isolate the causal factors, Baker et al. (1986) analyzed 211 R&D projects carried out between 1975 and 1982 by 21 companies and found that a positive answer to the following four questions is a likely sign of success.

- Has a relevant business need, problem, or opportunity been identified?
- Has an appropriate scientific need, problem, or opportunity been identified?
- Can the project results be transferred effectively to the internal user?
- How well can the internal user produce, market, distribute, and sell the resulting product or process?

Conversely, they found that a project has less of a chance to succeed if R&D personnel are unsure about its commercial potential, if the match between its technical and commercial aspects is vague, or the level of uncertainty on how the results are to be brought to the marketplace is high.

Much of this work corroborated the earlier findings of Balachandra (1984), who identified a set of 14 key variables shown to be highly correlated with project failure. The implied conclusion was that by evaluating changes in these variables periodically, the R&D manager would be better able to make the crucial decisions related to project initiation and termination.

In light of this research, Bard et al. (1988) developed a decision support tool to be used by the R&D manager to help update his or her portfolio at review time. In the remainder of this chapter we highlight their methodology and the ideas surrounding its implementation. Appendix 12A presents the results of a case study centering on a small computer firm specializing in peripheral equipment. Specific issues related to terminating a project once the decision has been made are detailed in Chapter 14.

12.6.1 Evaluating an Ongoing Project

To be useful to managers, quantitative methods must provide reliable results and fit within the existing decision-making framework. At a minimum, models should include those variables that managers feel are most important and for which they can provide hard data or firm opinions. As mentioned, Balachandra (1984) identified two groups of factors that strongly influence project outcomes. His work was based on a discriminant analysis of 114 R&D projects gleaned from 41 firms spanning heavy manufacturing, oil and gas, electromechanics and instrumentation, utilities, chemicals, and electronics. Table 12-2 summarizes the characteristics of the database. Each group is discussed below.

Critical factors.   The successful completion of an ongoing R&D project is closely linked to a number of critical factors. If it is determined that any one of the following has deteriorated significantly since the last review, immediate termination is implied.

1. Government regulations
2. Raw material availability
3. Market conditions
4. Probability of technical success

The first three, termed "exogenous critical factors," are generally outside the control of the firm. The fourth is assumed to be a function of the resources allocated to the project.

As an example of a negative change in government regulations (1), recall that the development of many diet foods based on saccharine had to be discontinued when the U.S. Food and Drug Administration affirmed their cancer-causing properties. With regard to raw material availability (2), we note that shortages are likely to have a damaging effect on market potential. In the 1970s, many Mexican pharmaceuticals had to discontinue research into the development of synthetic hormones from the barbasco root when the export market abruptly changed and the price of the plant soared.

Similarly, markets (3) may suddenly vanish as consumer tastes change or when substitutes appear to offer more immediate benefits. A good example of this was Polaroid's attempt to introduce instant movies (Polarvision). Unfortunately, the onslaught from video-cassette recorders was too great to contend with, and the product met a quick demise. The

TABLE 12-2   CHARACTERISTICS OF
DATABASE FOR DETERMINING
CRITICAL FACTORS

Item	Range
Number of employees	50–2000
Sales	$50M–$2B
R&D budget	$1M–$50M
Number of employees in R&D	10–50
Number of R&D projects	1–50
Project duration (years)	0.5–8

last critical factor is the probability of technical success (4)—a measure that is extremely difficult to assess (Rubenstein and Schroder 1977). In any event, if it is perceived to fall below some acceptable level, dependent projects must be set aside until the necessary technology materializes. In the early 1970s a number of computer firms had to shelve various bubble memory projects because CMOS chips did not become available on schedule. Today, however, many products using this memory device have found a niche as a result of belated technological advances.

If none of these critical factors has deteriorated significantly since the last review, the project would then be evaluated with respect to the key variables described below.

Key variables.   Variables in the second group are more volatile than those in the first but are not as strongly critical. A significant deterioration in a minority of them may not measurably affect outcomes. Thus project termination is implied only when a substantial majority have declined since the last review.

The key variables can be broadly categorized as environment related, project related, and organization related. Each subgroup is outlined below. An in-depth discussion is given by Balachandra (1984).

I. *Environment-related variables*
1. Positive chance event
2. Product-life-cycle stage

These two variables are outside the control of the organization but are very much influenced by the environment. A positive chance event (1) might be associated with the introduction of a complementary product into the marketplace that would enhance the desirability of a product currently in R&D. The development of 40+-megabyte tape backup units was abetted significantly when large-capacity hard drives were included on personal computers as standard features.

When a product is in the initial stages of its life cycle (2), the probability of false starts is greatest. Unfortunately, this is largely a function of the technological environment and is beyond the control of the R&D team. If a product quickly moves out of its infancy stage into its growth stage, R&D projects pertaining to the product are more likely to be successful.

II. *Project-related variables*
3. Pressure on project leader
4. R&D manager is project champion
5. Probability of commercial success
6. Support of top management
7. Project personnel commitment
8. Smoothness of technological route
9. End user market
10. Project champion appearing toward end

The eight variables in this subgroup are directly related to the project. A fraction of these depend on the subjective perceptions of the team managers and personnel. Specifically, it was found that positive pressure and feedback (3) from top management, as evidenced by the enthusiasm they show toward the project team, smooths the route to completion. If a project champion emerges (10), this can also strongly influence the chances for success. Without such a person, most desirably in the form of the R&D manager (4), organizational as well as technical barriers may become very difficult to overcome. The time when the project champion emerges also seems to make a difference.

The probability of commercial success (5) is the single most important variable in the group. To assess this measure a solid knowledge of the market and the costs associated with production and distribution are required. As the project evolves these factors become clearer to management. A product whose costs will be higher because of unanticipated technical and production problems is a serious candidate for termination. The probability of commercial success should increase or at least remain the same from one review period to the next.

The support of top management (6) and the commitment of project workers (7) are also highly correlated with success. The latter may decrease if problems such as poor leadership or snags in technology are perceived but not acknowledged.

The smoothness of the technological route (8), as viewed by the project leader and evidenced by delays in meeting deadlines, is another important variable. So is a limit on the number of end users (9). An increase in possible applications for a new product during its development may dilute the effort, resulting in delays and indecision. This, in turn, may lead to complicated redesign and subversion of the original goals.

III. *Organization-related variables*

   11. Company profitability
   12. Anticipated competition
   13. Presence of internal competition
   14. Number of projects in R&D portfolio

Each of the four variables in this subgroup is affected by conditions throughout the firm. In particular, it seems that the more profitable a company (11), the greater the chances of completing the project. This may be attributed to better managerial controls and better screening of new product ideas. If a product has no competition in the market (12), however, it is likely that the R&D team will take a more relaxed attitude toward its mission. This is a prelude to failure. On the other hand, if the competition is known to be working on a similar project, both pressure and motivation intensify.

In many cases, emergence of internal competition (13) for common resources can act as a catalyst. The existence of multiple demands for technicians and equipment enhances the motivation of the project team. Nevertheless, as the size of the portfolio grows (14), there is a greater chance of individual failures due to less management oversight and a proportional reduction in funding.

Monitoring scheme.   During the review process, if significant shortcomings in any of the four critical factors (i.e., regulations, raw materials, markets, and technology) are observed, the project is marked as a good candidate for scrapping. (Further investigation may be required before a final decision is made.) If no serious problems are found, the project is reviewed for negative changes in the key variables. A project score is computed by adding one point for each variable that has not deteriorated since the last review. A total of nine or more points indicates a high probability of success. Projects with scores between six and eight are deemed to be on the verge of failing and hence require an immediate and detailed evaluation. A score of six or less indicates a high probability of failure.

At any stage in the evaluation process it may be possible to save a marginally failing project by allocating additional resources to alter (5), (7), and (13), or by influencing the qualitatively controllable variables (3), (8), (9), and (14). For example, if the technological route is problematic or pressure on the project leader has declined, a commitment on the part of management may be all that is needed to bring a project score up to the desired level. A model that addresses this situation and takes into account the competition for resources among ongoing projects is developed in the next section. Due to the qualitative nature of most of the factors, an interactive approach is prescribed. This facilitates a timely assessment of the portfolio by allowing for on-line updates of performance data and the immediate disposition of marginal projects.

12.6.2 Analytic Methodology

At the beginning of a review period, each project is evaluated individually and collectively in accordance with the monitoring scheme outlined above. The first stage of this two-stage process involves the critical factors. If one or more of these is strongly negative, the project is terminated and its remaining resources redistributed. Next, the 14 key variables are evaluated. If the resulting score for a specific project equals or exceeds the threshold, T, it remains in the portfolio; if not, a judgment is made to see if the score can be raised to the desired level by altering one or more of the controllable factors. If this is not possible, the project is terminated and its resources reallocated.

These ideas are formalized in a three-step procedure using the following notation.

$$i = \text{index for projects}$$
$$j = \text{index for key variables}$$
$$n = \text{number of projects in the active portfolio}$$
$$\bar{n} = \text{total number of projects in the portfolio and on the candidate list}$$
$$n_{\text{max}} = \text{maximum number of projects to be included in the portfolio}$$
$$B = \text{total budget}$$
$$B_i = \text{current budget for project } i$$
$$b_i = \text{maximum funding allowable for project } i$$
$$p_i = \text{probability of technical success for project } i$$
$$P_i = \text{threshold value of } p_i$$
$$f_{ij}(t) = \text{value of key variable } j \text{ for project } i \text{ during review period } t$$
$$a_{ij} = \text{dependent zero–one scoring variable, indicating whether key variable } j \text{ for project } i \text{ is at an acceptable level}$$
$$T = \text{threshold value for project score}$$

- *Step 1*
  (a) Screen each project separately with respect to the three exogenous critical factors; terminate those with strong negative indicators.
  (b) Screen remaining projects in portfolio with respect to probability of technical success using threshold P_i; terminate those that cannot be improved sufficiently within budgetary guidelines.

- *Step 2*
  (a) Compute total score a_i for project i as follows: Let

  $$a_{ij} = \begin{cases} 1, & \text{if } f_{ij}(t) - f_{ij}(t-1) \geq 0, \\ 0, & \text{otherwise} \end{cases} \quad \text{for all } i \text{ and } j$$

  $$a_i = \sum_{j=1}^{14} a_{ij}, \quad i = 1,...,n$$

  (b) Compare the score obtained with threshold value T, and define a zero–one indicator variable $\hat{a}_i$ as follows:

  $$\hat{a}_i = \begin{cases} 1, & \text{if } a_i \geq T, \\ 0, & \text{otherwise} \end{cases} \quad i = 1,...,n$$

  (c) Determine the disposition of project i. If $\hat{a}_i = 1$, place the project in the portfolio; if not, evaluate the feasibility of increasing $\hat{a}_i$ from zero to 1 by increasing those a_{ij} values associated with the controllable key variables currently at zero. Terminate if not possible; otherwise, indicate whether or not additional effort will raise the score. Include the project in the portfolio if the response is positive.

- *Step 3.* Compute the amount of free resources, R, where

  $$R = B - \sum_{i=1}^{n} B_i \hat{a}_i$$

These three steps constitute the updating and qualitative evaluation of the current portfolio at the beginning of review period t. Some projects will be canceled outright; others will be further scrutinized by the decision maker to see if their condition can be improved.

To operationalize steps 1 and 2 in a manner that promotes consistency across projects and managers, two procedures are recommended. The first is to provide benchmarks for the interviewees in terms of background and reference data. With respect to market conditions, for example, the benchmark associated with a negative change might be determined by comparing sales figures for similar products over the last two quarters. The second procedure is aimed at building a consensus by soliciting responses from both the project leader and a subset of team members. Discrepancies can be fed back for reconsideration.

Model formulation.   At this point, we need to allocate the remaining funds, R, to the active projects, including those on the candidate list. A decision model is formulated for this purpose using the following additional notation:

V_i = present value of returns attributed to project i
y_i = additional amount of resources allocated to project i
x_i = total amount of resources allocated to project i
u_i = zero–one decision variable for continuing project i
$\bar{u}_i$ = zero–one decision variable for selecting project i from set of candidate projects to be in the portfolio

A project that is performing well at the beginning of a review period will not necessarily have its funding continued. Although normally this is not the case, it may be determined that the resources currently allocated to that project should be reduced and the difference reallocated to projects whose payoffs are potentially higher. Under such circumstances termination will occur if the probability of technical success, p_i, drops below its threshold, P_i. In general, p_i is assumed to be a function of the total budget, x_i, assigned to project i ($x_i = y_i + B_i$) and will be defined by one of the relationships shown in Fig. 12-4. Now, if the probability of commercial success is denoted by f_{is} (for simplicity the dependence of the critical factors on t will be dropped from the notation), we solve the following problem:

$$\max \sum_{i=1}^{n} V_i p_i(x_i) f_{is} u_i + \sum_{i=n+1}^{\bar{n}} V_i p_i(x_i) f_{is} \bar{u}_i \tag{12-1a}$$

subject to

$$\sum_{i=1}^{n} y_i u_i + \sum_{i=n+1}^{\bar{n}} x_i \bar{u}_i \leq R \tag{12-1b}$$

$$x_i = y_i + B_i \leq b_i, \qquad i = 1,...,n \tag{12-1c}$$

$$x_i \leq b_i, \qquad i = n+1,...,\bar{n} \tag{12-1d}$$

$$\sum_{i=1}^{n} u_i + \sum_{i=n+1}^{\bar{n}} \bar{u}_i \leq n_{\max} \tag{12-1e}$$

$$\sum_{j=1}^{14} a_{ij} \geq T u_i \qquad i = 1,...,n \tag{12-1f}$$

$$p_i(x_i) \geq P_i u_i, \qquad i = 1,...,n \tag{12-1g}$$

$$p_i(x_i) \geq P_i \bar{u}_i, \qquad i = n+1,...,\bar{n} \tag{12-1h}$$

$$x_i \geq 0, \quad y_i \geq -B_i, \quad u_i = 0 \text{ or } 1, \quad \bar{u}_i = 0 \text{ or } 1, \qquad \text{for all } i \tag{12-1i}$$

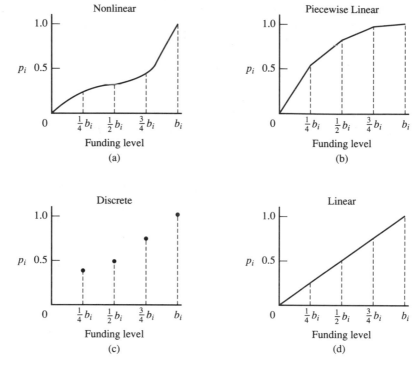

Figure 12-4 Relationships between probability of technical success and funding: (a) nonlinear; (b) piecewise linear; (c) discrete; (d) linear.

The objective function (12-1a) represents the expected return from the portfolio for both active and candidate projects. Constraint (12-1b) restricts the funding in period t to the remaining budget R. Constraints (12-1c) and (12-1d) place a limit on the amount allocated to a given project, while (12-1e) controls the maximum number of projects in the portfolio. The remaining structural constraints (12-1f) through (12-1h) assure that if a project is selected, its key variable score is at least equal the threshold value T (= 9), and its probability of technical success is at an acceptable level. This formulation permits resources to be removed from an active project as long as $p_i(x_i)$ does not fall below P_i.

Implicit in the construction of problem (12-1) is the assumption that additional resources allocated to project i will affect p_i as well as the three other quantitatively controllable key variables (5), (7), and (13). The functional relationship between x_i and the latter must be worked out on an individual basis. For example, the probability of technical success might be increased by the acquisition of better or more advanced laboratory equipment, while increasing worker commitment might be accomplished through the installation of a minicomputer to facilitate the project's data processing. The formulation above does not treat the key variables (5), (7), and (13) explicitly.

Implementation.   Problem (12-1) is a mixed nonlinear integer program (IP) whose degree of difficulty depends in part on the functional forms chosen to represent p_i. In the implementation, Bard et al. uses the discrete model in Fig. 12-4, so the problem reduces to a pure nonlinear integer program whose terms are at most quadratic in the decision variables x, u and $\bar{u}$. Such problems may be converted to integer linear programs by adding one variable and two constraints for each quadratic term (see Bard 1986). Because most R&D portfolios usually contain fewer than 30 projects, this type of transformation will yield a problem whose dimensions are well within the reach of current codes. If any of the other three models in Fig. 12-4 are used, different techniques may be required.

   To put the problem into a more manageable form, let us redefine the decision variables x_i such that x_{ik} equals 1 if project i is funded at level k, and 0 otherwise. Also, let p_{ik} be the probability of technical success associated with allocation b_{ik}, and K_i be the number of permissible funding levels for project i. This leads to the following pure zero–one linear formulation.

$$\max \sum_{i=1}^{N} V_i f_{i5}\left(\sum_{k=1}^{K_i} p_{ik} x_{ik} \right) \tag{12-2a}$$

subject to

$$\sum_{i=1}^{N} \sum_{k=1}^{K_i} b_{ik} x_{ik} \leq B \tag{12-2b}$$

$$\sum_{k=1}^{K_i} b_{ik} x_{ik} \leq b_i, \qquad i = 1,...,N \tag{12-2c}$$

$$\sum_{i=1}^{N} u_i \leq n_{\max} \tag{12-2d}$$

$$\sum_{k=1}^{K_i} p_{ik} x_{ik} - P_i u_i \geq 0, \qquad i = 1,...,N \tag{12-2e}$$

$$\sum_{k=1}^{K_i} x_{ik} \leq u_i, \qquad i = 1,...,N \tag{12-2f}$$

$$x_{ik} = 0 \text{ or } 1, \quad u_i = 0 \text{ or } 1, \qquad \text{for all } i \text{ and } k \tag{12-2g}$$

where $N \equiv \bar{n}$. Problem (12-2) assumes that if the score of project i is not at least at the threshold, T, then $u_i = 0$. This eliminates the need for constraint (12-1g). Also, the vector u has been redefined to include $\bar{u}$.

   In solving problems such as (12-2), it is important for the analyst to be able to enter data in a simple format and be able to change parameters easily while investigating various scenarios. Here the complete methodology was embodied in three separate modules: (1) a front-end, menu-driven routine for input and control; (2) a model generator for data formatting; and (3) a zero–one IP solver. All computer codes were written in BASIC for the

IBM-PC. Use of the methodology, along with the computations, are demonstrated in Appendix 12A.

TEAM PROJECT

Thermal Transfer Plant

Your team was invited to the CEO's office at TMS. At the meeting the CEO told you how impressed he was by the prototype rotary combustor project and expressed his confidence in your team leading the new waste management and recycling division. He expects this division to master the leading technology in waste disposal. To begin, you are asked to search the literature and to propose related high-tech R&D projects. The CEO would like you to present your proposal for the most appropriate such project at the next TMS board meeting.

It is clear that a detailed proposal addressing all aspects of the R&D project will be supported by the CEO. You are also aware of the once-in-a-lifetime opportunity that has been presented to you for recognition and advancement.

Prepare a proposal for the new R&D project, explaining the following:

- Why the project is the most appropriate for TMS
- What technology will be used
- The nature of the expected risks
- The proposed schedule and budget
- The approach you will take to maximize the probability that the project will be a success

DISCUSSION QUESTIONS

1. What characteristics distinguish R&D project management from conventional project management? What additional skills does an R&D project manager need?
2. Identify a few breakthrough technologies and the products that they spawned.
3. Pick a major U.S. industry, such as automobiles or computers, and discuss the lapses in technology and innovation on the domestic front that permitted foreign competitors to get a foothold and, in some cases, a dominant share of the market. Who or what do you think was to blame for this situation?
4. In the mid-1980s, General Motors undertook a $5 billion program to introduce robotics and computer-integrated manufacturing techniques into many of its assembly plants. The results were disappointing, to say the least. Enormous technical problems dogged the program from

the beginning and the ultimate gains in productivity were decidedly modest. What do you think went wrong? Why? From the long-term perspective, was the automation program a good idea?

5. Give a few examples where commercial success did not follow technical success with regard to new product introduction. What were the reasons for market failure?

6. Can you think of an example where lack of standards retarded the introduction of a new product or technology? Give some details.

7. Pick an industry and identify its base, key, and pacing technologies.

8. What are the differences between a strategic technical plan and an operational plan for an R&D project?

9. As the head of an R&D group contemplating the development of a notebook computer with a built-in fax machine, who would you like to have on your strategic planning team? Why?

10. Consider a new technology, such as superconductivity or magnetic levitation, and identify several (parallel) ways of realizing it on a commercial rather than a laboratory scale.

11. Identify a new technology where you feel that parallel development is not warranted. Explain you choice.

12. Why has simulation modeling, which is a descriptive technique, been preferred to mathematical programming, which is a prescriptive technique, for analyzing and providing help in the management of R&D projects?

13. Which of the critical factors and key variables described in Section 12.6 do you think would apply to conventional projects?

14. What are some of the shortcomings of the mathematical programming model presented in Section 12.6.2 to manage an R&D portfolio? Can you suggest ways of correcting or accommodating them?

15. What data are needed to run the mathematical programming model (12-2)? How would you go about collecting these data?

EXERCISES

12-1. Identify a new product based on an innovation in technology and draw up a strategic technical plan for its development. Be sure to discuss the risk factors at each stage and indicate how you would deal with each.

12-2. Assume that you are in charge of a round-trip mission to Mars. The goal is to spend 3 months on the plant surface performing experiments and collecting data that will be used to help set up a future colony. Construct a strategic plan for this mission.

12-3. The transonic airplane, now on the drawing board, is intended to be a commercial transport operating between continents at supersonic speeds. It will fly a ballistic trajectory and be able to reach Japan from the United States in only a few hours. Identify the base, key, and pacing technologies for this vehicle. Discuss the economic, political, social, and technical issues surrounding its development.

12-4. Select an industry such as semiconductors or consumer electronics, and go through the six stages of the strategic technical planning process listed in Table 12-1.

12-5. Consider the problem of trying to decide which tasks to fund in parallel to achieve a given technical objective. For example, the objective might be to develop a low-cost rechargeable battery to power an electric vehicle. The funding options might be the various types of battery technologies available (see Bard and Feinberg 1989). What are the decision variables and functional relationships associated with the problem? What data are required? Be specific with respect to probability functions and any other relationships that might exist.

12-6. Construct a mathematical programming model for the parallel funding problem discussed in Exercise 12-5 and Section 12.5.

12-7. Choose a new technology, describe its major features, and explain how you would apply the total quality management principles discussed in Chapter 6 to an R&D project aimed at commercialization. What is different about TQM applied to an R&D project versus a conventional project?

12-8. For each of the 14 key variables presented in Section 12.6.1, identify the internal and external data sources that can be used to ascertain their status.

12-9. *Computer Assignment.* Write an interactive computer program implementing the three-step procedure discussed in Section 12.6.2 for screening projects and computing project scores.

12-10. Using a commercial integer programming package, solve problem (12-2) initialized with the case study data presented in Appendix 12A.

REFERENCES

Project Selection

BAKER, N. R., "R&D Project Selection Models: An Assessment," *IEEE Transactions on Engineering Management*, Vol. EM-21, No. 4 (1974), pp. 165–171.

BARD, J. F., "A Multiobjective Methodology for Selecting Subsystem Automation Options," *Management Science*, Vol. 32, No. 12 (1986), pp. 1628–1641.

BARD, J. F., "Using Multicriteria Methods in the Early Stages of New Product Development," *Journal of the Operational Research Society*, Vol. 41, No. 8 (1990), pp. 755–766.

BARD, J. F., and A. FEINBERG, "A Two-Phase Approach to Technology Selection and System Design," *IEEE Transactions on Engineering Management*, Vol. EM-36, No. 1 (1989), pp. 28–36.

BARD, J. F., R. BALACHANDRA, and P. E. KAUFMANN, "An Interactive Approach to R&D Project Selection and Termination," *IEEE Transactions on Engineering Management*, Vol. EM-35, No. 3 (1988), pp. 139–146.

BU-BUSHAIT, K. A., "The Application of Project Management Techniques to Construction and Research and Development," *Project Management Journal*, Vol. 20, No. 2 (1898), pp. 17–21.

DEAN, B. V. (editor), *Project Management: Methods and Studies*, Elsevier Science Publishers, North-Holland, Amsterdam, 1985.

HIGGINS, J. C., and K. M. WATTS, "Some Perspectives on the Use of Management Science Techniques in R&D Management," *R&D Management*, Vol. 16, No. 4 (1986), pp. 291–296.

LIBERATORE, M. J., "An Extension of the Analytic Hierarchy Process for Industrial R&D Project Selection and Resource Allocation," *IEEE Transactions on Engineering Management*, Vol. EM-34, No. 1 (1987), pp. 12–18.

LIBERATORE, M. J., and G. J. TITUS, "The Practice of Management Science in R&D Project Management," *Management Science*, Vol. 29, No. 8 (1983), pp. 962–974.

SCHMIDT, R. L., and J. R. FREELAND, "Recent Progress in Modeling R&D Project-Selection Processes," *IEEE Transactions on Engineering Management*, Vol. 39, No. 2 (1992), pp. 189–200.

SOUDER, W. E. and T. MANDAKOVIC, "R&D Project Selection Models," *Research Management*, Vol. 29, No. 4 (July–August 1986), pp. 36–42.

WATTS, K. M., and J. C. HIGGINS, "The Use of Advanced Management Techniques in R&D," *Omega*, Vol. 15, No. 1 (1987), pp. 221–229.

Resource Allocation and Parallel Funding

ABERNATHY, W. J., and R. S. ROSENBLOOM, "Parallel Strategies in Development Projects," *Management Science*, Vol. 15, No. 10 (1969), pp. B486–B505.

BALTHASAR, H. U., R. A. BOSCHI, and M. M. MENLE, "Calling the Shots in R&D," *Harvard Business Review*, (May–June 1978), pp. 151–160.

BARD, J. F., "Parallel Funding of R&D Tasks with Probabilistic Outcomes," *Management Science*, Vol. 31, No. 7 (1985), pp. 814–828.

BOWER, J. L., *Managing the Resource Allocation Process*, Harvard University Press, Boston, 1970.

COOLEY, S. J., J. HEHMEYER, and P. J. SWEENEY, "Modeling R&D Resource Allocation," *Research Management*, Vol. 29, No. 1 (1986), pp. 40–45.

MILES, R. F., JR., "The SIMRAND Methodology: SIMulation of Research ANd Development Projects," *Large Scale Systems*, Vol. 7 (1984), pp. 59–67.

MOORE, L. J., and E. R. CLAYTON, *GERT Modeling and Simulation: Fundamentals and Applications*, Petrocelli-Charter, New York, 1976.

PRITSKER, A. A. B., *Modeling and Analysis Using Q-GERT Networks*, Second Edition, Wiley, New York, 1979.

TAYLOR, B. W., III, and L. J. MOORE, "R&D Project Planning with Q-GERT Network Modeling and Simulation," *Management Science*, Vol. 26, No. 1 (1980), pp. 44–59.

TYMON, W. G., and R. F. LOVELACE, "A Taxonomy of R&D Control Models and Variables Affecting Their Use," *R&D Management*, Vol. 16, No. 3 (1986), pp. 233–241.

WALLIN, C. C., and J. J. GILMAN, "Determining the Optimum Level for R&D Spending," *Research Management*, Vol. 29, No. 5 (1986), pp. 19–24.

New-Product Development

BALACHANDRA, R., "Critical Signals for Making the Go/No Go Decisions in New Product Development," *Journal of Product Innovation Management*, Vol. 2 (1984), pp. 92–100.

BALACHANDRA, R., *Early Warning Signals for R&D Projects,* Lexington Books, D.C. Heath, Lexington, MA, 1989.

COOPER, R. G., *Winning at New Products*, Gage Educational Publishing, Agincourt, Ontario, Canada, 1987.

COOPER, D., and C. CHAPMAN, *Risk Analysis for Large Projects*, Wiley, New York, 1986.

KIDDER, T., *The Soul of a New Machine*, Little, Brown, Boston, 1981.

NAYAK, P. R., and J. M. KETTERINGHAM, *Breakthroughs!*, Mercury Books, Philadelphia, 1987.

PETERS, T., "Get Innovative or Get Dead, Part One," *California Management Review*, Vol. 33, No. 1 (1990), pp. 9–26.

ROGERS, E. M., "New Product Adoption and Diffusion," *Journal of Consumer Research*, Vol. 2, No. 4 (1976), pp. 290–301.

ROSENBLOOM, R. S., and M. A. CUSUMANO, "Technology Pioneering and Competitive Advantage: The Birth of the VCR Industry," *California Management Review*, Vol. 29, No. 4 (1987).

SOUDER, W. E., *Managing New Product Innovations*, Lexington Books, D.C. Heath, Lexington, MA, 1987.

WHEELWRIGHT, S. C., and K. B. CLARK, "Creating Project Plans to Focus Product Development," *Harvard Business Review* (March–April 1992), pp. 70–82.

Critical Factors

BAKER, N. R., S. G. GREEN, and A. S. BEAN, "Why R&D Projects Succeed or Fail," *Research Management*, Vol. 29, No. 6 (November–December 1986), pp. 29–34.

BALACHANDRA, R., and J. A. RAELIN, "How to Abandon an R&D Project," *Research Management*, Vol. 18 (1980), pp. 24–29.

MELDRUM, M. J., and A. F. MILLMAN, "Ten Risks in Marketing High-Technology Products," *Industrial Marketing Management*, Vol. 20, No. 1 (1991), pp. 43–48.

PINTO, J. K., and D. P. SLEVIN, "Critical Factors in Successful Project Implementation," *IEEE Transactions on Engineering Management*, Vol. EM-34, No. 1 (1987), pp. 22–27.

RUBENSTEIN, A. H., and H. SCHRODER, "Managerial Differences in Assessing Probabilities of Technical Success for R&D Projects," *Management Science*, Vol. 24, No. 2 (1977), pp. 137–148.

SCHRODER, H. H., "The Quality of Subjective Probabilities of Technical Success in R&D," *R&D Management*, Vol. 6, No. 1 (1975).

Strategic Issues

ENGLERT, R. D., *Winning at Technological Innovation*, Chapter 3, "Strategic R&D Planning," McGraw-Hill, New York, 1990.

ERICKSON, T. J., J. F. MAGEE, P. A. ROUSSEL, and K. N. SAAD, "Managing Technology as a Business Strategy," *Sloan Management Review*, Vol. 31, No. 3 (1990), pp. 73–77.

GLUCK, F. W., S. P. KAUFMAN, and A. S. WALLECK, "Strategic Management for Competitive Advantage," *Harvard Business Review*, Vol. 58 (July–August 1980), pp. 154–161.

JAIN, R. K., and H. C. TRIANDIS, *Management of Research and Development: Managing the Unmanageable*, Wiley, New York, 1990.

ROSENBLOOM, R. S., and K. L. FREEZE, K, *Research in Technological Innovation, Management and Policy*, JAI Press, Greenwich, CT, 1985.

ROUSSEL, P. H., K. N. SAAD, and T. J. ERIKSON, *Third Generation R&D: Managing the Link to Corporate Strategy*, Harvard Business Press, Boston, 1991.

YATES, J. F. (editor), *Risk-Taking Behavior*, Wiley, New York, 1991.

APPENDIX 12A

Portfolio Management Case Study

Portable Solutions is a Texas-based company that has been distributing personal computers and peripherals since 1982. In May 1985 it expanded its operations and began to produce its own brand of tape backup units for the IBM-PC and compatibles. Since that time, the company has introduced to the market three different but related products, including a 10-megabyte self-threading backup system, a 20-megabyte streaming tape backup system, and a 40-megabyte streaming tape backup.

To maintain profitability and ensure its survival, Portable Solutions is now exploring two new ventures. The first involves vertical integration of its current line; the second centers on the development of new products to compete in complementary areas. To put these ideas into motion, the company has established an R&D portfolio that includes the following six projects.

1. A signature verification system with the capacity to store 145,000 signature files on an optical, nonerasable medium, and be able to access each within 3 seconds. The system is intended for use by banks and would replace current methods, which typically rely on microfiche as the storage medium. With respect to performance, optical technology offers far greater speed and reliability then do any of its competitors. In addition, the permanent nature of the files offers built-in security.

2. A signature verification system with the capacity to store 85,000 files on a hard drive and access each in 3 seconds or less. This system would incorporate most of the features of (1) but would use magnetic tape as the storage medium. The advantage here is lower development cost, while the disadvantage concerns record security (i.e., information can easily be altered).

3. A portable laser drive with the capacity to store 115 megabytes of information. This system should be fully portable without the hint of compatibility or installation problems. To date, speed and capacity have been limited by mechanical hard-drive technology. Optical technology, however, now permits storage of vast amounts of data in compact form, free of maintenance, and without the data integrity problems that have plagued magnetic media.

4. A small computer standard interface (SCSI) for existing and future product lines. This interface will be compatible with most mainframes as well as the Apple Macintosh and various workstations. At this time, the company's market is limited to the IBM-PC and compatibles. A SCSI will open up new opportunities throughout the industry.

5. A port extender interface for personal computers. This device permits the addition of peripherals when no extra slots for interface cards are available. It plugs directly into a floppy port, leaving the bus slots available for other applications. The proliferation of add-ons makes this an especially attractive product.

6. A combination 20-megabyte hard disk/20-megabyte portable backup system for military applications. This unit must pass stringent military quality and durability standards. The marketing department indicates that the demand for this product is high and that many channels exist for its promotion. Nevertheless, engineering has expressed serious doubts about achieving the required levels of performance within the target cost range.

In the course of marketing, management has recently identified four additional projects as potentially lucrative and is considering several funding alternatives. The new projects include:

7. A downloading peripheral that transfers data from a 9-inch mainframe magnetic tape to a 3.5-inch optical cartridge for use with personal computers. The mainframe standard for information archiving over the past 20 years has been the 9-inch tape. This medium requires constant maintenance to avoid data loss and consumes expensive CPU time during retrieval. Downloading tapes to optical media not only creates a maintenance-free environment but also permits easy access through personal computers.

8. A smart system capable of downloading 9-inch magnetic tapes to 3.5-inch optical cartridges. This system would be self-contained, not needing additional equipment to operate. It would differ from the system described in (7) by the incorporation of a central processing unit.

9. A record management system capable of storing up to 2 gigabytes of information in a 12-inch optical cartridge. As envisioned, a microcomputer would serve as processor, and up to 1,000 cartridges could be managed at once using a jukebox principle.

10. An image-scanning device capable of tracking pavement conditions on highways and determining when to record a damaged sector. In addition, it must be capable of classifying the damage and deriving a repair schedule based on severity of damage and equipment availability. The goal is to develop a system that will reduce highway repair costs by about 50%.

As is usually the case, Portable Solutions does not have the resources to fund all of these projects. Model (12-2) presented in Section 12.6 will be used to determine the best allocation of materials and personnel and to decide the level of activity for each project accepted.

Given the current demand for resources, management feels that at most seven projects should be undertaken at one time, and has imposed a $250,000 ceiling on the R&D budget. Table 12A-1 lists the input data for the 10 projects, and Table 12A-2 specifies the relationships between probability of technical success, p_{ik}, and funding level, b_{ik}. These data were derived from extensive interaction with the firm's four principal officers. They represent the consensus that emerged after two iterations of individual and collective discussions.

At the present time, five of the first six projects are actively being pursued at a cost of $240,000. The third project is not in the portfolio but is still considered a candidate. Table 12A-3 indicates the individual funding levels along with the total.

TABLE 12A-1 INPUT DATA FOR R&D CASE STUDY

Project	Probability of commercial success, f_{is}	Threshold probability, P_i	Present value of return, V_i	Maximum budget, B_i
1	0.75	0.35	$3.7M	$ 75K
2	0.82	0.40	8.2M	105K
3	0.67	0.45	7.5M	145K
4	0.92	0.35	4.1M	110K
5	0.55	0.30	5.1M	90K
6	0.88	0.35	7.8M	145K
7	0.68	0.40	3.5M	90K
8	0.75	0.35	9.0M	100K
9	0.67	0.30	7.5M	128K
10	0.94	0.45	8.6M	129K

TABLE 12A-2 RELATIONSHIP BETWEEN PROBABILITY OF TECHNICAL SUCCESS AND FUNDING LEVEL

Project	Level 1		Level 2		Level 3		Level 4	
	p_{i1}	b_{i1}	p_{i2}	b_{i2}	p_{i3}	b_{i3}	p_{i4}	b_{i4}
1	0.44	$22K	0.56	$34K	0.72	$54K	0.89	$ 72K
2	0.36	18K	0.45	26K	0.57	47K	0.82	64K
3	0.40	25K	0.58	52K	0.72	90K	0.95	130K
4	0.35	20K	0.50	38K	0.75	84K	0.94	100K
5	0.30	15K	0.55	40K	0.70	60K	0.90	90K
6	0.25	25K	0.50	50K	0.75	65K	0.98	120K
7	0.25	15K	0.56	40K	0.76	60K	0.89	82K
8	0.36	20K	0.49	40K	0.62	81K	0.82	94K
9	0.25	25K	0.54	50K	0.77	98K	0.95	125K
10	0.35	25K	0.50	48K	0.75	89K	0.94	129K

In the process of updating the portfolio, all current projects passed the critical factors test at step 1, and all but project 5 passed the key variables test at step 2. Running the model with the four remaining projects, project 3, and the four new ones, led to the selection of six projects, as shown in Table 12A-4. The total budget allocation accompanying this solution is $249,000 and the expected return is $19.56M. The specific funding levels are also shown in Table 12A-4.

The fact that projects 1, 3, and 7 were not chosen does not necessarily mean that they will be discarded but simply that they will be shelved until the next review or until additional funds become available. It is also possible that project 5 could be resurrected at a future time.

Regarding the computations, the solution was obtained in 15.22 minutes on an IBM-PC. This involved solving a 32-variable, zero–one integer programming problem with 11 constraints. In general, solution times grow exponentially with the total number of levels and projects. Note that constraints (12-2c) and (12-2d) can be handled implicitly, and that the nine project variables, u_i, can be eliminated by appropriately redefining (12-2d) and (12-2f).

TABLE 12A-3 FUNDING FOR BASIC PORTFOLIO

Project	Level	Funding
1	2	$ 34,000
2	4	64,000
4	1	20,000
5	3	60,000
6	3	62,000
		$240,000

TABLE 12A-4 RESULTS FOR UPDATED PORTFOLIO

Project	Level	Funding
2	2	$ 26,000
4	1	20,000
6	3	65,000
8	2	40,000
9	2	50,000
10	2	48,000
		$249,000

Assessment of Methodology

The presentation above demonstrates the facility with which an R&D manager can update his or her portfolio provided that all the pertinent data are available and that an accurate assessment of the key variables can be made. Updates can take place at any time but are commonly scheduled around the budgetary cycle. As some projects reach their critical stages, though, it may be desirable to increase the frequency with which the portfolio is reviewed.

After testing the methodology with a number of high-tech firms, it was found that most managers were less interested in the final results than in the process itself. The value for them was in systematically stepping through each project and assessing its status. As was often the case, the biggest stumbling block arose in the evaluation of changes in the key variables. The lack of up-to-date information often led to difficulties in making consistent judgments across the portfolio. In some instances, for example, not all managers were aware of a lapse in worker commitment or the critical need for a technological breakthrough.

Nevertheless, the information gathered at the interview sessions was prized as much for the insight it provided as for the confidence it instilled in the control process. By isolating the major components of the R&D management decision, the interactive dynamics enabled the participants to gain a better understanding of the forces at work.

13

Computer Support
for Project Management

13.1 INTRODUCTION

Project management is the process of achieving multidimensional goals related to on-time delivery, adherence to schedule, and cost minimization subject to resource availability, cash flow, and technological performance constraints, all in the presence of uncertainty. The tools that have been developed to assist project managers in their job were introduced earlier. Most of these tools are based on a model that transforms input data into some form of output that facilitates decision making. For example, scheduling by the critical path method transforms information about required activities, performance times, and precedence relations into a list of critical activities, available slack for noncritical activities, and an estimate of project completion time. Each tool is designed to handle a specific aspect of the project management process. However, a project manager frequently needs an integrated tool to handle several aspects of a project at once. This has led to the development of software packages for project management that now make it possible for different organizations to interact efficiently by standardizing procedures, reports, and data files.

Early software packages typically concentrated on scheduling and costs. Data input and processing were batch-oriented, and only a prespecified set of output reports were available. The introduction of new computers (principally PCs and workstations), rapid advances in software engineering, and reduced processing and computer memory costs led to the development of integrated software packages able to address a multitude of functions simultaneously.

The current trend in the area of software development is toward interactive, fully integrated systems. Many of these systems can handle the following aspects of project management throughout a project's life cycle:

- Configuration management
- Scheduling
- Budgeting
- Costs
- Resource management
- Monitoring and control

Potential users face two questions: (1) how to select the most appropriate software package for their needs, and (2) how to introduce the chosen package into their organization successfully. In the next section, guidance is offered to those charged with the responsibility of answering the first question. We use the software package SuperProject Expert by Computer Associates, furnished with the book, to illustrate concepts. This is followed by a discussion of the major criteria accompanying a benefit-cost analysis aimed at making the selection. The remainder of the chapter offers insights into smoothing the implementation of project management software in a new environment.

13.2 USE OF COMPUTERS IN THE PROJECT MANAGEMENT PROCESS

Project management requires the deliberate treatment of organizational processes, economic factors, and technological aspects, as well as the implementation of methodologies for planning, scheduling, and control. When choosing a project manager, it is important to consider leadership abilities, verbal communication skills, and motivation level. Today's computers cannot replace a skilled project manager because computers do not possess these attributes. However, they can support a project manager in certain decision-making processes if the problems at hand are well defined and amenable to quantitative or symbolic manipulation. Even if this is only partially the case, the computer's ability to store, retrieve, and process large quantities of data, along with its powerful communication capabilities, can help prepare information for the decision maker. Consequently, computers equipped with appropriate software can support many aspects of project management.

- *Work breakdown structure.* The initial step in project management is to define the project's content in the form of a work breakdown structure. The development of a WBS is greatly facilitated by computer packages whose input and presentation format reflect the underlying hierarchical structure of the project, that can assign appropriate codes to WBS elements at each level, and that can check for inconsistencies such as disconnected

WBS elements or lower-level WBS elements connected to more than one higher-level element. The division of the project/program into its basic building blocks is easier using a module that automatically assigns WBS codes and checks for inconsistencies as part of the data input process. Figure 13-1 illustrates a WBS diagram that corresponds to the seven tasks of the example project used throughout the book.

■ *Organizational breakdown structure.* The next step in project management is to develop the project's organizational breakdown structure. This structure depicts the communication lines for reports, work authorization, and so on. A module similar to the one for the WBS supports the creation of a clearly defined organizational structure. Integrating the WBS module with the OBS module generates a matrix that assigns each lower-level WBS element to a lower-level OBS element.

The OBS and WBS hierarchies allow for information processing through a roll-up mechanism. This mechanism transfers information from lower- to upper-level elements through the connections defined in the OBS–WBS matrix. The established relationships help to generate reports at several managerial levels.

One important aid to a project manager is a software package that supports the development, maintenance, and integration of the WBS and OBS. The process subdivides the project's work and allocates it among the groups participating. This division capability is also important in integrating individual efforts and helping update all groups on their share of the total effort as it relates to the entire project. The WBS and OBS modules not only generate reports at various managerial levels, but also keep these reports coherent and synchronous. By using the same OBS–WBS hierarchy throughout a project's life cycle, the plans developed during startup are used to execute, monitor, and control each stage of the project.

Once the organizational structure is defined and each participating unit is assigned tasks or a scope of work, it is possible to break down the project's work content further into activities and to estimate each activity's duration. This breakdown forms the basis for the next step in the project's planning cycle.

■ *Scheduling.* Scheduling starts by defining the time frame or calendar for the project. In the calendar definition phase, the project manager may select a current organizational calendar or develop one that is project specific. The calendar defines working days per week, daily working hours, scheduled holidays and vacations, and so on. One important decision is to select the minimal time unit. Some projects (such as machine tool maintenance) require a detailed schedule at the level of minutes or hours. For long-term construction projects, a minimum time unit of a day or a week may suffice. Figure 13-2 illustrates a

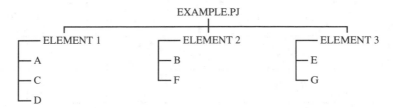

Figure 13-1   Work breakdown structure for the example project.

Calendar for: labor No. Units: 1 Std. Day: 8							
1994	Sun	Mon	Tue	Wed	Thu	Fri	Sat
Jan	02 WKND	03 HOLIDAY	04	05	06	07	08 WKND
Jan	09 WKND	10	11	12	13	14	15 WKND
Jan	16 WKND	17 HOLIDAY	18	19	20	21	22 WKND
Jan	23 WKND	24	25	26	27	28	29 WKND
Jan Feb	30 WKND	31	01	02	03	04	05 WKND
Feb	06 WKND	07	08	09	10	11	12 WKND
Feb	13 WKND	14	15	16	17	18	19 WKND
Feb	20 WKND	21 HOLIDAY	22	23	24	25	26 WKND
Feb Mar	27 WKND	28	01	02	03	04	05 WKND
Mar	06 WKND	07	08	09	10	11	12 WKND
Mar	13 WKND	14	15	16	17	18	19 WKND
Mar	20 WKND	21	22	23	24	25	26 WKND
Mar Apr	27 WKND	28	29	30	31	01	02 WKND
Apr	03 WKND	04	05	06	07	08	09 WKND
Apr	10 WKND	11	12	13	14	15	16 WKND
Apr	17 WKND	18	19	20	21	22	23 WKND
Apr	24 WKND	25	26	27	28	29	30 WKND

Figure 13-2   Calendar for the example project.

calendar for the example project, which is scheduled to start in the beginning of January 1994. The figure shows only the first four months since the rest of the calendar is defined similarly.

Based on the calendar and the estimated activity durations, activity scheduling can begin. This scheduling occurs in most software packages by defining precedence relations among activities. The first step is usually to enumerate finish-to-start precedence relations, that is, when an activity can start as a function of its immediate predecessors. The CPM (critical path method) logic is then applied and the early start, early finish, late start, and late finish dates are calculated for each activity. These dates are based on the calendar selected, with its predetermined holidays and vacations. The resulting schedule can be a table of activities with their corresponding dates and slacks, a Gantt chart, or a network model (AOA, AON). Some software packages include all three formats, while others include only a tabular report or a Gantt chart. Figure 13-3 presents the Gantt chart for the example project; Fig. 13-4 depicts the same activities as an AON diagram.

The basic schedule developed by the process described above is called an "unconstrained" schedule. The precedence relations among activities are assumed to be the only limiting factors. The next step is to introduce the time constraints imposed on activities or events (project milestones). Time constraints may require an activity to start or end on a given date, and may produce an infeasible schedule due to conflicts between the critical path's length and the imposed milestones. For example, a conflict occurs if the length of the critical path is 12 months but the contract calls for a delivery date 11 months after kickoff.

In some projects, managers can resolve conflicts by introducing other forms of precedence relations, such as start-to-start and finish-to-finish. Modeling the real situation with these alternatives may alleviate the problem. Some software packages support all types of precedence relations mentioned in Chapter 7, including those with built-in delays or lags. Understanding and controlling precedence relations are crucial in the scheduling process, and help develop a more realistic model. If managers cannot resolve conflicts, they must modify the project master plan until a feasible schedule is achieved.

In projects where a major concern is uncertainty, as evidenced by stochastic activity durations, PERT logic can be applied to analyze the effects of unanticipated disruptions on overall project length. Most commercial software packages do not handle stochastic activity durations. The few that do usually rely on Monte Carlo simulation, as explained in Chapter 7. Figure 13-5 presents a report on activity duration for the example project assuming that each duration is derived from the three time estimates associated with PERT. The corresponding standard deviations were computed by SuperProject Expert.

■ *Output.* The scheduling process occurs at the activity level, but it is desirable to be able to generate reports at any of the various OBS or WBS levels. Many software packages supporting the organization and work breakdown structures have this capability. In addition, some packages present schedules by mechanisms such as hammock activities or subnetworks. A hammock activity replaces a group of activities. This type of aggregation is suitable for high-level reports that do not require a single activity level of detail. The subnetwork facility is similar to the hammock concept but represents activity groups by two or more "aggregate" activities. Another possibility is to aggregate activities into tasks and

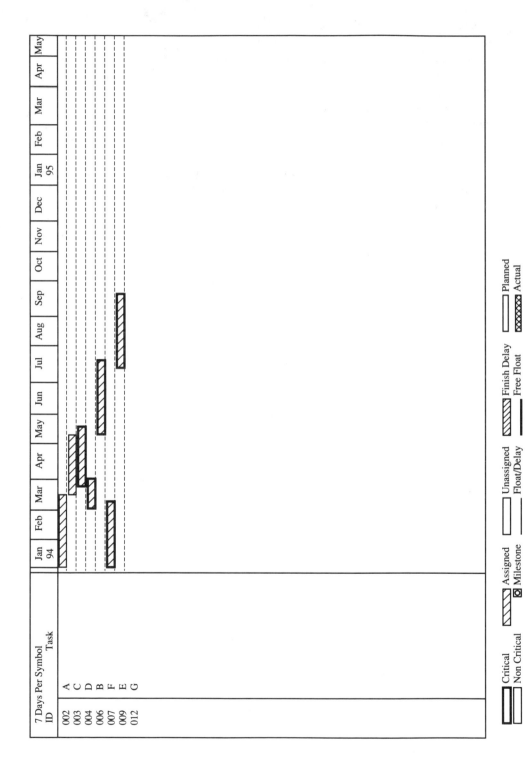

Figure 13-3 Gantt chart of the example project.

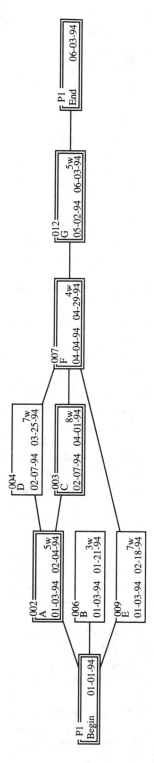

Figure 13-4 Activity-on-node network of the example project.

Task Name	Opt Dur	Likely Dur	Pess Dur	Dev
A	2w	5w	8w	1.00w
C	1w	5w	5w	0.67w
D	7w	7w	9w	0.33w
B	4w	4w	10w	1.00w
F	6w	6w	8w	0.33w
E	2w	6w	6w	0.67w
G	4w	5w	6w	0.33w

Figure 13-5   PERT-based report for the example project.

tasks into hammock tasks. Figure 13-6 represents each WBS element as a single hammock activity (or task) on the Gantt chart. These activities start with the earliest activity and terminate with the latest.

Charts designed for upper-level management normally present aggregated activities or tasks only; however, any mix of activities, tasks, subnetworks, and hammocks is possible. Figure 13-7 summarizes the schedule of the example project using both aggregated activities for WBS elements and detailed project activities. The schedule reports each activity's duration, scheduled start and finish, early start, late start, and float (in SuperProject Expert the term *task* is used for both low- and high-level activities).

■ *Resource planning.* In addition to time constraints, the project may be circumscribed by the availability of resources. Thus, the next step in the planning process is to add a resource dimension. The simplest approach is to assign resources to each activity, assuming that the same resource level is used throughout the activity's duration. Based on the earlier schedule, the required level for each resource type is calculated for each period. This approach helps identify time periods where resource requirements exceed resource availability. Project managers can reschedule activities to avoid resource overload or, if needed, they can try to acquire more resources.

Some sophisticated software packages allow for the uneven distribution of resource requirements over the activity's duration. When using such a package, a manager should specify each activity's required resource level during every period in which the activity is performed. Figure 13-8a gives the resource profile for the example project, accompanied by a Gantt chart assuming an early start schedule and unlimited resources.

Figure 13-8b gives the resource profile and the Gantt chart for the late start schedule. As can be seen by examining the two figures, the maximum of the resource profile is moved from the project's early phase in the early start schedule toward the project's middle phase in the late start schedule.

Software packages that support scheduling under resource availability constraints offer a large variety of decision support applications. One application is *resource allocation*, which specifies the availability level of each resource type for each calendar period. If resource requirements exceed resource availability for one or more resource types in a given period, the resource allocation procedure reschedules activities. Rescheduling may be limited to each activity's slack or be subject to a constraint imposed by a given project

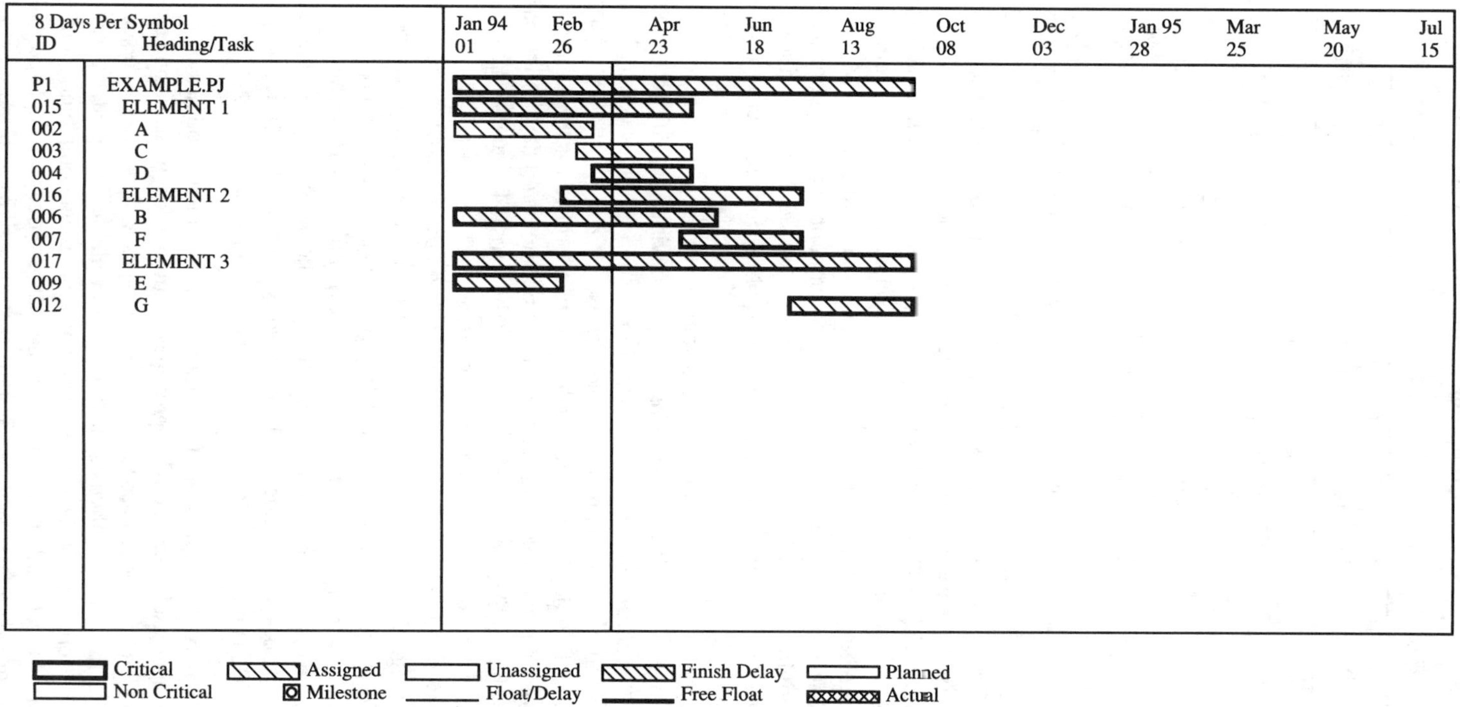

Figure 13-6 Gantt chart with hammock activities.

Task Name	Early Start	Early Finish	Late Start	Late Finish	Schd Start
A	01-01-94	03-14-94	01-04-94	03-14-94	01-04-94
C	03-14-94	04-11-94	03-15-94	04-11-94	03-15-94
D	03-14-94	05-02-94	04-06-94	05-24-94	03-15-94
B	01-01-94	02-07-94	03-02-94	04-06-94	01-04-94
F	05-24-94	08-06-94	05-24-94	08-26-94	05-24-94
E	01-01-94	02-16-94	04-11-94	05-24-94	04-11-94
G	08-26-94	10-26-94	08-26-94	10-26-94	08-26-94

Figure 13-7   Schedule summary report for the example project.

termination date. Priorities may be assigned to projects, or to activities within projects, so that high-priority activities receive scarce resources first.

Some packages allow for several types of resource capacities, such as overtime, second shifts, and subcontracting. The different capacity sources eliminate infeasibilities while minimizing a performance measure, such as resource costs. Moreover, some software packages offer the option of activity preemption. If this option is available, low-priority activities that are already started may be stopped if higher-priority activities compete for the same resource. When one or more of the higher-priority activities terminate, the preempted activities are resumed.

Another application available for resource management is *resource leveling*. Software packages with this option can reschedule activities to achieve a relatively constant use of one or more resources. As explained in Section 9.3, a leveled usage profile tends to decrease resource costs and increase resource use.

Figure 13-9 depicts a more leveled resource profile for the example project. Maximum resource requirements are reduced from 664 hours/49 days in the early start schedule, or 604 hours/49 days in the late start schedule, to 544 hours/49 days. Further leveling would necessarily delay the project's schedule.

Many resource allocation and resource leveling procedures are available. The resource management module of each software package is based on a specific algorithm. Due to the complexity of these scheduling problems, most commercial packages apply a *heuristic*, an algorithmic procedure that seeks a "good" feasible solution but does not guarantee that the optimum will be found. As a result, the performance of the resource leveling or resource allocation modules in different software packages varies with regard to computation (CPU) times and quality of resultant schedules.

■ *Resource management.* The resource management modules of commercial software packages use a variety of approaches to model the relationship between resource availability and a project's schedule. Some packages assume that all resources are renewable; that is, the same resource capacity level is available during each period. This assumption may be correct for a fixed workforce and equipment complement but not for subcontracting or materials. Other packages assume that resources are depleted with use, which is true for materials. Some packages assume that an activity's duration is a function of the

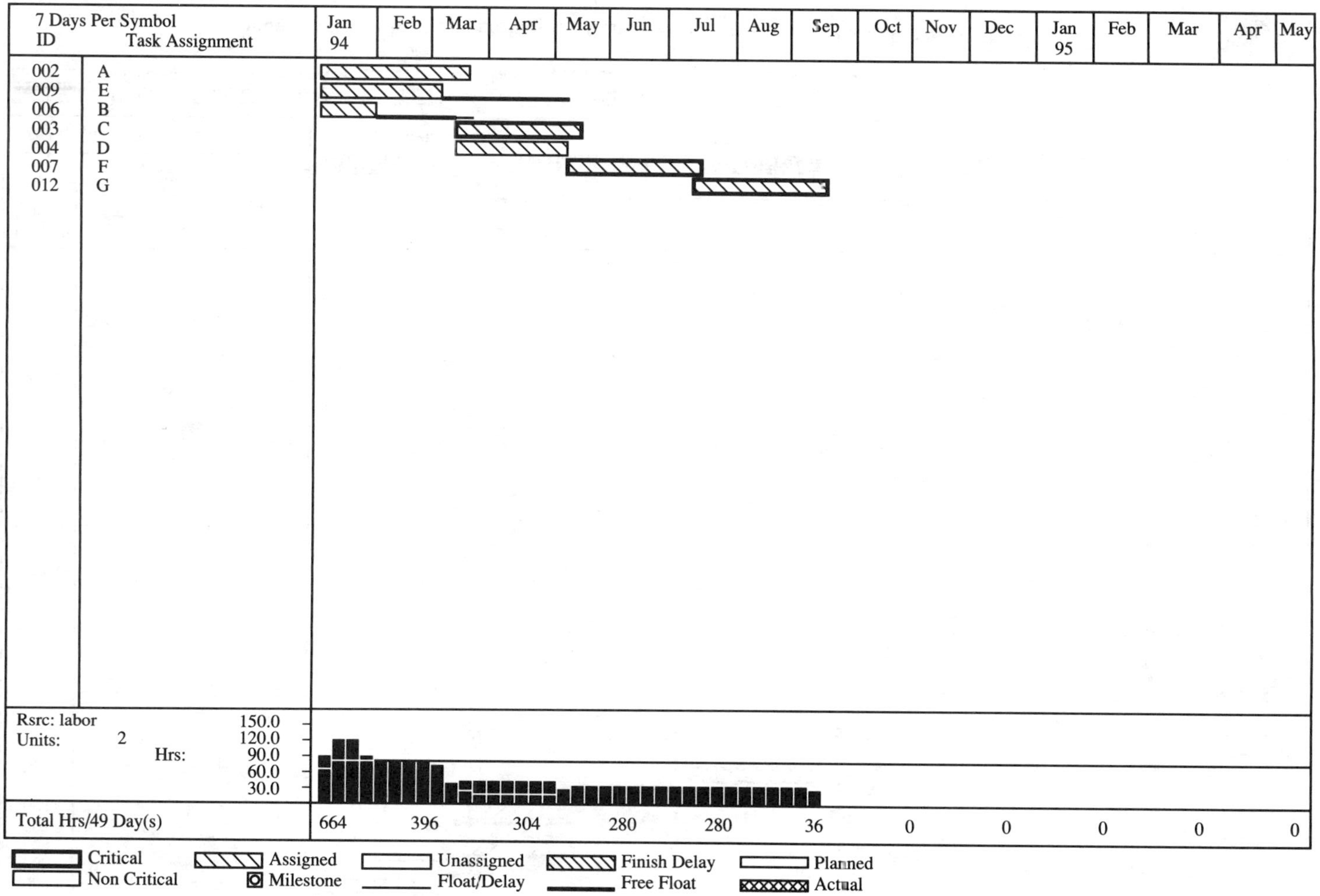

Figure 13-8(a) Resource profiles and Gantt charts for the example project: early start schedule.

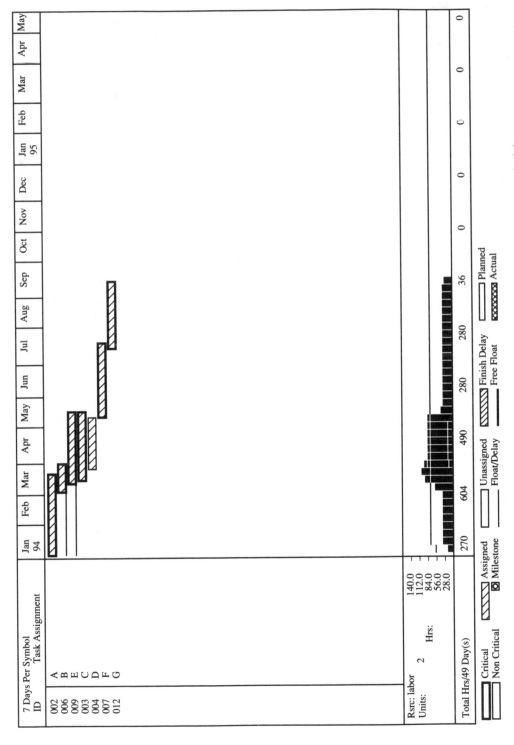

Figure 13-8(b) Resource profiles and Gantt charts for the example project: late start schedule.

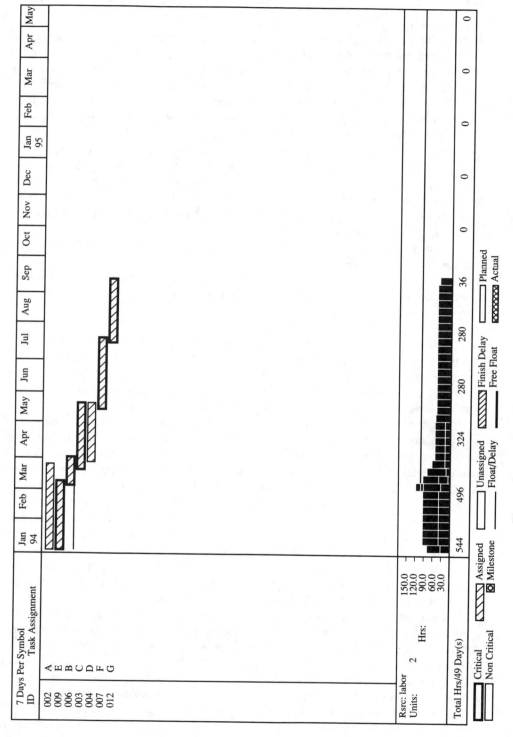

Figure 13-9 Leveled resource profile for the example project.

resources available to perform the activity (the time–resource trade-off). Other packages assume a fixed activity duration. Consequently, when selecting a software package for a specific project, a manager should carefully examine the type of resources the package can handle as well as the quality of resource leveling and resource allocation procedures.

The OBS–WBS matrix depicts the relationship between resources and functional management. Each resource is assigned to an organizational unit in the OBS and to activities related to WBS elements. This dual resource link provides traceability during the project's life cycle. Software packages that include the OBS–WBS matrix may support resource management by keeping track of the resources each OBS element uses to perform the activities on its assigned WBS components. Resource use is calculated by recording the actual effort associated with each resource in each period and comparing this effort to the resource's available capacity. Functional management and the project manager commonly use this calculation as a performance measure.

Software packages that support resource management are very helpful during the planning phase, where an important goal is to resolve conflicts. These packages are also helpful in the implementation phase, since uncertainty may cause changes in the original schedule leading to shifts in priorities. However, managing resources is important for other reasons as well: budgeting and cost management.

■ *Budget preparation.* The WBS, OBS, and schedule form the basis for project budgeting and costs. Cost estimating is performed at the activity level for direct labor and direct material costs. Indirect costs can be added at any OBS level. Each activity's direct costs should include its assigned resource costs. Therefore, managers should select a software package that can correctly represent these various components.

Some resource costs are based on an hourly rate and calculated as the rate times actual activity hours. Other resources, such as materials, have a per unit cost, where the measurement unit might be a pound or a cubic foot. Some resources may even have several rates; for example, overtime labor costs might be different from labor costs on regular time or a second shift. Overhead costs are charged against various baselines. Level-of-effort costs, such as those for project management, accrue as a passage of time, while apportioned-effort costs are based on a factor of a discrete effort, such as inspection. The initial project budget should specify other overhead costs, including facility operations and energy. Therefore, a software package's ability to communicate with the databases storing information about rates and actual costs is an important criterion for software selection. Figure 13-10 presents a cost–schedule report for the example project, which includes each activity's total costs along with its scheduled start and finish times.

Contractors responding to requests for proposals, in an effort to win contracts, must prepare cost estimates for each RFP. Thus the ability of a project management software package to support cost estimation may be an important aspect for such contractors. Cost estimates are based on a cost breakdown structure. (Refer to Chapter 10 for a discussion of cost component classification.) Information on the actual costs of previous projects stored in a user-friendly database is very helpful in preparing cost estimates.

Some software packages support a cost breakdown structure (CBS) and life-cycle cost analyses. The need for these functions becomes obvious when a client requires them

Task Name	Dur	Schd Start	Schd Finish	Total Cost
A	5w	01-04-94	03-14-94	1500.00
C	8w	03-15-94	05-09-94	3300.00
D	7w	03-22-94	05-09-94	4200.00
B	3w	03-01-94	03-22-94	3000.00
F	4w	05-09-94	07-11-94	6100.00
E	7w	01-04-94	03-04-94	5700.00
G	5w	07-11-94	09-09-94	7200.00

Figure 13-10   Cost report for the example project.

in a contract or an RFP. In some cases, the contractor may want to use an LCC analysis for its own benefit. By consistently using a CBS for each project, historical cost data are accumulated and bid preparation for future projects is more accurate and less time consuming.

In addition to budgeting, managers may need to forecast and manage the cash flow. Tying milestones and activities to cash flows makes it possible to schedule a project to achieve a desired cash flow, or to schedule several projects simultaneously under a variety of cash flow constraints. The schedule affects a project's budget and in many projects is subject to a host of monetary restrictions. The relationship between costs and schedule can be analyzed by time–cost models, where each activity's duration is assumed to be a function of the activity's costs. Recall that *crashing* is the term used to describe time–cost trade-offs. Not all commercial packages support this function, though, so managers with such needs should purchase a package that permits crashing or accommodates user-written subroutines.

■ *Configuration management.* In engineering projects, the technological aspects should be coordinated with the project management support system. The need to select a configuration baseline, to evaluate proposed engineering changes and their performance, cost, and schedule impacts, and to keep track of the current configuration translates into handling and controlling large data sets and transactions. Although software packages for configuration management are available on the market, most do not support the other facets of project management. Thus when selecting a software package, a project manager should consider the interaction between that package and the configuration management system and define the required interfaces.

Once a project's plan is established, the manager should examine the plan's sensitivity to changing conditions. Uncertainty plays a major role in most projects. Examples of sources of uncertainty are activity duration estimates, resource availability and costs, and lead time for material deliveries. The basic project planning models do not consider these aspects of uncertainty, and therefore we recommend a sensitivity analysis. This analysis frequently takes the form of "what if" questions. The project's plan is studied under various conditions. Following the analysis, risk management plans and contingency plans are developed for cases in which the original plan might lead to major disasters. A software package's ability to perform a "what if" analysis and to store several plans for the same project is therefore essential when significant levels of uncertainty are present. An acceptable project plan (1) outlines how the schedule, costs, and resource use fit within the imposed constraints, and (2) allocates the work in a feasible manner; the project is then ready to begin.

■ *Project monitoring.* During the production phase, the computer is useful to issue work orders and manage resources. A schedule for each resource that includes all the resource's assigned tasks and activities and their planned start and finish dates is a helpful tool. Figure 13-11 presents a labor schedule for the example project, in addition to information on each task's duration and number of assigned labor hours. Each task's number of assigned labor hours are calculated on a 10-hour labor-day.

To monitor progress, the actual accomplishments are input in terms of started and completed activities. In some cases, estimates of the percentage of work completed for ongoing activities are also used. Accompanying this information are accumulated data on resource use and expenditures. The analysis can take many different forms. Progress reports in the form of a Gantt chart, with completed activities marked, are very popular. Also common are tabular reports which indicate, from the project's start, actual progress versus planned progress for each task or activity during the current period or on a cumulative basis. While periodic reports are useful in exception identification, the cumulative reports are important for trend analysis. Figure 13-12 depicts a progress report based on a Gantt chart. The figure displays each task's status and original schedule, along with information on the percentage of work completed, the earned value, actual cost, and budgeted cost for each task. Figure 13-13 displays a resource-oriented progress report. For comparison, each task's actual start, actual finish, status and actual hours are reported, along with scheduled finish times. This type of report is helpful in monitoring resource use.

Information on actual resource usage by various OBS elements can serve as a basis for performance or management quality assessments, while actual cost information is the basis for project cost performance reevaluations. Although a software package's ability to store information on actual resource use and project costs is important, not all commercial packages can track both sets of data. Some packages can track actual resource use and then, assuming the original rate, calculate the estimated actual cost. Other packages can track actual costs but not actual resource use. Therefore, selecting a software package for a specific organization or project depends on the availability of other systems to perform these functions.

Task Name	Dur	Total Hours	Early Start	Schd Start	Early Finish	Schd Finish
A	5w	400	01-01-94	01-04-94	03-14-94	03-14-94
C	8w	240	03-14-94	03-15-94	05-09-94	05-09-94
D	7w	140	03-22-94	03-22-94	05-09-94	05-09-94
B	3w	120	01-01-94	03-01-94	01-24-94	03-22-94
F	4w	360	05-09-94	05-09-94	07-11-94	07-11-94
E	7w	350	01-01-94	01-04-94	03-04-94	03-04-94
G	5w	350	07-11-94	07-11-94	09-09-94	09-09-94

Figure 13-11   Detailed schedule for labor.

Task Gantt w/Actuals

Project: EXAMPLE.PJ
07-29-92

7 Days Per Symbol Task Name	Pct Comp	Earned (BCWP)	Actual Cost	Bdgt. Cost Work Schd.	Jan 94	Feb	Mar	Apr	May	Jun	Jul	Aug	Sep
A Completed	100	1500.00	1500.00	1500.00									
C Late Start	0	0.00	0.00	1072.50									
D Late Start	0	0.00	0.00	980.00									
B Completed	100	3000.00	2700.00	3000.00									
F Scheduled	0	0.00	0.00	0.00									
E Late Finish	50	2850.00	4500.00	5700.00									
G Scheduled	0	0.00	0.00	0.00									

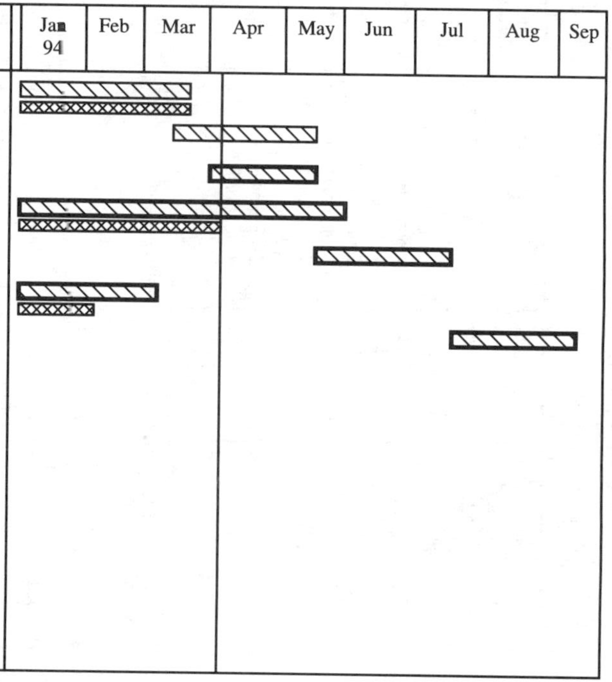

Critical   Assigned   Unassigned   Finish Delay   Planned
Non Critical   Milestone   Float/Delay   Free Float   Actual

Figure 13-12   Gantt chart–based progress report.

553

Heading/Task	Actual Start	Actual Finish	Status	Actual Hours	Schd Finish
EXAMPLE. PJ				546	09-14-94
ELEMENT 1			Late Start/Crit.	288	05-13-94
A	01-04-94	03-15-94	Completed	816	03-14-94
C			Late Start	0	05-09-94
D			Late Start/Crit.	0	05-13-94
ELEMENT 2			Late Start/Crit.	88	07-15-94
B	01-04-94	03-27-94	Completed/Crit.	472	03-25-94
F			Scheduled/Crit.	0	07-15-94
ELEMENT 3			Late Start/Crit.	170	09-14-94
E	01-05-94		Late Finish/Crit.	170	03-04-94
G			Scheduled/Crit.	0	09-14-94

Figure 13-13   Progress report for labor.

- *Project control.* Tracking progress, actual costs, and resource use form the basis of the project control system. Project control detects the deviations between the plans and the actual performances and analyzes trends. Deviations are examined to identify the problem's source and to forecast future trends. Based on the results, corrective measures are implemented. The control system compares planned and actual progress in several dimensions, including scope of work, schedule, expenditures, resource use, and technological performance. When a deviation is noted in any of these dimensions, the source is investigated to determine its influence on elements of the organizational and work breakdown structures.

One major problem with a control system based on a simple comparison between planned and actual values is the interaction between different dimensions of a project. For example, under some conditions, the interaction between costs and schedule makes it possible to shorten an activity's duration by increasing its direct costs. Similarly, technological changes may affect a project's costs, schedule, and resource requirements. A partial solution to the interaction problem is the DOD approach, summarized in the cost/schedule control systems criteria (C/SCSC). That approach, which is based on the earned value concept, is discussed in Chapter 11. In the case of a DOD project or a client who requires C/SCSC, one major issue when choosing a software package is compliance with the criteria. Even if compliance is not required, the control system's ability to detect deviations, to trace their source(s), and to forecast future performance from past accomplishments is an important consideration in selecting a project management system. Figure 13-14 presents a report predicated on earned value logic. The BCWP, ACWP, and BCWS values are enumerated for each task.

Another important aspect of project control is technological change control in the form of engineering change requests. This activity involves evaluating changes and deciding whether or not to accept them. A software package that supports configuration management activities is a useful tool in engineering project management.

- *Project life-cycle support.* A computer software package's ability to support decision making throughout a project's life cycle is an important factor in the software selection

Heading/Task	Pct Comp	Earned (BCWP)	Actual Cost	Bdgt. Cost Work Schd.
EXAMPLE. PJ	35	10850.00	8700.00	10797.75
ELEMENT 1	51	4590.00	1500.00	6269.66
A	100	1500.00	1500.00	1500.00
C	0	0.00	0.00	1072.50
D	0	0.00	0.00	960.00
ELEMENT 2	25	2275.00	2700.00	2203.16
B	100	3000.00	2700.00	3000.00
F	0	0.00	0.00	0.00
ELEMENT 3	25	3225.00	4500.00	4493.26
E	50	2850.00	4500.00	5700.00
G	0	0.00	0.00	0.00

Figure 13-14   Earned value–based progress report.

process. Analyzing a package's decision making capabilities involves answering the following questions:

- What does the package do?
- How does the package do it?
- What costs are involved in purchasing, using, and maintaining the software package?

In addition, how the package performs its functions relates to aspects such as the support level in the decision-making process, the time and effort required to learn the software, the human–machine interface, available logistics support for users, and hardware requirements.

The human–machine interface is a crucial factor affecting the costs of a software package's implementation and use, as well as its acceptance. New users more readily accept packages that can be learned quickly. User friendliness and easy learning are achieved with descriptive menus, on-line help screens, error-tolerant commands, and windows with pointing devices. A package that can access data from existing databases and can communicate with other management information systems is easier to introduce since some of the input formatting is already familiar.

■ *Report generation.* Report capabilities are another important issue to consider. Some project management computer packages contain only a standard set of tabular reports summarizing the results of the CPM analysis and, if applicable, resource allocation, resource leveling, and budgeting data. The problem with standard reports is that they may not use the organization's terminology and therefore may not be able to provide a specific answer to a specific question without additional work. Furthermore, it may not even be possible to derive the answer by scanning the output files.

Packages equipped with report generators are much more flexible and can produce reports for a given activity, for a given WBS or OBS element, and for a specific WBS or OBS level. Some generators even produce reports that integrate information from the project management system with information from other systems spreadsheets and word processors.

A user should consider a package's ability to design and produce graphical reports. Various types of charts and diagrams can summarize large amounts of data, including trends and correlations between different aspects of the project, on a single graph. Some packages contain a standard set of graphical reports, while more advanced packages allow the user to produce them from any data set in the project management system.

■ *Vendor support.* Apart from the software package's functions, the vendor's logistic support is also an important factor affecting the package's implementation success. A software package requires logistic support throughout its life cycle. In the early stages, learning the package is crucial. Vendors can provide numerous training options, including in-house training, training at the vendor's facility, tutorial programs, and manuals. A vendor can offer assistance in the early implementation stages, such as help with installation, data entry, and initial processing. Users may need more assistance during the operational phase since they may discover bugs or request specially tailored applications. Additional vendor support is helpful when the project management system is integrated with existing or new information systems.

■ *Hardware requirements.* Another issue to consider in the selection process is hardware requirements. Hardware costs, especially for personal computers, have decreased drastically since 1980. Software that can use larger amounts of RAM (random access memory) will run more quickly and so might be better than less expensive packages that perform numerous disk access operations to reduce RAM use. A software package's ability to support a variety of existing input/output devices is also important, as well as its ability to work in a network.

Well-designed and actively supported software packages can make routine tasks such as data collection, data processing, and data retrieval easier for a project manager. However, a package's successful implementation ultimately depends on how well the package fits the organization's needs. The following section lists helpful criteria for selecting the most appropriate software package for a specific application.

13.3 CRITERIA FOR SOFTWARE SELECTION

An organization is unlikely to find commercial software packages that provide 100% of the support it needs to manage its projects. Even if such a package is found, its cost may be prohibitive. Therefore, managers must systematically evaluate and select the most appropriate package. In so doing, three sets of criteria should be considered:

■ Operational criteria related to the software's capabilities and performance.
■ Information systems evaluation criteria applicable to any type of software package, not just project management software. These criteria are related to hardware requirements, software integrity, quality, and so on.
■ Life-cycle cost criteria.

The first set of criteria is based on the package's intended use and includes questions about the various functions, such as scheduling, budgeting, and project control. The second set is important in the selection of any management information system, and addresses questions related to the software's ability to function properly under different organizational and operational conditions. The third set is concerned with the cost of purchasing, installing, maintaining, and using a software package throughout its life cycle.

Although the specific criteria in each set depend on the package's intended applications and an organization's software needs, evaluators can develop a "generic" criteria list. Such lists frequently appear in articles evaluating and comparing software packages, a sampling of which is included in the reference section at the end of the chapter.

The level of sophistication of the various project management software packages varies considerably. The evaluator should adapt evaluation and selection criteria peculiar to the level of support needed. The set of functions available defines the scope of the package.

Unsophisticated users would be interested primarily in packages at the lowest level, mainly supporting the planning phase, which includes scheduling and budgeting activities. Some packages can also handle resources. In addition, they often are capable of producing a prespecified set of reports. Recall that a progress report is generated by replanning the project based on updated data. A comparison between the updated and the original plan forms the basis for project control.

Software packages at the second level support all the functions performed by first-level packages, as well as resource leveling, resource allocation, and project control. The corresponding modules identify cost and schedule variances and predict the budget at completion. Flexible report generators are also available at this level. These generators facilitate data presentation by permitting the user to select the most appropriate output formats.

Packages at the third level support organization and work breakdown structures for multiple projects. They can handle several projects competing for the same resources and can assign resources to projects based on predetermined priority rules. At this level, software packages allow users to write their own applications using a high-level programming language. Thus the packages can include applications such as configuration management and inventory and material management. Some of these packages have a graphical report generator and a relational database so that they can retrieve any data set to which the software has access and print it out using the report generator. This enables users to construct special reports to specific needs.

The following criteria, suggested by articles, books, and the authors' experience, have proven useful for selecting project management software packages. An appropriate subset can be adopted for each situation.

■ OPERATIONAL CRITERIA

Scheduling activities

   Number of activities per project
   Number of projects that can be analyzed simultaneously
   Types of precedence relations supported

Modeling of delays or lags within the precedence relations

Possible time units (hours, days, weeks)

Number of calendars that can be defined and saved

Critical path analysis

Computation of free and total slacks

External constraints on activity start and end dates

Support of milestones

External constraints on milestones

Support of hammock activities

Support of subnetworks

Network presentations as AOA

Network presentations as AON

Network drawings on screen, on a plotter, on a printer

Zooming capability on network drawings

Presentation of Gantt charts

Interactive editing of Gantt charts and of network drawings

Handling of stochastic activity duration: PERT or simulation analysis

Activity duration presented as a function of resource availability

Time–cost analysis

Automatic check of network logic for loops, disconnected activities

"What if" analysis

Budgeting, cost estimation, and cash flow

Support of several currencies

Handling of inflation rates

Connection between cost and activities, resources, milestones, organizations, work breakdown structure elements

Communication with existing cost accumulation, cost control, and cost estimation systems

Identification of direct versus indirect cost

Identification of cost categories, such as labor and material

Planning and budgeting the cost of materials and inventories

Support of cost breakdown structure

Support of statistical analysis of cost estimating relationships

Development of budgets and cash flows for a given schedule

Scheduling subject to budget constraints

Scheduling to minimize direct and indirect costs (PERT/Cost)

Support of LCC models and analysis

"What if" analysis

Resources

Number of different resources per activity

Number of different resources per project

Number of different resources for multiple projects

Handling of renewable resources (labor)

Handling of depleting resources (material)

Resource leveling

Resource allocation

Planning with alternative resources (e.g., subcontracting)

Preemption of activities

Definition of resource availability by dates, hours, organization

Allocation of resources among competing projects

Variable rate of resources (e.g., regular time versus overtime)

Variable usage of resources during the execution of an activity

"What if" analysis

Project structure

Definition of organizational structures: number of levels

Logical checks on completeness of OBS

Definition of work breakdown structure: number of levels

Logical checks on completeness of WBS

Integration of the OBS and WBS to form work packages

Drawing of OBS and WBS on screen, plotter, printer

Definition of communication lines and work authorization responsibility

Coding system for OBS–WBS matrix

Roll-up mechanism in OBS–WBS matrix for cost analysis

Limited access to data by passwords assigned to OBS units

Definition of a linear responsibility chart

Operation in a computer network

Configuration management

Definition of configuration items

Coding system for configuration items

Definition of baselines

Handling engineering change requests

Support of configuration identification

Support of configuration change control

Support of configuration status accounting

Support of configuration review and audits

Project control

Number of project baseline plans that can be handled and stored

Ability to define cost accounts and work packages

Ability to construct the budgeted cost of work scheduled (BCWS) at all WBS and OBS levels

Ability to accumulate, store, and retrieve the budgeted cost of work performed (BCWP or earned value) at all WBS and OBS levels

Ability to accumulate, store, and retrieve the actual cost of work performed (ACWP) at all WBS and OBS levels

Ability to calculate cost and schedule variances and indices at all WBS and OBS levels for each period and on a cumulative basis

Ability to forecast the estimated budget at completion based on actual progress (estimated by the earned value) and actual cost

Ability to compare actual progress with different baselines

Ability to signal cost and schedule deviations larger than predetermined thresholds

Ability to analyze trends in cost and schedule performances

Compliance with cost/schedule control systems criteria

Ability to control use of material and actual cost of material used

Ability to control use of resources and actual cost of these resources

Reporting

Standard reports available

Report generator

Graphical reports

Integration with word processor

Output to plotters

■ GENERAL SYSTEM CHARACTERISTICS CRITERIA

Friendliness: time to learn, help facilities, use of a menu, windows

Documentation: for operation, maintenance, installation

Security: data input, output, editing

Integrity of database

Communication with other information systems

Hardware requirements

Support available from vendor

User base: recommendations of current users

- LIFE-CYCLE COST RELATED CRITERIA

Purchase cost (per unit, quantity discounts)

Cost of hardware, facilities, and so on

Estimated cost of operation and maintenance

Expected service life

Cost of updating and new versions

Estimated value at phaseout time

The list above is generic and should be modified according to the specific needs of the project or organization. Appendix 13A presents an example of a criteria set developed by the Project Management Software Working Group in conjunction with the Belgian Building Research Institute. In Section 13.4 we demonstrate how a comprehensive list of criteria can be used to guide the software selection process.

13.4 SOFTWARE SELECTION PROCESS

Effective project management is a direct function of the tools available to support decision making at all levels of detail. An adequate software package facilitates the project manager's job by integrating different aspects of the project and simplifying routine data processing tasks. A software package that does not serve the project team's needs is of little value and may even prove a burden. Therefore, those responsible for choosing the software should approach their task advisedly.

The selection process begins by identifying data processing needs. This involves addressing the following questions:

- What is the project's expected size?
- How many different resources are needed?
- How many organizations will participate in the project?

The second step is to analyze the type of management decisions the software package will support. Should the package support configuration management? Should it support budgeting and cost estimates? Do existing systems already perform these functions satisfactorily?

Third, a criteria list should be constructed. This list is used, along with one of the selection/evaluation methodologies described in Chapters 3 and 4, to select and identify the most appropriate package. Since the evaluation techniques are subjective (relative importance or scores are subjectively assigned to each criterion), the selection decision should not be based on this analysis alone, nor should a final decision be made at this point in the process.

In fact, now is the time to analyze data from past projects and perhaps to construct a test project with attributes that reflect the current environment. The test project is simulated by first "planning" it using the software. Information on "actual" performance is added and reports are generated to support project "control." Since the simulation can be performed quickly, a 10-year project, for example, can be studied in 1 to 2 days and the results analyzed immediately. Allowing future system users to participate in the simulation helps them to better understand the software package and to identify and adjust for various package weaknesses. In any case, the package should only be approved for a trial period, and only after all intended users are satisfied with the simulated results.

We recommend trying out the package for a short period to test its suitability for the organization and upcoming projects. The selection team then decides whether to adopt the package or to investigate another based on the experience over the trial period.

To demonstrate the software package selection process using a scoring model, consider an organization that wishes to compare two project management software packages, A and B, using the generic criteria list presented above. Table 13-1 contains the relative weights assigned to each criteria type (set). Criteria related to software costs are not included, since a cost-effectiveness measure will be used as part of the selection process.

Next, a scale is developed for each criteria set that assigns a weight to each member of the set. The sum of the weights associated with each criterion in a set is 100. In the evaluation, each criterion is given a score between 0 and 10. This number is translated into points by multiplying it by the corresponding weight. Once the points are tallied for each criteria set and normalized by dividing them by 100, the total score for the package is calculated. This is done by forming the weighted sum using the weights in Table 13-1; the maximum package score is 100. The input data and the calculations for each package are summarized in Table 13-2.

The cost data for the two options is contained in Table 13-3; the results of the analysis are presented in Table 13-4. The results show that package B is better than package A in total score but ranks lower in scheduling, budgeting, and project structure. Package B's purchase cost is $6,000 higher than package A's and requires $300 more per year to update. Based on these results, management must now weigh B's higher cost against its superior

TABLE 13-1   RELATIVE WEIGHTS USED IN
THE SCORING MODEL

Criteria set	Weight
Activities and scheduling	20
Budgeting, cost estimation, and cash flow	15
Resources	15
Project structure	0
Configuration management	10
Project control	10
Reporting	10
General system characteristics	10
	100

performance and select the package most appropriate for the organization. Computing an "effectiveness/cost" ratio indicates that package A, with a score of 1.34 points per $1,000, is somewhat better than B, whose score is 1.29.

TABLE 13-2 CALCULATIONS FOR THE OPERATIONAL CRITERIA

Operational criteria	Weight	Package A score/points	Package B score/points
Activities and Scheduling			
Total Weight: 20			
Number of activities per project	5	8/40	6/30
Number of projects that can be analyzed simultaneously	5	7/35	7/35
Types of precedence relations supported	5	8/40	8/40
Modeling of delays or lags within the precedence relations	3	9/27	6/18
Possible time units (hours, days, weeks, etc.)	4	4/16	9/36
Number of calendars that can be defined and saved	5	5/25	4/20
Critical path analysis	5	8/40	5/25
Computation of free and total slacks	5	6/30	6/30
External constraints on activity start or end dates	5	2/10	6/30
Support of milestones	3	0/0	5/15
External constraints on milestones	2	0/0	3/6
Support of hammock activities	5	5/25	2/10
Support of subnetworks	1	3/3	5/5
Network presentation as AOA	5	6/30	0/0
Network presentation as AON	5	0/0	3/15
Network drawings on screen, on a plotter, on a printer	5	7/35	3/15
Zooming capability on network drawings	2	6/12	4/8
Presentation of Gantt charts	5	8/40	5/25
Interactive editing of Gantt charts, network drawings	2	6/12	4/8
Handling of stochastic activity duration PERT or simulation analysis	3	0/0	0/0
Activity duration presented as a function of resource availability	5	5/25	3/15
Time–cost analysis	5	7/35	6/30
Automatic check of network logic for loops, disconnected activities	5	10/50	5/25
"What if" analysis	5	7/35	5/25
Total	100	$\dfrac{565}{100} = 5.65$	$\dfrac{466}{100} = 4.66$

(Continues)

TABLE 13-2　(CONTINUED)

Operational criteria	Weight	Package A score/points	Package B score/points
Budgeting, Cost Estimation, and Cash Flow			
Total Weight: 15			
Support of several currencies	7	3/21	5/35
Handling of inflation rates	5	5/25	0/0
Connection between cost and activities, resources, milestones, organizations, work breakdown elements	10	7/70	5/50
Communication with the current budgeting and cost control systems	10	8/80	7/70
Identification of direct versus indirect cost	7	6/42	8/56
Identification of cost categories such as labor, and material	5	7/35	5/25
Management of the cost of materials and inventories	5	5/25	6/30
Support of cost breakdown structure	10	8/80	7/70
Support of statistical analysis of cost estimating relationships	5	5/25	0/0
Development of budgets and cash flows for a given schedule	10	8/80	6/60
Scheduling subject to budget constraints	8	6/48	6/48
Scheduling to minimize direct and indirect costs (PERT/Cost)	5	5/25	6/30
Support of LCC models and analysis	8	5/40	6/48
"What if" analysis	5	6/30	7/35
Total	100	$\frac{626}{100}$ = 6.26	$\frac{522}{100}$ = 5.22
Resources			
Total Weight: 15			
Number of different resources per activity	10	8/80	9/90
Number of different resources per project	10	6/60	8/80
Number of different resources for multiple projects	10	6/60	9/90
Handling of renewable resources (labor)	5	5/25	6/30
Handling of depleting resources (material)	8	6/48	6/48
Resource leveling	10	7/70	8/80
Resource allocation	10	6/60	6/60
Planning with alternative resources (e.g., subcontracting)	5	3/15	6/30
Preemption of activities	3	0/0	2/6
Definition of resource availability by dates, hours	8	5/40	4/32

Operational criteria	Weight	Package A score/points	Package B score/points
Allocation of resources among competing projects	8	6/48	5/40
Variable rate of resources (e.g., regular time versus overtime)	5	6/30	8/40
"What if" analysis	8	5/40	9/72
Total	100	$\dfrac{576}{100} = 5.76$	$\dfrac{698}{100} = 6.98$

Project Structure
Total Weight: 10

	Weight	Package A score/points	Package B score/points
Definition of organizational structures: number of levels	20	8/160	7/140
Definition of work breakdown structure: number of levels	20	8/160	6/120
Integration of the OBS and WBS to form work packages	20	9/180	7/140
Definition of communication lines and work authorization responsibility	15	6/90	5/75
Roll-up mechanism in OBS–WBS matrix for cost analysis	15	8/120	6/90
Limited access to data by passwords assigned to OBS units	10	10/100	8/80
Total	100	$\dfrac{810}{100} = 8.10$	$\dfrac{645}{100} = 6.45$

Configuration Management
Total Weight: 10

	Weight	Package A score/points	Package B score/points
Definition of configuration items	10	10/100	10/100
Definition of baselines	10	8/80	10/100
Handling engineering change requests	10	8/80	10/100
Support of configuration identification	15	10/150	10/150
Support of configuration change control	15	6/90	10/150
Support of configuration status accounting	20	7/140	10/200
Support of configuration review and audits	20	7/140	10/200
Total	100	$\dfrac{780}{100} = 7.8$	$\dfrac{1000}{100} = 10.0$

(Continues)

TABLE 13-2   (CONTINUED)

Operational criteria	Weight	Package A score/points	Package B score/points
Project Control			
Total Weight: 10			
Number of project baseline plans that can be handled and stored	10	5/50	6/60
Ability to define cost accounts and work packages	10	4/40	10/100
Ability to construct the BCWS at all WBS and OBS levels	10	5/50	7/70
Ability to accumulate, store and retrieve the BCWP (earned value) at all WBS and OBS levels	10	6/60	8/80
Ability to accumulate, store and retrieve the ACWP at all WBS and OBS levels	10	5/50	9/90
Ability to calculate cost and schedule variances and indices at all WBS and OBS levels for each period and on a cumulative basis	5	6/30	8/40
Ability to forecast the estimated budget to completion based on actual progress: earned value and actual cost	7	5/35	10/70
Ability to compare actual progress with different baselines	7	4/28	8/56
Ability to signal cost and schedule deviations larger than predetermined thresholds	7	5/35	6/42
Compliance with C/SCSC	8	6/48	8/64
Ability to control use of material and actual cost of material used	8	9/72	8/64
Ability to control use of resources and actual cost of these resources	8	6/48	8/64
Total	100	$\frac{546}{100} = 5.46$	$\frac{800}{100} = 8.0$
Reporting			
Total Weight: 10			
Standard reports available	20	8/160	9/180
Report generator	20	5/100	8/160
Graphical reports	20	3/60	8/160
Integration with word processor	20	5/100	5/100
Output to plotters	20	3/60	8/160
Total	100	$\frac{480}{100} = 4.8$	$\frac{760}{100} = 7.6$

Operational criteria	Weight	Package A score/points	Package B score/points
General System Characteristics			
Total Weight: 10			
Friendliness: time to learn, help facilities, menus, windows, etc.	20	5/100	7/140
Documentation for operation, maintenance, installation	5	8/120	6/90
Security, data input, output, editing	15	3/45	8/120
Integrity of database	20	5/100	7/140
Communication with other information systems	10	6/60	6/60
Hardware requirements	10	8/80	8/80
Support available from vendor	5	6/30	9/45
User base: recommendations of current users	5	5/25	9/45
Total	100	$\dfrac{560}{100} = 5.6$	$\dfrac{720}{100} = 7.2$

TABLE 13-3   COST DATA FOR SELECTION PROBLEM

Life cycle cost–related criteria	Package A	Package B
Purchase cost (per unit, quantity discounts)	$6,000	$12,000
Cost of hardware, facilities, etc.	$15,000	$15,000
Estimated cost of operation and maintenance	$2,500/yr	$2,500/yr
Expected service life	5yr	5yr
Cost of updating and new versions	$500/yr	$800/yr
Estimated value at phaseout time	0	0

TABLE 13-4   WEIGHTED SCORES FOR CRITERIA SETS AND RESULTS

Criteria	Weight	Package A score/points	Package B score/points
Activities and scheduling	20	5.56	4.66
Budgeting, cost estimation, and cash flows	15	6.26	5.22
Resources	15	5.76	6.98
Project structure	10	8.10	6.45
Configuration management	10	7.80	10.00
Project control	10	5.46	8.00
Reporting	10	4.80	7.60
General systems characteristics	10	5.60	7.20
Total points	100	48.40	56.10
Total cost		$36,000	$43,500
Relative cost (with respect to the lowest-cost package)		100%	121%
Effectiveness ratio (points/$1,000)		1.34	1.29

13.5 SOFTWARE IMPLEMENTATION

The successful implementation of project management software depends largely on the software's ability to support the project team's work. If, for example, the software can produce reports that were prepared manually in the past, the project team benefits from using the package. However, if top management requests new reports based on increased expectations of the software's capabilities, the extra work required to generate these reports may produce unanticipated slippage in the schedule.

The software evaluation team should involve all potential end users in the selection and implementation process. After all, the end users know which functions need support and what priorities should be assigned to each. Incorporating the end users into the decision-making process (criteria list development, assessment of each criterion's weight, and evaluation scale development) will go a long way toward smoothing acceptance of the selected package. In addition, including future users in package testing and project simulation allows them to contribute their insights and experiences to the selection process and, later, during implementation. Thus choosing the software should be a joint effort among information systems experts, analytic support personnel, and all potential system users.

Once the team selects a system, implementation starts with a training program in which future system users learn operational procedures and gain an understanding of each module's basic logic. This training program should precede actual system use to eliminate learning difficulties and unnecessary frustrations. We all have a limited tolerance for failure.

Only trained personnel should use the system, and the vendor or internal system experts should support initial applications since they can solve startup problems quickly. Management should begin implementation in areas where the system can help users perform current tasks and then expand it into other areas until all desired functions are included. In addition, when introducing the system, a manager should avoid assigning additional tasks to the project team. During the initial stages, the system should help users perform routine tasks efficiently. By performing routine tasks and alerting users early to potential problems, the system can free users' time to deal with exceptions and uncertainty. This will increase the chances for acceptance.

After a predetermined time, management should conduct a survey of users' opinions regarding the performance of the software. At this point, final procedures for system assessment, data updating, and data processing should be established. These procedures should support management's need for information while simplifying routine data-handling tasks.

Once again, we recommend a deliberate process that involves potential users in package selection, training, and phased implementation. Management should implement the software in stages to avoid overwhelming users with new and unfamiliar applications. Project management is greatly simplified when appropriate software tools are selected intelligently and introduced into the organization. Nevertheless, the best tools are useless if the project team does not accept them.

13.6 PROJECT MANAGEMENT SOFTWARE VENDORS

In 1991, International Data Corporation of Framingham, Massachusetts, estimated the market for project management software at $350 million (Wallace and Halverson 1992). Today, growth continues at a compounded annual rate of 18.6%. Since the market includes such a large number of products and since both products and vendors change constantly, an ongoing effort is required just to stay abreast. A list of vendors is provided in Appendix 13B. This list is based on a report issued by the Institute of Management & Administration (IMA 1992) and is not intended to be exhaustive. Additional products by the same vendors may also be available.

TEAM PROJECT

Thermal Transfer Plant

The R&D project that you have proposed in Chapter 12, coupled with your experience with the prototype rotary combustor has made your team the project management experts at TMS. Your success has motivated top management to introduce project management techniques throughout the organization. The management information systems (MIS) department has made a proposal to develop a project management application on a spreadsheet to support TMS's needs in this area. The department chief argues that since most TMS engineers and managers are familiar with Lotus 1–2–3, it would be much easier for them to learn and use an application based on this product. In addition, he foresees that full integration with existing databases and software will be achieved more quickly.

Because of your experience and expertise, TMS management has asked you to compare SuperProject Expert, the software used by your team, with the proposed 1–2–3 application. In so doing, explain what aspects of project management can be supported by a spreadsheet. Discuss the advantages and disadvantages of the MIS proposal.

Develop a software selection plan, including appropriate models that can be used to help in the selection process. Analyze SuperProject Expert's ability to satisfy the requirements and compare it to a tailormade spreadsheet program. Write a report and prepare a presentation that will summarize your analysis.

DISCUSSION QUESTIONS

1. For which aspects of project management are computers most useful? Why?
2. Are there aspects of project management for which computer support cannot be used?
3. Develop a list of criteria for a project management software package that will be used in a project management course.
4. What aspects of project management can be supported by a simple spreadsheet application?
5. Write a description of one application from the list you developed in your answer to Question 4.
6. What aspects of project management can be supported by a database system?
7. Write a description of one application from the list you developed in your answer to Question 6.
8. You are in charge of the selection and implementation of a new software package for project management in your organization. Develop a project plan and explain the details.
9. What risks are associated with the project discussed in Question 8?
10. Prepare a risk management plan for the project discussed in Question 8.
11. What improvements or changes in SuperProject Expert would you recommend?
12. To simplify SuperProject Expert, which features would you eliminate?

EXERCISES

13-1. Write a report on the software package SuperProject Expert, or another one with which you are familiar. Identify the advantages and disadvantages of the package as a tool for supporting the study of project management.

13-2. Develop a software selection methodology for the project "Design of a New Space Laboratory for Crystal Manufacture." Use the methodology to assess SuperProject's ability to support the management of this project.

13-3. Try to solve all the exercises in Chapters 7, 8, 9, and 11 using SuperProject Expert or some other package. Rewrite your answer to Exercise 13-1 based on your experience with the software.

13-4. Obtain a project management software package with which you are not familiar. For the example project in the book, determine how much time is required to learn its basic functions, including data entry, critical path analysis, and report generation.

13-5. Develop a spreadsheet application for project scheduling that calculates the early start, early finish, late start, late finish, total slack, and slack float for each activity.

13-6. Develop a spreadsheet application for resource management within an OBS–WBS framework. The application should present planned use, cost of resources, actual use and cost, and the deviations for each organizational unit.

13-7. Develop a PERT program on a spreadsheet based on the three time estimates for each activity. Your program should be able to calculate the probability of completing the project by a given date.

13-8. You have been asked to choose the project management software to be used as a teaching aid for a project management course.

(a) Write the software specifications for such a package.

(b) What might be the main differences in the software specifications associated with the following needs?

(1) A software package to be used as a teaching aid in a course

(2) A software package to be used for planning and controlling projects managed by your school

13-9. Develop a benchmark project to be used to evaluate the suitability of software packages as a teaching aid in project management studies.

13-10. Discuss the following comment and develop a numerical example to validate your remarks: "The purchasing price of a project management software package is a negligible issue as long as the price is no more than a few thousand dollars."

13-11. As part of your company's effort to select a project management software package, you have been asked to approach several other companies that presently use such packages.

(a) Develop a questionnaire to help collect the relevant information.

(b) Fill out two questionnaires, each representing a different software package.

(c) Compare the responses of the companies and select the best software of the two.

13-12. Read a recent article evaluating project management software packages. Such articles frequently appear in technical journals and magazines such as *Industrial Engineering* and *PC World*. Discuss the following issues, based on the article.

(a) What features do most of the packages have in common?

(b) What new features are starting to emerge?

(c) Which seem to be the leading packages, and what are the major reasons for this?

(d) Specify some of the more important criteria used in evaluating the packages.

(e) Suggest criteria for software evaluation other than those used in the article.

REFERENCES

ASSAD, A. A., and E. A. WASIL, "Project Management Using a Microcomputer," *Computers & Operations Research*, Vol. 13, No. 2 (1986), pp. 231–260.

DAUPHINAIS, B., and L. DARNELL, "Project Management One Step at a Time," *PC WORLD* (September 1984), pp. 240–250.

DAVIS, E. W., and R. D. MARTIN, "Project Management Software for Personal Computer: An Evaluation," *Industrial Management*, Vol. 27, No. 1 (January–February 1985), pp. 1–23.

DE WIT, J., and W. HERROELEN, "An Evaluation of Microcomputer-based Software Packages for Project Management," *European Journal of Operational Research*, Vol. 49 (1990), pp. 102–139.

FAWCETTE, J., "Choosing Project Management Software," *Personal Computing*, Vol. 8, No. 10 (1984), pp. 154–167.

FERSCO-WEISS, H., "High-End Project Managers Make the Plans," *PC Magazine*, Vol. 8, No. 9 (September 1989), pp. 155–195.

GIDO, J., *Project Management Software Directory*, Industrial Press, New York, 1985.

HECK, M., "Product Comparison-For That Really Big Project," *INFOWORLD* (February 1989), pp. 53–64.

IMA, *Choosing the Best PC-Based Project Management Software*, Engineering Management and Administration Report, Institute of Management and Administration, New York, April 1992.

LEVINE, A. H., "Computers in Project Management," in D. I. CLELAND and R. W. KING (editors), *Project Management Handbook*, Van Nostrand Reinhold, New York, 1988, pp. 692–735.

LISWOOD, W., "Project Management for Professionals Only," *PC WORLD* (November 1988), pp. 144–150.

RUBY, D., "Project Management: On the Critical Path," *PC Week*, Vol. 2, No. 3 (1985), pp. 51–59.

SMALLOWITZ, H., "Avoiding Pitfalls and Costly Detours," *Computer Decisions*, Vol. 16, No. 16 (1984), pp. 104–118.

SMITH, L., and S. GUPTA, "Evaluation of Project Management for Microcomputers for Production and Inventory Management Projects," *Production Inventory Management Review*, Vol. 5, No. 6 (1985), pp. 66–70.

WALLACE, R., and W. HALVERSON, "Project Management: A Critical Success Factor or a Management Fad," *Industrial Engineering*, Vol. 24, No. 4 (April 1992), pp. 48–53.

WHEELWRIGHT, J. C., "How to Choose the Project Management Microcomputer Software That's Right for You," *Industrial Engineering*, Vol. 18, No. 1 (January 1986), pp. 46–52.

WINSHIP, S., "High-End Project Manager Buyers Have High Expectations," *PC Week*, Vol. 7, No. 49 (1990), pp. 81–82.

APPENDIX 13A

BBRI Software Evaluation Checklist

The following checklist was developed by the Project Management Software Working Group (De Wit and Herruelen 1990) sponsored by the Belgian Building Research Institute.* Its purpose is to guide the potential user of project planning software through the time-consuming process of evaluating the multitude of products on the market. The weight to be assigned to the various criteria is applications dependent. Where necessary, the individual "check items" are accompanied by clarifying comments.

Check items	Comments
1. *Installation requirements and support*	
Does your PC meet the hardware requirements?	Check memory requirements, disk drive requirements (floppy disk drives or hard disk), need for color display, printer and plotter support (command language and speed).
Does your PC installation meet the software requirements?	Most packages run under MS-DOS version 2.0 or greater.
Does the user manual provide detailed installation guidelines?	Not all user manuals provide detailed setup steps, nor do they explain (DOS) commands and parameters in detail.
Are the installation and system diskettes copy protected? Can backup copies be made on diskette? On hard disk?	
Is accurate memory usage information provided on screen?	
Does the software vendor or dealer give technical support?	The presence of an experienced local dealer with sufficient project management knowledge is not always guaranteed.
Is the dealer familiar with project management?	
Is the dealer familiar with package details?	

(Continues)

*Members of the BBRI Working Group: W. Herroelen (Katholieke Universiteit Leuven, Chairman), J. De Wit (W.T.C.B.-Reporting Secretary), L. Deman (ABAY N.V.), Y. De Busschere (E.G.T.A. Contractors N.V.), M. De Wit (W.T.C.B.), G. Losseveld (C.F.E.N.V.), J. Pinchart (W.T.C.B.), J. Van Den Broeke (Maurice Delens N.V.).

Check items	Comments
Does the dealer provide installation support? Will sufficient information be made available on new package versions? Is a demo version available?	A realistic demo version should at least allow for the planning of projects with some 50 activities.
Are training sessions or courses available?	

2. *Documentation*

Is a user manual and/or a reference manual available?	User-friendly manuals explain the package fundamentals through a set of tutorial lessons.
Does the manual guide the user through all project management phases in a logical sequence?	
Does the manual contain detailed illustrations of all menu, input screen, and window displays?	If not, the user can obtain screen display printouts through the use of the **Print Screen** key.
Are all menu and input screen items and commands explained in detail?	
Does the manual have a detailed chart on the hierarchical organization on menus and screens?	
Is there a detailed table of contents? Is the index precise and reliable? Are all output reports explained in sufficient detail? Does the manual have a tutorial on project planning?	
Are all the calculations performed by the package explained in detail?	The manual should provide detailed information about the precise computational rules used in scheduling [processing of start-to-start (SS) and finish-to-finish (FF) relations where applicable, calculations of activity slack], cost, and resource monitoring.
Does the manual contain worked-out examples? Is sufficient information given about the generation of custom-made output reports?	

3. *Error detection and diagnostic*

Are error-diagnostic messages clear and informative?	The diagnostic message saying that a loop has been detected should contain detailed information about the activities involved in the loop. The message "A loop has been detected, please check your data" is *not* sufficient or very helpful.
Are corrective measures suggested and explained in sufficient detail?	

Check items	Comments
Does the program get blocked when erroneous input data are entered?	Entering a character instead of a number in a particular field may cause the user to be "removed" from the program, necessitating a program restart with accompanying loss of input data. A beep signal should be given when input data do not meet the required specifications.
Can the program cope with multiple errors at the same time?	
Does the program get blocked upon the first error?	
Is it possible to cancel commands and data input?	This should be possible through the use of the **Escape** key or the Functional keys.
Are the following errors signaled?	Some packages give diagnostics for each error. An overview of all errors committed may be helpful in preventing the user from going through a time-consuming correction process.
■ Loops	
■ Activities without predecessors and/or successors	Some packages report dangles (activity without entering or leaving precedence relations) as errors but do not require corrective action.
■ Logical errors in progress reporting	Upon reporting progress, it would be useful to have a list of all the activities that, based on the project progress information, should have been started but are not. To avoid activities without predecessor or successor relations, it is always advisable to have SS-relations in combination with FF-relations.
Can a list be obtained containing all data entered up to the occurrence of an error?	The **Print Screen** key may again be useful here.
Are timely indications given about the maximum number of activities that can still be entered?	It is very frustrating and time-consuming to discover in the middle of entering data for a a project that one is suddenly running out of memory space.

4. *Ease of use*

Is a walk-through tutorial available on disk?	
Is the menu structure clear? Are screen messages and commands clear and precise?	This is especially important during the learning phase.
Is an input buffer available (i.e., can data be entered and commands be generated before the full screen display appears)?	This capability has pros and cons; the user does not pay sufficient attention to the precise specifications on the screen, which may result in data-entry errors.

(Continues)

Check items	Comments
Are there limitations on the numbering procedure for network nodes?	Some packages will only accept increasing activity or node numbers, which is a serious inconvenience.
Are there limitations on the numbering of nodes connecting projects and subprojects?	
Can a data entry summary be on screen?	Programs offering this capability allow for considerable time savings during the project analysis stage.
Can the network logic be viewed on screen?	
Can a list be obtained of all activities and precedence relations ■ On screen? ■ On printer?	
Can project data be saved directly on diskette? Can the network diagram and the corresponding data be constructed on screen?	Drawing the network directly on screen can be very time consuming. For medium-sized and large networks it is advisable to prepare the network diagram on paper.
Can precedence relations be changed easily (e.g., can an FS-relation be changed quickly in a SS-relation)?	
Is it possible to change any data entry field without having to retype everything?	
Is it possible to use the **Insert** key?	
Is it possible to leave a menu or window when no data are entered?	Some packages do not allow this and require the system to be restarted again.
Does deletion of activities automatically lead to deletion of the precedence relations?	
To correct errors, it is necessary to leave the current menu?	
Are functional keys used?	For example, **F1** = **Help**.
Is time-phased acquisition of software possible?	

5. *Basic capabilities*

What is the planning system used? ■ Activity-on-arrow (PERT/CPM) ■ Precedence diagramming ■ Gantt chart	The precedence diagramming method is preferable within the building industry.
What is the maximum number of activities per project?	Some packages are misleading with respect to this item. A package with a capacity of 2,500 activities may in reality allow only for ±800 activities and 1,700 relations. In building projects, 1,000 activities and relations are common practice.
Does the program accept more than one start and finish activity within the same network?	
Is it possible to specify start and finish dates for each activity?	
Is it possible to denote negative float?	In some packages, negative float is used to indicate activity delay.

Check items	Comments
What type of precedence relations may be specified? ■ Start–start (SS)? ■ Finish–start (FS)? ■ Start–finish (SF)? ■ Finish–finish (FF)?	In building projects many overlapping activities occur. As a result, it is not advisable to use software based on the exclusive use of FS-relations. The SF-relation is useful for the direct coupling of an order to an activity, so reducing the allowable float.
Is it possible to have (per activity): ■ Several SS relations? ■ Several FF relations? ■ Several FS relations? Is it possible to specify hammock activities?	This is a must but the limitation of one FF-relation per activity is not critical. The duration of hammock activities is automatically updated as the times of the nodes to which they are connected are realized. Hammock activities are useful for specifying summary' activities.
Is it possible to specify milestones? Is it possible to specify activity codes for the responsibility, the work breakdown structure, and the work areas? What is the maximum number of incoming precedence relations for an activity? Is it possible to create subprojects? What type of calendar is used? ■ Working days ■ Calendar days ■ Both Is it possible to use more than one calendar?	Some programs allow for only three to five incoming relations. A minimum of five incoming relations is a necessity. Some packages allow for more than one calendar per project. In most cases one calendar is sufficient. It is important to check whether the number of working days in a week can be changed (working weeks of 3, 4, 5, 6, or 7 days).
How many holidays and nonworking days can be entered? Is it possible to use multiple formats for entering calendar dates? How many different codes can be used in generating output reports? Is it possible to sort the activities in nonnumeric order?	In the building industry this is an important issue. For long projects taking from 2 to 3 years, it should be possible to enter from 80 to 100 nonworking days. Packages requiring dates to be entered as month–day–year are inconvenient for European users. Packages often exaggerate the number of code selections. Identification and retrieval of network activities is rather complicated with this option. If the option is available, the software should allow for at least four identification characters.
How many characters may be used for describing: ■ projects? ■ activities? ■ nodes?	For activities, a minimum of 32 characters is reasonable. For nodes, a minimum of four characters is acceptable.

(Continues)

Check items	Comments
Is the time scale: ■ Hours? ■ Shifts? ■ Days? ■ Weeks? ■ Other?	Some packages may automatically convert durations from one scale to another.
Is it possible to have several activities with the same start and ending node number?	Packages having this option should not be used.
Is it possible to change the number identification of an existing activity?	
Is it possible to enter fixed starting times?	This option is a necessity.

6. *Resource and cost monitoring*

Does the package allow for the monitoring of resources?	Although most users in the building and construction industry are interested primarily in the time aspect of planning, competitive gains can be generated through effective resource and cost monitoring.
How many resource types can be entered?	The possibility to specify some 20 resource types should be sufficient for most building industry projects.
Are resource requirements to be specified as: ■ A constant amount per time unit for the entire activity duration? ■ A constant amount per time unit through part of the activity duration? ■ Any quantity to be arbitrarily divided over the activity duration?	
Does the package allow for: ■ Resource leveling? ■ Resource constraints?	The procedures for resource leveling (smoothing the resource requirements for a given project duration by scheduling activities somewhere within their available float) and resource-constrained scheduling (scheduling the activities subject to resource constraints) used by many packages are extremely weak, very fuzzy and almost never explained in detail. Regrettably, the use of these procedures is not recommended when the user does not have the know-how to evaluate the effects of the procedures used.
Is the package sufficiently precise about the details of the heuristic procedures used?	
Does the package allow for the monitoring of costs?	
Is the package precise and clear on its procedure to estimate costs?	
Can the user override the estimations by entering specific cost data?	

Check items	Comments
What are the specific capabilities for updating the project plan with actual cost/schedule information to compare actual versus planned costs?	Many packages now have this important capability and generate reports on time and cost variance. None of the packages tested are of specific help during the process of project bidding, nor during the process of performing time–cost trade-off analyses.
7. *Progress reporting* Is progress reporting to be done by: ■ Effective start time? ■ Effective finish time? ■ Remaining duration? ■ Percentage complete? ■ Percentage remaining to complete?	These are interesting options for analyzing the project as built. For some packages the actual start and finish times should be used to ensure correct results, especially in FF- and SS-relations.

Source: Adapted from J. De Wit and W. Herroelen, "An Evaluation of Microcomputer-Based Software Packages for Project Management," *European Journal of Operational Research*, Vol. 49 (1991), pp. 102–139.

APPENDIX 13B

List of Software Vendors

Product	Vendor	Environment	Price	Comments
ABT Project Workbench/Standard	Applied Business Technologies, Inc. New York, NY	IBM PC/XT/AT	$750	Small to midsize projects
Action-Newton w/ Project Query Language	Information Research Corp., Charlottesville, VA	IBM PC/XT/AT	1,498	Project tracking budget management, multiproject control
Artemis	Lucas Management Systems, Fairfax, VA	IBM MVS, VMHP	5,000	CPM project management database
Artemis I/CSCS	Lucas Management Systems, Houston, TX	IBM PC/XT/AT, PS/2	3,500	"What-if" analysis, resource leveling
Harvard Project Manager	Software Publishing Corp., Mountain View, CA	IBM PC/XT/AT, PS/2	595	WBS, project control management; PERT/Gantt charts
InstaPlan 5000	InstaPlan Corp., Mill Valley, CA	IBM PC/XT/AT	495	Handles 16,000 activities and resources
Micro Planner	Micro Planning Software, San Francisco, CA	IBM PC/XT/AT	495	Planning, resource management
Micro Planner Plus	Micro Planning Software, San Francisco, CA	IBM PC/XT/AT, PS/2, Unix, Xenix, VMS, Apple Macintosh	495	Planning, resource management
Micro Trak	Softrak Systems, Salt Lake City, UT	IBM PC/XT/AT, PS/2, Unix, Xenix, VMS	595–895	Project scheduling, control, resource management using CPM
MicroMan II	Poc-It Management, Santa Monica, CA	IBM PC/XT/AT	2,895	Tracks, manages information systems activity
Microsoft Project	Microsoft Corp., Redmond, WA	IBM PC/XT/AT	495	Resource planning, cost analysis, Gantt charts

Product	Vendor	Environment	Price	Comments
Multitrak	Multisystems, Cambridge, MA	IBM MVS, DOS, CIC, IBM PC	Contact vendor	Resource allocation
N1100	Nichols & Company, Culver City, CA	IBM PC, HP 150, Wang PC	1,800	Critical path analysis, project control
OpenPlan	Welcom Software Technology, Houston, TX	IBM PC/XT/AT, PS/2	4,200	Critical path analysis, resource scheduling
Opera	Welcom Software Technology, Houston, TX	IBM PC/XT/AT, PS/2	2,200	Risk analysis, extensin scheduling
PAC Micro	AGS Management Systems, King of Prussia, PA	IBM PC/XT/AT	900	Multiple project scheduling, CPM
Parade	Primavera Systems, Bala Cynwyd, PA	IBM PC/XT/AT, PS/2	2,000	Performance measurement, earned value analysis
Pertmaster Advanced	Pertmaster Int'l, Santa Clara, CA	IBM PC/XT/AT, PS/2	1,500	Plan, schedule, manage project with interrelated tasks
Plantrac	ComputerLine, Inc., Pembroke, MA	IBM PC/XT/AT	695+	Planning, tracking tool
Plot Trek	SofTrak Systems, Salt Lake City, UT	IBM PC/XT/AT, PS/2, Unix	295–495	Micro Trakgraphics, network diagrams, Gantt charts
PMS-11	North America Mica, San Diego, CA	IBM PC/XT/AT	1,295	
PMS-80-Advanced	Pinnell Engineering, Portland, OR	IBM PC/XT/AT	2,500 (base) 1,500 (graphics option)	
PREMIS	K&H Professional Management Services, Wayne, PA	IBM OS, VS	Contact vendor	Time analysis, resource management
Primavera Project Planner	Primavera Systems, Bala Cynwyd, PA	IBM PC/XT/AT, PS/2	795–5,000	Cost analysis and control, activity coding
Project Alert	CRI, Inc., Santa Clara, CA	DEC VAX, HP3000, Apollo	10,000+	PERT and Gantt chart scheduling and monitoring
Project OUTLOOK	Strategic Software Cambridge, MA	IBM PC/XT/AT, PS/2	495	Interactive, MS-Windows, build networks on screen with mouse

(Continues)

Product	Vendor	Environment	Price	Comments
Project Scheduler 4	Scitor Corp., Foster City, CA	IBM PC/XT/AT, PS/2, Wang PC, Macintosh	685	Handles multiple projects
Project Workbench Advanced	Applied Business Technologies, New York, NY	IBM PC/XT/AT, Wang PC	1,150	Complex projects; Gantt charts, resources screens
Prothos	New Technology Association, Evansville, IN	IBM PC/XT/AT, DEC VAX, Unix Sys V	600+	
Quick-Plan II	Mitchell Management Westborough, MA	IBM PC/XT/AT	250	Planning, resource scheduling
Skyline	Applitech Software Cambridge, MA	IBM PC/XT/AT, PS/2	295	Project outliner, uses CPM
SSP's PROMIS	Strategic Software Cambridge, MA	IBM PC/XT/AT, PS/2	2,995	Subnetworking for large projects
SuperProject Expert	Computer Associates Garden City, NY Mountain View, CA	IBM PC/XT/AT, PS/2	695	PERT, Gantt charts, CPM, probability analysis
Task Monitor	Monitor Software, Los Altos, CA	IBM PC	695	
Time Line	Symantec Corp., Novato, CA	IBM PC/XT/AT, PS/2	495	Project planning, resource allocation, cost tracking
Top down Project Planner	Ajida Technologies, Santa Rosa, CA	IBM PC/XT/AT, PS/2	495	Top-down approach WBS
Qwiknet Professional	Project Software & Development, Inc., Cambridge, MA	IBM PC/XT/AT, PS/2, DEC VAX/VMS	1,495 18,900	CPM, multiproject scheduling
ViewPoint	Computer Aided Management, Inc. Petaluma, CA	IBM PC/XT/AT, PS/2	1,995	Top-down planning scheduling
VISIONmicro	Systonetics, Fullerton, CA	IBM PC/XT/AT	995	Mouse-driven interface; project management and graphics
Vue	National Information Systems Cupertino, CA	IBM, DEC, UNIX, HP, Honeywell	1,995	Scheduling, uses CPM

14

Project Termination

14.1 INTRODUCTION

Project termination is an important, yet often mismanaged phase in a project's life cycle. At some point, management must decide to terminate the project. However, this can be a difficult and agonizing activity, since projects tend to develop a life and constituency of their own. Team members, subcontractors, and other support personnel often become effective advocates for continuing a project long after its useful life has expired. Nevertheless, all projects must end and it is up to management to see that their concluding phase is smooth, timely, and as painless as possible.

The reality is that team members frequently overlook or try to delay termination to the last possible moment. Such delays can have serious consequences because they create unnecessary stress and are costly for both the organization and the project personnel. Therefore, a successful project must include a well-planned and executed termination phase that saves time and money and avoids unnecessary conflict.

Managing project termination revolves around two central questions concerned with when and how to terminate the project. The answer to the first question seems obvious: Terminate the project when its mission is accomplished. Some projects, though, are perforce canceled before this criterion is met because of changing market conditions, organizational shakeups, cost overruns, or technical difficulties. However, if a manager is convinced that a project will produce results, he or she may be predisposed to slant cost and performance data in the most favorable direction. Sometimes when managers realize that a project is in real trouble, rather than accept failure, they may choose to invest more resources. As a general rule, though, premature termination should be considered only when the probability of success is clearly too low to justify further investment in the project.

How to terminate the project requires a clear set of procedures for reassigning materials, equipment, personnel, and other resources. A project manager with good leadership skills can decrease anxiety levels within the organization and among the outside participants by carefully planning and executing the project's termination.

14.2 WHEN TO TERMINATE A PROJECT

In the section above we presented a simplistic answer to the question of when to cancel a project. Judging when a project's mission is accomplished is difficult because the degree of success or failure associated with most projects is tricky to measure at any specific time. The success (or failure) level tends to increase at a decreasing rate, implying that change is less visible with the passage of time. But since detecting a partial success or failure is not easy, management is wont to delay termination until the outcome is clearer or more information is available. This "wait and see" attitude can be very expensive. Project costs may escalate, and in most failed projects, these costs cannot be recovered. In many cases the project manager is forced to act subjectively without full confidence in the decision.

On the other hand, a project's termination costs may be a stumbling block to what objectively looks like the best course of action. When the initial decision to start a project is made, managers rarely know, or even consider, what the closing costs and salvage value of the project will be if it is terminated prematurely. New projects are supposed to succeed, not fail. It would be psychologically disturbing to think or plan otherwise. So when management is faced with a huge bill for project termination, the decision might be to continue spending money with the hope that the situation will improve, despite the evidence to the contrary. At the end of the Cold War, the United States was faced with just such a situation. The reality of canceling tens of billions of dollars in defense contracts meant skyrocketing unemployment in the aerospace and shipbuilding industries and huge financial penalties to buy out extant contracts. To cite one example, in 1992 the U.S. Congress decided to go ahead with a $3 billion program to build a prototype of the next-generation nuclear attack submarine to avoid closing down General Dynamic's Electric Boat Division in Groton, Connecticut. Politics and the severe short-term economic effects that the local community would probably have experienced were the determining factors.

Economics and politics alone, though, do not always drive the termination decision. The L1011 Tri-Star program of Lockheed is a prime example. For more than a decade the aircraft accumulated enormous losses and, in fact, was never really expected to earn a profit. But the program was Lockheed's reentry into commercial aviation and became a symbol that broadened the company's image beyond simply being a defense contractor (Staw and Ross 1987).

This suggests another difficulty in reaching consensus on the exact termination point of a project: namely, defining the mission. For example, consider a construction project in a residential neighborhood. The project may accomplish its mission as soon as the houses are built, as soon as they are sold and tenants move in, or possibly, as soon as the 1-year contractual warranty period expires. The situation may be even more difficult when R&D projects such as the development of a space station are concerned. In this example, the design team is likely to make engineering changes throughout the station's construction, assembly, and even operation. Members of the R&D team may be assigned to other parts of the organization (NASA) or may continue as a team involved in related projects and activities. Here, project termination is almost impossible to define. A third scenario involves an engineering team designing a new product intended for mass production, such as a notebook computer. When a prototype is successfully developed, the team may be integrated into the parent company as a division to manufacture, support, and improve the new product.

Meredith and Mantel (1989) propose three approaches to project termination: extinction, inclusion, and integration.

1. *Termination by extinction* occurs when the project stops because its mission is either a success or a failure. In either case, all substantial project activities cease at the time of assessment. The project team or special project termination team conducts the phaseout. Either team's aim is to reassign resources, close out the books, and write a final project report. This is discussed in Section 14.5.

2. *Termination by inclusion* occurs when the project team becomes a new part of the parent organization. Resources are transferred to the new organizational unit, which is integrated into the parent organization. This type of project termination is typical for organizations with a project/product structure.

3. *Termination by integration* occurs when the project's resources, as well as its deliverables, are integrated into the parent organization's various units. This approach is very common in a matrix organization because most people involved in a project are also affiliated with one or more functional units. When the project terminates, team members are reintegrated into their corresponding units.

As noted earlier, most projects do not reach clear success or failure points. Therefore, management should monitor each project vigilantly to look for signs that call its continuation into question. Monitoring is facilitated by the project control system discussed in Chapter 11. In addition, it is a good idea to conduct regular evaluations and audits to evaluate the status of critical milestones during a project's life cycle. Unlike the project control system, which is operated by the project team, an external organizational unit not directly involved with the project should conduct the audits to assure a more objective analysis. The client may also require formal evaluations and audits as each phase ends. These results may be presented at the preliminary design review, the critical design review, or other milestone dates.

The financial audits commonly used in organizations concentrate on their financial well-being and economic status. By contrast, the project audit covers a large number of aspects, including:

- The project's *current status versus stated goals* as related to schedule, costs, technology, risk, human relations, resource use, and information availability

- *Future trends*, that is, forecasts of total project costs, expected completion time, and the likelihood that the project will achieve its stated goals
- *Recommendations* to change the project's plans or to terminate the project if success seems unlikely

If performed conscientiously, the audit report will be more objective than the project control system reports. However, due to auditing costs, audit reports are not issued regularly. Termination decisions, then, frequently result from information from the control system. If the cumulative information indicates that success is remote, an audit team may be assembled to evaluate the situation more closely. We note here that a decision against initiating the termination phase (i.e., the "do nothing" decision) should be based on a project's satisfactory performances, not on a lack of alarm signals. For assistance in this matter, the project manager must rely on the control system throughout the project's life cycle. The information that it provides can trigger an audit to support the termination decision.

Assuming that the control system functions well and that current information is available, management needs a methodology for reaching a termination decision. Project management researchers have developed lists of questions designed to address this issue (Buell 1967). Although most studies have focused on R&D projects, the following list is appropriate in the majority of circumstances. The questions may be difficult to answer, requiring a special audit to obtain the necessary information.

- Did the organization's goals change sufficiently so that the original project definition is inconsistent with the current goals?
- Does management still support the project?
- Is the project's budget consistent with the organizational budget?
- Are technological, cost, and schedule risks acceptable?
- Is the project still innovative? Is it possible to achieve the same results with current technology faster and at lower cost without completing the project?
- How is the project team's morale? Can the team finish the project successfully?
- Is the project still profitable and cost-effective?
- Can the project be integrated into the organization's functional units?
- Is the project still current? Do sufficient environmental or technological changes make the project obsolete?
- Are there opportunities to use the project's resources elsewhere that would prove more cost-effective or beneficial?

Based on the answers to these questions, perhaps obtained with the help of the economic analysis and project evaluation/selection techniques discussed in Chapters 2, 3, and 4, management should be able to decide whether it is time to cancel the project. Once a termination decision is made, the question then becomes how to minimize the likely disruption that such action would cause.

As mentioned, management should repeatedly consider whether to continue or to terminate a project throughout its life cycle. In addition, an outside group should be asked to provide input to the decision, since the project manager and team members have a vested interest that may compromise their candor. The outside analysis should be a part of the project audit effort which should be designed to yield an objective evaluation of the project's status.

Since project success (or failure) is multidimensional, the evaluation should cover at least the following:

- *Economic evaluation.* Given the costs of all project efforts to date, is project continuation justified?
- *Project costs and schedule evaluations.* Given the current costs, schedule, and control system's trend predictions, should the project be canceled?
- *Management objectives.* Given the organization's current objectives, does the project serve these objectives?
- *Customer relations and reputation.* If premature termination is justified, how will this affect the organization's reputation and its customer relationships?
- *Contractual and ethical considerations.* Is project termination possible given current client and supplier contracts? Is project termination ethical?

In conjunction with these questions, the auditing process should consider a multitude of factors as well as their impact on the organization. Building on the work of Buell, Balachandra, and Raelin (1980), identify the following quantitative and qualitative factors:

Quantitative factors

Probability of commercial success
Anticipated annual growth rate
Capital requirements
Project use
Investment return
Annual costs
Probability of technical success
Amount of time actual project costs equaled budgeted project costs

Qualitative factors

Degree of consumer acceptance of the project's outcome
Probability of government restrictions
Ability to react successfully to competition
Degree of innovation

Degree of linkage with other ongoing projects

Degree of top management support

Degree of R&D management support

Degree of the project leader's commitment

Degree of the project personnel's commitment as perceived by top management, R&D management and project leaders

Presence of persons with sufficient influence to keep the project going

One methodology supporting a project termination decision is the early termination monitoring system (ETMS) designed to generate an overall index of a project's viability (Meredith 1988). By using input from the project's control system, ETMS reports the effects of an early termination on the organization's image, the project team's performance, the marketplace economics, and the penalty costs that will be incurred.

Finally, Table 14-1 enumerates 10 critical reasons identified by Dean (1968) in a study of 36 companies for premature R&D project termination. Taken together with the lists above, we begin to see why this life-cycle phase is so difficult to manage. The difficulty stems from the many factors involved in the decision to begin phaseout, as well as in the complexity of termination planning and execution.

TABLE 14-1 PRINCIPAL REASONS FOR CANCELING R&D PROJECTS

Factors	Reporting frequency
Technical	
Low probability of achieving technical objectives or commercial results	34
Available R&D skills cannot solve the technical manufacturing problems	11
R&D personnel or funds required for higher-priority projects	10
Economic	
Low investment profit or return	23
Individual product development too costly	18
Market	
Low market potential	16
Change in competitive factors or market needs	10
Other	
Too much time to achieve commercial results	6
Negative effects on other projects or products	3
Patent problems	1

14.3 PLANNING FOR PROJECT TERMINATION

Like any other phase in the project's life cycle, termination planning tries to maximize the project's probability of success. Once management approves cancellation, the following action should be taken:

- Set project termination milestones.
- Establish termination phase target costs and budget allocations.
- Specify major milestone deliverables.
- Define desired organizational structure and workforce after termination.

Although each project may have a different set of goals, some activities are required in almost all cases. Archibald (1976) suggests the following activity termination list.

Project office (PO) and project team (PT) organization

Conduct project closeout meetings.
Establish PO and PT releases and reassignments.
Carry out necessary personnel actions.
Prepare a personal performance evaluation for each PT member.

Instructions and procedures

Terminate the PO and PT.
Close out all work orders and contracts.
Terminate the reporting procedures.
Prepare the final report(s).
Complete and dispose of the project file.

Financial

Close out the financial documents and records.
Audit the final charges and costs.
Prepare the final project financial report(s).
Collect the receivables.

Project definition

Document the final approved project scope.
Prepare the project's final breakdown structure and enter it into the project
file.

Plans, budget, and schedules

> Document the actual delivery dates of all contractual deliverable end items.
>
> Document the actual completion dates of all other contractual obligations.
>
> Prepare the project's final and task status reports.

Work authorization and control

> Close out all work orders and contracts.

Project evaluation and control

> Assure the completion of all action assignments.
>
> Prepare the final evaluation report(s).
>
> Conduct the final review meeting.
>
> Terminate the financial, personnel, and progress reporting procedures.

Management and customer reporting

> Submit the project's final report to the customer.
>
> Submit the project's final report to management.

Marketing and contract administration

> Compile the final contract documents, including revisions, waivers, and related correspondences.
>
> Verify and document compliance with all contractual terms.
>
> Compile the required proofs of the shipment and customer acceptance documents.
>
> Officially notify the customer of the contract's completion.
>
> Initiate and pursue any claims against the customer.
>
> Prepare and conduct the defense against the customer's claims.
>
> Initiate public relations announcements regarding the contract's completion.
>
> Prepare the final contract status report.

Extensions—new business

> Document the possibilities for project or contract extensions or other related new business.
>
> Obtain an extension commitment.

Project records control

> Complete the project file and transmit it to the designated manager.
>
> Dispose of other project records as required by established procedures.

Purchasing and subcontracting (for each purchase order and subcontract)

> Document compliance and completion.
>
> Verify the project's final payment and proper accounting.
>
> Notify the vendor/contractor of the project's completion.

Engineering documentation

> Compile and store all engineering documents.
>
> Prepare the final technical report.

Site operations

> Close down all site operations.
>
> Dispose of all equipment and materials.

Based on this list and on additional (project specific) activities, management can perform a project scheduling analysis of the termination phase. The results obtained from the analysis form the basis for budgeting and staffing during phaseout. Spirer (1983) suggests the use of diagrams, matrices, and checklists as analytical tools for the management of project termination. In addition, he suggests a work breakdown structure, as shown in Fig. 14-1, to identify the problems that are likely to arise in the process.

The project termination phase has a significant emotional impact on the people involved. Four types of groups may be identified: end users, customers, team members and producers, and consultants and maintenance personnel. The following example clarifies the differences among the groups. A company that manufactures elevators is the producer, its customer is the builder, the end users are the tenants who are going to occupy the building. Each of the four groups is involved and affected differently by project termination. Therefore, it is extremely important to identify the nature of the impact and be able to treat any untoward consequences. Although the contractor is the immediate customer of the elevator manufacturer, the end users and the other interested parties, such as the maintenance crew and the consultants, represent future customers that should be taken into account. The immediate customer may want to terminate the project as soon as possible, even if the unit installed has not been tested sufficiently under normal operating conditions. However, if this unit does not meet the expectations of the end users, costly rework may be required and the reputation of the elevator company may be damaged.

The following list identifies typical problems that employees working on a project may face during the termination phase:

- Loss of interest in the project
- Insecurity regarding their prospect to get new jobs
- Insecurity regarding the uncertainty involved in a new project
- Problems in handling the project to the customer

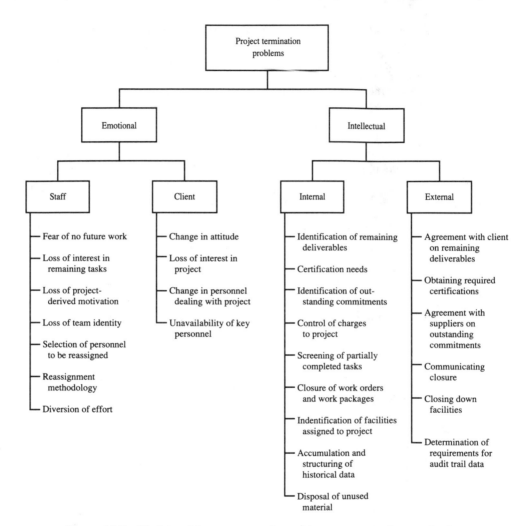

Figure 14-1 Work breakdown structure for problems accompanying termination.

From an emotional point of view, project termination has a separation effect. Each project team member faces the following troublesome questions:

- What, if any, are my plans after the project?
- What is my future role in the organization?
- What is my next assignment?

The project manager should consider specific answers as well as the best way to communicate these answers to the team members. Furthermore, the project manager may worry about his or her own future after the project. Planning ahead how to resolve these personal problems and fears will help to reduce high levels of individual anxiety among all team members.

During phaseout, due to the natural feelings of uncertainty, project team members may experience low morale, lose their interest in the project, or try to delay its termination. The frequency and intensity of conflicts tend to increase, and even successful projects may leave many members feeling angry, upset, or both. To minimize these effects, management should try to reduce the members' uncertainty levels. Suddenly canceling a project may be disastrous. Team members may find it difficult to terminate the project effectively if they face sudden unexpected changes requiring them to invest their time and energy developing adaptive strategies. Consequently, management's sensitivity, thoughtful planning, and consideration of members' emotions can reduce the negative effects of cancellation and support a project's successful closing.

14.4 IMPLEMENTING PROJECT TERMINATION

Once management decides to cancel a project and develops a project termination plan, a termination phase leader must be chosen. Project managers are natural candidates, but if they are emotionally unsettled and uncertain about their own futures, they might not be able to do a reliable job. A second candidate is a professional project termination manager who may be unfamiliar with the project's substance but experienced and well trained in closing down projects efficiently and effectively. The choice depends on the answers to the following questions:

- Did the project achieve its mission?
- Is the project manager assigned to a new project? If yes, when will the new assignment begin?
- Is the client satisfied?
- Is an experienced project termination manager available?

If the project is completed successfully, the client is satisfied, and the project manager knows his next assignment, the project manager is the best candidate to head up the termination effort. Otherwise, appointing an experienced alternative is a wiser choice because the current project manager may not be motivated to do the job conscientiously.

The termination leader should implement the termination plan by notifying all project team members of the decision to cancel the project. Communicating with team members and laying out a road map for their futures reduce their uncertainty levels. Once this is accomplished, the next step is to reduce and eventually eliminate the use of all resources, while implementing procedures that will facilitate a smooth transition of all personnel to their next assignments.

Throughout project termination planning, implementation, and execution, management should be extremely sensitive to the various aspects of human relations. The need for cooperation in future projects should guide all interactions with current team members, the client, suppliers, and subcontractors. The termination phase is a bridge to future projects. One cornerstone of this bridge is the project's final report.

14.5 FINAL REPORT

The basis of total quality management is continuous improvement. Since each project has a limited lifetime, TQM requires improvement from one project to the next. To facilitate this notion, one important outcome of the termination phase is the project's final report, documenting activities at each stage of the project's life cycle. Such a report emphasizes weak points in the planning and implementation phases in order to improve organizational procedures and practices. The report also explains working procedures developed during the project's life cycle that contributed to its success, and proposes adopting these procedures in future projects. The report helps management to plan future projects and to train future managers and team members. Thus the report forms the basis for improving organizational–project–management practices and developing new and improved working procedures.

To accomplish these goals, the final report begins by stating the project's mission. Next, it discusses in detail the plans developed to achieve that mission, the trade-off analyses conducted, and the planning tools used. Finally, the report compares the project's original mission and plans with the actual results and deviations, and explains why such deviations occurred.

Based on this analysis, the report evaluates the project's specific procedures and tools for planning, monitoring, and control. Details should be furnished on any new procedures and analytical methods developed during the project, and recommendations should be made regarding their adoption if it is believed that they can be implemented successfully by the entire organization. Recommendations on the future uses of, or modifications to, existing procedures should also be cited. Next, the report evaluates resource use and the performance of vendors and subcontractors, judging specifically whether or not they should be included in future projects. Finally, the report evaluates and documents the performance of project team members, auxiliary personnel, and functional unit managers.

Developing a standard format for final reports allows an organization to store the information collected in a database, making it accessible for future projects. Many standard formats are designed around one of the following:

- *Standard WBS*, such as the one suggested by MIL-STD-881A. Using a standard WBS allows management to retrieve information on relevant WBS elements in past projects.
- *Standard CBS*. Storing cost information in a standard cost breakdown structure allows cost estimators and life-cycle cost analysts easy access to this type of data for future project use.
- *Standard SOW*. Storing work statements in a standard format makes responding to future requests for proposals easier, since similar SOWs from past projects can serve as a basis for new proposals.

A well-structured final project report can help an organization improve and learn from its experience. Submitting the report to management is the last step in any well-managed project.

TEAM PROJECT

Thermal Transfer Plant

The rotary combustor was assembled, tested and successfully delivered to the client organization. TMS management wants to learn from your experience with the project and has requested a final report. This report should be a prototype for future project teams at TMS to use.

Explain in your report the plan for phasing out the rotary combustor project. Present a schedule to execute this task and list resources required. Comment on the experience you have gained, the lessons you have learned, and the mistakes you have made, and how this information can be used guide others in future projects. Include in your report a chronological review of recommendations regarding project management tools and techniques used throughout the project's life cycle, and all the data that might be helpful in TMS's future development activities.

DISCUSSION QUESTIONS

1. Develop a flow diagram that shows how project termination decisions should be made.
2. Explain the difference between termination by integration and termination by inclusion, using an example for each process.
3. In what ways does the termination phase of a project differ from the close-down of a failed company? What are the similarities?
4. What is the difference between the input of the project control system and the input of an audit team to the project termination decision process?
5. In what way should the planning of the project termination phase be affected by personnel considerations?
6. Why do some projects that are clearly "losers" seem to go on forever? Can you identify a few at the national level? State level? Local level?
7. What is the most important information that a final report should contain?
8. In 1969, the U.S. Congress canceled funding for a supersonic commercial transport (SST). Do you think that was a good decision? Can you imagine what the pros and cons were?

9. Assume that you are working for a computer manufacturer as a software engineer and that you are told abruptly that your project will be canceled within 4 weeks. List the questions that you would have for management. After absorbing the shock, what would you do?

10. Identify the closeout costs for a big project such as the U.S. Space Station or superconducting super collider, which is about half complete.

11. Many people in and out of government have proposed sunset laws for all projects and agencies. That is, after a fixed amount of time, a project or agency would be closed down unless sufficient justification to continue its activities were offered. Why is such a law needed? What might constitute "sufficient justification"?

12. List the political and sociological reasons why a project might continue to be supported even though it cannot be justified economically. Can you identify such a project in your private life?

EXERCISES

14-1. Develop guidelines for writing a project final report.

14-2. Write a job description for a project termination manager.

14-3. Develop a "generic" project termination plan based on the list of activities presented in the chapter. What are the precedence relations among these activities? Develop a linear responsibility chart for the termination phase.

14-4. Develop a cost breakdown structure for the termination phase, assuming that only activities not related to the substance of the project are performed at that phase.

14-5. A flagrant example of a program that has outlived its mission is the Rural Electrification Program started in the 1930s by the Roosevelt administration. Its original goal was to bring electricity to all U.S. communities. Today, with its mission long accomplished, the program, budgeted at $5 billion annually, provides subsidies to such unneedy giants as MCI, Houston Lighting and Power, and GTE Sprint. The Office of Management and Budget has tried periodically to shut this program down, but has never been able to prevail over its powerful beneficiaries. Nevertheless, anticipating the emergence of more rational heads, you have been asked to write a final termination report for this program. The report should document its beginnings, its successes, and the reasons why it has flourished for so long, as well as the more traditional information associated with termination.

14-6. Identify two projects in which you have been involved recently.
  (a) Describe each project briefly.
  (b) Suggest criteria that may have been used to identify the start of the termination phase of each project.
  (c) Give two examples of activities that were performed poorly during the termination phase of either project, and suggest measures that might have been taken to improve the situation.

14-7. Develop a questionnaire to capture the importance of various activities that should be performed during the termination stage.
  (a) Administer the questionnaire to a sample of project managers.
  (b) Summarize and analyze the results.

14-8. Identify two projects (local or national) that were terminated prematurely.
  (a) Analyze the reasons why each was canceled.
  (b) Compare the results of the two cases.

14-9. Discuss the following statement made by a project manager: "We have already spent 70% of the budget required to complete the project and it would be a waste of money to abandon it at this stage."

REFERENCES

ARCHIBALD, R. D., *Managing High Technology Programs and Projects*, Wiley, New York, 1976.

BALACHANDRA, R., and J. A. RAELIN, "How to Decide When to Abandon a Project," *Research Management*, Vol. 23, No. 4 (1980), pp. 24–29.

BUELL, C. K., "When to Terminate a Research and Development Project," *Research Management*, Vol. 10, No. 4 (1967), pp. 275–284.

DEAN, B. V., *Evaluating, Selecting, and Controlling R&D Projects*, American Management Association, New York, 1968.

DEUTCH, M. S., "An Exploratory Analysis Relating the Software Management Process to Project Success," *IEEE Transactions on Engineering Management*, Vol. 38, No. 4 (1991), pp. 365–375.

MEREDITH, J. R., "Project Monitoring and Early Termination," *Project Management Journal*, Vol. XIX, No. 5 (1988).

MEREDITH, J. R., and S. J. MANTEL, JR., *Project Management: A Managerial Approach*, Second Edition, Wiley, New York, 1989.

NORTHCRAFT, G. B., and G. WOLF, "Dollars, Sense, and Sunk Costs: A Life Cycle Model of Resource Allocation Decisions," *Academy of Management Review*, Vol. 9, No. 2 (1984), pp. 225–234.

PINTO, J. K., and S. J. MANTEL, JR., "The Causes of Project Failure," *IEEE Transactions on Engineering Management*, Vol. 37, No. 4 (1990), pp. 269–276.

PINTO, J. K., and J. E. PRESCOTT, "Variations in Critical Success Factors over the Stages in the Project Life Cycle," *Journal of Management*, Vol. 14 (1988), pp. 5–18.

SHAFER, S. M., and S. J. MANTEL, JR., "A Decision Support System for the Project Termination Decision," *Project Management Journal*, Vol. XX (1989), pp. 23–28.

SPIRER, H. F., "Phasing Out the Project," in D. I. CLELAND and W. R. KING (editors), *Project Management Handbook*, Van Nostrand Reinhold, New York, 1983, pp. 245–262.

STAW, B. M., and J. ROSS, "Knowing When to Pull the Plug," *Harvard Business Review*, Vol. 65, No. 2 (1987), pp. 68–74.

TADISINA, S. K., "Support System for the Termination Decision in R&D Management," *Project Management Journal*, Vol. XVII, No. 5 (1986), pp. 97–104.

THAMHAIN, H. J., and D. L. WILEMON, "Conflict Management in Project Life Cycles," *Sloan Management Review*, Vol. 17 (1975), pp. 31–50.

APPENDIX

Short Guide to SuperProject Expert*

A.1 INTRODUCTION

This guide gets you started with a hands-on look at the basic operations of SuperProject. An on-line tutorial for more intensive instruction is included in the SuperProject program package. In this introductory overview, you will learn to start the program, give commands, create projects, and analyze task relationships over time. You will also preview some of the more advanced features available for your use.

SuperProject's features include:

- A choice of sequential or concurrent task dependence relationships (Finish-to-Start, Start-to-Start, Finish-to-Finish, with lead/lag)
- Baseline scheduling for comparison with current schedules
- Resolution of conflicts through resource leveling; leveling between projects

*SuperProject Expert is a product of Computer Associates, Inc., 1240 McKay Drive, San Jose, CA 95131, telephone: (408) 432-1727, 1-800-342-5224. The material in this appendix is an edited version of "SuperProject Expert 10-Minute Guide." Use of the SuperProject software is restricted to those who have purchased the book and the accompanying disk.

- User control over program functions—or virtually complete automation, as needed
- Numerous project views for accurate management perspectives: PERT Charts, Gantt Charts, Work Breakdown Structure Outlines and Charts, plus individual task and resource summaries (including Calendars)

In the sections that follow you will explore these features by using SuperProject to schedule activities and assign resources to a project with several task levels.

A.2 GETTING STARTED

A.2.1 Copying Diskettes

Use the **DOS COPY** command to copy files from the original diskette to your working diskette.

Note

During this procedure you will be making a new diskette. To avoid confusion, the diskette that came with this book is called the *original* diskette; the diskette you are making is called the working *copy*.

For this procedure you will need:

- Your DOS diskette if your system does not have a hard drive
- The original SuperProject diskette
- A formatted empty diskette

To copy all files from the original to the working copy:

1. The DOS command prompt (>) should still be on the screen. If it is not, go through the DOS activation process to bring it up. The cursor should be beside the DOS prompt. If any diskettes are in either drive, remove them.
2. Place write-protect tabs on the *original* diskette.
3. Insert the original diskette in drive A and the working copy (the formatted blank diskette) in drive B.
4. Type the following command (make sure that you leave a space between the second asterisk and the *B*):

```
COPY A:*.* B:
```

Check that you have typed it correctly; press **RETURN**.

5. As files are copied from one diskette to the other, their names appear on the screen. When the process is completed, a message appears indicating the total number of files copied, and the DOS command prompt appears with the cursor beside it.

You have now created your working copy of the SuperProject diskette. Remove the original diskette from drive A and the working copy from drive B. You should store your SuperProject original diskette in a safe place. You should not need it again for normal use. Leave the DOS prompt on the screen and the disk drive empty. Proceed to the section "Starting SuperProject."

A.2.2 Installing SuperProject on a Hard Disk

SuperProject can be run on IBM PC, XT, AT, or compatible hard-disk systems which have at least 512K of random access memory (RAM) and DOS 2.0 or higher. It can also be activated for network use with 3Com, IBM PC Network, and Novell systems running under DOS 3.1. An OS/2 version is available from Computer Associates.

Below are instructions for transferring SuperProject from the original diskette (or working copy) to your hard disk. The instructions assume that you have a basic knowledge of those features of DOS that pertain to hard-disk operations. You can find this information in your DOS computer manual.

The following procedures assume that your hard disk has been formatted and is drive C. You can copy SuperProject into one of the following:

1. The root directory
2. An existing subdirectory
3. A new subdirectory

We recommend option 3 (putting the project files into a new subdirectory), on which the following procedures are based. You will create one new directory called *\SPJ* (or any other name that you choose) on drive C to contain program files. If you prefer option 1 or 2, you must adjust the procedures accordingly.

Copy the program diskette

1. Be sure that you are logged onto the root directory of your hard disk. Check that you have at least 1.5 MB of free memory available on your hard drive. Enter

   C:

2. Make the SPJ directory. Enter

   MD SPJ

3. Log onto the SPJ directory. Enter

   CD SPJ

4. Now copy all the files from all the original diskette to the SPJ directory. Insert the SuperProject diskette in drive A. Enter

COPY A:*.* C:

When the process is complete, you see the DOS prompt (C>) again.

You have just copied SuperProject onto your hard disk. Before proceeding further, put the original SuperProject diskette away in a safe place. You should not need it again unless something damages or erases the information on your hard disk.

A.2.3 Installing Your Printer

To use the Print Screen, View, and Reports options from the Output menu, you must first install your character printer.

1. From any screen, enter:

/Output Character Printer Install (type /OC)

2. Select your printer from the menu using the arrow keys.
3. Finally, SuperProject prompts you to save the configuration. Respond *Yes* to write the configuration selection to disk; SuperProject will use this configuration for subsequent sessions. The file that is created on disk is called *SYSPREF.SPJ*. Respond *No* to have the configuration active for this session only.

SuperProject directs output for the character printer through the DOS printer port. If you want to change the printer port, you must do so outside SuperProject (before you start a session). Consult your DOS manual for specific instructions.

A.2.4 Installing Your Graphics Device
(Plotter, Dot Matrix, or Laser)

To use the Plot/Graphs option from the Output menu, you must first install your graphics device.

1. From any screen, enter:

/Output Graphics Device Install [] (type /OG)

2. From the submenu select the type of graphics device you have:
   (a) B/W printers
   (b) Color printers
   (c) Plotters

3. From the next submenu, select the model of your graphics device.

4. Next, SuperProject provides the Port dialogue box. Set the Port options to match your graphics output device. Press **F10** to accept an option.

5. Finally, SuperProject prompts you to save the configuration. Respond *Yes* to write the configuration selection to disk; SuperProject will use this configuration for subsequent sessions. Respond *No* to have the configuration active for this session only.

A.2.5 Starting SuperProject

Your diskette contains a self-running demo program (DEMO.XQT); we suggest that you view it before continuing with this guide. It will introduce you to the main features of SuperProject—what they do and how they work. To run the demo program, follow the startup instructions below. At step 4, type **SPJ /MDEMO**, then press **RETURN**.

This section shows you how to setup DOS and start SuperProject. The examples we show are typical for DOS and SuperProject but you can alter them to match your computer and preference.

1. Before starting SuperProject you should do the following from DOS.
   (a) Set the system date (and time, optional). SuperProject reads the system date when you start it and bases its calculations on that date.

Note

SuperProject does not change the current date once it is running. Thus if you leave SuperProject running over midnight, the current date will be wrong.

   (b) On a hard-disk system, you should set the path to the SPJ directory. Enter

   PATH = \SPJ

   Tip: Put the PATH command in your AUTOEXEC.BAT file to set it each time you start your computer. Note that if you already have a PATH = statement that includes other directories, add \SPJ to that statement.

2. You can significantly increase the speed of SuperProject by increasing the number of DOS buffers (DOS defaults to 2). We suggest that you try 12. Place the line BUFFERS = 12 in your CONFIG.SYS file. Every DOS buffer reduces memory available for your projects by 528 bytes. Thus, when you increase from 2 to 12 buffers, you lose 5.28K project space.

3. Log onto the SuperProject program drive and directory.
   (a) For diskette systems, insert the diskette in drive A and log onto that drive.

   A:

(b) For hard disk systems, log onto your hard disk and change to the SuperProject directory.

C:
CD \SPJ

4. Load SuperProject. Enter

SPJ

You see the SuperProject Title Screen:

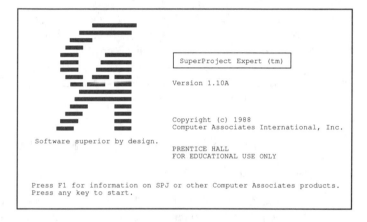

5. Press any key to continue. You see a PERT Chart with the Project Beginning/Ending headers.

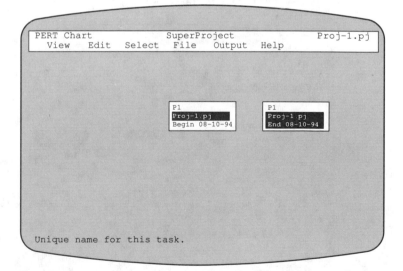

You are now in SuperProject. The bar across the top of the screen is called the *menubar;* you'll learn more about it later. Notice that "Proj-1.pj" appears at the top right of the bar. Proj-1.pj is the default name for the active project.

A.3 BASIC INFORMATION

A.3.1 Output Preferences Files

The Preferences file contains your settings at the Setup Preferences menu and the Character Printer and Graphics Device drivers. When you install either a Character Printer or a Graphics device, you are prompted whether to save that installation to disk.

The default Preference file is *SYSPREF.SPJ* in the current logged directory. SuperProject automatically reads this file when you start a session. When you save your device installation to disk, you can specify another name. Then you can use the /S command line option to use that preference file when you start SuperProject. For example, suppose that you save a configuration for a HP II LaserJet as *HP.SPJ*. You can then start SuperProject with the following command:

SPJ /SHP.SPJ

Thus you can easily maintain different preferences files for different output devices.

A.3.2 Loading SuperProject

At the DOS prompt, type **SPJ** and press **RETURN**—boldface type means "type this key or these characters." You see an initial screen with the name, version number, type of computer, and serial number of your program. The screen also contains copyright information. At the bottom of the screen you see a prompt to "press any key to continue." Do so, and the PERT chart appears, shown here with its View menu pulled down:

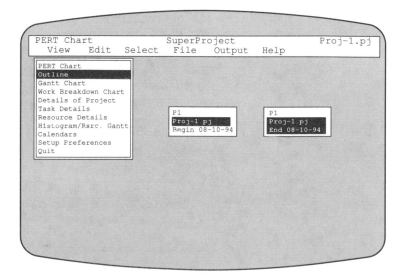

Menubar.   At the top of each screen is the *menubar*. The top line lists the current screen, program, and current project names—from left to right. The second line displays these menu names: View, Edit, Select, File, Output, Help.

Making menu selections.   Each name on the second line of the menubar represents a pulldown menu. These menus are much like window shades—pull them down when you need them, let them "roll" back up when you're done.
     There are two ways to access the SuperProject menus. Use whichever one you find most convenient:

- Press and hold down **F9**, then move the cursor over the menu names with the **RIGHT** and **LEFT ARROW** keys. The **UP** and **DOWN ARROW** keys move the cursor through the menu selections. Either press **RETURN** or remove your finger from **F9** when the cursor is on the desired selection.
- Press / followed by the first letter of the menu name. For example, /**E** brings up the **Edit** menu. Then either press the first letter of your menu selection, or use the arrow keys to highlight the desired selection and press **RETURN**.

A.3.3 Command Line Options

You can specify a project filename on the command line for SuperProject when you start a SuperProject session. For example, the following command starts SuperProject and loads the project file *BUILDING.PJ* in one step. Note that you do not need to specify the default *.PJ* extension.

```
                          SPJ BUILDING
```

     SuperProject has several command line options that allow you to configure the program for special circumstances. A command line option is entered following the project filename on the DOS command line and must begin with a slash. The options are:

/**K** Disables the command key ALT and CTRL. You can then use SuperProject with many keyboards that are not 100% IBM compatible; however, you must use menus and slash commands instead of key commands.

/**L** Indicates a plasma or LCD monitor without a dim attribute. Chosen fields are indicated by underlining rather than highlighting. Use this option when SuperProject "thinks" it is using a color monitor when in fact it is not.

/**M** Executes the macro file specified.

/**S** Loads the Preferences from the path\filename specified. Use this option when the preferences file (*SYSPREF.SPJ*) is not in the default directory. Note that when you save preferences, SuperProject creates the file *SYSPREF.SPJ* in the default directory. Use this option when the file has been renamed and/or moved to another directory.

/V Detects and sets the maximum number of rows for EGA and VGA monitors. For example, use this option to detect and use the 43 lines/screen mode of the EGA. ALT-W toggles between the EGA/VGA expanded mode and the standard (25 lines/screen) mode.

/VnXn Sets the screen size (rows × columns). No detection of the screen mode is attempted with this option. The screen must be set to the proper size at DOS prior to starting SuperProject. SuperProject uses the Preference option for Mono or Color with this option.

Examples:

```
SPJ MYPROJ /K
SPJ MYPROJ /MJANUARY
SPJ /SHPLASER
```

A.3.4 Leaving the Program

To leave SuperProject and exit to DOS, you can:

- Enter **/View Quit** (type **/VQ**).
  The program asks, "Are you sure you want to quit? Y or N." Type:

  Y

  then press **RETURN**. After you press **RETURN** (if you have changed anything) it will ask: "Proj-1.pj has been modifed. Save it? Y or N?" Press **N**.
  If you are using a two-diskette system, you may see this message:

  Insert COMMAND.COM in drive A
  and strike any key when ready.

  If so, put your DOS diskette in drive A and press **RETURN**. The DOS prompt reappears.
- Use the key command **ALT-(minus)**, where minus equals the - next to **0** on the top line of the keyboard.
- Press **CTRL-ALT-Q**. This option promptly leaves SuperProject with *No Questions Asked*. No confirmations are provided.

A.3.5 On-Line Tutorial

The SuperProject on-line tutorial is an excellent way to learn how to use the program. To begin, log onto the directory containing all program files and the *XQT* files. Then invoke the tutorial macro when you start SuperProject typing

SPJ /MTUTORIAL

A.4 BUILDING A PROJECT: MAIN FEATURES

In this section we discuss basic SuperProject scheduling operations: creating, linking, and deleting tasks; viewing and interpreting the primary work screens. You will begin working in Beginner mode. When you get SuperProject, it is set this way so that you can learn its main features with a less complicated feature set. Later in this guide you will learn how to switch experience modes to access more advanced functions—for now, just be aware that you are seeing only part of the total program.

Suppose that you are director of a project to develop a personal backpack helicopter and put it into production. You can imagine how eager your financial supporters are to see the project completed: no more crowded freeways or jammed buses and commuter trains, millions of potential purchasers waiting with cash in hand. So you swing into action and plan the project. Begin by defining:

- *What* you need to accomplish
- *Who* can participate
- *How* much money is in the budget
- *When* the project must be completed

With these goals in mind, you can consider some additional information before you begin:

- Project phases
- Activities (tasks) within the phases
- Approximate task durations
- Available resources (personnel, facilities, equipment, materials)
- Estimated resource costs
- Deadlines and milestones
- Task sequences and dependencies

You can summarize your information on paper, or you can use SuperProject for "brainstorming" with little formal preparation—just start with the project objective and build from there on SuperProject's Outline screen. For example, you could start with notes like these:

```
                              NOTES
                    Two main phases:
                    A. Phase I-Design the product
                    B. Phase II-Prepare a production facility

Phase I                              Phase II
a. Define product specifications     a. Define facility needs
b. Complete preliminary design       b. Hire and assign design group
c. Prepare prototype                 c. Approve design
d. Test                              d. Complete construction
e. Approve for manufacture           e. Buy/install machines
                                     f. Approve for startup
```

Naturally, this is only a skeleton outline. You will have gathered information on the expected duration of these tasks, will have personnel assignments in mind, and undoubtedly will have layers of subtasks beneath these primary tasks, but for our purpose this list is enough.

SuperProject offers four main View screens for project building:

- *PERT Chart:* network model of the project
- *Outline screen:* spreadsheet model and a database of the project
- *Task Gantt:* Gantt chart with dates and task durations
- *Work Breakdown Chart:* work breakdown structure of the project

Each screen presents the project from a different perspective—you'll choose to display each one, depending on the information you need to enter or review. The screen you see after typing **SPJ** at the prompt is the PERT Chart. The label *PERT Chart* appears in its upper left corner. In SuperProject, the PERT Chart allows you to visualize the sequence relationships between tasks in your project. It also displays the *critical tasks* in a project. Delaying a critical task delays the completion date of the whole project.

A.4.1 Outline Screen: Project Summary

While you can use the PERT Chart for many project-building operations, the Outline screen is SuperProject's main screen for data entry and review. To view the Outline screen, type **/V** to pull down the View menu. Then, type **O** to display the Outline screen:

```
Outline                    SuperProject                    Proj-1.pj
    View    Edit    Select    File    Output    Help

 ┌──────────────┬──────┬────┬─────┬────────┬────────┬──────┬────┬───────┐
 │ Heading/Task │ Task │ Pr │ Dur │  Schd  │  Schd  │ Allc │ Un │ Total │
 │   Resource   │  ID  │    │     │  Start │ Finish │      │    │ Hours │
 ├──────────────┼──────┼────┼─────┼────────┼────────┼──────┼────┼───────┤
 │ Proj-1.pj    │ P1   │    │  0  │08-10-94│08-10-94│      │    │   0   │
 │              │      │    │     │        │        │      │    │       │
 └──────────────┴──────┴────┴─────┴────────┴────────┴──────┴────┴───────┘
```

Its data-entry fields let you store and display a variety of project, task, resource, and assignment information, such as:

- *Project/heading/task name:* the default name can be changed or left alone.
- *Resources:* personnel, tools, materials, or other resources allocated to each task.
- *Task priority:* used in automatic scheduling.
- *Duration:* number of workdays required to complete the task (elapsed calendar time, not person-time). You can type the number followed by **h** to enter duration in hours, or type **w** for weeks.
- *Scheduled start and finish dates:* calculated from other project entries.

Some field entries are calculated, while you can enter others. All fields in white or bright type are "editable"—you can enter information there. You can also enter information into fields with gray or dim type *if* the background turns blue or inverse video when the cursor is on that field. Noneditable fields have gray or dim type with an amber or underlined cursor block. Their information is always calculated.

The Outline screen supports virtually all program operations—you can create tasks and resources, enter task and assignment durations, assign resources to tasks, enter dependency relationships among tasks, establish project structure, view scheduled and planned dates, and enter actual hours and costs.

Help.   To learn more about this screen, press **F1** for help with the entry fields, then **F1** again for general Help information. After reviewing the Help screen, press **ESC** to return to the Outline screen. All the screens have help available when you press **F1**. Also, as you move from field to field, check the Help line at the bottom of the screen. It changes with each cursor movement to guide your use of SuperProject.

Data entry.   You'll be entering the Phase I tasks of your project. You will change task names, assign resources, and enter durations. Ultimately, you'll see how these changes are reflected throughout the program—on the PERT, Gantt, and Work Breakdown Charts as well as the Outline screen.

Notice that the cursor is on the Project Name field. We'll change this name to make it distinctive and easy to remember. Type **HELIPROJ** and press **RETURN**. The cursor moves to the next field, the Project ID. Notice that *.PJ* has been added to the name. This is the default ending that SuperProject assigns to all new projects unless you set another one on the Preferences screen.

Now, we'll create the first task, shown on the sample project notes as "Define product specification." Use the **ARROW** keys to move the cursor to the line just below the project name. Type the abbreviated task title, **Define Prod Spec**, at the cursor and use the **DOWN ARROW** to move to the next line. Task ID *001* and a duration of *5*, the default set by the program, are displayed in the row to the right of the task name. For now, do not change anything.

Notice that the program has computed a start date (the current date) and a finish date (5 working days after the start). In its computations, SuperProject uses the start date, the duration, and a set of calendars that contain the weekends and each resource's days off. The project duration has changed from *0* to *5* and has the same Scheduled Start and Finish dates as Task 001:

```
┌─────────────────────────────────────────────────────────────┐
│ Outline              SuperProject              HELIPROJ.PJ   │
│    View    Edit    Select   File   Output   Help            │
├─────────────────────────────────────────────────────────────┤
│ Heading/Task    Task │ Pr │ Dur │ Schd    │ Schd    │Allc│Un│Total│
│   Resource       ID  │    │     │ Start   │ Finish  │    │  │Hours│
│                                                              │
│ HELIPROJ.PJ      P1  │    │  5  │08-10-94 │08-14-94 │    │  │  0  │
│   Define Prod Spec 001│   │  5  │08-10-94 │08-14-94 │    │  │  0  │
│                                                              │
│                                                              │
│                                                              │
│                                                              │
│                                                              │
└─────────────────────────────────────────────────────────────┘
```

Add the next task in the same way; place the cursor directly under the first task name and type **Prelim Design**, an abbreviation.

On our notes, this task is listed as "Complete preliminary design." We estimate that it will take about 20 days for this design task. Use the **RIGHT ARROW** to move the cursor to the duration field. Notice that Task 002 displays in the ID column, and the default *5* displays for the duration. The duration is shown in bright type, so you can change it—type **20** over the *5* and press **RETURN**.

To create the third task, "Prepare Prototype," type **Prototype** directly below *Prelim Design*. Press return and the **RIGHT ARROW** to move to the Duration field. Then replace the default with **30**. Notice that all three tasks are scheduled to start on the same day but have different Scheduled Finish dates. The project's Scheduled Finish date is the same as for Task 003, the longest task.

Tasks 001 and 002 have blue, or dim, ID numbers while Task 003's ID is in bright type. This indicates that Task 003 is on the *critical path*—if that task is delayed, it will delay the entire project.

Now, let's assign a resource to each of these tasks. Move the cursor up to the first task name, *Define Prod Spec*. Then type /**E** to pull down the Edit menu. Type **A** to select the third option, Add Assignment. A resource appears indented below the task. Type over the default name to show that **Don** is assigned to the first task.

Notice the fields to the right of the resource name—these give information about the task-resource *assignment*—its priority, duration, scheduled start and finish, allocation type (how resource hours are calculated), the number of resource units (a single person or a three-person department, for example), and finally, the number of hours assigned (8 hours per day × 5 days).

Now, put the cursor over the other task names and type /**EA** to assign resources to them—assign **Anne** to *Prelim Design* and **Mike** to *Prototype*. Press **!** (that is **SHIFT-1**) to recalculate the project schedule. Your screen should look like this, with different dates:

Heading/Task Resource	Task ID	Pr	Dur	Schd Start	Schd Finish	Allc	Un	Total Hours
HELIPROJ.PJ	P1		30	08-10-94	09-18-94			440
Define Prod Spec	001		5	08-10-94	08-14-94			40
Don	001	50	5	08-10-94	08-14-94	dayx	1	40
Prelim Design	002		20	08-10-94	09-04-94			160
Anne	002	50	20	08-10-94	09-04-94	dayx	1	160
Prototype	003		30	08-10-94	09-18-94			240
Mike	003	50	30	08-10-94	09-18-94	dayx	1	240

Menu bar:
Outline   SuperProject   HELIPROJ.PJ
View   Edit   Select   File   Output   Help

A.4.2 Linking Tasks

You have created the first three tasks of your project and have assigned resources to them. There is still one more step to accomplish before you can begin to evaluate the project schedule—you must link dependent tasks to show the sequence relationship between the finish of one and the start of the next.

Because the product specifications must be defined before the product can be designed, we need to link Task 001 to Task 002. First, move the cursor over Task 002 and press **F4**. (This has the same effect as selecting Link Tasks from the Edit menu.) These prompts appear at the bottom of the screen:

> **Link from:Prelim Design   To:Prelim Design**
> **<ARROWS OR TAB>=Points to task, <RET>=Enter, <BS>=Backup**

The program has automatically inserted the name of the current task into both link fields, and the cursor is on the "Link from:" field. You can use any of these methods to identify both "ends" of the link:

- Type the task ID number at the cursor, or
- Type the task name at the cursor, or
- Use the arrow keys to move from one task to another until the appropriate name appears at the cursor.

Note: To accept an entry, you press **RETURN**. Use **BACKSPACE** to move between the link fields.

In this case you will have either *001* or *Define Prod Spec* in the "Link from:" field and *002* or *Prelim Design* in the "To:" field. Notice that the Task 002 Scheduled Start and Finish dates have changed to show that Task 002 does not start until Task 001 is finished.

The type of link between these two tasks is called Finish-to-Start, which means that the first task must *finish* before the second can *start*. Other types of links are allowed in SuperProject. These are Start-to-Start, one task's start depends on the start of a second but they can finish in any order. Finish-to-Finish links show that one task's finish depends on the finish of another, regardless of when they started. You can only use these link types if you are using SuperProject in Intermediate or Expert mode.

Now, link Task 002 to Task 003 to show that the preliminary design must be completed before prototyping can begin. Do this by pressing **F4** and responding to the prompts, as before. Notice that the project duration has changed to 55 days, to reflect the link sequence.

A.4.3 Links on the PERT Chart

You can see links best on the PERT Chart. Select /View **PERT** Chart (type **/VP**, which is the same as selecting PERT Chart from the View menu). If you'd like to know more about the PERT Chart, press **F1**, and **ESC** when done.

Type **/EA** (the same as selecting Arrange PERT Chart from the Edit menu). Respond **Y** to the prompt about arrangement order and press **RETURN**. The tasks appear as boxes in a horizontal row across the middle of the screen. Use the **ARROW** keys to move the viewing window to the right and left over the task boxes, so you can view them all. The box at the cursor is highlighted to show that it is now the "current task." To reduce the boxes so that they are all visible, type **/ER** (for /**E**dit menu **R**educe View). Your screen will look like this, if Task 003 is the current task (the small rectangle for Task 003 is highlighted):

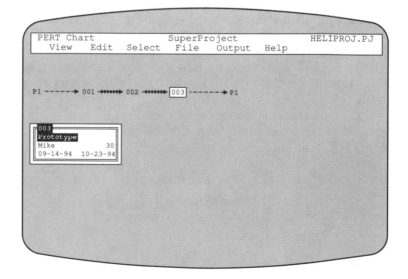

The task in the lower left corner is the current task, at the cursor. You can use the **TAB** and **BACKTAB (SHIFT-TAB)** to move forward and backward through the reduced task boxes to display different tasks in the corner (Arrow Keys Left and Right will also work). Notice that the PERT task boxes show the Task ID in their upper left, with the task name, resource, duration, and Scheduled Start and Finish dates displayed within the box.

The boxes with thick borders and *P1* in the corner are the Project Heading boxes. They "enclose" all their subordinate tasks. Press **CTRL-V** (the **CTRL** key plus **V**) to remove the heading boxes from the screen. Then, press **/EE** (/Edit Enlarge View) to restore the task boxes to their normal size.

Critical path, critical tasks.   Notice that Tasks 001, 002, and 003 are connected by a string of diamond symbols (or dots) (red on color monitors). These diamonds indicate that the tasks are on the *critical path*, where a delay or change in any task on this path affects the project's schedule as a whole.

The task boxes have a double line around them:

This shows that they are critical tasks, or tasks on the critical path.

Noncritical tasks.   Suppose that you have thought of a minor task not listed on the project notes—"Prepare marketing specification sheet." It is a logical outgrowth from Task 001, "Define Prod Spec," and will be performed by the same resource—Don.

You can create that task on the PERT Chart. All the View screens are interactive—whenever you make a change on one, it affects all the others. To do this, you can type **/EC**, which is the same as selecting Create Task from the PERT Edit menu. Or, you can use a function key—**F3**—which works the same as **/EC** but with fewer keystrokes. Try it, press **F3**.

A task box for Task 004 displays in the center of the screen, with the cursor at the default task name. Type the new name, **Mkt Spec Sheet**, over the default and press **RETURN**. Enter **Don** as the resource in the next field and press **RETURN**. The default duration of *5* is correct, so no other fields need to be changed.

Type **/EA** (for Edit menu, Arrange PERT Chart). Respond **Y** to the prompt about arrangement order and press **RETURN**. Task 004 now appears below the other tasks.

Next, you need to link Task 001 to Task 004 to show that specifications must be defined before the "spec" sheet can be written. Press **F4** to display the link prompt and be sure that Task 001's name or number shows after *From:*, with Task 004 following *To:*. Press **RETURN** to create the link.

Notice that Task 004 is now scheduled to start after Task 001 has finished. You can rearrange the screen to make the task time sequence clearer: type **/EA** and respond **Y** as before. Your screen should look like this, with different dates:

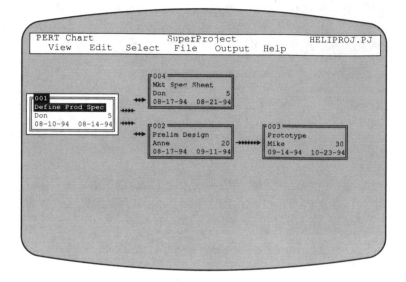

    The new task is connected to Task 001 with a dotted line of ovals, blue on color monitors. The new task box has only a single line around it. The link symbols and single-line task box show that the new task is not on the critical path. Based on the task's duration and resource assignment, SuperProject has decided that it is noncritical. If it is delayed, the entire project will not be delayed as well. Task 004 is directly over Task 002 because they have the same start date.

    To see your project from a different perspective, view the Task Gantt.

A.4.4 Task Gantt: Task Duration

While the PERT Chart is ideal for showing the sequences of tasks and critical paths, task relationships over time are not as obvious. To review these, you can look at the Task Gantt. This chart is another way of viewing the same information entered and displayed on the PERT Chart.

    The Task Gantt, like the other SuperProject screens, is available through the View menu. To display the Task Gantt, select Gantt Chart from the View menu (type **/VG**).

```
┌──────────────────────────────────────────────────────────────────┐
│ Task Gantt              SuperProject              HELIPROJ.PJ       │
│    View    Edit    Select    File    Output    Help                │
├──────────────────────────────────────────────────────────────────┤
│ 1 Day Per Symbol                            │September 94          │
│ ID    Heading/Task      Resource │10   17   24   31   07   14   21 │
│ P1   HELIPROJ.PJ                                                   │
│ 001     Define Prod Spec                                           │
│ 001                Don                                             │
│ 002   Prelim Design                                               │
│ 002                Anne                                            │
│ 003   Prototype                                                   │
│ 003                Mike                                            │
│ 004   Mkt Spec Sheet                                              │
│ 004                Don                                             │
└──────────────────────────────────────────────────────────────────┘
```

The Task Gantt displays tasks in rows with the duration and scheduled dates shown in a time-scale bar graph. Beneath each task is a line showing the resource assigned to that task and the amount of time committed to the task. As you can see, this is a time-scale representation of the information you have just entered on the Outline screen and PERT Chart. It is easy to see when each task begins and ends, as well as their sequence.

The duration of the project may be longer than what you see on the Task Gantt. With **SCROLL LOCK** on, you can use the arrow keys to bring other portions of the schedule into view. To see if **SCROLL LOCK** is on, look for the word *SCRL* in the lower right corner of your screen. Experiment with the cursor keys, if you wish.

Gantt symbols.   The Task Gantt uses a set of symbols to show task status. For now, just note that:

- Small dots represent time available for work, but not yet scheduled.
- Tasks on the critical path have graph bars shown with heavy shading, red on a color monitor.
- Noncritical tasks are shown with lighter shading, blue on color monitors.

You can press **F1** or type **/HL** for more on Gantt symbols. When you are done, press **ESC** to return to the Gantt Chart.

Task Gantt Edit menu.   To add and alter information displayed on the Task Gantt, see the Edit menu. Type **/E** to pull it down. It displays these options: Create, Delete, Link Tasks, Unlink Tasks, Enlarge View, Reduce View, Modify Days Per Symbol, Goto Date, Add Assignments, View Options [], Hide lower level (−), Show next level (+), and Position [].

Note that you can do many of the same things that were available through the PERT Chart Edit menu.

You may find *Goto Date* helpful now, if you scrolled through the screen and want to return to your original starting point. Type **G** or move the cursor to that selection and press **RETURN**. Then enter today's date or whatever date you entered into DOS when you started the program, if other than today's. After you press **RETURN**, the Gantt bars will scroll back to their original positions. This feature is useful for viewing parts of long projects.

For now, though, you can move to a different screen to display your resource assignments from another viewpoint.

Histogram/Resource Gantt.   Press **/VH** to select Histogram/Rsrc.Gantt from the View menu. The Resource Histogram displays:

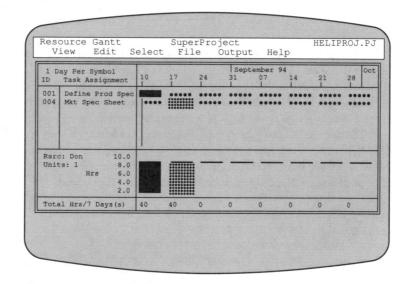

If Don's Histogram is not on-screen, press **TAB** until you see *Don* following *Rsrc:* at the left middle of the screen in the Histogram Window. The Histogram Window shows that his second assignment begins immediately after his first, that he is scheduled for 8-hour days without overscheduling, and that he has unassigned time beginning 10 working days after the start of his assignments. You can see that one of his assignments is critical, while the other is not. The Resource Histogram is helpful when you are adjusting the schedule of individual resources.

To review assignments for all resources, press **CTRL-V** to display the Resource Gantt:

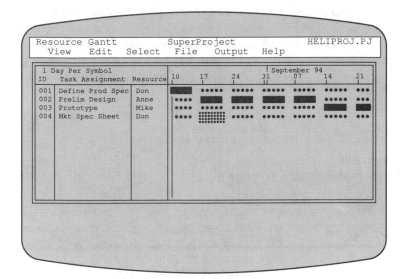

The Resource Gantt shows the amount of time scheduled for each resource per task, and whether the tasks are critical or noncritical. If any of the resources were overscheduled, with more hours assigned than they are scheduled to work in a day, their chart symbols would blink to alert you.

And now, back to the Outline screen . . . Type **/VO** to return.

A.4.5 Deleting a Task

As you study the Outline screen, you realize that you are not ready to add Task 004. Perhaps development and manufacturing preparations should be farther along before your company prepares marketing literature. Possibly marketing preparations should be a third development phase by themselves. So you decide to delete the new task.

Place the cursor over the task name, "Mkt Spec Sheet." Then, press **F5**, which functions as a Delete key throughout the program. (This is the same as selecting Delete from the Edit menu, or type **/ED**.) Respond **Y** to the confirmation prompt when it appears at the bottom of the screen. When you press **RETURN**, the selected task will disappear.

At this point, you might want to save your project—either to take a break, or protect against unexpected data loss.

A.4.6 Saving the Project File

You need to save your projects for two reasons: to prevent data loss during a work session (in the event of a power failure or other unanticipated problem) and to end a SuperProject work session. There is a separate selection on the File menu for each reason to save.

Begin by pulling down the File menu (**/F**). You see these options: Save and Continue, Load Project, Create Project, Put Away Project, Discard Project, Include, View Memory Layout, recalculate (!).

You can type **S** to Save and Continue or **P** to Put Away Project. You will probably want to use Save and Continue in the middle of a work session because the project stays in memory and remains on your screen ready for action. Use Put Away Project to clear the screen for the next project or as preparation for leaving the program.

For now, type **P**. This message, or one similar to it on a two-drive computer, will appear:

Save project as: C:\SPJ\HELIPROJ.PJ

Press **RETURN** to save the project under that name.

A.5 BASIC OPERATIONS SUMMARY

At this point you have viewed the PERT Chart, Task and Resource Gantts, Outline Screen, and Resource Histogram by selecting the appropriate item on the View menu (by typing **/V** and the first letter of the desired screen). You have learned to:

- Pull down menus from the menubar and make selections from them
- Create tasks by selecting Create from the Edit menu (**/EC**) or by pressing the **F3** key
- Link tasks by selecting Link from the Edit menu (**/EL**) or by pressing the **F4** key
- Delete tasks by selecting Delete from the Edit menu (**/ED**) or by pressing the **F5** key
- Distinguish critical from noncritical tasks on the PERT Chart (critical tasks have a diamond link line and double lines around the task box) and Task Gantt (critical tasks have heavily shaded, red graph bars; noncritical tasks have lightly shaded, blue bars)
- Save projects in the middle of and at the end of a work session by typing either **/FS** or **/FP**

In the next section you will learn some additional SuperProject operations, how to:

- Use the Project and Resource Calendars to establish projectwide holidays and personal schedule adjustments
- Promote and demote tasks to form a task hierarchy or Work Breakdown Structure
- Change preference settings to customize program functions
- Display the remaining SuperProject View screens

A.6 FINISHING THE PROJECT: ADVANCED FEATURES

Now it is time to finish the project. You need to:

- Enter the remaining tasks, assign resources to them
- Enter the phase headings, link the phases

For the moment, we'll postpone entering more tasks and proceed to some other SuperProject activities. If you put your project away (**/FP**) at the end of the last main section of this Guide, begin by reloading it into memory. If the project name (*HELIPROJ.PJ*) still displays in the upper right corner of your screen, skip to the next section, "Creating Headings."

Loading the project.   Type **/FL**. When the dialogue box appears, move the cursor over *HELIPROJ.PJ* and press **RETURN**. The project will soon appear, in the PERT view. Type **/VO** to display the Outline screen.

A.6.1 Creating Headings

SuperProject supports Work Breakdown Structures (task hierarchies) with up to nine task levels. The project itself is the first level. In SuperProject, tasks with subordinate tasks are called *headings*. You can create headings by promoting and demoting tasks on the Outline screen, Task Gantt, and Work Breakdown Chart. Heading durations and dates are calculated from the tasks beneath them.

Our notes for the backpack helicopter project call for three basic project levels: the project itself, the phases, and the tasks. So far, you have entered three Phase I tasks. Now it is time to create the Phase I heading.

Use the **ARROW** keys to move the cursor to the project name. Press **F3** to create a task. Then, name the task **Phase I**, by typing over the default name. With the cursor over the phase name, press **SHIFT-LEFT ARROW**. SuperProject automatically indents the remaining tasks below *Phase I*. The phase name displays in yellow or bright type to show that it is a heading. The other tasks have become subordinate to it.

To create the Phase II heading, place the cursor over *Phase I* and press **F3**. A new task appears below the indented tasks, at the same level as *Phase I*. Name the new task **Phase II**. Notice that it appears in the same type as the other tasks, with the default duration (5). It is not a heading yet, because it has no subordinate tasks.

At this point, your screen should look like this with different dates:

```
Outline                     SuperProject              HELIPROJ.PJ
   View    Edit    Select    File    Output    Help
```

Heading/Task Resource	Task ID	Pr	Dur	Schd Start	Schd Finish	Allc	Un	Total Hours
HELIPROJ.PJ	P1		55	08-10-94	10-23-94			440
Phase I	005		55	08-10-94	10-23-94			440
Define Prod Spec	001		5	08-10-94	08-14-94			40
Don	001	50	20	08-10-94	08-14-94	dayx	1	40
Prelim Design	001		20	08-17-94	09-11-94			160
Anne	002	50	30	08-17-94	09-11-94	dayx	1	160
Prototype	003		30	09-14-94	10-23-94			240
Mike	003	50	5	09-14-94	10-23-94	dayx	1	240
Phase II	006			08-10-94	08-14-94			0

Now, create the first Phase II task: "Define facility needs." Place the cursor below *Phase II* and type the new name, **Define Fac Needs**; enter a duration **10**. You can demote the task, to make it subordinate to *Phase II*. Place the cursor on its name—*Define Fac Needs*—and press **SHIFT-RIGHT ARROW**. The new task is now indented to the level of the Phase I tasks and *Phase II* displays in emphasized type. Notice that the duration of *Phase II* has changed to *10*, with the same Scheduled Start and Finish dates as *Define Fac Needs*.

Once you have created headings, you can "hide" their tasks to show only the heading. Place the cursor on Phase I, then press the −, on the numeric keypad. All of the Phase I tasks will disappear, and a "+" displays in front of the phase name. To show the tasks again, press + (by the numeric keypad). You can hide and show resources as well. Try it—place the cursor on a task and press −. A plus-or-minus sign (±) is displayed before the task. Press + with the cursor over the task name to show its resource again.

The structure of your project is starting to shape up. To see the division into project phases from another perspective, display the Work Breakdown Chart.

A.6.2 Work Breakdown Chart: Project Structure

SuperProject supports up to nine levels of tasks, with the project itself as the first level. The Work Breakdown Chart shows the project Work Breakdown Structure, or task hierarchy, in the form of an "organization chart."

Type **/VW** to display the Work Breakdown Chart from the View menu. The headings and tasks display in the form of boxes, similar to those on the PERT Chart. Headings have wide shaded borders, while critical tasks have double-line borders. You can select Reduce View from the Work Breakdown Edit menu (**/ER**) to fit all the tasks on screen at once. Press **CTRL-V** to display the chart four different ways. Here is the Left Right view, reduced:

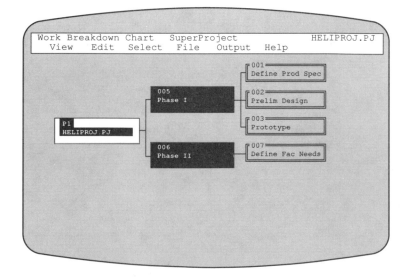

The Top Down view, shown here reduced one more time, gives a clear picture of the tasks that "report to" each heading:

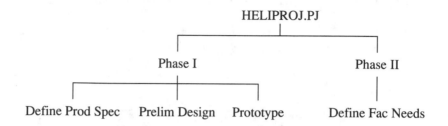

A.6.3 Linking Phases

SuperProject's headings have a unique relationship to their subordinate tasks. Changes made at the heading level—to default settings, for example—filter down to the tasks under that heading. Also, when a heading is linked to another heading, or a task, the effect is the same as if each task were linked individually. Return to the Outline screen to demonstrate this (type **/VO**).

Suppose we decided that all Phase I tasks must be completed before Phase II could start. At present, Phases I and II are scheduled to start on the same day and the project has a duration of 55 days. Now, press **F4** to display the link prompt. Enter the phase heading ID numbers or names—**Phase I** after *From:*, and **Phase II** after *To:*.

Look at the Outline screen—Phase II starts after Phase I has finished, and the project duration has increased to *65*.

Deletions work the same way. You have the option of deleting all subordinate tasks with their headings. Grouping tasks by heading is a convenient way to handle a large number of related jobs or projects with many tasks.

From here, you can go to create the remainder of your project's tasks, assign resources, and fine-tune the schedule. Feel free to do that on your own. For now, though, we'll take a quick tour through the rest of SuperProject's View screens: the Calendars, Preferences, and the three Details screens—Project, Task, and Resource.

A.6.4 Calendar Adjustments

SuperProject's Project and Resource Calendars contain information on scheduled work-days and workday lengths for the entire project and individual resources.

Project holidays.   To schedule nonworking days for all project resources at once, go to the Project Calendar by selecting Calendars from the View menu (/**VC**):

```
Project Calendar          SuperProject              HELIPROJ.PJ
    View     Edit    Select    File     Output    Help

Calendar for: All Project Resources

 1994   Sun    0 | Mon    8 | Tues   8 | Wed    8 | Thu    8 | Fri    8 | Sat    0

 Aug    09  WKND | 10     8 | 11     8 | 12     8 | 13     8 | 14     8 | 15  WKND

 Aug    16  WKND | 17     8 | 18     8 | 19     8 | 20     8 | 21     8 | 22  WKND

 Aug    23  WKND | 24     8 | 25     8 | 26     8 | 27     8 | 28     8 | 29  WKND

 Aug    30  WKND | 31     8 | 01     8 | 02     8 | 03     8 | 04     8 | 05  WKND
 Sep

 Sep    06  WKND | 07        | 08     8 | 09     8 | 10     8 | 11     8 | 12  WKND
                 | Labor Day
```

The numbers next to the day names indicate the standard workday length for that day, throughout the duration of the project. If *0* is displayed, the day is labeled *WKND*, for "Weekend." To declare a projectwide holiday, move the cursor to the appropriate date block and press **F3**, a shortcut way to give the Calendars Edit menu command, Create (/**EC**). Press **RETURN**; the word *Holiday* appears for that date in blue or dim type.

You can name holidays as well. Move to another date block and type a holiday name. Press **RETURN** to create that holiday. To delete a project holiday, move the cursor to the date block and press **F5** (or select /**E**dit **D**elete).

Holidays on the Project Calendar are automatically assigned to all resources. You can schedule resources to work on a company holiday using the Resource Calendar. Or, you can schedule individual nonworking days (vacation, family emergencies) on normal project workdays. To view the first Resource Calendar, press **TAB**. Note that a Resource Calendar is automatically created by SuperProject for each resource assigned to a project. The name of the resource appears in the upper left corner.

```
┌──────────────────────────────────────────────────────────────────────┐
│ Resource Calendar          SuperProject              HELIPROJ.PJ       │
│      View     Edit    Select    File     Output     Help              │
├──────────────────────────────────────────────────────────────────────┤
│ Calendar for: Don          No. Units: 1        Std. Day: 8            │
├───────┬────────┬────────┬────────┬────────┬────────┬────────┬────────┤
│ 1994  │ Sun  0 │ Mon  8 │ Tues 8 │ Wed  8 │ Thu  8 │ Fri  8 │ Sat  0 │
├───────┼────────┼────────┼────────┼────────┼────────┼────────┼────────┤
│ Aug   │ 09 WKND│ 10     │ 11   8 │ 12   8 │ 13   8 │ 14   8 │ 15 WKND│
│       │        │Vacation│        │        │        │        │        │
├───────┼────────┼────────┼────────┼────────┼────────┼────────┼────────┤
│ Aug   │ 16 WKND│ 17   8 │ 18     │ 19     │ 20     │ 21     │ 22 WKND│
├───────┼────────┼────────┼────────┼────────┼────────┼────────┼────────┤
│ Aug   │ 23 WKND│ 24     │ 25     │ 26     │ 27     │ 28     │ 29 WKND│
├───────┼────────┼────────┼────────┼────────┼────────┼────────┼────────┤
│ Aug   │ 30 WKND│ 31     │ 01     │ 02     │ 03     │ 04     │ 05 WKND│
│ Sep   │        │        │        │        │        │        │        │
├───────┼────────┼────────┼────────┼────────┼────────┼────────┼────────┤
│ Sep   │ 06 WKND│ 07     │ 08     │ 09     │ 10     │ 11     │ 12 WKND│
│       │        │Labor Day│       │        │        │        │        │
└───────┴────────┴────────┴────────┴────────┴────────┴────────┴────────┘
```

Project holidays show in blue or dim type, while resource nonworking days show in red or bright type. You create resource holidays by pressing **F3** (this can only be done on the resource calendar). No work will be scheduled for a resource on either project holidays or nonworking days unless you request an override. Move the cursor to a holiday and type a number between 1 and 24, to indicate the number of hours to be worked on that day. Press **RETURN** to schedule the resource for that number of hours on that day.

You can also schedule partial days that way—go to a regular 8-hour day and enter a number less than 8, say **4** to allow for a morning appointment. The new number appears in the upper and lower right corners of the date block; if you try to schedule the resource for more than that number of hours for that day, SuperProject will indicate that the resource is overscheduled.

A.7 ADDITIONAL SCREENS

A.7.1 Changing Experience Modes: Preferences Screen

The remaining View screens are only available in Intermediate or Expert mode. Type **/VS** (for **V**iew menu, **S**etup Preferences). The Preferences screen will display in Beginner mode:

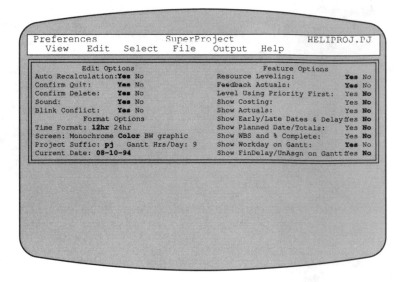

```
  Preferences               SuperProject           HELIPROJ.PJ
      View    Edit    Select    File    Output    Help

        Edit Options                        Feature Options
  Auto Recalculation:Yes No        Resource Leveling:          Yes No
  Confirm Quit:      Yes No        Feedback Actuals:           Yes No
  Confirm Delete:    Yes No        Level Using Priority First: Yes No
  Sound:             Yes No        Show Costing:               Yes No
  Blink Conflict:    Yes No        Show Actuals:               Yes No
        Format Options             Show Early/Late Dates & DelayYes No
  Time Format: 12hr 24hr           Show Planned Date/Totals:   Yes No
  Screen: Monochrome Color BW graphic  Show WBS and % Complete: Yes No
  Project Suffic: pj  Gantt Hrs/Day: 9   Show Workday on Gantt:  Yes No
  Current Date: 08-10-94           Show FinDelay/UnAsgn on GanttYes No
```

This screen shows all the settings you can change within Beginner mode to customize the way SuperProject works. The highlighted fields show which settings currently
are selected. You can change them using the arrow keys and **RETURN**. Now, though, we
need to pull down the Edit menu and select Expert mode.

Type **/E** to display the menu, with these options: Beginner Mode, Intermediate
Mode, Expert Mode, Save as Default, and Date & International Preferences. Then press **E**
to change to Expert mode. You see an expanded Preferences screen that lets you take advantage of every SuperProject feature. Many of the available preference settings determine
which fields are available for data entry and review on the Details screens. Type **/VD** (for
View menu, Details of Project) to display the Project Details screen.

A.7.2 Project Details Screen

The Project Details screen contains a variety of information about the project—descriptive
material in the upper left portion, schedule data, default settings to be passed down to resources and tasks, and totals which summarize project Scheduled, Planned, and Actual
activity—including costs:

```
Project Details          SuperProject           HELIPROJ.PJ
      View     Edit    Select    File    Output    Help

Name: HELIPROJ.PJ
ID: P1   Directory:c:\SPJ\
  Author:             Leader:          ┌──Defaults──┐  ┌──Totals──┐
WBSMask:11.22.33.44.???? Freeze:Yes No  Hours:    40   Var:   11000.00
Created:08-11-94 Revised: 08-14-94 # 3  Fixed:   0.00  Fixed:     0.00
  Start:08-10-94                         Rate:   25.00  Total: 11000.00
 Finish:10-27-94     Dur: 56  0h         Rate Mult: 1.00 Actual:    0.00
   Late:10-27-94     Dev:0.00            Duration:   5  Hours: 440 Ovr:  0
                                         Allc: dyx Pr :50 Tasks: 5 Rsrc: 3

 Worday  Start Finish Hrs  121 2 3 4 5 6a7 8 9 1011121 2 3 4 5 6p7 8 9 1011
 Sunday    8:00a  8:00a  0  --------bb------------bb------------bb------
 Monday    8:00a  5:00p  8  --------bb------------bb------------bb------
 Tuesday   8:00a  5:00p  8  --------bb------------bb------------bb------
 Wednesday 8:00a  5:00p  8  --------bb------------bb------------bb------
 Thursday  8:00a  5:00p  8  --------bb------------bb------------bb------
 Friday    8:00a  5:00p  8  --------bb------------bb------------bb------
 Saturday  8:00a  8:00a  0  --------bb------------bb------------bb------
```

Much of the work schedule information at the bottom of the screen is included on the Project Calendar. As you saw, you can make changes there when you are working in Beginner mode and cannot access the Project Details screen.

SuperProject lets you schedule resources across projects. You can link those projects through the Project Details Edit menu. Press **CTRL-V** to show the Linked Projects view.

A.7.3 Task Details Screen

Type **/VT** to display the first Task Details screen—there is one for each task in the project:

```
Task Details             SuperProject           HELIPROJ.PJ
      View     Edit    Select    File    Output    Help
┌ID: 001
│Name: Define Prod Spec

 Duration:5      Actual Dur: 0   ┌─Start─┐ ┌─Finish─┐ ┌──Totals──┐
 Strt Del:0      Finish Del: 0   Erly:08-10-94 08-17-94 Var: 1000.00
  Float:0        Free Float: 0   Late:08-10-94 08-17-94 Fix:    0.00
 Pct Comp:  0  BCWP:       0.00  Schd:08-10-94 08-17-94 Tot: 1000.00
    Type:ASAP ALAP Must Span     Actl:                  Act:    0.00
  WBS:01.01.00.00.0001  Acct: 0  Plan:                  Pln:    0.00
                                 Scheduled/Crit. Priority: Hrs: 40
                                 Dev:0.00                 Pln: 0 Ovr: 0

 Resource │ Hrs │ Allc │ Un │  Predecessors  │  Successors        │ Lag
 Don      │  40 │ dayx │ 1  │                │ 002 Prelim Design FS│ 0h
```

At the bottom of the screen is information about the resource assignments for this task—assignment data are summarized into the Task Totals in the upper right portion of the screen.

A.7.4 Resource Details Screen

To view the first Resource Details screen, type **/VR**:

```
┌─────────────────────────────────────────────────────────────────────┐
│ Resource Details          SuperProject              HELIPROJ.PJ       │
│     View    Edit   Select   File    Output    Help                   │
├─────────────────────────────────────────────────────────────────────┤
│ Rsrc Name: Don                          ┌─Defaults─┐   ┌─Totals─┐    │
│ Work Code:       Accrue: Strt Prorate End │Hours:      40│ Var:  1000.00│
│ Total Overscheduled:   0 Rate Mult: 1.00  │Fixed:    0.00│ Fix:     0.00│
│    Calendar Variance: -8 No. Units: 1     │Rate:    25.00│ Tot:  1000.00│
│     Workday:   Sun Mon Tue Wed Thu Fri Sat│Standard Day:  8│ Act:    0.00│
│ Start: 8:00a   0   8   8   8   8   8   0   │Allocation: dayx│ Hrs: 40│
├────┬────────────────┬─────┬─────┬──────┬────┬─────┬─────┬─────┬──────────┬──────────┤
│ ID │     Task       │ Dur │ Hrs │ Allc │ Un │ Ovr │Actl │ Pr  │  Start   │  Finish  │
├────┼────────────────┼─────┼─────┼──────┼────┼─────┼─────┼─────┼──────────┼──────────┤
│001 │Define Prod Spec│  5  │ 40  │ dayx │  1 │  0  │  0  │     │ 08-11-94 │ 08-17-94 │
│    │                │     │     │      │    │     │     │     │          │          │
└────┴────────────────┴─────┴─────┴──────┴────┴─────┴─────┴─────┴──────────┴──────────┘
```

This screen is similar to the Task Details screen—at the bottom it contains information on the resource's task assignments, and this assignment information is summarized in the Resource Totals. Much of the resource work schedule information at the left just above the assignment information is also included on the Resource Calendar, where you can change it in any mode (Beginner, as well as the others).

A.7.5 Printing Screen Views

You can print any screen accessible through the View menu by selecting the first or second item from the Output menu. The Beginner mode contains these options: Screen, View, Graphics Device Install, Character Printer Install. In Intermediate and Expert modes, you can do graphic plots of PERT and Work Breakdown Charts and the Gantts using a graphic printer. SuperProject supports many dot matrix, laser, and plotter devices.

A.8 OUTPUT

A.8.1 Cost/Schedule Status Report Fields

The Cost/Schedule Status Report is required by many U.S. government contractors, and contains fields that compare the current estimates to both the original plan and the actual work to date. The following fields are available on SuperProject.

- *Budgeted cost of work scheduled (BCWS):* calculates how much the work scheduled through today should have cost. If the Planned Start of a task is later than the current date, this value is zero. If the Planned Finish is earlier than the current date, this value equals the Planned Cost. If the Planned Start is earlier than or equal to the current date and the Planned Finish is later than or equal to the current date, then

$$\text{BCWS} = \frac{\text{planned workdays to date}}{\text{planned workdays}} \times \text{planned cost}$$

Planned Start/Finish Hours are ignored in this calculation. The current date is included when calculating planned days. Partial work on any day is rounded up to be one full day.

- *Schedule variance (SV):* difference between earned value (BCWP) and budgeted cost of work scheduled (BCWS); a negative number indicates that the project is behind schedule.
- *Cost variance (CV):* earned value (BCWP)—actual cost; a negative number indicates that the project is behind schedule and over budget.
- *Variance at completion (VAC):* planned cost—total cost; a negative number indicates that the current plan exceeds the original budget.
- *Cost remaining:* total cost—actual cost.
- *Hours remaining:* total hours—actual hours.

A.8.2 Matrix Reports

If you are using SuperProject in Intermediate or Expert mode you can select Reports from the Output menu. A selection screen appears listing numerous types of information available through SuperProject. You can build custom reports by selecting the exact information needed.

A.9 SUMMARY

In the latter half of this guide you learned to:

- Demote tasks to subtasks using **SHIFT-RIGHT ARROW** on the Outline screen; promote them to headings using **SHIFT-LEFT ARROW**

- Display the Work Breakdown Chart with **/VW**; reduce and enlarge its views by typing **/ER** and **/EE**; and display alternative views with **CTRL-V**
- Hide subtasks by pressing –; show them again with +
- Change experience modes and Preferences settings by typing **/VS** to view the Preferences Screen and **/E** to pull down the Edit menu
- Display the Project and Resource Calendars (**/VC**), and the Project (**/VD**), Task (**/VT**), and Resource (**/VR**) Details screens for information entry and review

A summary of the View screens and their contents is shown in Table A.1. The **ALT** key plus a number is another way of displaying View screens:

TABLE A.1 SUMMARY OF VIEW SCREENS

View	Display keys	Contents
PERT Chart	**ALT-1, /VP**	Task sequence network
Outline	**ALT-2, /VO**	Overall information summary
Task Gantt	**ALT-3, /VG**	Durations graphed over time
Work Breakdown Chart	**ALT-4, /VW**	Task structure network; levels of tasks
Project Details	**ALT-5, /VD**	Overall project schedule and cost definition information
Task Details	**ALT-6, /VT**	Schedule, assignment, and cost information for each task
Resource Details	**ALT-7, /VR**	Schedule, assignment, and cost information for each resource
Histogram/Resource Gantt	**ALT-8, /VH**	Horizontal and vertical assignment bar graphs
Calendars	**ALT-9, /VC**	Project and personal days off
Preferences	**ALT-0, /VP**	Program function settings

Final Comments

This guide only covers a small fraction of SuperProject's capabilities. It should be clear, though, after experimenting with the sample project that SuperProject is a powerful, yet easy-to-use tool. If you want to know more about its features, the on-line help function can always be accessed by pressing **F1** to provide further information.

Index